Use of Landscape Sciences for the Assessment of Environmental Security

NATO Science for Peace and Security Series

This Series presents the results of scientific meetings supported under the NATO Programme: Science for Peace and Security (SPS).

The NATO SPS Programme supports meetings in the following Key Priority areas: (1) Defence Against Terrorism; (2) Countering other Threats to Security and (3) NATO, Partner and Mediterranean Dialogue Country Priorities. The types of meeting supported are generally "Advanced Study Institutes" and "Advanced Research Workshops". The NATO SPS Series collects together the results of these meetings. The meetings are co-organized by scientists from NATO countries and scientists from NATO's "Partner" or "Mediterranean Dialogue" countries. The observations and recommendations made at the meetings, as well as the contents of the volumes in the Series, reflect those of participants and contributors only; they should not necessarily be regarded as reflecting NATO views or policy.

Advanced Study Institutes (ASI) are high-level tutorial courses intended to convey the latest developments in a subject to an advanced-level audience

Advanced Research Workshops (ARW) are expert meetings where an intense but informal exchange of views at the frontiers of a subject aims at identifying directions for future action

Following a transformation of the programme in 2006 the Series has been re-named and re-organised. Recent volumes on topics not related to security, which result from meetings supported under the programme earlier, may be found in the NATO Science Series.

The Series is published by IOS Press, Amsterdam, and Springer, Dordrecht, in conjunction with the NATO Public Diplomacy Division.

Sub-Series

A.	Chemistry and Biology	Springer
B.	Physics and Biophysics	Springer
C.	Environmental Security	Springer
D.	Information and Communication Security	IOS Press
E.	Human and Societal Dynamics	IOS Press

http://www.nato.int/science
http://www.springer.com
http://www.iospress.nl

Series C: Environmental Security

Use of Landscape Sciences for the Assessment of Environmental Security

edited by

Irene Petrosillo
Department of Biological and Environmental Sciences and Technologies,
University of Salento, Lecce, Italy

Felix Müller
NATO CCMS Co-Director, Ecology Centre, University of Kiel, Germany

K. Bruce Jones
U.S. Geological Survey, Geography Discipline, Reston, Virginia, U.S.A

Giovanni Zurlini
Department of Biological and Environmental Sciences and Technologies,
University of Salento, Lecce, Italy

Kinga Krauze
International Centre for Ecology, Polish Academy of Sciences, Lódz, Poland

Sergey Victorov
Research Institute of Remote Sensing Methods for Geology, St. Petersburg, Russia

Bai-Lian Li
University of California, Riverside, California, U.S.A

William G. Kepner
NATO CCMS Co-Director, U.S. Environmental Protection Agency, Office of Research and Development, Las Vegas, Nevada, U.S.A

Springer

Published in cooperation with NATO Public Diplomacy Division

Results of the NATO/CCMS Pilot Study on the
Use of Landscape Sciences for Environmental Assessment
2001–2006

A C.I.P. Catalogue record for this book is available from the Library of Congress.

ISBN 978-1-4020-6589-7 (PB)
ISBN 978-1-4020-6588-0 (HB)
ISBN 978-1-4020-6594-1 (e-book)

Published by Springer,
P.O. Box 17, 3300 AA Dordrecht, The Netherlands.

www.springer.com

Printed on acid-free paper

CONTENTS

FOREWORD

The assessment of land use and land cover is an important activity for contemporary land management. Human land-use practices are the most significant factors influencing environmental management at local, regional, national, and global scales. In the past, environmental policies have often reflected a reactive response to environmental perturbations with management efforts focused on short-term, local-scale problems such as pollutant abatement. Currently, environmental management philosophy is evolving toward examination of critical environmental problems over larger spatial scales and assessment of the cumulative risk resulting from multiple problem sources. Today's environmental managers, urban planners, and decision-makers are increasingly expected to examine environmental and economic problems in a larger geographic context that crosses national boundaries and scientific disciplines. Secondly, contemporary policy-makers have also been challenged on how they view security. The conventional definition of national security has been expanded to include environmental threats resulting from resource scarcity and overpopulation and it is recognized that environmental factors may have an impact in creating conflict and world instability. Thus the working definition of security has been broadened beyond relying on militaristic aspects alone and has evolved to include the environment.

In 1969, the North Atlantic Treaty Organization (NATO) established the Committee on the Challenges of Modern Society (CCMS) partly in response to examine the link between environmental issues and security. CCMS was created for the purpose of addressing problems affecting the environment of the member nations and the quality of life of their citizens. The key goal of the NATO/CCMS has been to utilize a science framework to promote international cooperation and to meet the challenges associated with evaluating environmental problems related to land use at multiple spatial-temporal scales, understanding the consequences of environmental change, and to examine non-traditional threats to security, the CCMS initiated a Pilot Study on the *Use of Landscape Sciences for Environmental Assessment* in March 2001 (http://www.nato.int/science/pilot-studies/lsea/lsea-index.htm). Specifically, the Pilot Study was developed to explore the potential of quantifying and assessing environmental condition, processes of land degradation, and subsequent impacts on natural and human resources (including security) by combining the advanced technologies of remote sensing, geographic information systems, spatial statistics, and process models with landscape ecology theory. The Landscape Sciences Pilot Study is co-chaired by the United States and Germany and is designed to enhance the ability of senior-level decision-makers, environmental managers, and the public to 1) address a range of environmental problems that have inherently different scales; 2) evaluate cumulative impacts to ecological and hydrological resources; 3) provide a framework for large-scale assessment in which to put surrounding communities in perspective; 4) communicate analysis and assessment results to a wide range of technical and non-technical audiences; and 5) develop products, such as regional and watershed assessments, analysis tools, digital maps, and databases, for a variety of international audiences. The project uses landscape ecology, i.e., the study of the distribution patterns of communities and ecosystems, the ecological processes that effect those patterns, and changes in both pattern and process over time, as its foundation.

In this book a new perspective of environmental security is proposed, considering and integrating two different viewpoints: ecological and social (perception). This starts from the recognition that human decisions, producing environmental change, depend on the perception and awareness that people have of their environment.

The research and implementation agenda of the Pilot Study was accomplished through the completion of multiple national studies throughout Europe and the United States which emphasizes thematic areas related to landscape characterization, land cover change detection, landscape indicators, landscape assessment, and landscape theory and models that, in some study areas have been linked with the assessment of perception of environmental security.

The results from the Pilot Study and from the Editorial Board meetings illustrate the utility of adopting a landscape sciences approach in public health and environmental decision-making, natural resource management and planning, and ecological preservation/restoration projects. The landscape assessment framework and methodologies should provide a number of key benefits to environmental managers and the public in regard to determining how different land-use choices impact ecological integrity and subsequently, environmental security.

Thus this CCMS Pilot Study on the *use of landscape sciences* provides a multi-lateral forum for cooperation, information exchange, and dialogue among the environmental, development, foreign and security policy communities. Additionally, it provides an “enabling environment” to facilitate joint work programs and further advances the NATO goal of utilizing a science framework to promote international cooperation and peace.

William G. Kepner
Felix Müller
Giovanni Zurlini
Bai-Lian Li
K. Bruce Jones
Kinga Krauze
Sergey Victorov
Irene Petrosillo

CONTRIBUTIONS OF LANDSCAPE SCIENCES TO THE DEVELOPMENT OF ENVIRONMENTAL SECURITY

FELIX MÜLLER*
Ecology Centre, University of Kiel, Germany

K. BRUCE JONES
USGS Headquarters, Reston Virginia, USA

KINGA KRAUZE
Polish Academy of Sciences, Institute of Hyrology, Lodsch, Poland

BAI-LIAN LI
Ecological Complexity and Modeling Laboratory, Department of Botany and Plant Sciences, University of California, Riverside, USA

SERGEY VICTOROV
Russian Academy of Sciences, Institute of Remote Sensing Methods for Geology, St. Petersburg, Russia

IRENE PETROSILLO AND GIOVANNI ZURLINI
Laboratory of Landscape Ecology – Department of Biological and Environmental Sciences and Technologies – University of Salento, Lecce, Italy

WILLIAM G. KEPNER
US Environmental Protection Agency, Office of Research and Development, P. O. Box 93478, Las Vegas, Nevada 89193, USA

1. Environmental threats as objects of security

The growing knowledge about environmental change, stress, and degradation has increased the visibility of environmental condition as an important

*Corresponding Author: Dr. Felix Müller, Ecology Centre, University of Kiel, Olshausenstrasse 75 D 24118 Kiel (Germany); e-mail: fmueller@ecology.uni-kiel.de.

I. Petrosillo et al. (eds.), Use of Landscape Sciences for the Assessment of Environmental Security, 1–17.

determinant of security, especially given the emergence of new political, economic, social, and environmental challenges since the end of the cold war. The relationship between environment and security now is a common interest among both the scientific and policymaking communities, supported by the fact that the traditional security concepts based on territorial integrity and political sovereignty have been revisited following the changes in the geopolitical landscape at the end of the last century.

Security generally is related to both a perception of freedom from risk and freedom from anxiety or fear. Security aims at providing expected services, safety, and protecting valuable assets from harm, even during times of increased threat or risk. Security is achieved through both prospective (preventative) and retrospective (mitigation) actions on the part of governments, agencies, and people. Perceptions of security by individuals, communities, and societies are strongly linked to human well-being and to the satisfaction of the population.

The notion of environmental security has been historically linked to international conflicts caused by environmental degradation, e.g. through overuse of renewable resources, pollution, or impoverishment in the space of living (Tuchel, 2004; Herrero, 2006; Liotta, 2006). In this context, the concept of environmental security has been developed mainly by international policy researchers and has focused on the role of the scarcity of renewable resources such as cropland, forests, water, and fish stocks. Statistical data demonstrate that agriculture and natural resource availability plays an important role in many events of acute violence, which often occur in rural areas (De Soysa et al., 1999). Therefore, attention has been devoted to the theoretical analysis of possible pathways that lead to loss of environmental security, beginning with scarcity and leading to outbreaks of violence. Thus, environmental security has been discussed as a concept of international security policy (Brauch, 2006).

This debate began in the late 1980s and has been quite intense. Recently, environmental security issues received increasing worldwide interest by governments, scientific institutions, intergovernmental organizations, and nongovernmental groups, calling for greater attention to the potential threats to security posed by environmental problems (Dabelko, 2004; Matthew et al., 2004, UNEP, 2004).

The decrease in quantity and quality of resources, rapid global population growth, and unequal access to resources are the basic drivers behind increasing environment-related security risks. Notably renewable resources like water and land are crucial factors in security issues, especially with respect to instability and migration between and within countries or regions. Moreover, environmental degradation often results in changes in important ecological and landscape processes that can have irreversible impacts to critical renewable resources such as water, fiber, food, and clean air. This can lead to a relatively permanent loss of land that can be inhabited by people and human societies, as in the case of desertification (Kepner, 2006). The subsequent scarcity of nonrenewable resources can contribute to national and international instability (Dabelko et al., 2000).

Results of the increasing human appropriation of regional landscapes can have a variety of ecological effects directly drawing on the state of environmental security. Consequently, focusing on environmental security as a theme is an essential step in the study of interactions between humans and the environment, and it is critical in understanding how humans create and respond to environmental change.

2. The significance of environmental security

Concerns about the adverse impact of human activities on the environment, the direct and indirect effects of various forms of environmental change on national and regional security, focus on the potential consequences of triggering, intensifying, or generating conflict, and instability relevant to conventional security thinking. Some examples for such impact processes that are discussed throughout this book are summarized in Box 1. The pathways and results highlighted in Box 1 suggest that the environment is going through massive changes that can significantly alter environmental security. These changes can result in negative or even catastrophic affects on environmental conditions which in turn can pose significant social and political threats.

Environmental and human systems develop various significant interrelations, and consequently they have to be analyzed and managed in an integrative manner, taking into account structural and process-based linkages. From the perspective of human systems (e.g. economic stability and quality of life), an optimization is being strived for trying to utilize natural systems as intensively as possible without disturbing focal environmental provisions or provoking future catastrophes. A security-based, environmental management approach should therefore reduce the loss of important goods and services upon which society depends while providing options for restoring services that have been lost. To underline the significance of the environmental security discussion, a recent comprehensive overview of the environmental security field (MEA, 2005; see also Zurlini and Müller, i.p.) additionally observes that:

- The environment is the most transnational issue, and its security is an important dimension of peace, national security, and human rights
- Over the next 100 years, one third of current global land cover will be transformed, hence the world will be facing increasingly hard choices among consumption, ecosystem services, restoration, preservation, or degradation

- Environmental security is central to national security, comprising the dynamics and interconnections among humans and natural resources
- To achieve political, economic, and social security, environmental security has to be taken into account; overall security does not prevail without considering environmental security

Box 1. Some pathways between environmental stresses and socioeconomic risk situations.

- Soil degradation (e.g. energy conversion → acid rain → soil pH → loss of nutrients → loss of soil fertility → loss of agricultural and forestry potentials → decreasing provision of food and other environmental goods)
- Eutrophication (e.g. agriculture → fertilizers → algal growth in aquatic systems → decrease of fish populations → loss of fishing potentials → decreasing provision of food)
- Toxification (e.g. industry → pollutants → ground water → reduced quality of drinking water → increasing health problems for people and ecosystem components)
- Water shortage and quality (e.g. watershed management and water usage in international catchments → international conflicts about changing water balances and changing water quality)
- Desertification (e.g. agriculture/livestock → overgrazing → erosion → loss of agricultural potential → loss of productivity → decreasing provision of food, irreversibility)
- Global climate change (e.g. energy usage → CO_2 production → global temperature rise → loss of various potentials → loss and displacement of goods and services; sea level rise → existence of countries in flat coastal regions is endangered; extreme weather conditions)
- Urbanization (e.g. loss of habitats by land conversion → loss of water storage capacity → peak runoff → decoupling of production and utilization of goods and services → critical dependencies between cities and their surroundings)
- Demographic change (e.g. loss of soil due to overgrazing, but also consequences of global change → loss of productivity → displacement of services → migration; aging and shrinking in Europe)
- Landscape loss and fragmentation (e.g. agriculture intensification → habitat isolation → reduce species and gene flows and exchanges and fluxes of materials and energy; deforestation, land conversion)
- Biodiversity decrease (e.g. species loss and extinction → reduce the functions of these species → decrease ecosystem service)

Based on these assumptions, there are many different ways to define environmental security, most of which originate from international policy debates. Some of these definitions are highlighted in Table 1. The AC/UNU Millennium Project states that the term "environmental security" refers to several general categories, but most of these arise from two general society concerns:

- About the adverse impact of human activities on the environment (the emphasis here is on the security of the environment as a good in itself, for the sake of future generations, as the context for human life)
- About the direct and indirect effects of various forms of environmental change which may be natural or human-generated on national and regional security (here the focus is on environmental change triggering, intensifying, or generating the forms of conflict and instability relevant to conventional security thinking)

TABLE 1. Some definitions of environmental security.

AC/UNU Millennium Project[†]	Environmental security is the relative public safety from environmental dangers caused by natural or human processes due to ignorance, accident, mismanagement, or design and originating within or across national borders.
AC/UNU Millennium Project[1]	Environmental security is the state of human–environment dynamics that includes restoration of the environment damaged by military actions, and amelioration of resource scarcities, environmental degradation, and biological threats that could lead to social disorder and conflict.
Barnett (1997)	Environmental security is the proactive minimization of anthropogenic threats to the functional integrity of the biosphere and thus to its interdependent human component.
Belluck et al. (2006)	By ensuring environmental security we mean guarding against environmental degradation in order to preserve or protect human, material, and natural resources at scales ranging from global to local.
AC/UNU Millennium Project[1]	Environmental security is the maintenance of the physical surroundings of society for its needs without diminishing the natural stock.
"Environmental Security of Russia", Issue 2, The Security Council of the Russian Federation, Moscow (1996), p. 55	Environmental security is protectedness of natural environment and vital interests of citizens, society, the state from internal and external impacts, adverse processes and trends in development that threaten human health, biodiversity, and sustainable functioning of ecosystems, and survival of humankind.
"On Principles of Environmental Security in the Commonwealth States", December 4, 1997	Environmental security is the state of protection of vital interests of the individual, society, natural environment from threats resulting from anthropogenic and natural impacts on the environment.

[†] http://www.acunu.org/millennium/env-sec1.html

The Millennium Project concludes that the condition of environmental security is one in which social systems interact with ecological systems in sustainable ways, all individuals have fair and reasonable access to environmental goods, and mechanisms exist to address environmental crises and conflicts. NATO formulates a similar position, which is cited in Box 2.

Box 2. NATO's position on environmental security.

The Environment and Security
(From the NATO online library)

The whole notion of security as traditionally understood – in terms of political and military threats to national sovereignty – must be expanded to include the growing impact of environmental stress – locally, nationally, regionally, and globally.

World Commission on Environment and Development, 1987

"Security has traditionally been defined as individual or collective safety achieved through the protection of territorial integrity, political sovereignty and national interests. However, the understanding of security has evolved over time. It has been recognised that environmental factors have an impact on conflicts and levels of stability.
Even though the causes of conflict and insecurity are often complex, evidence suggests that environmental degradation and resource depletion are a source of tension in many regions of the world. Land degradation, climate change, water quality and quantity, and the management and distribution of natural resources (e.g. oil, forests, minerals) are factors that can contribute directly to conflict or be linked to them by exacerbating other causes such as poverty, migration, infectious diseases, poor governance and declining economic productivity. In sum, environmental problems can threaten human livelihoods and contribute to social and economic inequalities.
Concern for the environment has grown and the need to integrate environmental policies into security measures has therefore been and continues to be a priority. The scarcity of renewable resources and the cross-border character of environmental issues have led the international community to take an active role in initiating environmental projects."

http://www.nato.int/docu/environment/html_en/environment_02.html

In order to introduce environmental security with reference to social–ecological systems, it is useful to refer to the definition provided by Wolfers in 1962, stating that "*environmental security, in an objective sense, measures the absence of threats to acquired values, in a subjective sense, the absence of fear that such values will be attacked.*" In short, the previous statements could be tentatively reformulated to define environmental security in the context of landscape sciences, according to the following: *Environmental security, in an objective sense, aims to evaluate the level of threats to acquire and sustain landscape values in terms of ecosystem goods and services at multiple scales and, in a subjective sense, represents the level of fear that such values will be attacked and possibly lost.*

3. A landscape approach to environmental security

In this volume, environmental security is described from the viewpoint of landscape sciences. This approach seems to be particularly suitable because landscapes are comprised of the abiotic and biotic ecological structures and processes of an area *and* their interrelations with the components of the human society (International Association for Landscape Ecology [IALE], see: http://www.landscape-ecology.org/web link). Thus, IALE states that landscape ecology can be portrayed by several of its core themes, such as:

- The spatial pattern or structure of landscapes, ranging from wilderness to cities
- The relationship between pattern and process in landscapes
- The relationship among human activities and landscape pattern, process, and change
- The effect of scale and disturbance on the landscape

Landscapes are human–environmental systems, and social–ecological landscapes (SEL) are the real-world objects of interest from the perspective of environmental security (Zurlini and Müller, i.p.). Applying landscape ecology thus stimulates the integration of various disciplines to develop adequate tools and methodologies for addressing environmental security issues at relevant scales.

Landscapes change constantly, triggered by natural and anthropogenic drivers. Land use and land cover changes as a result of human actions have been identified as a primary effect of humans on natural systems. Changes in land use and land cover underlie fragmentation and habitat loss, which are great threats to biodiversity and ecosystem services (MEA, 2005). Consequently, landscape change is one of the foremost themes underlying landscape research, comprising several potential feed backs with societal implications.

Assessing environmental security on the landscape scale requires consideration of two complex, yet interactive systems: the social subsystem, characterized by human intent, and the ecological system that results from a deep evolutionary history. The conceptual distinction of these components has led to a very schematic and reductionist way to address the complexity of social–

ecological landscapes in the past, since those components have been strongly interacting historically. Society strongly determines and is influenced by the landscape ecological components of the whole system (Antrop, 2006; Berkes and Folke, 1998). Taking into account the relevant security items as mentioned above, it is therefore necessary to consider that:

- Security of social–ecological landscapes must be assessed both objectively and subjectively, because security is meaningless unless there is somebody perceiving it.
- Security is value laden, and what we consider values is related to our normative systems that nowadays recognize concepts like ecosystem functions and services, ecosystem integrity and sustainability as fundamental values for the survival and well-being of mankind.
- Humans have historically threatened these values from local to global scale, but natural hazards and disasters also pose significant threats, especially given human-induced changes in landscape composition and pattern.

Within these considerations, environmental security has to do with the risks or fragilities (or vulnerabilities) of loosing ecosystem goods and services as well as the perception of those risks (Zurlini et al., 1999; Petrosillo et al., 2006; 2007). Thus, risk and fragility are deemed as multilayered, multi-scale, and complex, existing in both the objective physical and social realms, as well as in the subjective realm. The perception of security is fundamental at all levels of human organization, from the individual to the government. Therefore, perception is one of the most important aspects when addressing environmental phenomena. As Simoni and Allen (1998) say "People interact on the basis of the meanings they assign to the world around them. Individuals respond to each other in terms of their perceptions" and "Even if people are "mistaken" in their interpretation of a situation, such interpretations, nevertheless, have a "real" consequences". Following this idea, since the late 1960s, a substantial body of literature has been developed (Rohrmann, 1999) to analyze people's perception of risk, even if most of the studies refer to the perception of natural hazards.

Regarding environmental security in the subjective sense, the "threats" are of an abstract nature, in the domains of feelings and cognition. The level of fear that such values will be attacked and possibly lost largely depends on the correct information and the consciousness of the role and significance played by ecosystem goods and services. Given that both objective and subjective measures can provide reliable estimates of environmental security through indicators and monitoring programs, it is interesting to compare the concordance between "objective" and "subjective" evaluations (Figure 1).

Environmental security		
	Objective	
Subjective	++ (a)	+ – (b)
	– + (c)	– – (d)

Figure 1. Possible combinations between "objective" and "subjective" evaluations (After Zurlini and Müller, i.p.).

As an example, the environmental security of the same location might be evaluated differently in objective and subjective terms (Figure 1). In cases (a) and (d) there is concordance between objective and subjective evaluations; for case (a) both agrees on positive (high) environmental security, whereas for case (d) both agrees on negative (low) environmental security. In contrast, cases (b) and (c) are discordant; in the first case there is no fear that values such as ecosystem goods and services will be attacked and possibly lost, while the objective evaluation says just the opposite. This situation is the most dangerous but also the most common in the real world because people are often unaware of the environmental degradation they cause. In contrast, in case (c) there is fear that ecosystem goods and services will be attacked and possibly lost (low environmental security) but there is no objective reason for such fear. For example, certain Mediterranean beaches are often naturally covered by seaweed leaves (*Posidonia oceanica*), that is an indicator of good coastal ecosystem quality. However, most tourists wrongly perceive beaches covered by leaves as "dirty" and "insecure", so leaves are removed.

These differences may be focal starting points for environmental conflicts. Consequently, *the focal task of the landscape security approach is to identify and assess the overall risk due to the difference between subjective and objective environmental security and to figure out possible ways to minimize them. This science-based activity is a precondition for a rational decision-making process to be fulfilled by a satisfactory state of information and environmental knowledge.* The practical goal therefore is to develop landscape assessment approaches, including indicators and vulnerability concepts that increase our understanding of the interrelationships between humans, ecosystems, and services, and decrease the differences between perceived and actual vulnerabilities and risks to security.

4. A Framework to assess environmental security

In order to assess and manage the determinants of environmental security, a conceptual model can be adopted which is based on the DPSIR approach of the

European Environmental Agency (1999). Following this conception, the agents and factors influencing environmental security can be represented in terms of drivers, pressures, states, impacts, and responses. These factors are arranged with respect to progressions that lead to modifications of environmental security (Figure 2): Land use provokes a change of landscape states (including potential states with increasing risks), and this modification effects changes of ecosystem or landscape services, potentially including reduced provisions of goods and regulatory functions. As a consequence, several items of human well-being will be changed, leading to new societal demands, drivers, and motivations, including security relevant developments in the society. These dynamics will influence environmental decision processes in association with external constraints, and finally new opinions, how to optimally cope with the environment, will be realized in political response processes. Such an approach is fundamental to the concept of adaptive management.

The focal linkage between the ecological features of the model and the societal components is provided at the interface between ecosystem services and the components of human well-being. This is the position at which security-related values are at risk. These include:

- *Provisional services*: The products obtained from ecosystems, including, for example, genetic resources, food, fiber, and freshwater.
- *Organizational services*: The benefits obtained from the regulation capacity of ecosystem processes, including the regulation of climate, water, and some human diseases. To distinguish this class from the regulation services of the Millennium Ecosystem Assessment (MEA, 2005), also synergetic features are included, such as buffer capacities, fragilities, or resilience.
- *Cultural services*: The nonmaterial benefits people obtain from ecosystems through spiritual enrichment, cognitive development, reflection, recreation, and aesthetic experience, including, e.g. knowledge systems, social relations, and aesthetic values.

From a synoptic look at the categories of ecosystem services, one fact becomes obvious: all services are strongly dependent on the performance of the organizational (regulation) functions (e.g. Müller, 2004). The correlated processes do not only influence processual linkages, but on the long run they also determine the potentials of ecosystems to provide material products and cultural services. Therefore, the performance of all service types is based upon *landscape integrity* (Müller, 2005; Petrosillo et al., 2005). All of these provisions influence the features of human well-being, as shown in Box 3. At this position of the information cycle in Figure 2, security relevant decisions are met: Does the environmental situation support human well-being or is there a decrease? If this is the case, the overall degree of security for a certain group of humans might be endangered.

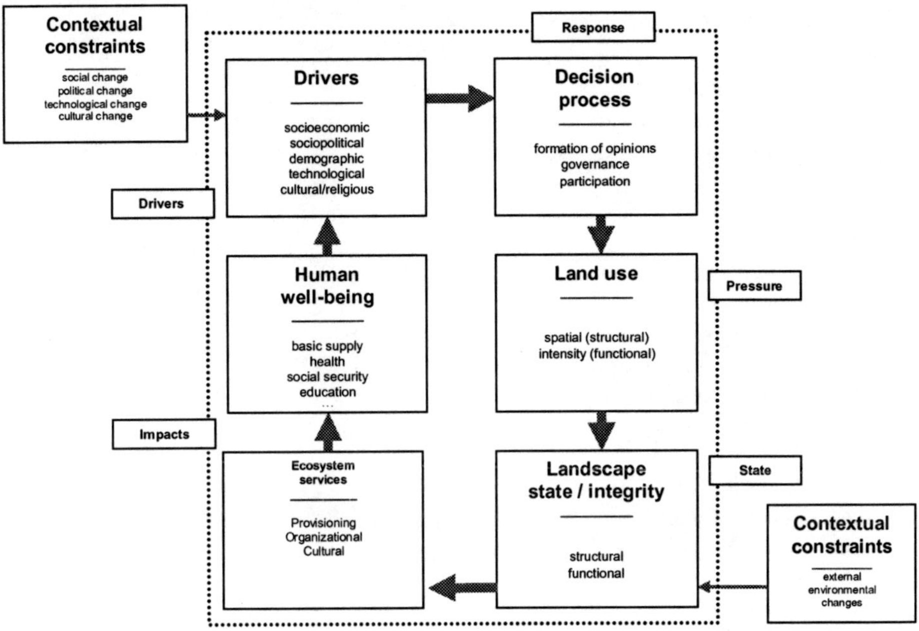

Figure 2. A conceptual DPSIR model of human–environmental systems, focusing on the risk of loosing ecosystem goods and services (after Müller, 2004).

The ecological key variables within this human–environmental framework are landscape integrity and self-organization. The former represents a complex systems approach that is mainly based upon functional variables of energy and matter flows and budgets and structural features of whole ecosystems (Barkmann, 2002). Taking into account the focal ideas of the security concept, it is possible to use an alternative formulation for the ecological components of sustainable development in this context: "meet the needs of future generations" means "keep available ecosystem services on a long-term, inter-generational, and broad scale, intra-generational level". As self-organization represents the integral component of the organizational services, it becomes clear that the benefits to environmental security are strictly dependent on the degrees and the potentials of self-organization processes (Kay, 2000; Müller, 2004).

To maintain these services, the ability for future self-organizing processes within the respective system has to be preserved for the social–ecological system as a whole. Applying this viewpoint, we can define ecological integrity as a "political target for the preservation against non-specific ecological risks that are general disturbances of the self-organizing capacity of ecological systems. Thus, the goal should be a support and preservation of those processes and structures which are essential prerequisites of the ecological ability for self-organization" (Barkmann et al., 2001).

Box 3. Some features of human well-being. (Source: Millennium Ecosystem Assessment, 2005[‡]; Burkhard and Müller, i.p.)

Material well-being
- Adequate livelihoods
- Sufficient nutritious food
- Shelter and accommodation
- Access to goods
- Job security, labor

Health
- Life expectancy
- Strength
- Feeling well
- Access to clean air and water

Political stability and security
- Personal safety
- Secure resource access
- Security from disasters

Education and information

Political freedom of choice and action

Social relations
- Social cohesion
- Mutual respect
- Gender equality
- Family life

In the context of environmental security, it is crucial to address integrity in very broad sense and at different hierarchical levels in the panarchy, integrating terms of quality of life, cultural identity, and self-organization potential of the whole social–ecological landscape in focus, meanwhile considering constraints and conditionings from upper and lower hierarchical levels, reducing the gap between the real and the perceived state of the environment, by targeted environmental information and educational activities. This integrated multilevel approach could help to reach a development linked with an enhancement of self-organizing processes leading to a corresponding enhancement of environmental security.

To realize that strategy, some theory-based requirements have to be fulfilled (see also Petrosillo et al., this volume), including the concept of social–ecological landscapes, the concept of self-organization (Joergensen and Müller, 2000), the ecosystem approach (CBD, 2005, see www.biodiv.org/), hierarchy theory (O'Neill et al., 1986; Gunderson and Holling, 2003; Zurlini et al., 2006;

[‡] see http://earthtrends.wri.org/text/index.php

this volume), and the approaches of resilience, fragility, and buffer capacities. Applying these features, it has to be realized that breakdown, decay, and novelty are intrinsic features of system development processes. An evaluation of the respective consequences is an important task of adaptive management (see Petrosillo et al., this volume). Basing on these conceptual items, the environmental security assessment techniques described in this book follow a logical sequence, laid out by Müller et al. (2003):

- Landscape identification and characterization, providing the basic and initial information of the area in question
- Land cover change detection and analysis of landscape dynamics, providing information about the changes of important landscape features in time
- Landscape indicators, providing important and representative variables for the aggregated characterization of landscape structures and functions
- Landscape theory, providing basic ideas and concepts for an improved understanding of landscape development and potential vulnerabilities and risks
- Landscape scenarios and modeling, providing synthesized understanding of structures and processes and delivering basic information about management consequences for decision-makers
- Landscape assessment and adaptive management, providing concepts for evaluation and manipulation of landscapes in an iterative manner

5. Security assessment in this book

Taking into account these security related items and concepts, this book follows three main objectives:

- To demonstrate the potentials of landscape science in addressing environmental security
- To identify significant landscape components and their organizational structure related to environmental security
- To show methods to evaluate landscape security at different scales and to introduce this concept into landscape management

The focal starting point of the resulting landscape security concept is to reduce differences between the perceived risks and the actual (realistic) threats of a certain situation. The degrees of subjective and objective security differ, i.e. due to a lack of intersubjective information. Thus, the task of landscape science and adaptive management is to minimize this gap and to contribute to decision-making processes by providing knowledge, methodology, and information. This target will be discussed referring to the following list of questions:

- What indicators, models, or other monitoring and assessment approaches capture important aspects of environmental security at landscape scales? To what degree do they capture each of the five categories of the DPSIR framework?
- Which goods and services are critically related to environmental security? How well do existing data, indicators, and models capture these? Which indicators and models capture multiple environmental themes and/or goods and services?
- Which landscape properties indicate potential vulnerability? Can these be measured from existing data, models, and approaches?
- What methods and approaches can be used to relate components of the DPSIR framework? How well do these approaches address issues of scale?
- How can landscapes be optimized to increase environmental security and reduce hazards and risks?

Moreover, the landscape approach described in this book uses a different frame of reference and organizing principles than traditional environmental assessments. With regards to provisional, regulatory (organizational), and cultural services, and environmental security it explicitly recognizes the importance of spatial variation in land use and landscape condition, and the interrelationships across spatial domains. It explicitly relates the spatial pattern and interactions of humans and environmental features as they relate to ecological services and risks to environmental security. By understanding these spatial relationships and interactions, it is possible to develop a set of management recommendations to improve and protect ecological services upon which society depends. This is the unique and primary contribution of the landscape sciences described in this book.

Sections in this book address these questions and suggest a number of landscape sciences approaches that provide a unique perspective to various aspects of environmental security assessments. In Part II of this book, some general features of environmental security will be presented, and the implications of the security approach for environmental management and policy will be described and exemplified. Part III highlights the applicability of landscape ecological techniques to assess environmental security. In Part IV, various approaches are presented for the indication of landscape states and their dynamics. Part V provides different case studies of landscape analyses of watersheds at a range of spatial scales. These papers also demonstrate the potential of models for the understanding and management of landscapes processes within watershed analysis units. In Part VI of the book, human influences on socio–ecological landscapes are the primary focus. Among several land use-based conflicts, the potential consequences of demographic change on the ecological states of landscapes are important subjects of these articles. Part VII includes papers on landscape impact assessments, as well as a set of summary comments, conclusion, and recommendations related to the application of the landscape sciences toward understanding and achieving environmental security.

References

Antrop, M., 2006. Sustainable landscapes: contradiction, fiction or utopia? Landscape and Urban Planning 75, 187–197.

Barkmann, J., 2002. Modellierung und Indikation nachhaltiger Landschaftsentwicklung – Beiträge zu den Grundlagen angewandter Ökosystemforschung. Diss. University of Kiel.

Barkmann, J., Baumann, R., Meyer, U., Müller, F., and Windhorst, W., 2001. Ökologische Integrität: Risikovorsorge im Nachhaltigen Landschaftsmanagement. Gaia 10(2), 97–108.

Barnett, J., 1997. 'Environmental Security: Now What?' Seminar, Department of International Relations, Keele University.

Belluck, D.A., Hull, R.N., Benjamin, S.L., Alcorn, J., and Linkow, I., 2006. Environmental security, critical infrastructure and risk assessment: definitions and current trends. In: B. Morel and I. Linkow (Editors), Environmental Security and Environmental Management. Springer, Dordrecht, pp. 3–16.

Berkes, F. and Folke, C. (Editors), 1998. Linking Social and Ecological Systems: Management Practices and Social Mechanisms for Building Resilience. Cambridge University Press, Cambridge.

Brauch, H.G., 2006. Desertification: a new security challenge for the Mediterranean? In: W.G. Kepner, J.L. Rubio, D.A. Mouat, and F. Pedrazzini (Editors), Desertification in the Mediterranean Region: A Security Issue. Springer, Dordrecht, pp. 11–85.

Burkhard, B. and Müller, F. (in press). Indicating ecosystem health and integrity. In: A. Denhard (Editor), Sustainability in River Basins – A Question of Governance. Springer, Dordrecht.

Dabelko, G.D., 2004. The next steps for environment, population, and security. Environmental Change and Security Project Report 10. Woodrow Wilson International Center for Scholars, Washington, DC, pp. 3–7.

Dabelko, G., Lonergan, S., and Matthew, R., 2000. State-of-threat review on environment, security and development cooperation. Working Paper. OECD DAC Working Party on Development Cooperation and Environment, OECD, Paris. Online: http://www.oecd.org/dataoecd/8/51/2446676.pdf.

De Soysa, I., Gleditsch, N.P., Gibson, M., Sollenberg, M., and Westing, H., 1999. To cultivate peace: agriculture in a world of conflict. PRIO Report 1/99. PRIO, Oslo. Online: http://www.futureharvest.org/peace/report.pdf.

EEA [European Environment Agency], 1999. Information for Improving Europe's Environment. EEA, Copenhagen.

Gunderson, L.H. and Holling C.S. (Editors), 2003. Panarchy: Understanding Transformations in Human and Natural Systems. Island Press, Washington, DC.

Herrero, S.T., 2006. Desertification and environmental security: the case of conflicts between farmers and herders in the arid environments of the Sahel. In: W.G. Kepner, J.L. Rubio, D.A.

Mouat, and F. Pedrazzini (Editors), Desertification in the Mediterranean Region: A Security Issue. Springer, Dordrecht, pp. 109–132.

Joergensen, S.E. and Müller, F., 2000. Ecosystems as complex systems. In: S.E. Joergensen and F. Müller (Editors), Handbook of Ecosystem Theories and Management, CRC Press, Boca Raton, FL, pp. 5–20.

Kay, J.J., 2000. Ecosystems as self-organised holarchic open systems: narratives and the second law of thermodynamics. In: S.E. Joergensen and F. Müller (Editors), Handbook of Ecosystem Theories and Management, CRC Press, Boca Raton, FL, pp. 135–160.

Kepner, W.G., 2006. Introduction: desertification and security. In: W.G. Kepner, J.L. Rubio, D.A. Mouat, and F. Pedrazzini (Editors), Desertification in the Mediterranean Region. Springer, Dordrecht, pp. 3–9.

Liotta, P.H., 2006. The Poseidon prairie. In: W.G. Kepner, J.L. Rubio, D.A. Mouat, and F. Pedrazzini (Editors), Desertification in the Mediterranean Region: A Security Issue. Springer, Dordrecht, pp. 87–108.

Matthew, R.A., Brklacich, M., and McDonald, B., 2004. Analyzing environment, conflict, and cooperation. In: UNEP (United Nations Environment Programme), Understanding Environment, Conflict and Cooperation, UNEP, Division of Early Warning and Assessment (DEWA), Nairobi, pp. 5–15.

Millennium Ecosystem Assessment [MEA], 2005. Ecosystems and Human Well-being: Synthesis. Island Press, Washington, DC.

Müller, F., 2004. Ecosystem indicators fort he integrated management of landscape health and integrity. In: S.E. Joergensen, R. Costanza, and X. Fu-Liu (Editors), Ecological Indicators for Assessment of Ecosystem Health, Taylor & Francis, Boca Raton, FL, pp. 277–303.

Müller, F., 2005. Indicating ecosystem and landscape organization. Ecological Indicators 5(4), 280–294.

Müller, F., Kepner, W., and Cäsar, K., 2003. Landscape sciences for environmental assessment. Proceedings from a CCMS pilot study meeting at Salzau Castle, Germany. EcoSys Bd. 10.

O'Neill, R.V., De Angelis, D.L., Waide, J.B., and Allen, T.H.F., 1986. A hierarchical concept of ecosystems. Monographs in Population Ecology 23, Princeton University Press.

Petrosillo, I., Zurlini, G., Kepner, W., and Müller, F. (Editors), 2005. Linkages among landscape assessment, quality of life and environmental security. EcoSys Bd. 11.

Petrosillo, I., Zurlini, G., Grato, E., and Zaccarelli, N., 2006. Indicating fragility of socio-ecological tourism-based systems. Ecological Indicators 6, 104–113.

Petrosillo, I., Zurlini, G., Corlianò, M.E., Zaccarelli, N., and Dadamo, M., 2007. Tourist perception of recreational environment and management in a marine protected area. Landscape and Urban Planning 79, 29–37.

Rohrmann, B., 1999. Risk Perception Research – Review and Documentation. Research Center Juelich, RC Studies 68.

Simoni, M. and Allen, P., 1998. Public perception of risk management in environmental controversies, a U.K. case study. Risk, Health, Safety & Environment 9, 309.

Tuchel, D., 2004. Romanians warn of Danube Delta ecological disaster. Environmental News Service, September 28, 2004.

UNEP [United Nations Environment Programme], 2004. Understanding environment, conflict and cooperation. UNEP, Division of Early Warning and Assessment (DEWA), UNEP, Nairobi. Available from http://wwics.si.edu/topics/pubs/unep.pdf.

WCED [World Commission on Environment and Development], 1987. Our Common Future. Oxford University Press, New York.

Wolfers, A., 1962. Discord and Collaboration. Essays on International Politics. John Hopkins University Press, Baltimore, MD.

Zurlini, G., Amadio, V., and Rossi, O., 1999. A landscape approach to biodiversity and biological health planning: the map of Italian nature. Ecosystem Health 5, 296–311.

Zurlini, G., Riitters, K., Zaccarelli, N., Petrosillo, I., Jones, K.B., and Rossi, L., 2006. Disturbance patterns in a socio-ecological system at multiple scales. Ecological Complexity 3, 119–128.

Zurlini, G. and Müller F. (in press), Environmental security and system ecology. In: Encyclopedia of Ecology. Amsterdam.

PART I INTRODUCTION

ENVIRONMENTAL SECURITY AND LANDSCAPE ECOLOGY

GIOVANNI ZURLINI
Laboratory of Landscape Ecology – Department of Biological and Environmental Sciences and Technologies – University of Salento, Lecce, Italy

The relationship between environmental change, stress, biodiversity loss, and environmental degradation relative to the issue of security has garnered increased importance as new challenges have emerged since the end of the cold war. Results of the increasing human appropriation of regional landscapes can have a variety of ecological effects directly drawing on the notion of environmental security. Focusing on environmental security is the essential step to be developed in the study of interactions between humans and the environment in social–ecological landscapes (SEL), and it is critical in understanding how humans create and respond to environmental change. The real challenge for environmental security is dealing with systems that are not only cross-scale but also dynamic, whereby the nature of cross-scale influences in the linked social–ecological system changes over time, creating fundamental problems for division of responsibility among centralized and decentralized agents. Because the self-organizing properties of complex social–ecological systems and associated management systems seem to cause uncertainty to grow over time, understanding should be continuously updated and adjusted, and each management action viewed as an opportunity to further learn how to adapt to changing circumstances.

This challenge gives the scientist a new role in decision-making from being an objective and detached specialist expected to deliver knowledge to managers to becoming one of several actors in the learning and knowledge-generation process.

LANDSCAPE ECOLOGY AND ENVIRONMENTAL SECURITY: BASIC CONCEPTS AND REGIONAL APPLICATIONS FOR THE MEDITERRANEAN IN THE 21ST CENTURY

HANS GÜNTER BRAUCH[*†]
Free University of Berlin, Otto-Suhr-Institute of Political Science Peace Research and European Security Studies (AFES-PRESS), Mosbach, Germany

Abstract. This chapter offers a framework of analysis for studying linkages between landscape ecology and environmental security with a special focus on environmental threats, challenges, vulnerabilities, and risks on the common and differentiated trends on both shores of the Mediterranean between 1950 and 2050. The chapter reviews the debate on reconceptualizing of security for different worldviews of analysts, mind-sets of policymakers, scientific schools, and research programs in the social sciences. The PSR model by OECD and the DPSIR model by EEA will be developed further into a PEISOR model that will be illustrated for the Mediterranean region by addressing both common geo-ecological trends (air, water, soil) and major differences on either shore (population, agriculture, urbanization) by referring to common environmental and different socioeconomic challenges in the Mediterranean. Extreme outcomes will be addressed such as hazards, migration, environmental crises, and conflicts. The chapter concludes with contributions of landscape science to vulnerability mapping and early warning.

Keywords: Security; PSR; DPSIR and PEISOR model; Mediterranean; global environental change; climate change; natural hazards; landscape ecology

1. Introduction: disciplines and research questions

Landscape ecology as a discipline in the geosciences focuses on interactions between living communities and their environmental conditions. While geo-

[*]Corresponding Author: Dr. Hans Günter Brauch, Otto-Suhr-Institute of Political Science Peace Research and European Security Studies (AFES-PRESS), Alte Bergsteige 47, 74821 Mosbach, Germany; e-mail: brauch@onlinehome.de

[†]This chapter relies on previous work by Brauch (2000, 2001, 2003, 2003a, 2003b, 2003c, 2005, 2005a, 2005b, 2006, 2006a, and on forthcoming books (Brauch et al., 2007, 2007a, 2008). The copyright for all figures remains with the author and Springer Heidelberg.

I. Petrosillo et al. (eds.), Use of Landscape Sciences for the Assessment of Environmental Security, 21–42.

ecology (Huggett, 1995) deals with the abiotic subsystem within the ecosystem, landscape ecology aims at a holistic approach that combines abiotic, biotic, and anthropogenic interactions. Troll has merged the analysis of landscapes (physic geography) with ecology (biology). Remote sensing and geographical information systems have become important tools for analyzing anthropogenic land use changes that are used in manifold practical applications in spatial planning and nature conservation and protection. Landscape ecology has become a multidisciplinary research programme and a practical approach (Lexikon der Geowissenschaften, 3, 2001: 232–233; Lexikon der Geographie, 2, 2002: 308). The analysis of security concepts and problems has been a topic of many disciplines in the social and human sciences, from philosophy, international law to sociology and political science and of its programs: policy analysis and international relations. Peace, security, development, and environment form a conceptual quartet whose linkages have been analyzed from the respective work programs of security studies, peace and conflict research, development and environmental studies.

This chapter focuses on the linkage between the environment – or more specifically the segment covered by landscape ecology – and security, that have been discussed as "environmental" or synonymously as "ecological security". It first introduces into the debate on security concepts, reviews four models on the relationship between nature and humankind before introducing the author's own PEISOR model. Two aspects: six drivers of global environmental change and climate-related natural hazards are illustrated for the Mediterranean region and the potential contribution of landscape ecology for enhancing environmental and human security is discussed.

2. Security analysis: worldviews, mind-sets, schools, and programs

As a scientific concept, "*security* is ambiguous and elastic in its meaning" (Art, 1993: 820). Wolfers (1962) pointed to two sides of the security concept: "Security, in an *objective sense*, measures the absence of threats to acquired values, in a *subjective sense*, the absence of fear that such values will be attacked." For social constructivists security is "what actors make of it" (Wendt, 1992), thus adding an *intersubjective* meaning.

The *perception* of security threats, challenges, vulnerabilities, and risks (Brauch, 2003, 2005) depends on the worldviews or traditions of the analyst and on the mind-set of policymakers. Three basic views have been distinguished by the English school (Bull, 1977) that of: (a) *Hobbesian* pessimism (realism) where *power* is the key category; (b) *Kantian* optimism (idealism) where *international law* and *human rights* are crucial; and (c) *Grotian* pragmatism where *cooperation* is vital (Brauch, 2003, 2005). This Eurocentric typology refers to three "ideal type" approaches to security that also exists in other cultures and philosophical traditions. Booth argued that "old mind-sets" have often distorted the assessment of "new challenges", and that many old mind-sets have survived the global contextual change of 1989–1990 unchanged (Booth, 1998: 28).

Influenced by these worldviews and mind-sets, security is a key concept of competing schools of (a) *war*, *military*, *strategic*, or *security studies* from a Hobbesian or realist perspective, and (b) *peace and conflict research* that has focused on conflict prevention from a Grotian pragmatist and/or Kantian idealist view. Since 1990, the distance between both schools had narrowed, but since 2002 it has again widened. Since the end of the cold war many authors have observed a "*widening*" of the security concept in OECD countries to five dimensions (Buzan et al., 1998), a "*deepening*" (from the "state" to other "referents" and levels of analyses), and a "*sectorialization*" (energy, food, health, water, etc.), while in other countries a narrow military security concept has prevailed (Brauch et al., 2003).

This *widening* and *deepening* of the security concept since 1990 is illustrated in Table 1 that refers horizontally (⇨) to the five security dimensions and vertically (⇩) to five levels of interaction (or reference points) with a special focus on the environmental security dimension where the complex interaction between human beings and humankind (🡫) and the global and planetary level of nature (🡩) has been indicated by arrows.

TABLE 1. Widening and deepening of security concepts. (Source: Brauch, 2003, 2005.)

Security dimension ⇒ Level of interaction (reference point) ⇓	Military	Political	Economic	Social	Environmental 🡫 (longer-term environ-. mental challenges)
Human ➔					Cause and victim
Societal/Community					🡫🡩
National (short-term threats)	Middle East discourses on "security dilemma"				🡫🡩 „survival dilemma"
International/Regional					🡫🡩
Global/Planetary ➔					GEC

Table 2 responds to three terms (reference object, value at risk, sources of threat) and questions (security of whom, of what, and from whom or what?) posed by Wolfers (1962) distinguishing between five security concepts: the classic *national security* and the more recent *societal*, *human*, *environmental*, and *gender security* concepts that have been widely used in the social sciences and in political discourses since 1990 relying on the categorization that was suggested by Møller (2003) and Oswald (2001).

TABLE 2. Expanded concepts of security. (Source: Møller, 2003; Oswald, 2001.)

	Reference object (security of whom?)	Value at risk (security of what?)	Source(s) of threat (security from whom or what?)
National security (political, military dimension)	The state	Sovereignty, territorial integrity	Other states, terrorism (substate actors)
Societal security	Nations, societal groups	National unity, identity	(States) Nations, migrants, alien cultures
Human security	Individuals humankind	Survival, quality of life	State, globalization, GEC, nature, terrorism
Environmental security	Ecosystem	Sustainability	Humankind
Gender security	Gender relations, indigenous people, minorities	Equality, identity, solidarity	Patriarchy, totalitarian institutions (governments, religions, elites, culture), intolerance

Landscape ecology focuses on the environmental security dimension where the human beings and humankind have acted as drivers of global environmental change (GEC) and become victims of extreme weather events: storms (winter, hurricanes, cyclones), flash floods and flooding, drought, and forest fires whose numbers and intensity has grown due to climate change (IPCC, 2001).

But differences existed in the "referent object" of environmental security analyses. While the first three phases of environmental security research has been "state-centred" by adding a new "mission" to the US national security agenda (Ullmann, 1983; Mathews, 1989; Myers, 1989), others suggested that its research agenda should be addressed from a "people-centred" perspective (Barnett, 2001). UNDP (1994) saw environmental security as one of seven components of "human security". While GECHS (1999) focused on the linkages between global environmental change and human security, UNU-EHS (Bogardi and Brauch, 2005; Brauch, 2005, 2005a) will develop the environmental dimension of human security by addressing water-related hazards. Oswald (2001) has combined HUman, Gender and Environmental security (HUGE).

3. Hard security threats and environmental security challenges

This "widening" of the security concept from its narrow political and military core (hard security) during the cold war, to enclose economic, societal, and environmental dimensions (soft security), was complemented by a "deepening" by shifting from the nation state ("national security") and military alliances ("alliance security") to the people or humankind ("human security"). Furthermore, a "sectorialization" of security could be observed as food, health, and water security that are also relevant for landscape ecology for the analysis of overgrazing, deforestation, soil erosion, desertification, and drought in arid and semiarid regions (Brauch, 2006b).

Three different approaches to security analysis are distinguished: objective (absence of threats), subjective (absence of fear), and intersubjective. From the perspective of the Copenhagen school securitization refers to a "discursive and political process through which an intersubjective understanding is constructed within a political community to treat something as an existential threat to a valued referent object, and to enable a call for urgent and exceptional measures to deal with the threat" (Waever, 2007). According to this constructivist theory the "referent object" (that is threatened and holds a general claim on "having to survive") points to the state or the environment or humankind. The "securitizing actor" makes the claim (in a speech act) "pointing to an existential threat to a referent object thereby legitimizing extraordinary measures," and the "audience" (have to be convinced in order for the speech act to be successful in the sense of opening the door to extraordinary measures). According to this theory a key question is: "How does the politics of a given issue change when it shifts from being a normal political issue to becoming ascribed the urgency, priority and drama of 'a matter of security'" (Waever, 2007).

For the analysis of the environmental dimension of human security this author distinguished between four concepts with regard to (objective) security dangers and (subjective) security concerns: threats, challenges, vulnerabilities, and risks that are often used synonymously but at the same time they have, e.g. the "vulnerability concept", many different meanings in the development, environment, global environmental change, in the global change, hazard, and early-warning communities. This author suggested referring to "threats" only for hard security and to "challenges" for nonmilitary or soft security (objective) dangers and (subjective) concerns (Brauch, 2005, 2006).

With regard to landscape ecology, the question is which anthropogenic changes in the interaction between humankind and nature, or human beings living in a specific landscape pose challenges, vulnerabilities and risks to other human beings, to their livelihood, well-being, and survival. According to Table 2, the "referent object" of environmental security is the ecosystem, or more specifically those parts that are environmentally vulnerable, the "value at risk" is sustainability and the "source of the threat or challenge" is humankind that is also a victim of natural hazards. However, those who cause climate change and the primary victims of climate-related hazards are not identical.

4. Model: global environmental change and fatal outcomes

Instead of a simple stress-response model that claims direct links between stress factors, environment change and societal responses, the *Pressure-State-*Response (PSR) model of OECD (1994, 1998, 1999, 2001, 2001a) assumes that human activities put pressure on nature that leads to environmental changes (climate change, water and soil degradation, biodiversity loss) to which the state and society respond with ecological and economic measures and programs.

The OECD's PSR model distinguished between "pressure" (P), "state of the environment" (S), and "response" (R) indicators. Among "*pressure*" key factors are listed (population growth, consumption, poverty), while "state" refers to the environmental conditions that emerge from this pressure (air pollution, deforestation, degradation) that influence human health and well-being, and "*response*" points to the manifold activities of society to avoid, prevent, and reduce negative impacts on the environment and to protect natural resources from these effects. Among the *pressures* are human activities in the energy, transport, industry, and agricultural sector on natural resources (air, water, soil, organisms) to which the state, society, business, and international actors respond. Between these three elements of the PSR model there are many complex interactions (resource transfers, information, decisions).[‡]

The *UN Commission for Sustainable Development* (UN-CSD) used with its DSR (*Driving Force-State-Response*) model a slightly modified framework.[§] The *European Environment Agency* (EEA, 1998) has developed a framework that distinguishes "*D*riving Force–*P*ressure–*S*tate–*I*mpact–*R*esponse" (DPSIR)[**]

[‡] See: Jesinghaus (n.d.), at: <http://esl.jrc.it/envind/theory/handb_03.htm >: "The PSR model was developed in the 1970s by the Canadian statistician Anthony Friend, and subsequently adopted by the OECDs State of the Environment (SOE) group; for an illustration see at: <http://www.virtualcentre.org/en/dec/toolbox/Refer/EnvIndi.htm>.

[§] See: UN Commission for Sustainable Development: "Indicators of Sustainable Development", at: <http://www.un.org/esa/sustdev/isd.htm>.

[**] See: Jochen Jesinghaus: "European System of Environmental Pressure Indices", at: < http://esl.jrc.it/envind/theory/handb_03.htm>.

that offers a mechanism for the analysis of environmental problems and for the development of environmental indicators (Figure 1).[††]

A different model was used as a framework for the *Millennium Ecosystem Assessment* (MA, 2002; MEA, 2005) that distinguished between direct and indirect drivers of change that directly affect human well-being and ecosystem services. In this framework besides the material minimum for a good life, health, and good social relations, security is considered as one of the key elements of human well-being that influence the freedom of choice. Security has been defined as: (a) the ability to live in an environmentally clean and safe shelter, and (b) the ability to reduce vulnerability to ecological shocks and stress (MEA, 2005; Leemans, 2007). In the words of the *Human Security Commission* (CHS, 2003) this refers to two basic principles: "protection" and "empowerment".

These four models have served a specific purpose: to develop environmental indicators and to guide the ecosystem assessment. But they did not focus – from a security perspective – on the linkages between processes of global environmental change (GEC) and natural hazards and their sociopolitical consequences.

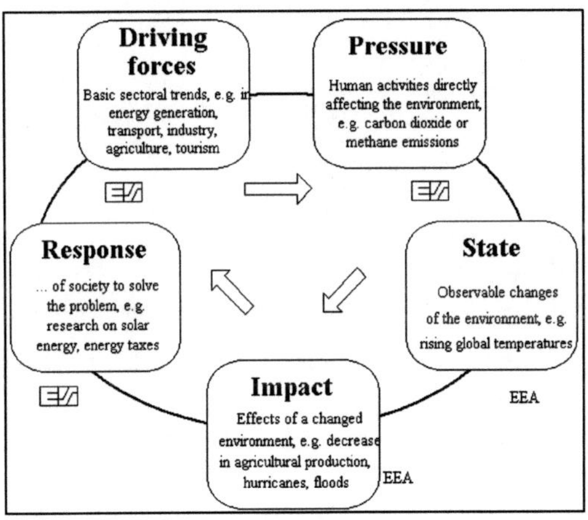

Figure 1. DPSIR model of EEA. (Source: Jesinghaus (n.d.), at: <**http://esl.jrc.it/envind/ theory/ handb_03. htm**>.)

This is the goal of the PEISOR model (Brauch, 2005; 2006) where:

- *P* (*pressure*) refers to six drivers of global environmental change (*survival hexagon*)

[††]See: European Commission: "Towards Environmental pressure Indicators for the EU", at: <http://www.e-m-a-i-l.nu/tepi/firstpub.htm>.

- *E* to the *effects* of the linear, nonlinear, or chaotic interactions within the "hexagon" on environmental scarcity, degradation, and stress
- *I* to extreme or fatal *impacts* of human-induced and climate-related natural hazards (storms, flash floods, flooding, land slides, drought)
- *SO* to *societal outcomes*: internal displacement, migration, urbanization, crises, conflicts, state failure
- *R* to *response* by the society, the business community, the state where both traditional and modern technological knowledge can make a difference

While hazards cannot be prevented, their impact in terms of deaths, affected people, economic and insured damages can be reduced by a combination of policies and measures that link protection with empowerment of the people to become more resilient.

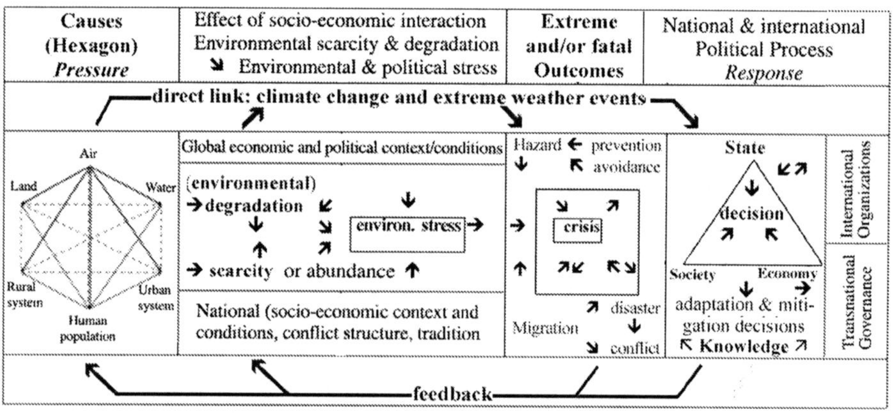

Figure 2. PEISOR Model. (Source: Brauch, 2005.)

Figure 2 refers in the first column under causes or pressure to six key factors contributing to GEC, three supply or environmental factors (land, air, and water), and three demand or human factors (population living and working in rural or urban systems). These six factors interact in a nonlinear or chaotic way and pose pressure on the political and societal context where they may trigger, impact, or affect socioeconomic interactions either causing or contributing to *environmental degradation* (of water, soil, air) or *scarcity* (of water and soil). The interaction among these two processes may result in *environmental stress* that can cause various extreme and in a few cases even fatal impacts. However, there may also be a direct impact of climate change resulting in an increase in hydrometeorological hazards. Environmental stress may increase the impact of hazards (especially for those with a high degree of social vulnerability) and cause or contribute (with natural hazards and conflicts) to internal displacement, urbanization, and to transboundary migration. Whether these factors result in domestic crises, disasters, and in a few worst cases in violent conflicts or whether these can be avoided depends on many specific factors and activities resulting from the interaction between the three actors representing the state, the society, and the business community but also on the use of both traditional and modern technical and organizational knowledge and knowledge-based response strategies, by international organizations and transnational societal and economic organizations (governance).

The PEISOR model will be illustrated next for the impacts of the six factors of global environmental change on the environmental dimension of human security for the Mediterranean space during the 20th and 21st centuries that will have severe repercussions on landscape ecology by posing new challenges to environmental and human security.[‡‡]

5. Mediterranean space: geo-ecological commonalities versus differences

There is no accepted definition of the Mediterranean – its space, region, climate, and way of life – that combines Europe, Africa, and Asia. There is a consensus that it is a region that was the cradle of several civilizations and of three monotheistic religions. The Mediterranean is characterized by *unity* and *diversity*, by periods of cooperation and conflict, of tolerance and violent intolerance, by intensive cultural exchange and cultural clashes, of close economic cooperation, interdependence but also by exploitation, unequal exchange, and dependence. These contradictions challenge the uncritical application of prevailing security concepts and social-science approaches (Brauch, 2001, 2003, 2003a).

The only existing pan-Mediterranean regime emerged from the *Convention for the Protection of the Marine Environment and Coastal Region of the Mediterranean* (*Barcelona Convention*) that entered into force on 12 February 1978 (Haas, 1990, 1993). From an environmental perspective, this Barcelona regime refers to the narrow and truly "Mediterranean space".

After the end of the cold war several political and military strategists have pointed to new uncertainties, challenges, risks, or even threats confronting Europe or the Euro-Atlantic region. The nature of the challenge seems to have grown with the distance of the observers from the Mediterranean region and for some these "new risks" have been used to legitimize a missile defence system. Others, primarily in southern Europe, have pointed to new *soft security challenges*, e.g. illegal migration and environmental stress.

‡‡ Brauch 2001; 2003; 2006; see for speeches by the author in Lecce (2004), Bonn (2005), Israel, Mexico and Thailand (2006), at: <http://www.afes-press.de/html/download_hgb.html>.

This author (Brauch, 2000, 2001, 2003, 2003b) has pointed to six fundamental long-term challenges that will confront the Mediterranean in the 21st century:

- Different levels of population growth on the northern, southern, and eastern shores
- The regional impact of global climate change due to projected increases in temperature, decline in precipitation, and rise of the sea level
- The scarcity of water for drinking and irrigation
- The decline in self-sufficiency in food production and the increasing need for imports of cereals ("virtual water", Allan, 2003)
- The progressing soil erosion and desertification confronting all countries around the Mediterranean
- The progressing urbanization and pollution in major coastal cities in the MENA

These six factors form a "*survival hexagon*" (Figure 2; Brauch, 2000: 305; 2003). None of these factors poses a direct military security threat but both the *eco-systemic* (climate, water, soil) in conjunction with the *anthropogenic* (population, urban, and rural systems) drivers of GEC and of regional environmental change in the Mediterranean pose environmental security challenges, and create many vulnerabilities and risks for the people living in the basin and especially along the coastline and in the delta regions (Nile, Po).

For the analysis of "environment" and "security" linkages a focus on the interactions of natural processes in the Mediterranean geographical space – including it specific landscapes – with human activities and their possible impacts on the political region is needed. Within the Mediterranean these are common features:

- *Climate change*, and weather extremities (droughts, floods) do not distinguish between Turino or Tanger, Almeria or Algiers (Brauch, 2002).
- *Water scarcity* will affect Andalusia and Murcia in the same way as it will affect semiarid and arid regions in the south in North Africa and in the Middle East.
- *Soil erosion and desertification* is a challenge not only for North Africa, the Middle East, but also for large parts of Spain, Sicily, and southern Italy (Brauch, 2003, 2006b).

But there are major differences among the other three factors:

- *Demography*: increase pressure in the south, relieve in the north

- *Food scarcity and surplus*: the gap will become more severe
- *Urbanization and pollution*: urgency and the sources to cope with them are different in the north and south

The Mediterranean environment within the natural boundaries of the Sahara desert, Europe, and the Atlantic Sea does not respect national borders.

5.1. COMMON ENVIRONMENTAL CHALLENGES IN THE MEDITERRANEAN UNTIL 2100

The *Mediterranean climate* implies complex interactions between global climate change and regional impacts that requires an analysis of global processes and regional activities. For the progressing desertification an analysis of local activities, of anthropogenic and natural processes, and their interaction with climate change is essential. Both climate change and desertification may constrain the water supply and food production. But for the analysis of population policy and urbanization national policies set the framework. Population growth contributes to the increasing demand for water and food.

While for the analysis of "environmental security" issues an environmental spatialization of the Mediterranean applies, for policies avoiding extreme societal outcomes, including conflicts emerging from medium-term implications of long-term structural causes that transgress national borders, early and joint action by the countries participating in the wider Euro-Mediterranean space of the Barcelona EMP framework is needed.

In its first three assessment reports (IPCC, 1990, 1996, 2001) and in a regional report (1PCC, 1998) the *Intergovernmental Panel on Climate Change* has not offered an integrated assessment of the climate change impacts for the Mediterranean region. According to the IPCC (2001) global average temperature has increased during the 20th century by an average of 0.6°C and is projected to rise between 1990 and 2100 between 1.4°C and 5.8°C, and the sea level has risen by 10–20 cm and is projected to rise during the 21st century between 9 and 88 cm.

Parry (2005), a cochairman of WG II of the IPCC pointed to these trends for Europe: the summers will get drier in southern Europe, and the winters will become wetter in northern Europe with an increase of the intensity of rainfall what will lead to increasing risks of flooding in northern Europe and in the Alps. In Europe, hot summers will double in frequency by 2020, and will be ten times as frequent by 2080. Does climate change in Europe pose challenges for environmental and human security, especially for sensitive coastal regions or Alpine landscapes? Table 3 illustrates potential impacts of climate change posing security treats, challenges, vulnerabilities, and risks.

TABLE 3. Compilation of environmental threats, challenges, vulnerabilities, and risks.

Environmental causes, stressors, effects, and natural hazards pose	Natural and economic factors		Societal impact factors	
	Substantial threats for	Challenges affecting	Vulnerabilities for	Risks for
		Security objects (for what or whom?)		
Climate change – Temperature increase (creeping, long term)	– Human health – Agriculture (yield decline) – Biodiversity – Desertification	– Tourism – Food security – Fisheries – Government – Economy	– Inf., disease – Crop damage – Nat. systems – Water scarcity – Forest fire	– Human populations – The poor, old people, children in heat waves
– Sea level rise (creeping, long term)	– Small island states – Marine ecosyst. – Indigenous communities, – Industry, energy	– Deltas – Coastal zones – Marine, freshwater ecosystems	– Coastal cities, habitats, infrastructure, jobs – Cities, homes, jobs	– Livelihood – Poor people, – Insurance, – Financial services
Abrupt climate change (cooling in central, northern Europe, North America	– Countries and people in northern Europe, benefiting from Gulf stream	– Livelihood – Survival	– Agriculture – Habitat – People	– Human life and animals, property – Forced migration of people
Climate change – Extreme weather events: hurricanes, cyclones, winter storms	– Habitat, technical infrastructure, transportation etc.	– Forests (health of trees) – Food security	– Coastal ecosystems – Forests, settlements – Electricity transmission	– Human life and property – Insurance – Financial services
Climate change on: – Extreme weather events: floods	– Habitat, technical infrastructure and people	– vulnerable, flood-prone areas	– persons living in flood-prone areas	– human life and property
– Extreme weather events: drought	– Availability of water and food, survival of people	– Decreased crop yield and water quality and quantity	Arid and semiarid zones, agriculture – Forests (tree health)	– Human life animals, property

Climate change has direct implications on soil erosion and desertification and on the precipitation pattern that have significantly changed in the Mediterranean during the last 30–50 years and that are projected to change unequally, causing negative impacts for agriculture in some landscapes and climate zones while potentially benefiting others. Temperature increase also raises evapotranspiration and thus increases the water needs of plants and may result in a yield decline concerning many cereals.

The environmental challenges posed by climate change, soil erosion, and desertification as well as by increasing water scarcity and stress in the Mediterranean will not distinguish between southern Europe and the MENA region and they will affect the two referents of environmental (ecosystems) and human security (human beings) but also the affected nation states. "Sustainability" and "survival" of some plants and human livelihoods will be at risk. The sources of these threats and challenges to environmental and human security may be humankind, nature, the state, and progressing globalization.

5.2. DIFFERENT SOCIOECONOMIC CHALLENGES IN THE MEDITERRANEAN UNTIL 2100

However, the demographic and urbanization trends in southern Europe and in the North Africa region have differed during the 20th century and will further disintegrate until 2030 (urbanization) and 2050 (population growth).[§§] According to the 2004 Population Review of the UN Populations Division the population of the five North African countries has been projected to grow by 102 million persons between 2000 and 2050 (thereof in Egypt alone by 58.6 million) the population of the five south European countries may decline by 1 million taking a significant immigration from the south into account (UN, 2005). In North Africa the urbanization rate will increase from 2000 to 2030 by 15% from 48.4% (2000) to 63.4% (2030) while it will rise in the same time period in southern Europe by only by 8.7%.

While the cities with more than 1 million inhabitants will decline in southern Europe, they are all projected to grow significantly until 2015 in North Africa. While the five south European countries will not be seriously affected by the projected decline in the production of cereals, North Africa and the Middle East will be resulting from two opposite trends, declining yields due to climate change and water stress and increasing demand due to rapid population growth. Between 1995 and1997 and 2030 the self-sufficiency in cereals has

[§§]These trends have been presented in the author's keynote speech in Lecce on 5 September 2004 that may be accessed at: <http://www.afes-press.de/pdf/Brauch%20Lecce%20.pdf>.

been projected by FAO (Bruinsma, 2003; Brauch et al., 2003: 974) to decline from 65% to 56% and the import needs for virtual water to rise by 150%. These projected trends pose severe threats to economic security and for the foreign debt of these affected countries.

5.3. FATAL OUTCOMES: HAZARDS, MIGRATION, CRISES, AND CONFLICTS IN THE MEDITERRANEAN

These changes in climate will also impact on water-related extreme weather events, on the probability of intensive storms, flash floods, flooding, and drought. According to the Munich Reinsurance company (MunichRe, 2006), between 1950 and 2005 of all major natural hazards globally, 71% were water related, causing 45% of the 1.75 million deaths and 69% of the economic damage. According to the international disaster database (EMDAT) of the Centre for Research on the Epidemiology of Disasters (CRED) in Louvain (Belgium) for the years 1974–2003 all reported water-related hazards caused 71% of a total of 2.066.273 persons who died and 98% of 5.076.494 (Hoyois and Guha-Sapir, 2003). Both the direct and indirect effects of climate change have killed and affected millions of people during the 20th century thus posing severe threats to the survival of many people confronting them with a survival dilemma.

With regard to the dominant hazard types, the Mediterranean has been atypical due to its high seismic and volcanic activity and its earthquake-prone regions where the African, European, and Asian plates regularly collide (Brauch, 2003a). The differences in the number of reported flood victims between 1975 and 2001 in Algeria (17 events, 1,201 deaths, 141,765 affected), Morocco (11 events, 873 deaths, 155,757 affected), and Egypt (5 events, 673 deaths, 229,868 affected) compared with France (30 events, 143 deaths, 372,125 affected), Italy (16 events, 319 deaths, 67,622 affected), Spain (13 events, 198 deaths, 741,300 affected), Greece (8 events, 78 deaths, 10,150 affected), and Portugal (4 events, 99 deaths, 47,220 affected) reflect the different levels of economic development, social vulnerability, and resilience within the society and disaster preparedness. With regard to the number of people affected by hazard types in southern Europe those affected by drought prevailed (due to a single event in Spain in 1995), followed by wind storms, earthquakes, and floods while in North Africa those affected by earthquakes were followed by the victims of floods, drought, and to a lesser extent of storms.

5.4. CONTRIBUTIONS OF LANDSCAPE SCIENCE TO VULNERABILITY MAPPING

Across the Mediterranean there has been a significant vulnerability gap for the people in North Africa and in southern Europe that have been affected by environmental (and water) scarcity, degradation, and stress as well as by natural hazards. The number of deaths and of affected people from weather- and water-related hazards has been much higher in North Africa than in southern, central, and northern Europe due to flash floods and droughts (Brauch, 2003a; 2003c), resulting from a higher degree of environmental and social vulnerability, but partly also due to poverty and to the fact that many informal houses in the rapidly growing urban slums have been built in flood-prone areas beyond the control of local governments. Environmental stress and climate-related natural hazards have forced people, and especially the young, to leave their rural homes and livelihoods trying to find an income in the urban centers or across the sea in Europe. In some rare cases, water and food scarcity and increase in food prizes have resulted in food and hunger riots in Egypt (1977), Tunisia (1984), Algeria (1988), and Morocco (1984, 1990).

What has been and should be the contribution of landscape ecology to achieving a higher degree of environmental, societal, and human security in the Mediterranean and elsewhere? To which extent can the DPSIR model of EEA and this author's PEISOR model guide the practical work of landscape ecologists? Which contribution can the conceptual thinking on new security concepts make for landscape ecologists in guiding the practical applications of their methods to enhance the human security of potential victims living in environmentally vulnerable ecosystems and landscapes?

6. Environmental scenarios and fields for environmental assessment

The Mediterranean environment within the natural boundaries of the Sahara desert, Europe, and the Atlantic Sea does not respect national borders. This ecological region has been highly vulnerable to manifold natural and anthropogenic challenges and multiple complex interactions between nature and human beings in different sensible landscapes. Two editions of the *Blue Plan* (Grenon and Batisse, 1989; Benoit and Comeau, 2005), a report of the EEA for UNEP (2006) on *priority issues in the Mediterranean environment* and a report of the German Advisory Council on Global Change (WBGU, 2006) on the future of the seas have neither employed any of the five models referred to above nor have they specifically focused on the role of landscape ecology. But these four reports refer to many challenges the sensitive landscapes in the Mediterranean – and thus landscape ecologists – will be confronted with.

Three groups of activities and events should be distinguished that pose manifold threats, challenges, vulnerabilities to the environmental, societal and economic security dimensions, to human and in some cases to national and regional security as well as for water and food security, and to a lesser extent for health security of the people affected:

- Direct impacts of incremental environmental (soil and water) stress resulting from the interaction of environmental scarcity (water and soil) leading directly (overgrazing, pollution) to the degradation of soil (erosion, desertification)
- Slow-onset (creeping) changes posed by temperature increase and sea-level rise due to global climate change and regional impacts
- Rapid-onset climate and water-related natural hazards (storms, flash floods, flooding, landslides, forest fires, drought, etc.)

The first Blue Plan (Grenon and Batisse, 1989) has focused on the economic activities in six areas: (a) agro-food, (b) industry, (c) energy, (d) tourism, (e) transport, and (f) urbanization, and on their impact on the environment until 2025 based on different scenarios. The second Blue Plan (Benoit and Comeau, 2005) addressed six sustainability issues for the three strategic sectors (water, energy, transport) and their impacts on urban, rural, and coastal areas.

In its four scenarios on the *Future of the Mediterranean Basin*, the first Blue Plan (Grenon and Batisse, 1989) has pointed to the manifold environmental challenges the Mediterranean riparian states will be confronted with until 2025 (Brauch, 2001). The low- and the high-growth scenarios have both demonstrated the difficulties of development especially of the southern and eastern Mediterranean riparian countries. The Blue Plan suggested two alternative scenarios: A1 with strong north–south cooperation, and A2 with predominant south–south cooperation and suggests a combination of both scenarios as the "most propitious" over time. However, since 1985 both alternatives were not pursued: economic south–south cooperation even among the Arab states was minor (e.g. the AMU is still paralyzed due to the unsolved Sahara conflict) and north–south cooperation was not sufficient to cope with the challenges.

In many suggestions the Blue Plan team called for cooperation on the Mediterranean level oriented at the goal of a "sustainable development" in the areas of (a) control of urbanization, (b) the supervision of productive or transport activities, (c) waste-water treatment plants. Furthermore, they called for (1) the progress of knowledge, (2) cooperation on management and the environment (e.g. in coastal, urban, water resources, forest, of protected areas, and of marine living resources), (3) a move from collaboration to Mediterranean solidarity, and (4) a program for the young generations.

To conclude, to cope with these environmental challenges in the Mediterranean in the decades to come, already now cooperation on these nonmilitary challenges to life and survival of the whole region, a close political and economic cooperation is crucial. The Euro-Mediterranean partnership (Barcelona EMP process since 1995) offers the only political "institution" where competence, legitimacy, and resources are present. Too often persisting conflicts (in the Middle East) but also lack of understanding, courage, and vision have paralyzed or impeded this only institution that could change the negative trends referred to by the Blue Plan team that have not been reversed since 1989.

The diagnosis in the second Blue Plan "confirmed the gloomy trends projected 15 years ago", and it noted that "urbanization, especially in the coastal zones, has been even more extensive than projected" and that "unsustainable consumption and production patterns are continuing to spread in the developed countries." In the environment real, many risks have been confirmed, "especially the greater stresses on water, soils and coastal zones and degradation of natural resources and ecosystems. The social and economic repercussions are becoming increasingly obvious" (Benoit and Comeau, 2005: 359).

The second Blue Plan has discussed the increased vulnerability of *coastal urban areas* (built-up coasts, loss of agricultural land, valuable natural habitats, degradation of ecosystems, coastal landscapes) where most large cities and the national wealth are concentrated to natural hazards. *Rural areas*, especially in the MENA region, are confronted with other challenges (poverty, desertification, environmental degradation, forest fires, and degradation of agricultural landscapes) due to the concentration of development in specific areas. *Coastal overdevelopment* ("littoralisation") and the abandonment of the hinterland has resulted in environmental degradation that has affected primarily poor people who live in a bad or polluted environment without access to health water (Benoit and Comeau, 2005: 24–25).

According to the new baseline scenario of the second Blue Plan soils would be especially affected by desertification, salinization, wind and water erosion, and deforestation and the degradation of grazing and cultivated land. Desertification would further increase, and the "water stress would continue to increase in the southern and eastern Mediterranean with 63 million people having less than 500 m^2 per capita per year". In addition the baseline scenario assumes for 2025 that "the effects of sudden natural hazards… would have increasingly serious consequences" (Benoit and Comeau, 2005: 363). The second plan foresees the following for the baseline scenario.

The accumulation of multiple environmental pressures and impacts would inevitably result in degradation of very valuable common goods on a global scale, including the loss of terrestrial (including agricultural) and coastal biodiversity, the degradation of the Mediterranean landscapes, and historical and cultural heritage (sites and monuments). The annual global costs of these multiple negative impacts, which already amount to several percentage points of GDP, would rise (Benoit and Comeau, 2005: 364). To avoid these projected trends in an alternative scenario the second Blue Plan aimed at four major objectives:

- Protecting the natural and cultural heritage and preventing natural risks
- Decoupling economic growth from pressures on the environment
- Reducing internal disparities and promoting in-depth development of countries
- Creating added value by enhancing specific Mediterranean assets, in particular the wealth and diversity of its heritage, and exploiting synergies between activities (Benoit and Comeau, 2005: 365)

To achieve these objectives, the second Blue Plan suggests a combination of measures from (a) enhancing water supply by managing demand, (b) energy efficiency, (c) breaking the vicious cycle of transport, (d) lifestyle changes in the cities, (e) better water management and soil conservation in agriculture, (f) a better reconciliation of tourism and the environment, (g) reducing domestic disparities and relieving coastal zones, and (h) enhancing the strong points of the Mediterranean region (Benoit and Comeau, 2005: 366–378).

While the WBGU Report (2006) has focused on the impact of global warming on marine ecosystems in the oceans, the EEA Study (2006) for UNEP reviewed the impact of human activity (sewage, urban runoff, solid waste, industrial effluents, urbanization, Euthrophication, sand erosion, marine transport, biological invasions, harmful algal blooms, exploitation of marine resources, expansion of aquacultures, and natural hazards) around the Mediterranean and its impact on the Mediterranean Sea.

To realize the alternative scenario of the second blue plan and the goal of the EEA study for UNEP, this requires manifold cooperative policy decisions for which the Barcelona Declaration of 1976 and the Barcelona Declaration of 1995 can offer a policy and a functional framework. At the Barcelona summit in November 2005, the 35 participating states agreed in their Five-Year work program (2006–2010).

Promote environmental sustainability and implement the Mediterranean Strategy for Sustainable Development. To develop as soon as possible a road map for the de-polluting the Mediterranean by 2020, based on the recommendations of Euromed Environment Ministers using inter alia the MSSD and the UNEP Mediterranean Action Plan toward this end, while providing adequate financial and technical assistance to this end. The goal should be to tackle all the major sources of pollution including industrial emissions, municipal waste ad particularly urban wastewater. Exchange experience on sustainable development in the Baltic Sea, the Mediterranean, and the Black Sea.[***]

Which contribution can landscape ecology as a "technical tool" make to counter the effects of the business as usual scenario and by shifting to a more sustainable development strategy for the whole Mediterranean basin and to protect the Mediterranean Sea? Landscape ecologists could contribute to both public awareness and specific spatial coastal, rural, and urban planning to reduce environmental and social vulnerability by developing vulnerability indicators (Birkmann, 2006) and contribute to vulnerability mapping as a tool for urban and city planning. Within the PEISOR model landscape ecologists can contribute to the development of landscape specific knowledge on environmental scarcity, degradation, and stress on the one hand and on contributing to improved disaster preparedness by vulnerability mapping as a tool of rural and urban spatial planning on the other.

References

Allan, J.A., 2003. Water security in the Mediterranean and the Middle East. In: H.G. Brauch, P.H. Liotta, A. Marquina, P. Rogers, and M.E.-S. Selim (Editors), Security and Environment in the Mediterranean. Conceptualising Security and Environmental Conflicts. Springer, Berlin, Heidelberg, pp. 705–718.

Art, R., 1993. Security. In: J. Krieger (Editor), The Oxford Companion to Politics of the World. Oxford University Press, Oxford, New York, pp. 820–822.

Barnett, J. (Editor), 2001. The Meaning of Environmental Security. Ecological Politics and Policy in the New Security Era. Zed, London, New York.

Benoit, G. and Comeau, A. (Editors), 2005. A Sustainable Future for the Mediterranean. The Blue Plan's Environment & Development Outlook. Earthscan, Sterling, VA, London.

Birkmann, J. (Editor), 2006. Measuring the Unmeasurable. UNU Press, Tokyo.

Bogardi, J. and Brauch, H.G., 2005. Global environmental change: a challenge for human security – defining and conceptualising the environmental dimension of human security. In: A. Rechkemmer (Editor), UNEO – Towards an International Environment Organization – Approaches to a Sustainable Reform of Global Environmental Governance, Nomos, Baden-Baden.

Booth, K., 1998. Cold wars of the mind. In: K. Booth (Editor), Statecraft and Security. The Cold War and Beyond, Cambridge University Press, Cambridge, pp. 29–55.

Brauch, H.G., 2000. Partnership building measures to deal with long-term non-military challenges affecting north-south security relations. In: H.G. Brauch, A. Marquina, and A. Biad (Editors), Euro-Mediterranean Partnership for the 21st Century, St. Martin's Press, New York, Macmillan, London, pp. 281–318.

Brauch, H.G., 2001. The Mediterranean "Space" beyond geopolitique and globalization. Common space – divided region. In: A. Marquina and H.G. Brauch (Editors), The

[***] See at: <http://ec.europa.eu/comm/external_relations/euromed/summit1105/index.htm> and at: <http://ec.europa.eu/comm/external_relations/euromed/summit1105/five_years.pdf>.

Mediterranean Space ant Its Borders. Geography, Politics, Economics and Environment, Collection Strademed 14, AFES-PRESS, UNISCI – Mosbach, Madrid, pp. 109–144.

Brauch, H.G., 2002. Climate change, environmental stress and conflict - AFES-PRESS report for the Federal Ministry for the environment, nature conservation and nuclear safety. In: Federal Ministry for the Environment, Nature Conservation, and Nuclear Safety (Editor), Climate Change and Conflict. Can Climate Change Impacts Increase Conflict Potentials? What is the Relevance of this Issue for the International Process on Climate Change? Federal Ministry for the Environment, Nature Conservation and Nuclear Safety, Berlin, pp. 9–112.

Brauch, H.G., 2003. Security and environment linkages in the Mediterranean: Three phases of research on human and environmental security and peace. In: H.G. Brauch, P.H. Liotta, A. Marquina, P. Rogers, and M.E.-S. Selim (Editors), Security and Environment in the Mediterranean. Conceptualising Security and Environmental Conflicts, Springer, Berlin, Heidelberg, pp. 35–143.

Brauch, H.G., 2003a. Natural disasters in the Mediterranean (1900–2001): from disaster response to disaster preparedness. In: H.G. Brauch, P.H. Liotta, A. Marquina, P. Rogers, and M.E.-S. Selim (Editors), Security and Environment in the Mediterranean. Conceptualising Security and Environmental Conflicts, Springer, Berlin, Heidelberg, pp. 863–906.

Brauch, H.G., 2003b. Towards a fourth phase of research on human and environmental security and peace: conceptual conclusions. In: H.G. Brauch, P.H. Liotta, A. Marquina, P. Rogers, and M.E.-S. Selim (Editors), Security and Environment in the Mediterranean. Conceptualising Security and Environmental Conflicts, Springer, Berlin, Heidelberg, pp. 919–954.

Brauch, H.G., 2003c. Urbanization and natural disasters in the Mediterranean – population growth and climate change in the 21st century. In: A. Kreimer, M. Arnold, and A. Carlin (Editors), The Future of Disaster Risk: Building Safer Cities. December 2002. Conference Papers, World Bank, Washington, DC, pp. 149–164.

Brauch, H.G., 2005. Freedom from Hazard Impact: Developing the Environmental Dimension of Human Security. Intersections, 2, UNU-EHS.

Brauch, H.G., 2005a. Threats, Challenges, Vulnerabilities and Risks in Environmental Human Security, Source 1/2005, UNU-EHS, Bonn.

Brauch, H.G., 2006. Impact of global warming and non-conventional water resources: a conceptual contribution to conflict resolution and long-term conflict avoidance. In: S. Hillel and D. Hassan (Editors), Proceedings of the Second Israeli-Palestinian Conference Water for Life in the Middle East, held in Antalya, Turkey, 10–14 October 2004 (Jerusalem: IPCRI, 2006), pp. 567–689.

Brauch, H.G., 2006a. Potential of solar thermal desalination to defuse water as a conflict issue in the Middle East - Proposal for functional cooperation in the Gulf of Aqaba. In: I. Linkov and M. Benoit (Editors), Role of Risk Assessment in Environmental Security and Emergency Preparedness in the Mediterranean Region, Springer, Dordrecht, pp. 25–48.

Brauch, H.G., 2006b. Desertification – a new security challenge for the Mediterranean? Policy agenda for recognising and coping with fatal outcomes of global environmental change and potentially violent societal consequences. In: W. Kepner, J.L. Rubio, D. Mouat, and F. Pedrazzini (Editors), Desertification in the Mediterranean Region. A Security Issue, Springer, Dordrecht, pp. 11–85.

Brauch, H.G., Grin, J., Mesjasz, C., Krummenacher, H., Chadha, B.N., Chourou, B., Oswald Spring, U., and Kameri-Mbote, P. (Editors), 2007a. Facing Global Environmental Change: Environmental, Human, Energy, Food, Health and Water Security Concepts, Springer, Berlin.

Brauch, H.G., Liotta, P.H., Marquina, A., Rogers, P., and Selim, M.E.-S. (Editors), 2003a. Security and Environment in the Mediterranean. Conceptualising Security and Environmental Conflicts, Springer, Berlin, Heidelberg.

Brauch, H.G., Liotta, P.H., and Selim, M.E.-S., 2003. Security concepts for cooperation in the Mediterranean: conclusions and outlook for the 21st century. In: H.G. Brauch, P.H. Liotta, A. Marquina, P. Rogers, and M.E.-S. Selim (Editors), Security and Environment in the

Mediterranean. Conceptualising Security and Environmental Conflicts, Springer, Berlin, Heidelberg, pp. 955–990.
Brauch, H.G., Mesjasz, C., Grin, J., Oswald Spring, U., Kameri-Mbote, P., Chourou, B., Hayek, B., and Birkmann, J. (Editors), 2008. Coping with Global Environmental Change, Disasters and Security – Threats, Challenges, Vulnerabilities and Risks, Springer, Berlin, Heidelberg.
Brauch, H.G., Selim, M.E.-S., and Liotta, P.H., 2003. Security concepts for cooperation and partnership in the Mediterranean: conclusions and outlook for the 21st century. In: H.G. Brauch, P.H. Liotta, A. Marquina, P. Rogers, and M.E.-S. Selim (Editors), Security and Environment in the Mediterranean. Conceptualising Security and Environmental Conflicts, Springer, Berlin, Heidelberg, pp. 955–990.
Bruinsma, J. (Editor), 2003. World Agriculture. Towards 2015/2030. An FAO Perspective. Earthscan, London.
Bull, H. (Editor), 1977. The Anarchical Society. A Study of Order in World Politics. Columbia University Press, New York.
Buzan, B., Wæver, O., and Wilde, J. (Editors), 1998. Security. A New Framework for Analysis. Lynne Rienner, Boulder, London.
CHS [Commission on Human Security], 2003. Human Security Now. Commission on Human Security, New York.
EEA, 1998. Europe's Environment. The Second Assessment. EEA, Copenhagen.
EEA, 2006. Priority Issues in the Mediterranean Environment. Rev. version. EEA Report No. 4/2006, EEA, Copenhagen.
GECHS, 1999. Global Environmental Change and Human Security. GECHS Science Plan, IHDP Report No. 11, IHDP, Bonn.
Grenon, M. and Batisse, M. (Editors), 1989. Futures for the Mediterranean Basin. The Blue Plan. Oxford University Press, Oxford.
Haas, P. (Editor), 1990. Saving the Mediterranean: The Politics of International Environmental Cooperation. Columbia University Press, New York.
Haas, P., 1993. Epistemic communities and the dynamics of international environmental cooperation. In: V. Rittberger (Editor), Regime Theory and International Relations, Clarendon Press, Oxford, pp. 168–201.
Hoyois, P. and Debarati, G.-S., 2003. Three decades of floods in Europe: A preliminary analysis of EMDAT data. Working paper (draft). CRED, Brussels.
IPCC [Intergovernmental Panel on Climate Change], 1990. Climate Change. The IPCC Impacts Assessment. WMO; UNEP; IPCC, Geneva.
IPCC [Intergovernmental Panel on Climate Change], 1996. Climate Change 1995. The Science of Climate Change. Contributions of Working Group I to the Second Assessment Report of the Intergovernmental Panel on Climate Change. Cambridge University Press, Cambridge.
IPCC [Intergovernmental Panel on Climate Change], 1998. The Regional Impacts of Climate Change. An Assessment of Vulnerability. Cambridge University Press, Cambridge.
IPPC [Intergovernmental Panel on Climate Change], 2001. Climate Change 2001. The Scientific Basis. Cambridge University Press, Cambridge.
IPPC [Intergovernmental Panel on Climate Change], 2001a. Climate Change 2001. Impacts, Adaptation and Vulnerability. Mitigation. Cambridge University Press, Cambridge.
Leemans, R., 2007. The millennium ecosystem assessment: securing interactions between ecosystems, ecosystem services and human well-being. In: H.G. Brauch, J. Grin, C. Mesjasz, H. Krummenacher, N. Chadha Behera, B. Chourou, U. Oswald Spring, and P. Kameri-Mbote (Editors), Facing Global Environmental Change: Environmental, Human, Energy, Food, Health and Water Security Concepts. Hexagon Series on Human and Environmental Security and Peace, Vol. 4, Springer, Berlin, Heidelberg.
Lexikon der Geographie, Vol. 2, 2002 (Spektrum Akademischer Verlag, Heidelberg, Berlin).
Lexikon der Geowissenschaften, Vol. 3, 2001 (Spektrum Akademischer Verlag, Heidelberg, Berlin).

Mathews, J.T., 1989. Redefining Security. Foreign Affairs 68, 162–177.

MEA [Millenium Ecosystem Assessment], 2002. Ecosystems and Human Well-being: A Framework for Assessment. Islands Press, Washington, DC.

MEA [Millenium Ecosystem Assessment], 2005. Ecosystems and Human Well-Being. Islands Press, Washington DC.

Møller, B., 2003. National, societal and human security: discussion – a case study of the Israeli-Palestine conflict. In: H.G. Brauch, P.H. Liotta, A. Marquina, P. Rogers, and M.E.-S. Selim (Editors), Security and Environment in the Mediterranean. Conceptualising Security and Environmental Conflicts, Springer, Berlin, Heidelberg, pp. 277–288.

MunichRe, 2006. Topics – Annual Review of Natural Catastrophes 2005, MunichRe, München.

Myers, N., 1989. Environment and Security. Foreign Policy 74, 23–41.

OECD, 1994. Environmental Indicators: OECD Core Set. OECD, Paris.

OECD, 1998. Towards Sustainable Development: Environmental Indicators. OECD, Paris.

OECD, 1999. OECD Environment Monograph. OECD, Paris.

OECD, 2001. Environmental Indicators – Towards Sustainable Development. OECD, Paris.

OECD, 2001a., Key Environmental Indicators. OECD, Paris.

Oswald Spring, U., 2001. Sustainable development with peace building and human security. In: M.K. Tolba (Editor), Our Fragile World. Challenges and Opportunities for Sustainable Development. Forerunner to the Encyclopedia of Life Support Systems, EOLSS Publishers, Oxford, pp. 873–916.

Parry, M., 2005. Impacts of Climate Change on Agriculture in Europe. Informal Meeting of EU Agriculture and Environment Ministers, 11 September 2005, London, at: <http://www.fco.gov.uk/Files/kfile/Parry.pdf>.

Ullman, R., 1983. Redefining Security. International Security 8, 129–153.

UN, 2005. World Population Prospects. The 2004 Revision. Highlights Population Division, Department of Economic and Social Affairs, UN Secretariat, New York.

UNDP, 1994. Human Development Report 1994. Oxford University Press, New York.

Waever, O., 2007. Societal security. In: H.G. Brauch, J. Grin, C. Mesjasz, P. Dunay, N. Behera, B. Chourou, U. Oswald Spring, and P. Kameri-Mbote (Editors), 2006. Globalisation and Environmental Challenges: Reconceptualising Security in the 21st Century, Springer, Berlin, Heidelberg.

Wendt, A., 1992. Anarchy is What States Make of It: The Social Construction of Power Politics. International Organization 46(2), 391–425.

Wolfers, A., 1962. National security as an ambiguous symbol. In: A. Wolfers (Editor), Discord and Collaboration. Essays on International Politics, John Hopkins University Press, Baltimore, pp. 147–165.

LANDSCAPE MANAGEMENT FOR ENVIRONMENTAL SECURITY: SOME PERSPECTIVES OF ADAPTIVE MANAGEMENT APPROACHES

IRENE PETROSILLO* AND GIOVANNI ZURLINI
Laboratory of Landscape Ecology – Department of Biological and Environmental Sciences and Technologies – University of Salento, Lecce, Italy

FELIX MÜLLER
Ecology Center, University of Kiel, Germany

Abstract. Managing Socio-Ecological Landscapes (SELs) for environmental security implies the recognition that social systems interact with ecological systems, and requires new strategies to improve environmental policies, in light of the mismatch between existing global management capacity and likely threats to environmental resources. This because mismanagement is recognized as one of the main causes that generates environmental insecurity. Thus, decision-makers should consider whether a course of action is reversible and should utilize procedures to evaluate the outcomes of actions, learning from them. These are the principles of adaptive management, based on the ability to adapt to change considered as an intrinsic feature of SELs. In the conclusion we present some general requirements to enhance environmental security through adaptive management approach, highlighting that managing SELs for environmental security demands for a new strategy which includes uncertainty and the dynamics of systems.

Keywords: adaptive management; environmental security; socio-ecological landscape; stakeholders' participation

*Corresponding Author: Dr. Petrosillo Irene, Laboratory of Landscape Ecology – Department of Biological and Environmental Sciences and Technologies Ecotekne – University of Salento, Prov.le Lecce-Monteroni, 73100 Lecce (Italy); phone: +39 0832 298896; fax: +39 0832 298626; e-mail: irene.petrosillo@unile.it.

I. Petrosillo et al. (eds.), Use of Landscape Sciences for the Assessment of Environmental Security, 43–56.

1. Introduction

It is not long time ago when the focal keywords in ecology and environmental management are stability, equilibrium, and control, and along these basic ideas ecologists have believed in growth and development. If there is any change of all in ecological natural system, it will lead to a more mature, complex, and elaborated state.

It has turned out in the 1990s, when Heraclitus' knowledge, from 500 B.C., has been rediscovered: nothing is as permanent as change. According to his theory everything is flowing and a man never is able to jump into the same river twice, because meanwhile both man and river are changed.

For ecosystem managers this shift of basic ideas implies that:

- Stability is an illusion
- Ecosystems never return to the old stage again
- There is not equilibrium but all processes are based on a nonequilibrium
- Many dynamic processes cannot be controlled

With the concept of the Resilience Alliance, these basic ideas can be accomplished by even more radical ideas:

- Breakdown, catastrophe, and chaos are intrinsic processes of systems adaptation, and they cannot be avoided.
- The ecosystem organizations according to hierarchies, whereby all single process operate on different spatial–temporal scales. It is not possible to keep all of them under control.
- Environmental systems are complex and can be characterized by many nonlinear linkages, because of high number of components and high number of linkages. Thus, their behavior can only hardly be foreseen.

In addition, reciprocal influences among humans and the climate, biota, and ecological goods and services of the world have become both stronger and more widely recognized, there has been also the acknowledgment that in the majority of ecosystems, structure, and function are now determined primarily by human interactions, perceptions, and behaviors, so that nowadays new ideas of management are necessary (Kay, 1999). This produces consequences for security because nature does not work deterministically but hierarchically generating high level of uncertainty, and management for security could only operate at the level of constrains. So management for security has to take into account that slow natural processes and fast human processes are integrated and coevolving, with the recognition that adaptation of natural systems is difficult for the time required. So that the important task of adaptive management to sustain security is to recognize all these distinct development velocities.

Sustainable development is a key aspect of environmental security and should be promoted through a wide range of mechanisms including education, steering national economies with tax and subsidy programs, promoting green technology innovation, and multilateral environmental agreements. Although sustainable development and environmental security are mutually reinforcing concepts for policy, they are not the same thing. Sustainable development focuses on environmentally sound development that is economically, financially, socially, and environmentally sustainable. On the other hand, the notion of environmental security has been historically linked to environmentally induced conflicts, e.g. caused by overuse of renewable resources, pollution, or impoverishment in the space of living. The notion has been developed mainly by international policy researchers and focused on the role of the scarcity of renewable resources such as cropland, forests, water, and fish stocks. Attention has been devoted to the theoretical analysis of the possible pathways, beginning with scarcity and leading to outbreaks of violence.

Developing theory for sustainable, more secure environmental future options requires a model of how human and ecological processes interact across space and time. Viewing environmental security from such perspective implies that self-organization of ecological systems establishes the arena for evolutionary change, but under the increasingly dominating human influence in space and time, thus defining limiting constraints at "higher scales" and altering the natural functioning of ecological processes in absence of human influences. Therefore, after all, the self-organization of human institutional patterns establishes the arena for future sustainable, environmentally secure opportunities.

These challenges grow since there are strong linkages between natural and human system, since ecosystems are essential for human survival and well-being through their provisioning, regulating, cultural, and supporting service (Figure 1). Preserved and unpreserved, agricultural and even urban areas provide a variety of vital ecosystem services to humans. Evidence in recent decades of escalating human impacts on ecological systems worldwide raises concerns about the consequences of ecosystem changes for human well-being. Natural system produces resources that are useful for human use system, but when this use turns out into the overexploitation of resources environmental hazards are possible.

2. Basic components of adaptive management

Adaptive management to cope with the multiple problems must be driven by explicit goals, executed by policies, protocols, and practices, and made adaptable by monitoring and research based on the best understanding of the ecological interactions and processes necessary to sustain ecosystem composition, structure, and function and human well-being. Usually management approaches have often focused on maximizing short-term yield and economic gain rather than long-term sustainability.

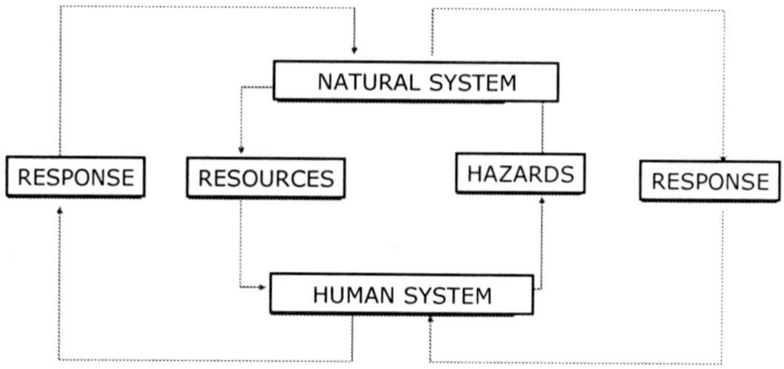

Figure 1. The linkages between natural and human systems.

There are several reasons that contribute to this discrepancy, including: inadequate information on the function and dynamics of ecosystems; the openness and interconnectedness of ecosystems across scales that transcend management boundaries; a prevailing human motivation that the immediate economic and social value of supposedly renewable resources outweighs the risk of future ecosystem damage or the benefits of alternative management approaches. The goal of adaptive management is to overcome these obstacles, including the following strategies:

1. Adaptive management requests the establishment of measurable management goals that specify future processes and outcomes.
2. Adaptive management recognizes that biological diversity and structural complexity strengthen human–environmental systems against disturbance and provide ecosystem goods and services necessary to adapt to long-term change.
3. Adaptive management recognizes that change is an intrinsic feature of human–environmental systems that are dynamic in space and in time, so that it avoids keeping human–environmental systems in a particular state or configuration. Successful adaptive management requires, on the one hand, institutions that are adaptable to changes in ecosystem characteristics and, on the other hand, the interaction among scientists, managers, and the public. Communication must flow in both directions, and scientists must be willing to prioritize their research with regard to critical management needs.

4. Adaptive management recognizes that ecosystem processes operate over a wide range of spatial and temporal scales, and their behavior at any given location is greatly affected by surrounding systems.
5. Adaptive management values the active role of stakeholders in achieving sustainable management goals.
6. Adaptive management acknowledges that management approaches must be viewed as hypotheses to be tested by research and monitoring programs.
7. Adaptive management addresses the uncertainty characterizing systems, by combining democratic principles, scientific analysis, education, and institutional learning to increase our understanding of ecosystem processes and the consequences of management interventions, and to improve the quality of data upon which decisions must be made.

3. Adaptive management and environmental security

Adaptive management is not a rejection of a biocentric for a totally anthropocentric worldview. Rather it is management that acknowledges the importance of human needs but, at the same time, recognizes the existence of a mismatch between existing global management capacity and likely threats to environmental resources and services (environmental security).

The condition for an increase of environmental security is that the socio-economic system interacts with ecological system in sustainable ways, all individuals have fair and reasonable access to environmental goods and services, and mechanisms exist to address environmental crises and conflicts.

In the real geopolitical world, it is possible to identify two distinguished cases that can be interested by environmental insecurity: within a country and transborders. Similarly, the main reasons which can lead to environmental security can be classified into at least three groups: by ignorance and/or mismanagement, by intention, and by a mix of natural and human actions. Combining these criteria it is possible to classify some actual examples of environmental insecurity (Table 1).

4. Adaptive management in socio-ecological landscapes

Anthropogenic disturbances such as changes in land use are determined by the social components of Socio-Ecological Landscapes (SELs) which consist of groups of people organized in a hierarchy at different levels (e.g. household, village, county, province, region, and nation). In face of resource use and

TABLE 1. Within a country and transborders examples of environmental insecurity caused by ignorance and/or mismanagement, by intention a by a mix of natural and human actions.

	By Ignorance and/or Mismanagement	By Intention	By a mix of Natural and Human Actions
Within a Country	Groundwater contamination	Chemical attacks and draining marshes in Iraq	Floods
	Hazardous wastes	Poisoning or diversion or misuse of water resources	Famines
	Soil erosion	Sarin gas attack in Tokyo subway	Salinization
			Introduction of exotic species
Transborders	Rain forest depletion	Burning oil fields in Kuwait	Global Warming
	Rivers usage	Dam construction	Desertification
	Chernobyl Nuclear Accident	Biological weapons	Population Growth
			Rich–Poor Gaps

environmental hazards both human and natural systems react through more or less appropriate responses. A social–ecological perspective of environmental security stresses adaptability and learning through thoughtful probing. Emphasis needs to be placed both on dealing with threats and hazards, and on human response to risk considering both social (perception) and economic aspects. Encouraging risk adverse behavior or discouraging risk-prone behavior is more effective than simplistic schemes intended to reduce hazard. The reason is that people tend to engage in more risky behavior, if they perceive a more secure environment (risk homeostasis).

Typical responses coming from human systems aim to guarantee natural resources in a more or less sustainable way. However, because human–environmental systems are complex, and exhibit chaotic behavior, to cope with environmental threats optimal solutions do not exist, as it is difficult to learn general lessons from several local cases, whereas we need to know trends at different scales. Both holistic solutions and cultural change are necessary to reduce risk and improve environmental security.

In the context of management the term "adaptive management" arose over 20 years ago, and its concepts and applications have been widely discussed in the resource-oriented literature (Holling, 1978; Walters, 1986; Halbert, 1998), e.g. catchment management of fish resources (Olsson and Folke, 2001), management of rural areas (Chambers, 1985), of disturbed areas (Dale et al.,

1998), of land use (Dale et al., 2000), management of resources based on participatory approach (Walker et al., 2002). Adaptive management is a process for addressing the uncertainties of resource management policies by implementing the policies experimentally and documenting the results. This does not mean that simply changing management direction in the face of failed policies (crisis) constitutes adaptive management, and in this case it is better to speak of "maladaptive management". On the contrary adaptive management is a planned approach to reliably learn why policies, or critical components of policies, succeed or fail. In the literature three main types of adaptive policy evaluation can be identified: (1) "reactive" management, when a policy change results from public criticism or legal challenges; (2) "passive" management is the deliberate evaluation of one policy alternative at a time through formal monitoring; and (3) "active" management is the deliberate, experimental evaluation of several policy alternatives by implementing them at the same time and comparing their outcomes.

The definition of adaptive management currently accepted stated that "[adaptive management] is a systematic process for continually improving management policies and practices by learning from the outcomes of operational programs. Its most effective form is 'active' adaptive management that employs management programs that are designed to experimentally compare selected policies or practices, by evaluating alternative hypotheses about the system being managed" (Gunderson and Holling, 2002).

This kind of management is in conflict with the typical will of managers that seek to command and control processes of change in simplified landscapes in an attempt to stabilize ecosystem outputs and sustain consumption patterns (Holling and Meffe, 1996; Carpenter and Gunderson, 2001). Humans control agricultural pests through herbicides and pesticides; convert multispecies variable-aged forests into monocultures of single-aged plantations; hunt and kill predators to produce a larger, more reliable supply of game species; suppress fires and pest outbreaks in forests to ensure a steady lumber supply; clear forests for pasture development to achieve constantly high cattle production. Fire management provides a good example: the suppression of fire in ecosystems that have evolved with fire as an integral part of their environments is successful in the short term, but the consequence is an accumulation of fuel, over large areas, which eventually generates large fires that may fundamentally change the state of the ecosystem (Holling and Meffe, 1996) and the whole SEL.

Social–ecological systems may have different dynamics when compared to the ecological component alone because the social domain contains the element of human intent. Thus, management actions can deliberately avoid or seek the crossing of actual and perceived thresholds. It is still unclear whether a common framework of system dynamics could be used to examine and explain both social and ecological systems. Conditions like those in Europe could be the best to test models because, e.g. European landscapes are the result of consecutive reorganizations of the land for a long time to adapt uses and spatial structures to meet changing societal demands (Antrop, 2005).

If self-organization of ecological systems establishes the arena for evolutionary change, self-organization of human institutional patterns establishes the arena for future sustainable opportunity. Consequently, the possible management strategies and the self-organization of ecological systems are not independent because the characteristic scales of particular phenomena like anthropogenic changes should entrain and constrain ecological processes, and be related to the scales of human interactions with the biophysical environment.

The diversity of biotic systems across scales, from genes to landscapes, and the ecosystem goods and services they generate, provides the basic foundation on which social and economic development depends. Therefore, a major challenge is to manage our interconnected environmental assets in a fashion that secures their capacity to support societal development for a long time into the future (Costanza et al., 2000).

5. Theoretical requirements

All landscapes are exposed to gradual changes in climate, nutrient loading, habitat fragmentation or biotic exploitation. Nature usually responds to gradual change in a smoothly homogeneous way, but sudden drastic switches to a contrasting state can interrupt smooth change, with serious social and economic consequences. Although stochastic events like storms or fire can trigger off such shifts, recent studies of rangelands, coral reefs, forests, lakes, and oceans show that loss of resilience usually is the way for a shift to an alternate state (Scheffer et al., 2001), typical characteristic of complex adaptive systems like Socio-Ecological Landscapes (SELs).

Assessing and evaluating environmental security in the context of SELs requires a shift in thinking and perspective (Ludwig et al., 2001); the notion of security, as traditionally understood, must be expanded to include the subjective perspective, in terms of the level of awareness and fear that ecosystem goods and services will be eroded and possibly lost (risk perception). Nevertheless, the objective sense of environmental security needs also to be extended because the previous worldview of nature and society as systems near equilibrium is being replaced by a dynamic view, which emphasizes complex multi-scale and nonlinear relations between entities under continuous change and facing discontinuities and uncertainty from complexes or suites of synergistic stresses and shocks.

For relevant and successful environmental management and development planning however, it is crucial that decision-makers be able to get a clear picture of local situations and adjust their plans accordingly. Studying the factors that make people feel safe and secure gives us a deeper understanding of their day-to-day thinking when making decisions that affect their livelihood (Petrosillo et al., 2007). It also puts focus on the strategies that work well and could be further developed and encouraged from a management and planning perspective (Haag and Hajdu, 2005).

The policy or management changes made to address problems and opportunities related to ecosystems and their services, whether at local scales or national or international scales, need to be adaptive and flexible in order to benefit from past experience, to hedge against risk, and to consider uncertainty. The understanding of ecosystem dynamics will always be limited, socio-economic systems will continue to change, and outside determinants can never be fully anticipated. Decision-makers should consider whether a course of action is reversible and should incorporate, whenever possible, procedures to evaluate the outcomes of actions and learn from them. Debate about exactly how to do this continues in discussions of adaptive management, social learning, and the precautionary principle. But the core message of all approaches is the same: acknowledge the limits of human understanding, give special consideration to irreversible changes, and evaluate the impacts of decisions as they unfold (Millennium Ecosystem Assessment, 2003).

Systems with high adaptive capacity are able to reconfigure themselves without significant declines in crucial functions in relation to primary productivity, hydrological cycles, social relations, and economic prosperity. A consequence of a loss of resilience, and therefore of adaptive capacity, is loss of opportunity, constrained options during periods of reorganization and renewal, an inability of the system to do different things.

The adaptive capacity in ecological systems is related to genetic diversity, biological diversity, and the heterogeneity of landscape mosaics. In social systems, the existence of institutions and networks that learn and store knowledge and experience, create flexibility in problem solving and balance power among interest groups play an important role in adaptive capacity. Resilience is one key to enhance adaptive capacity. It is possible to identify and expand on some critical factors that interact across temporal and spatial scales and that seem to be required for dealing with natural resource dynamics during periods of change and reorganization.

Summarizing we can say that the concept of adaptive management is based on a system comprehension of socio-ecological units. This implies some theoretical requirements:

- The management objects are comprehended as dynamic systems, consisting of observer-defined components and their interrelationships. From this perspective, structures as well as functions and processes have to be used as indicators, and the situation analysis should follow the methodological steps of systems analysis. This procedure will make sure that all significant influences are included and that the mutual linkages between the components are considered as significant parts of the procedure.
- The management objects are self-organized units. They provide a high potential for self-regulation, which is mainly influenced by exterior constraints. Thus, management should change the prevailing constraints and rely upon the self-organization capacities of the systems themselves.
- All components are dynamic. Decrease, decay, and destruction are important parts of their dynamics, which often can not be avoided. Therefore, not only growth and development have to be taken as management guidelines. If adaptation to changing constraints is the management target, we have to realize that the break down of a certain nonadaptive (sub)system can be the optimal strategy to adapt to the new situation. Of course, managers have to evaluate carefully if the respective loss of structures can be accepted.
- The elements and subsystems are operating in a functional hierarchy. Under steady-state conditions slow processes with broad extents are constraining fast and small-scaled processes. If we manage social–ecological systems for environmental security, it is necessary to consider these spatial and temporal distinctions, i.e. referring to different velocities of (the often very rapid) human changes and the (often much more slow dynamics) of natural components.
- All components have distinct behavioral features. Their resilience, fragility and buffer capacity vary in a wide range and are changing continuously. Furthermore, we have to realize, that change is a permanent feature of environmental features. Thus, there are differences between a management for conservation and a management for adaptation, which have to be realized.

Furthermore, different management conceptions have to be taken into account if adaptive management strategies are used to optimize environmental security:

- Sustainable development is closely linked with the target of environmental security. It strives for a long-term functionality of the systems in a global perspective. The respective sustainability programs thus also are valid for security management.

- The adoption of an ecosystem comprehension implies that the guidelines of the ecosystem approach of the CBD should be realized. Their focal innovations are related to:
    - A focus on the relationships and processes within ecosystems (e.g. stores and flows of energy, water, nutrients)
    - The mutual increase of benefit-sharing (e.g. by including concepts as environmental security, sustainability, or the valuation of ecosystem goods and services)
    - The usage of adaptive management practices (e.g. managing uncertainties, understanding management as a learning process, introducing flexibility to ecosystem management, adopting a long-term orientation with intersectoral cooperation)
- Ecosystem and landscapes should be managed in a way that increases the health and integrity of the respective ecological units. These concepts are consequent applications of the ecosystem approach. Their central objective is to increase the self-organizing capacity of the environmental units.
- The final target of adaptive management is the optimization of human well-being and the long term increase of human quality-of-life. Therefore the optimal provision of ecosystem services has to be a focal target. Consequently, management has to find the best suited land use structure, thereby realizing the multifunctionality of socio-ecological landscapes.
- Management should not be carried out without the contribution of stakeholders. This integration of interest groups will lead to a good consideration of local knowledge on the one side and will guarantee a highly motivated execution of the proposed management measures on the other.
- Adaptive management should use modern instruments for landscape analysis, landscape modeling, and scenario building.

6. Adaptive management methodology

Adaptive management is a six steps cycle (Figure 2). The first step (Assess Problem) consists on the definition of both spatial–temporal scale of problem, and sensitivity of resource values (risk assessment). On the basis of results the definition of measurable management objectives and the list of potential management actions are produced. For each objective a set of key indicators is identified: the choice of indicators is relevant because they have to be relevant to objectives and responsive to management actions; it is necessary to take into account the cost and practicality of measuring each indicator; then it is important to select some indicators that respond in the short term, some in the medium term, and some in the long term. Select indicators that respond at different spatial scales (e.g. site, landscape, region).

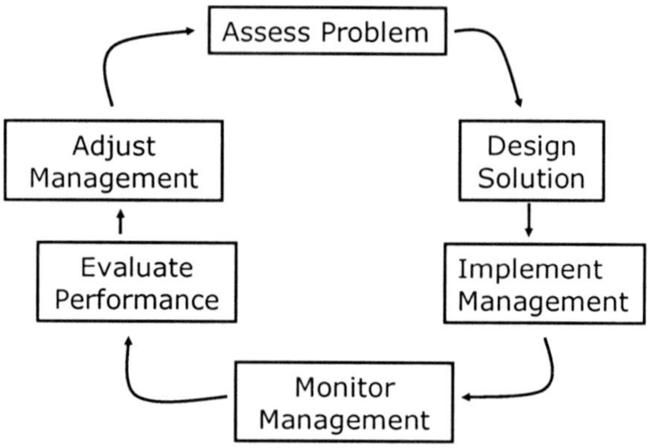

Figure 2. The adaptive management scheme.

The second step (Design Solution) has the purpose to design a management plan and monitoring program that are informative and provide reliable feedback. In this context several alternative management options are compared also testing a range of options at a pilot scale, before testing one or more at a larger scale. Then, one management option will be chosen for the implementation.

Usually, after the last step the plan has to be followed, but in some circumstances, it may be necessary to deviate from the original plan (Implement Management step), deciding when and what type of deviations are acceptable.

The Monitor step allows assessing how actions actually affect indicators and evaluating the effectiveness of alternative actions, and taking appropriate corrective action. Monitoring may also detect "surprising" events.

In the fifth step (Evaluate Performance), data are analyzed and actual results are compared to forecasts made in Step 1. The evaluation should explain why results occurred and include recommendations for future action. Negative or unexpected outcomes can be as informative as positive, predicted outcomes.

During the last step (Adjust Management) information gained through the preceding five steps should be used to verify or update the models used to make the initial forecasts, and adjust management actions as necessary. Objectives should be reviewed and adjusted to ensure that they remain consistent with overall goals and values.

Usually an effective application of Adaptive Management requires integration among managers, scientists, and modelers: managers with scientists identify the problem, then with modelers they develop a computer simulation model that they then use to explore various "what if..?" scenarios and evaluate potential outcomes of different management actions. In addition, significant benefits are derived from the process of building the model. This approach intends to encourage debate about system response to management actions and to stimulate a creative search for new solutions, rather than build consensus around a single solution. Scenarios building allow managers to: project potential effects over time and space; forecast potential effects of cumulative management actions; and it provides a consistent basis for participants to discuss and evaluate management options.

7. Conclusions

Managing SELs for environmental security demands for a new strategy which includes uncertainty and the dynamics of systems. The respective adaptive management concept therefore has to consider the following items:

- The relevant objects of environmental security in the real geopolitical world are social–ecological landscapes (SELs) at all hierarchical levels.
- Adaptive management has to address the risks of loosing ecosystem goods and services, as well as the subjective perception of those risks, by developing holistic risk assessments.
- To optimize adaptive management retrospective studies have to be implemented to learn from the past system trajectory.
- Managers must learn to cope with change and uncertainty as intrinsic features of SELs.
- Diversity should be nurtured to optimize resilience.
- Adaptive management, based on learning by doing, should be supported at all hierarchical levels of governance.
- Institutional structures and function should be modified from a top-down, hierarchical approach to a bottom-up process of policy development and refinement, including risk perception.
- Communication should be improved including education, innovation, and the combination of different types of knowledge for learning.
- Trust through stakeholder involvement should be optimized at all stages.
- Opportunity for self-organization should be created towards social–ecological sustainability.

References

Antrop, M., 2005. Sustainable landscapes: contradiction, fiction or utopia? Landscape and Urban Planning 75(3–4), 187–197.

Carpenter, S.R. and Gunderson, L.H., 2001. Coping with collapse: ecological and social dynamics in ecosystem management. BioScience 51, 451–457.

Chambers, R. (Editor), 1985. Managing Rural Development: Ideas and Experience from East Africa. Longman, London.

Costanza, R., Daly, H., Folke, C., Hawken, P., Holling, C.S., McMichael, A.J., Pimentel, D., and Rapport, D., 2000. Managing our environmental portfolio. BioScience 50, 149–155.

Dale, V.H., Brown, S., Haeuber, R.A., Hobbs, N.T., Huntly, N., Naiman, R.J., Riebsame, W.E., Turner, M.G., and Valone, T.J., 2000. Ecological principles and guidelines for managing the use of land. Ecological Applications 10, 639–670.

Dale, V.H., Lugo, A.E., MacMahon, J.A., and Pickett, S.T.A., 1998. Ecosystem management in the context of large, infrequent disturbances. Ecosystems 1, 546–557.

Gunderson, L.H. and Holling, C.S. (Editors), 2002. Panarchy: Understanding Transformations in Human and Natural Systems. Island Press, Washington, DC.

Haag, F. and Hajdu, F., 2005. Perspectives on local environmental security, exemplified by a rural South African village. Environmental Management 36(4), 483–494.

Halbert, C.S., 1998. How adaptive is adaptive management? Implementing adaptive management in Washington state and British Columbia. Reviews in Fisheries Science 1, 261–283.

Holling, C.S. (Editor), 1978. Adaptive Environmental Assessment and Management. Wiley, London.

Holling, C.S. and Meffe, G.K., 1996. Command and control and the pathology of natural resource management. Conservation Biology 10, 328–337.

Kay, J.J., Regier, H., Boyle, M., and Francis, G.R., 1999. An ecosystem approach for sustainability: addressing the challenge of complexity. Futures 31(7), 721–742.

Ludwig, D., Mangel, M., and Haddad, B., 2001. Ecology, conservation, and public policy. Annual Review of Ecology and Systematics 32, 481–517.

Millennium Ecosystem Assessment, 2003. Ecosystems and Human Well-being: A Framework for Assessment. Island Press, Washington DC, pp. 1–25.

Olsson, P. and Folke, C., 2001. Local ecological knowledge and institutional dynamics for ecosystem management: A study of Lake Racken watershed, Sweden. Ecosystems 4, 85–104.

Petrosillo, I., Zurlini, G., Corlianò, M.E., Zaccarelli, N., and Dadamo, M., 2007. Tourist perception of recreational environment and management in a marine protected area. Landscape and Urban Planning 79, 29–37.

Scheffer, M., Carpenter, S.R., Foley, J., Folke, C., and Walker, B., 2001. Catastrophic shifts in ecosystems. Nature 413, 591–696.

Walker, B., Carpenter, S., Anderies, J., Abel, N., Cumming, G., Janssen, M., Lebel, L., Norberg, J., Peterson, G.D., and Pritchard, R., 2002. Resilience management in social-ecological systems: a working hypothesis for a participatory approach. Conservation Ecology 6(1), 14. [online] URL: http://www.consecol.org/vol6/iss1/art14

Walters, C.J. (Editor), 1986. Adaptive Management Of Renewable Resources. McGraw-Hill, New York.

THE POLICY FRAMEWORK GMES AS A GUIDELINE FOR THE INTEGRATION OF ENVIRONMENTAL SECURITY RESEARCH AND LANDSCAPE SCIENCES

THOMAS BLASCHKE*, PETER ZEIL, STEFAN LANG,
STEFAN KIENBERGER, AND ANNELENE KAMMER
Z_GIS Centre for Geoinformatics, Salzburg University, Austria

Abstract. "Environmental security" is a relatively new concept. It might be regarded as a subset of the wider term "human security". In this section we commence from a short theoretical discussion on the underlying concepts. Then we hypothesise that monitoring capacities for environmental security are increasingly needed. We argue that these monitoring tasks require not only technical means but a sound methodology to analyse evaluate and predict changes to a citizen's secure and lively environment. To understand what a "secure and lively environment" for (European) citizen may mean, socio-political aims are to be defined. We use the political goals of the European Commission drawn up in Gothenburg in 2001 and the subsequently evolving GMES programme as a guideline and juxtapose this somewhat positivistic political view to the state of the art of the monitoring capacities and underlying methodologies. In practical terms, GMES is today a joint initiative of the European Commission and the European Space Agency, designed to establish a European capacity for the operational delivery and use of information in support of Environment and Security policies. Despite advances made in sensor technology, observing networks, data evaluation techniques and information technology, production of the information needed for effective policy decisions on environment and security issues often remains below its full potential. Politically, it is of particular concern that Europe does not possess the independent capacity to assess the key drivers and impacts of change in these areas, or to evaluate its policy responses. Still, there are a very limited number of independent, critical, and scientifically sound investigations on GMES beyond

*Corresponding Author: Prof. Dr. Thomas Blaschke, Z_GIS Centre for Geoinformatics, Salzburg University, Hellbrunner Str. 34, A-5020 Salzburg, Austria; Phone: +43 662 80445210; Fax: +43 662 8044525; e-mail: Thomas.Blaschke@sbg.ac.at

I. Petrosillo et al. (eds.), Use of Landscape Sciences for the Assessment of Environmental Security, 57–73.

somewhat positivistic "official" EU publications. This paper investigates the linkages and future consequences of this programme to environmental security research and the role of landscape sciences. Building on ongoing work within the GMES Network of Excellence – GMOSS and on studies within the NATO working group on landscape sciences we elucidate the technical capacities of Earth observations and GIS. We conclude that the scientific challenges are mainly methodological. Finally, we sketch what a future framework for the integration of environmental security research and landscape sciences should take into consideration.

Keywords: Environmental security; landscape ecology; earth observation; landscape sciences

1. (Re-)conceptualising environmental security

The term security has various connotations and has gone through different conceptual shifts in history. Introduced by Cicero and Lucretius referring to a philosophical and psychological state of mind, or the subjective feeling of freedom from sorrow, the current understanding of security is triggered through major political developments in the past. Since the global turn of 1989 and 1990 and since the terrorist attack of September 11, 2001 in many countries a change in the concept of security has occurred away from the narrow military, political and economic security of the cold war towards a wider scope that has also included societal and environmental dimensions (Brauch, 2005). As Brauch (2005) further states, since the early 1990s, influenced by the concerns for "human development" (UNDP, 1994), a shift in the referent object of the security concept has taken place from an exclusive focus on the "nation state" to "human beings and humankind" or from the prevailing "national security" to "human security". Since the late 1990s two parallel debates have taken place on "environmental security" and on "human security" both in the social sciences and within international organisations that have also been stimulated by several international commissions and high-level expert panels.

Dalby (2002) and Brauch (2003) identified three different stages (ranging from the 1970s to the 1990s) within the evolvement on environmental security which has been influenced by different research groups and "schools". According to Brauch (2005), many studies in the environmental security debate since 1990 have ignored or failed to integrate the contributions of the global environmental change community in the natural sciences. To a large extent the latter have also failed to integrate the results of this debate.

Following Brauch (2005), while the academic debate on environmental security influenced the policy agenda of several international organisations, the human-security concept used by UNDP (1994) triggered a global and ongoing scientific debate. Since then, human security has been referred to as: (a) a level of analysis; (b) as a human-centred perspective (Annan, 2001); and (c) as an encompassing concept (UNDP, 1994). For the first approach, the individual human beings or the persons affected by environmental stress and its outcomes (hazards, migration, crises, and conflicts) are the referent object; for the second a normative orientation is essential while the third is a combination of all five dimensions and five levels of a widened security concept.

Two main causes for the re-conceptualisation of "security" in the 1990s can be summarised (Brauch, 2003):

1. Fundamental changes in the international political order resulted in new hard security threats, soft (environmental) security challenges, in new vulnerbilities, and risks that are perceived and interpreted differently depending on worldview, mind-set, and models by the analyst.
2. There exists an increasing perception of new challenges triggered by global environmental change (GEC) and processes of globalisation that may result in fatal outcomes (hazards, migration) that escalate into political crises and violent conflicts.

Therefore, Brauch (2003, 2005) suggested a fourth phase of research on Human and Environmental Security and Peace (HESP) that should combine structural factors from the natural (climate change, water, soil) and human dimenions (population growth, urbanisation, pollution, agriculture, and food) based on the expertise from the natural and social sciences with outcomes and conflict constellations.

2. The GMES framework and the GALILEO and INSPIRE initiatives

GMES is a joint initiative of the European Commission and the European Space Agency, designed to establish a European capacity for the operational delivery and use of information in support of Environment and Security policies. Up until now, investment in data capture and the subsequent generation of information, products, and services to meet the needs of policy development and appraisal for environment and security in Europe has been piecemeal and without coordination. Despite advances made in sensor technology, observing networks, data-evaluation techniques and information technology, production of the information needed for effective policy decisions on environment, and security issues often remains below its full potential. At a time when command of information has important geo-strategic implications, it is of particular concern that Europe does not possess the independent capacity to assess the key drivers and impacts of change in these areas, or to evaluate its policy responses.

GMES can also be seen as a political instrument to reach policy aims as defined in the Cardiff process. At the European Council meeting in Barcelona 2002, a sustainable development strategy was adopted that emphasised the integration of environmental concerns into sectoral policies. The European Council in Seville (June 2002) approved a conflict prevention programme that aimed both at short-term prevention and at the root causes of conflict, in its development cooperation with poverty reduction. It also stressed "the importance, in the context of sustainable development, of maintaining the objective of food security as a basic component of the fight against poverty, as the World Food Summit in Rome has just reiterated".

The GMES capacity is based on four interrelated components as represented by the GMES diamond (see Figure 1):

- Services
- Observations from space
- In situ (including airborne) observations
- Data integration and information management capacity

From the users' perspective, the priority component is the provision of services to fulfil the policy and users' needs. The range of services available by 2008 will be developed progressively.

The provision of services relies on the space and in situ components that capture the required data. In addition, access to socio-economic and other statistical data will be important in order to provide the maximum added value to the services foreseen.

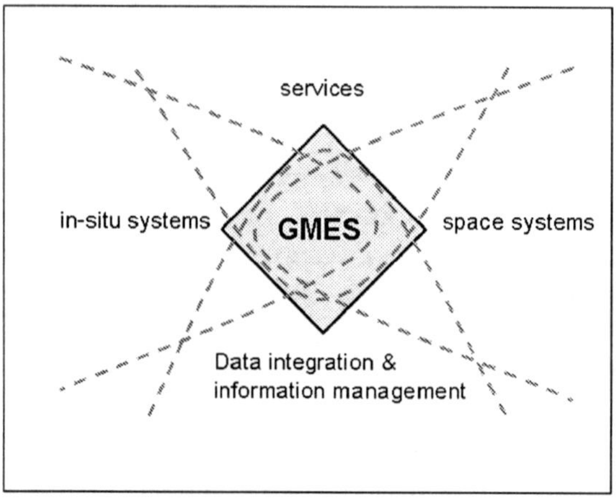

Figure 1. The GMES diamond. (Source: http://www.gmes.info, modified.)

The data integration and information management will enable user access and the sharing of information. These components will be developed in conjunction with the set of services that are required. Such a "build-as-you-need" approach requires GMES to retain a modular open system approach that can easily accommodate new elements.

The objective of the GMES Action Plan's Initial Period was to prepare proposals for the Implementation Period (2004–2008), jointly between EC and ESA, for the establishment of a European capacity for Global Monitoring for Environment and Security by 2008. The strengths and the weaknesses of the current European capacity for monitoring and information production were identified, as well as the needs for improvements, in the scientific, technical, legal, economic, and institutional domains.

In November 2005, the European Commission adopted a new Communication on Global Monitoring for Environment and Security (GMES): From concept to reality (COM(2005)565 final). This Communication plans to introduce the first three Earth observation services: emergency management, land monitoring and marine services. After GALILEO, the European navigation satellite system, the Global Monitoring for Environment and Security (GMES) has become the second EU flagship in space policy. Natural and man-made catastrophes in Europe, America, Asia, and Africa, coupled with increased security needs, have further reinforced the case for improved monitoring systems. It is argued that GMES gathers relevant Earth observation data, for example concerning environmental pollution, floods, forest fires, or earthquakes, and supplies them in reliable and user-friendly services in support of public policymakers' needs. GMES shall ensure that crisis situations can be better anticipated and managed.

Starting with emergency management, land monitoring and marine services as three fast track services (land, marine, emergency) GMES will be developed in steps by the end of 2008. The **Emergency Management service** aims to reinforce the European capacity to predict and respond to crises and emergencies associated with natural and man-made disasters such as:

- Meteorological-driven hazards, e.g. storms, fires, floods
- Geophysical hazards, e.g. earthquakes, tsunamis, volcanic eruptions, landslides
- Technological disasters, whether deliberate or accidental, e.g. urban fires, chemical accidents, on industrial sites
- Humanitarian disasters generated by sudden natural, technological, and weather-driven disasters.

The **Land Monitoring service** will deliver timely, important information on land use and land cover changes for a number of identified areas at European, national and local scale. In the short to medium term the proposed service will focus on the mapping of land cover and land use for Europe's:

- Spatial development perspective
- Urban environment strategy
- Implementation, review, and monitoring of EU policies

The proposed **Marine services** will provide data, information products, and indicators on the marine environment. The objective is to provide general information, structured at the European level, on the condition of the seas, including:

- Their monitoring
- Indicators for global and regional mapping of ocean climate and variability
- Operational ocean analyses and forecasts (e.g. ocean current, temperature, salinity fields)
- Operation, validation, and maintenance of in situ observing networks

GMES and the related INSPIRE initiative will substantially increase the ability of geospatial information to support a variety of EU policies. Geoinformation plays an ever increasing role in environmental monitoring (Blaschke, 2001). A major future challenge is monitoring global climate change and the environmental commitments made within the framework of international treaties such as the Kyoto protocol and – globally – to control its causes.

GMES encompasses information from the environment as well as security. Even if one focuses on an environmentally sustainable situation for European citizens, the planning and management of land use that contributes to sustainable development involves the analysis of a number of complex issues. As a result, the skills and knowledge required is often fragmented, so that the capacity of both professional and non-professional groups to resolve sustainable land use issues is often limited (Nowotny et al., 2001). This statement rings even more true when bridging environmental monitoring and monitoring the security of European citizens. Before trying to overcome these barriers, it is important to lay the foundations of such a bridge in terms of scientific definitions and methodologies.

From a global perspective, there seems to be a slight overlap between GMES and GEO. As a result of the Earth Observation Summit in Washington DC, 2003, an ad hoc Group on Earth Observations (GEO) was established to prepare a 10-year implementation plan for a coordinated, comprehensive, and sustained Earth observation system of systems. Since the citizen's security and environmental aspects are not yet secured at a high level in the global market economy, it is a clear European Union challenge to develop a methodology to consider environmental and security aspects in land use planning, environmental protection, security applications, and consequently in policy. This is interpreted as being one of the reasons for these parallel strategies of GMES and GEO. The future challenges are to achieve increased interoperability of acquisition systems, to harmonise and to foster standardisation of data structures and interfaces, to overcome policy barriers to data sharing, to design quality assurance mechanisms, to achieve fusion of data from different levels and to provide innovative, user-friendly services that are cost effective and sustainable. In the context of the above, NATO CCMS working group is speculating whether these developments should become part of a coherent, single overall programme. In this case we see a potential danger that pragmatic questions and investments override scientific aspects of environmental monitoring and – as discussed later – of the integration of landscape sciences in the monitoring of environmental security.

Interdisciplinary and coordinated research is of critical importance to generate both the technologies and the understanding that Europe needs to strengthen its security monitoring capability. The definition and perception of security threats, challenges, vulnerabilities and risks are part of a socio-political process resulting from a complex policy assessment in EU partner countries by governments, societal groups, research institutions, and the media.

Information derived from remote sensors can contribute to an early recognition and warning of such threats and can enable policymakers to prevent the emergence of conflicts or to reduce their impact. Technology itself, however, cannot guarantee security, but security can be greatly enhanced by technological support. Space technology and geo-data provide context information on security problems and contribute to strengthen preparedness. Using technology effectively can enable policy-makers to enhance the coping capacities and can – at least in principle – reduce the vulnerability of states and thereby foster stability.

2.1. GALILEO

As stressed in the European Commission White Paper on European transport policy for 2010, the European Union needs an independent satellite navigation system. GALILEO is Europe's contribution to a Global Navigation Satellite Infrastructure (GNSS).

There are at present two radio navigation satellite networks in the world, one American (GPS), and one Russian (Glonass). Both were designed as military systems. Since the Russian system seems to have not succeeded in generating any significant civil applications, GALILEO offers a real alternative to the establishment of a de facto monopoly in favour of GPS and American industry.

- GALILEO has been designed and developed as a non-military application, while nonetheless incorporating all the necessary protective security features. Unlike GPS, which was essentially designed for military use, GALILEO therefore provides, for some of the services offered, a very high level of continuity required by modern business, in particular with regard to contractual responsibility.
- It is based on the same technology as GPS and provides a similar – and possibly higher – degree of precision, thanks to the structure of the constellation of satellites and the ground-based control and management systems planned.
- GALILEO is more reliable as it includes a signal "integrity message" informing the user immediately of any errors. In addition, unlike GPS, it will be possible to receive GALILEO in towns and in regions located in extreme latitudes.
- It represents a real public service and, as such, guarantees continuity of service provision for specific applications. GPS signals, on the other hand, in recent years have on several occasions become unavailable on a planned or unplanned basis, sometimes without prior warning.

Nonetheless, GALILEO also complements GPS insofar as:

- Using both infrastructures in a coordinated fashion (double sourcing) offers real advantages in terms of precision and in terms of security, should one of the two systems become unavailable.
- The existence of two independent systems is of benefit to all users since they will be able to use the same receiver to receive both GPS and GALILEO signals.

The European Commission and the European Space Agency attach great importance to the complementary and interoperable relationship between GALILEO and GPS in order to provide improved and safer services to the users worldwide. This can be seen from the EGNOS programme (European Geostationary Navigation Overlay Service) which makes significant improvements to the services offered in Europe by the GPS and Glonass satellite constellations. EGNOS, which has been developed since 1993, increases the number of GPS signals, applies a differential correction and adds an integrity message. EGNOS is also set to be incorporated into GALILEO. The way that GALILEO complements GPS is useful in presenting Europe as a credible partner to the

USA. By the same token, Europe would have put itself out of the running had it abandoned the GALILEO programme.

2.2. INSPIRE

The general situation on spatial information in Europe is one of fragmentation of data-sets and sources, gaps in availability, lack of harmonisation between data-sets at different geographical scales and duplication of information collection. These problems make it difficult to identify, access and use data that is available. Fortunately, awareness is growing at national and at EU level about the need for quality geo-referenced information to support understanding of the complexity and interactions between human activities and environmental pressures and impacts The INfrastructure for SPatial InfoRmation in Europe initiative (INSPIRE) aims at making available relevant, harmonised and quality geographic information for the purpose of formulation, implementation, monitoring and evaluation of Community policy. The INSPIRE initiative is therefore timely and relevant but also a major challenge given the general situation outlined above and the many stakeholder interests to be addressed.

The INSPIRE Proposal for a Directive was adopted by the Commission in July 2004. This was a major milestone for the use of Geographic information in Europe as a contribution to environmental policy and sustainable development. It was the first step in a co-decision procedure that should lead to the formal adoption of the INSPIRE Directive, which then has to be implemented in every EU Member State. INSPIRE is complementary to related policy initiatives, such as the Commission proposal for a Directive on the reuse and commercial exploitation of Public Sector Information. The initiative intends to trigger the creation of a European spatial information infrastructure that delivers to the users integrated spatial information services. These services should allow the users to identify and access spatial or geographical information from a wide range of sources, from the local level to the global level, in an interoperable way for a variety of uses. The target users of INSPIRE include policymakers, planners and managers at European, national and local level, and the citizens and their organisations. Possible services are the visualisation of information layers, overlay of information from different sources, spatial, and temporal analysis, etc.

3. Monitoring environmental security

Despite undoubted strengths in environmental data acquisition from remote sensing, from in situ monitoring and from field surveillance, there is both scope and need for investment in improved monitoring and surveying in Europe. The use of Earth observation (EO) for environmental monitoring has reached operational status in meteorology, climatology and climate change research, sea surface monitoring as well as vegetation and land cover mapping. Elsewhere, the use of EO however remains limited because it is unable to deliver useful information at a required accuracy or with sufficient resolution, timeliness, and continuity.

Next to the NATO CCMS pilot study this paper partially builds on the EU 6th Framework Network of Excellence GMOSS (*Global Monitoring for Security and Stability*, http://gmoss.jrc.it). The aim of this network is to integrate Europe's civil security research to acquire and nourish the autonomous knowledge and expertise base Europe needs in order to develop and maintain an effective capacity for global monitoring using satellite EO. The science involved can be divided into three broad classes (1) generic technologies that are common to nearly all security applications of Earth observation such as feature recognition, (2) specific applications such as population monitoring, and (3) socio-political studies that aim to assess the threats that need to be monitored and identify the stakeholders needs' for information.

Alternatively to the sectoral view described above we can attempt to dissect the complexity of the wide field of environment and security into application fields or disciplines. The GMES Service elements are structured according to this manner, but for the purpose of this paper we would rather follow a process-orientated view and utilise the widely agreed disaster management cycle as an example (see Figure 2).

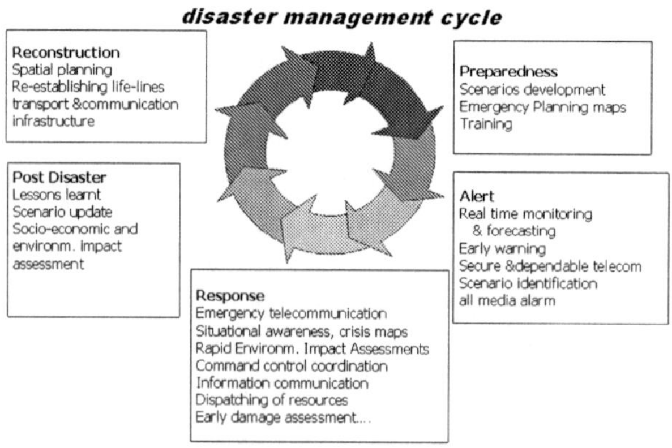

Figure 2. The disaster management cycle and the processes associated with it.

Technically, the vision for the future is that space technology-based methods such as satellite communication, satellite-based navigation, or remote sensing can provide a significant contribution in environmental security monitoring. Regarding monitoring aspects, information systems based on mobile devices using the Global Navigation Satellite Systems/Global Positioning System (GNSS/GPS) augmented by complementary systems and/or sensors, for example, could assist rescue teams by determining their position and navigating within difficult and unfamiliar terrain. These systems could also safely guide mountaineers and hikers through mountain regions by combining their positional data with hiking routes stored in a mobile computer, e.g. Personal Digital Assistant (PDA). With receivers and transmitters for broadband technology, such navigation tools could support the environmental monitoring tasks with online safety information such as avalanche warnings or weather forecasts and transmit SOS messages in case of accidents to emergency call centres.

Remote sensing methods may also be used to assess the conditions and development of remote or difficultly accessible regions. Applications include deforestation as a result of storms, calamities, erosion damage due to overuse, path construction on steep slopes, distribution of settlements, assessment, and monitoring of surface deformations, mass movements, etc. The Geographic Information System (GIS)-based combination of remote sensing results with other spatial information such as geological maps or Digital Terrain Models (DTM) allows for determining the hazard potential and vulnerability of the region. (D)INSAR (Differential interferometry) technologies designed to detect damages on houses and infrastructures are yet another example of how space technologies can contribute to safety in mountain regions. GIS assist decision making experts, the media and the public in visualising areas prone to natural hazards and are important tools for planning purposes as well as for the management of rescue operations. During the last decade there has also been an increase in the awareness in the links between disaster impact and environmental conditions (Kelly, 2005). GIS can therefore also play an important role in the identification of perceived salient environmental issues related to immediate relief and recovery operations (Kelly, 2005).

Coordinated research and education is of critical importance to generate both the technology as well as the understanding which we need to adjust to environmental security threats. Technology provides information on threats, contributes to protection, and, where appropriate, enables society to counter them. But a prerequisite is that a consensus exists within society on the nature of these dangers to environmental security, as well as the political will and the means to put preventive activities and counter measures into place. The terminology used herein bridges environmental and security fields. Some terms used such as, for instance, "counter measures" may be associated with a military origin. The terminology is not agreed as yet; it needs some adaptation and – as laid out in the next section – an integration of landscape science concepts.

4. Integrating landscape sciences

4.1. LANDSCAPE SCIENCES, LANDSCAPE ECOLOGY, APPLIED ECOLOGY

The NATO working group has the overall vision to establish and exchange information about landscape science approaches that are useful for environental assessment and to transfer landscape assessment technologies among the study group participants for use in environmental protection and preservation programmes. Land use and land cover characterisation and the use of landscape indicators is extensively used throughout this book. Environmental assessment is usually defined as a process by which scientific evidence and technological information are analysed for the purpose of evaluating present conditions or forecasting the outcomes of alternative future courses of actions. The assessments are directed towards specific ecological resources and socially relevant end points such as watershed conditions (water quality, quantity, and vulnerability to flooding), landscape resilience (ability to sustain ecological goods and services when subjected to conditions of anthropogenic and natural stress), and biodiversity (wildlife habitat).

For landscape sciences to progress, it must incorporate both its scientific foundations and the ability to develop and integrate novel methods and ideas into ecological theory and research (Starzomski et al., 2004). As stated in the first sections, "environmental security" includes the human perspective. A "catastrophe" is ecologically defined as environmental stochasticity of an extreme form that affects the entire meta-population, causing local extinctions in a density-independent manner. Catastrophes are therefore separated from "normal" environmental fluctuations. Fluctuations that are correlated (synchronised) over a scale that encompasses several local populations are a mixture of catastrophe and "normal" variability. In turn, a tsunami or an earthquake causes immense loss to human life and property. A tsunami may have caused untold environmental damage, with reports of destroyed coral reefs and uprooted mangrove forests, but these consequences are regarded as secondary compared to the human losses. This fact is neither "good" nor "bad" but has to be regarded in a future framework of "environmental security". Wu and Overton (2002) have coined the term crises-oriented ecology (see Blaschke, 2005b). Kelly (2001) points out that disaster relief assistance can only be effective if managers and decision-makers are aware of and include critical factors such as the environmental impacts of the disaster or the relief action. Environmental damage and degradation often threaten livelihoods and can cause or aggravate tensions. This can then lead to the risk of secondary disasters which in turn can lead to even greater vulnerability and further conflict (Organisation for Economic Co-operation and Development, 2005).

Sustainable solutions to complex social, economic, and environmental problems and the long-term success of GMES are unlikely to be the result of a narrow, technical investigation. Owens and Cowell (2002: 49) argue that: "Increasing recourse to technical rationality, for example, might be seen as part of an 'analytical arms race', in which intangible environmental qualities must be rendered material in the planning process to match the apparently 'objective' status of other claims". These authors go on to trace the evolution of assessment tools from "Environmental Impact Assessment" to "Strategic Environmental Assessment". Future research needs to enhance the demonstration of assessment tools that apply to land use and land use change, not just for the evaluation of major political initiatives, but also to examine longer-term effects of predictable changes and changes that are policy driven. No formal disaster-focussed environmental impact assessments existed before the 1990s and the few disaster-related environmental impact assessments done were generally based on normal environmental impact assessment (EIA) procedures, or addressed multiyear disasters such as the locust crisis in Africa in the late 1980s and early 1990s (Kelly, 2005). The procedures followed in a normal EIA do not, however, provide useful and reliable results in disasters, particularly quick-onset events where the time to do an impact assessment is rather limited. A collaborative project between the Benfield Greig Hazard Research Centre and CARE International has recently developed, tested, and disseminated a rapid environmental impact assessment (REA) process for emergency assessment, planning, and response (Kelly, 2001). There exists, however, a great opportunity for better integration of the spatial and geographic components for further analysis of the data and information collected during the REA process.

Generic tools for monitoring the environment and security are currently being developed amongst others in various EU FP6 projects such as GMOSS, SENSOR, ORCHESTRA, OASIS, and various GMES Service elements. Similarly, several Coordinated Actions (CA) and Specific Targeted Research Projects (STREPS) focus on vulnerability and impact assessment tools. For instance the CA ATLAS centres on sustainable impact assessment tools, while another initiative, EASY-ECO, aims at organising training courses and conferences on sustainability issues.

4.2. SPATIAL ARRANGEMENT MATTERS

Blaschke (2005a) juxtaposes two different paradigms in landscape ecology and landscape planning, the natural capital paradigm and the patch–matrix–corridor model, as discussed later. His critique develops from the suggestion by Potschin and Haines-Young (2005) that if sustainability on a landscape scale is mainly viewed in terms of maintaining the output of "goods and services", then probably many different landscape configurations can be regarded as sustainnable. Thus, planning solutions merely have to be "adequate" rather than "optional", so that the key task for Landscape Ecology is to understand the extent of the choice-space within which decisions can be made.

Blaschke (2005a) argues that spatial pattern matters fundamentally because context can have a fundamental influence of meaning and value. Moreover, landscape structure is important in its own right, because different structures have different implications for processes. By implication, therefore, planning in the context of sustainability must not only take account of the outputs of goods and services, but also the nature of landscape patterns as an issue in its own right. If we only focus on outcomes in terms of benefits how, he asks, are we to help people with the issue of where to place things?

Transferring this sustainability discussion to environmental security and placing the people in the central view one can argue that, in contradistinction to the anthropocentric arguments developed in favour of the natural capital paradigm (Haines-Young, 2000), the patch–matrix–corridor model of Forman (1995) still offers much that is of value to Landscape Ecology and landscape sciences. This may provide the key to understanding land use systems and land use changes through the development of structural or spatial indicators that can sit alongside other sustainability measures that cover the economic, social, and cultural aspects of sustainability.

Blaschke (2005a) claims that in almost all cases spatial pattern is relevant. Landscape structure often has a fundamental connection with process, and the analysis of structure is often fundamental to any understanding of ecosystem integrity and how ecological systems can be maintained so that they continue to deliver benefits to society. The main point of difference it seems depends on the question of whether in the context of a multifunctional landscape, there is an optimal pattern or spatial arrangement of landscape elements that science seeks to discover (Forman, 1995) or a larger set of landscape configurations that are "adequate" or "sufficient" in terms of the outputs they can maintain. So, from a critical discussion Potschin and Haines-Young (2005) conclude that "whatever position we take on this question, it seems inescapably the case that spatial pattern does matter".

If this holds true, we can hypothesise that this "spatialisation" may become the most important factor in the environmental security research concept.

5. Conclusions and research needs

This chapter discusses the scientific foundations of a proposed integration of environmental security with landscape science concepts. The underlying vision is a "spatially enabled framework" for environmental security. We have briefly outlined the integrating potential of such a spatial framework and used the European INSPIRE directive as a proof of concept. As a first step towards this foresight, numerous issues relating to the timely provision of comprehensive, consistent, and readily accessible information services were discussed and technical developments were juxtaposed to methodological challenges and recent achievements, both in the arising trans-disciplinary field of environmental security and in landscape sciences.

Conceptually, this chapter discusses that:

- Very few areas of European policy do not interact in some measure with environment and security. GMES may be a general environmental policy framework within which the EU operates. Its security definition may serve as a guideline for scientific research.
- Research towards "Environmental security" is not simply environmental research plus security research. Rather, we suggest following the upcoming concept/paradigm of vulnerability to include other areas of policy and legislation which need to be considered.
- The information needed is extremely diverse, cutting across policy domains, environmental media and themes. It depends upon a variety of technologies and data sources, many of which are not environment or security specific. We demonstrate that the spatial domain may be the key to integrate these various data-sets through the recent means of GIS and remote sensing tools.
- Better integration between the various data sources is needed to facilitate the flow of information between applications but also from regional and national bodies to the European level and to overcome isolated sectoral views.
- Despite the diversity of needs, there is considerable scope to design a framework for information gathering designed for multiple use across different areas of application. By empirical examples we illustrate the necessity to strike a balance between the provisions of information that is as specific as possible to any particular application, while exploiting every opportunity to draw on common sources.

Continued investment is needed to improve the availability and supply of consistent, high resolution basic geographical data. Rapid adoption of the INSPIRE Directive on the establishment of a European Spatial Data Infrastructure is important in this respect and action should be taken, in the context of this Directive, to improve the supply and maintenance of this information, by establishing EU-wide mapping standards for topographic and other basic geographic data, and by applying these to the development of consistent European spatial data-sets, as identified in the INSPIRE priority themes.

References

Annan, K.A., 2001. Foreword. In: R. McRae and D. Hubert (Editors), Human Security and the New Diplomacy. Protecting People, Promoting Peace. Ithaca McGill Queen's University Press, Montreal, Kingston, London, pp. xix.

Blaschke, T., 2001. Environmental monitoring and management of protected areas through integrated ecological information systems – an EU perspective. In: C. Rautenstrauch and S. Patig (Editors), Environmental Information Systems in Industry and Public Administration. Idea Group Publishing, Hershey, London, pp. 75–100.

Blaschke, T., 2005a. The role of the spatial dimension within the framework of sustainable landscapes and natural capital. Landscape and Urban Planning 75, 198–226.

Blaschke, T., 2005b. Crisis-oriented landscape ecology research: the need for a new research framework. IALE European conference 2005; CD-ROM.

Brauch, H.G., 2003. Security and environment linkages in the Mediterranean: three phases of research on human and environmental security and peace. In: H.G. Brauch, P.H. Liotta, A. Marquina, P. Rogers, and M.E. Selim (Editors), Security and Environment in the Mediterranean. Conceptualising Security and Environmental Conflicts, Springer, Berlin, Heidelberg, pp. 35–143.

Brauch, H.G., 2005. Threats, Challenges, Vulnerabilities and Risks in Environmental and Human Security, Source No. 1/2005, UNU-EHS. ISBN: 3-9810200-5-7 (electronic version).

COM(2005)565 final. COMMUNICATION FROM THE COMMISSION TO THE COUNCIL AND THE EUROPEAN PARLIAMENT. Global Monitoring for Environment and Security (GMES): From Concept to Reality, Brussels.

Dalby, S., 2002. Security and ecology in the age of globalization. In: Wodrow Wilson International Centre for Scholars (Editor), Environmental Change & Security Project Report, Issue No. 8 (Summer), pp. 95–108.

Forman, R., 1995. Land Mosaics: The Ecology of Landscapes and Regions. Cambridge University Press, Cambridge.

Haines-Young, R., 2000. Sustainable development and sustainable landscapes: defining a new paradigm for landscape ecology. Fennia 178(1), 7–14.

Kelly, C., 2001. Benfield Greig Hazard Research Centre, University College London. Rapid Environmental Impact Assessment: A Framework for Best Practice in Emergency Response. Presented at: Sharing Experiences on Environmental Management in Refugee Situations: A practitioner's workshop, Geneva, 22–25 October 2001.

Kelly, C., 2005. Getting better with Age? Rethinking Rapid Environmental Impact Assessment in Disasters. IAIA Meeting, Boston.

Nowotny, H., Gibbons, M., and Scott, P., 2001. Rethinking Science: Knowledge and the Public in an Age of Uncertainty. Blackwell, London.

Organisation for Economic Co-operation and Development, 2005. Developing Assistance Committee (DAC), Mainstreaming Conflict Prevention, Issues brief: Overview of links between the environment, Conflict and Peace. [online] URL: http://www.oecd.org/dataoecd/26/49/35785489.pdf

Owens, S. and Cowell, R., 2002. Land and Limits: Interpreting Sustainability in the Planning Process. Routledge, London.

Potschin, M. and Haines-Young, R., 2005. "Rio+10", sustainability science and landscape ecology. Landscape and Urban Planning 75(3), 162–174.

Starzomski, B.M., Cardinale, B.J., Dunne, J.A., Hillery, M.J., Holt, C.A., Krawchuk, M.A., Lage, M., McMahon, S., and Melnychuk, M.C., 2004. Contemporary visions of progress in ecology and thoughts for the future. Ecology and Society 9(1), 14. [online] URL: http://www.ecologyandsociety.org/vol9/iss1/art14/

UNDP, 1994. Human Development Report 1994. Oxford University Press, New York.

Wu, J. and Overton, C., 2002. Asian ecology: pressing problems and research challenges. Bulletin of Ecological Society of America 83(3), 189–194.

PART II INTRODUCTION

LANDSCAPE SCIENCE METHODOLOGIES TO ASSESS ENVIRONMENTAL SECURITY

BAI-LIAN LI
Ecological Complexity and Modeling Laboratory, Department of Botany and Plant Sciences, University of California, USA

WILLIAM G. KEPNER
U.S. Environmental Protection Agency, Office of Research and Development, Nevada, USA

Landscape science recognizes that spatial patterns and heterogeneous environments are critical factors in understanding how systems work across a range of scales. There are three chapters in this section. In the first one by Stoyan et al., the authors investigated the landscape pattern in a protected area in Bulgaria, using satellite images, GIS analyses tools, and a set of landscape metrics to evaluate the level of threats to particular ecosystem services connected with biodiversity and further to contribute for the objectives of environmental security assessment. The second chapter by Lang et al. highlighted that an active development of urban green spaces is the key for sustainable urban ecosystems, presenting two approaches for advanced urban green mapping in Arizona and Austria. In the third chapter by Li et al., the authors outlined a new conceptual approach to the problem of environmental security, based on the study of the self-sustainable or self-organizing properties of the natural biota which reflects the ability to maintain all environmental characteristics in a state optimal for living systems. They showed that the bases of environmental stability in natural ecological systems can also be analyzed on allometric scaling.

INVESTIGATING LANDSCAPE PATTERNS IN PROTECTED AREAS USING ASTER IMAGES

STOYAN NEDKOV * AND MARYIANA NIKOLOVA
Bulgarian Academy of Sciences, Institute of Geography, Sofia, Bulgaria

TIMO KUMPULA
University of Joensuu, Department of Geography, Joensuu, Finland

Abstract. Understanding the spatial structure of the areas with preserved nature is important for the support of their habitat function, which provides ecosystem values the maintenance of biological and genetic diversity. This study uses satellite image interpretation and GIS analysis tools to determine the main land cover classes, to evaluate their conservation value and the spatial structure of the habitats. The classification of the ASTER TERRA VNIR image results in the differentiation of the territory into 11 land cover types with particular distribution among the previously defined two landscape regions. Three land cover classes with high conservation value were assessed, another three have middle value, while the other five have low or no conservation value. Different landscape metrics were used to evaluate the spatial structure of the habitats and land cover classes with high conservation values. The results show that beech–hornbeam forests provide the best habitat conditions for interior species, while in oak-flowering ash and dry meadow classes they are under a higher threat. The results can be used to evaluate the level of threat to the habitat ecosystem services and further to contribute to the improvement of the environmental security.

Keywords: Landscape pattern; remote sensing; biological conservation, landscape metrics

*Corresponding Author: Dr. Nedkov Stoyan, Bulgarian Academy of Sciences, Institute of Geography, Akad. G. Bonchev St., Bl. 3, 1113 Sofia, Bulgaria; Phone: +3592 979 33 71; e-mail: snedkov@abv.bg

I. Petrosillo et al. (eds.), Use of Landscape Sciences for the Assessment of Environmental Security, 77–92.

1. Introduction

Bulgaria is a fairly small country, with rich biodiversity. It is due to the unique geographical location, between three natural zones. Another reason is the varied relief, which ensures rapid change of the ecological conditions within a small distance. The biodiversity is considered to be one of the most important national values of the country. The areas with preserved biodiversity are key sources for the genetic material that ensures the future development of the nature as well as the biological resources vital to human life. Providing habitat for wild plant and animal species is one of the four main functions of natural ecosystems formulated by de Groot et al. (2002). This function provides ecosystem services like the maintenance of biological and genetic diversity. Supporting healthy habitats is a necessary precondition when providing these services. It can be realized by setting these areas under special protection regime (reserves, national parks etc). From that point of view the biological conservation (focused on the biodiversity protection) can contribute to environmental security, which is described, in an objective sense, with the aims to evaluate the level of threats to acquired and sustained ecosystem values in terms of services on a multiple scale. The analyses of spatial structure of species distribution and habitat loss may be used to provide a measure of ecological safety for a habitat network system on a regional scale (Chen et al., 2005, 2006).

Biological conservation is focused on different levels of organization ranging from genes to biomes. At the landscape level, the persistence of biodiversity is chiefly affected by human activities, including the fragmentation of natural habitat and the degradation of the quality of the matrix in which natural areas are embedded (Knight and Landres, 2002). Landscape ecology can be useful for nature conservation because it takes into consideration the spatial arrangements of the habitats and consequently considers structures and processes as perceived by the different species (Farina, 2000). Remote Sensing and GIS provide appropriate tools to investigate landscape patterns for the objectives of biological conservation and the management of the protected areas. Land cover classification based on satellite images is one of the most popular methods. In the past, the classifications of remote sensing images have been criticized to be too general and inaccurate. In most cases they were based on field measurements and visual interpretation of the images, whereby image characteristics like resolution (pixel size), spatial coverage, and spectral channel effects could limit the usability of the images in vegetation classification (Kumpula et al., 2005). The images provided by the ASTER (Advanced Spaceborne Thermal Emission Radiometer) could satisfy the needs of relatively high resolution (15 m) with reasonable cost, which is the reason for their choice in this investigation. During the last decade a lot of GIS tools for landscape analyses have been developed. Klug and Zeil (2004) present 13 different software packages working independently or as extensions of popular GIS programs like the ArcView and the ArcGIS. Among them is the Analytical Tools Interface for Landscape assessment (ATtILA) which was chosen for the present study. It is very flexible easy to use, compatible with widely available software, able to accept a wide range of data as input for calculation of various metrics and allows easily reproducible results.

The nature protection system in Bulgaria includes 6 categories of protected areas ranked according to their limitation regime as: (1) Reserve; (2) National Park; (3) Nature Landmark; (4) Supported Reserve; (5) Nature Park; (6) Protected Site. Nature Parks, as specific protection areas, are founded with the aim: (1) to support ecosystems and their biodiversity; (2) to give opportunity for scientific, educational and recreational activity; (3) to ensure sustainable use of renewable natural resources and to preserve the traditional livelihood of the local communities. Within its borders it is possible to have settlements and resorts as well as production activities, which do not pollute environment. Other protected areas within the park borders preserve their regime according to the regulations. The existence of different kinds of land use with different kinds of activities and objectives often leads to conflicts between the involved sides, for example, the park administration and forestry directorate. On the other hand, the legislative framework, which ensures both conservation and some economical activities within the park area, gives the opportunity to develop Nature Parks as models for sustainable interaction between nature and society. The realization of such a plan can lead to particular reduction of the level of threats to ecosystem values, especially to the preservation of biodiversity.

The main objectives of this work are: (1) to detect the possible land cover classes using ASTER images; (2) to evaluate their conservation value; (3) to evaluate the spatial structure of the habitats using the GIS analyses tools. The achievement of these objectives can be useful to the evaluation of the level of threats to refugium ecosystem function, which ensures the important service of the maintenance of biological and genetic diversity and further contributes to the objectives of environmental security assessment.

2. Research area

Nature Park Sinite Kamani occupies 113.08 km^2 situated on the south slopes of Slivenska Planina, which is a part of the Balkan Range. The relief is predominantly low-mountain and only the highest parts lie in the middle mountain belt. The altitude varies from 290 m in the southern part to 1,181 m around Bulgarka Peak, which is the highest in Eastern Balkan Mountains

(Figure 1). The great biodiversity of the area is determined by a unique combination of relief, climate, rocks and soil conditions. Totally 1,024 higher plant species are found in the park territory, 938 of them are seminal, 19 fern species, and 67 moss species (Stoeva et al., 2004). There are 89 plant species with conservation value within the park territory, 42 of them are in the Bulgarian Red book, 39 are regional endemic (Balkan peninsular), 16 are national endemic, 6 are included in the Bern Convention, 10 are on the European Red list. There is also a great variety of animal species – 244 vertebrates are found within the park territory and 45 of them are in the Bulgarian Red book.

The park of 66.8 km^2 was announced as protected area on 28 November 1980. According to the former legislation it belonged to the category of the so-called People's Park. Kutelka reserve (6.45 km^2) was created within the park area in 1983 to ensure a strong protection of the habitats of rare and endangered predatory and scavenger birds like some eagle and vulture species. According to the new nature protection legislation of Bulgaria adopted in 1998, the area was transformed into Nature Park. This category is less strict and allows some kinds of economical activities as was mentioned above. The park territory was extended to the east and reached its present size in 2002.

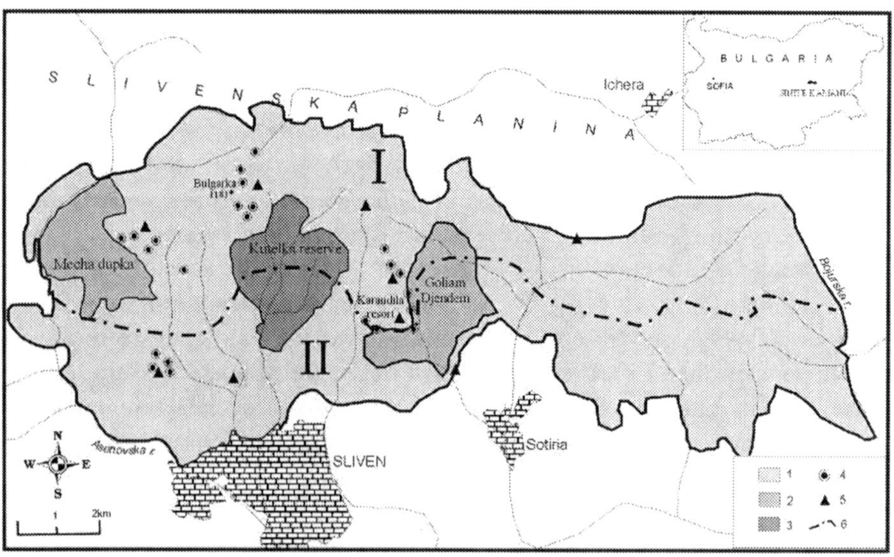

Figure 1. Location of the research area Sinite Kamani Nature Park: 1 – park territory; 2 – areas with special protection regime; 3 – reserve area. Field work points: 4 – training sites; 5 – rectification points. Landscape features: 6 – border between landscape regions; I – Landscape Region 1; II – Landscape Region 2 (more details in the text).

3. Materials and methods

3.1. DATA ACQUISITION

The ASTER TERRA VNIR scene used in this study was acquired in May 2003. Its sensor has a 15 m spatial resolution and it is able to measure the reflection from the ground surface ranging from the visible to near infrared VNIR (wavelengths 0.52–0.86). The swath width of the ASTER TERRA is 60 km at a 16-day repeat interval (Abrams, 2000). ASTER TERRA also has a 15 m resolution NIR along-track stereo-band looking 27.6° backwards from nadir. The stereo-band (3B) covers the same spectral range of 0.76–0.86 μm as the nadir band (Kaab, 2005). From the ASTER stereo data it is possible to generate Digital Elevation Model (Welsh et al., 1998).

The field data was collected in September 2005. It consists of field training sites for the land cover classification and points for the image georectification (Figure 1). The field training sites were selected randomly to represent the main vegetation types of Sinite Kamani Nature Park. The main tree species, field layer vegetation, and mineral ground coverage were inventoried on each site and the locations were marked with GPS. The total number of the inventoried sites is 23. For the georectification of the satellite image nine clearly visible crossings from different parts of the ASTER TERRA image were marked with GPS.

3.2. LANDSCAPE CHARACTERIZATION

The study of spatial variation in landscapes on multiple scales is the main objective of landscape ecology. The spatial and hierarchical aspects of landscape heterogeneity are investigated on two levels of the scale.

The first level represents the landscape differentiation caused by the influence of main flows of energy and matter. According to the classification scheme of Bulgarian landscapes made by Velchev et al. (1992), the area of Sinite Kamani belongs to two landscape types. They are differentiated as a result of changes in the hydro-climatic conditions caused by the elevation (zonal altitude). The first type, called *low mountain subhumid submediterranean with oak forests and shrubs*, occupies the lower (below 800 m) part of the park situated on the southern slopes of the mountain. The second – *mid mountain temperate humid with beech forests*, is located in the higher part of the mountain. These land units were considered as landscape regions according to the concept of Forman (1995). They were mapped using a regression model of interpolation based on the correlation between hydro-climatic coefficients and the altitude (Gikov and Nedkov, 2005).

On the second level of the hierarchy the heterogeneity is investigated on the basis of land cover. The differentiation here is a result of the combined interaction of large-scale hydro-climatic influence and human activity. The land cover classification was based on the ASTER TERRA satellite image (see above). This is the main operating level for landscape structure investigation using the concept of patches, corridors, and matrix introduced by Forman and Gordon (1986).

Image interpretation for the needs of the landscape classification was done using ERDAS imagine 8.7 software. According to the field data and interpretations, the ASTER TERRA image was classified into 10 main land cover classes and three subclasses. The supervised maximum likelihood classifier was used as a classification scheme. Due to the misclassifications, the class number was diminished to eight main vegetation classes and three subclasses. The image classification was imported to grid format in the ArcGIS software. The fragmentary of the classification map was reduced using a smooth 2 × 2 grid-majority filter. Three additional classes were added after manual digitalization directly from the image. They include urban places used mainly for recreation purposes, paved roads and sports facilities such as ski tracks. Most of them occupy a limited area, which impedes their correct delineation by automatic classification but on the other hand allows manual digitalization. This data in form of shape-files was incorporated into the land cover grid using the AGWA Land Cover Modification tool.

3.3. INDICATORS AND ASSESSMENT

The indicators used in this work were divided into two main objectives: (1) to evaluate the conservation value of the land cover classes; (2) to explore the spatial aspects of the landscape pattern in order to evaluate its usability for the conservation functions according to the park's regulations. The conservation value of the land cover classes was evaluated using the number and distribution of the endangered, endemic, and relict plant species. The main source for information was a floristic investigation in Sinite Kamani Nature Park (Stoeva et al., 2004). It contains a list of all the plant species observed in the territory of the park, which are included in the Bulgarian Red Book and other national and international lists of protected species. The list includes 90 species with some kind of conservation value. We have inventoried their habitats and distributed them among the land cover classes delineated from the satellite image. The conservation value of every class was evaluated according to the number of species; their protection status and the number of unique species for a particular land cover class.

The landscape metrics were calculated for the exploration of the landscape pattern. They were derived using the ATtILA tool. The core input data is the land cover grid of the Sinite Kamani Nature Park, derived from the ASTER image. The classification was adjusted to the program using the so-called land cover reclassification table, which has four precoded systems (including NLCD, ANDERSON I, II, and SAA). There is also an option to incorporate new classification that was used in our work. The process is handy and flexible, which enables repetitive adjustment for every calculation (if necessary). Three main types of metrics were derived. (1) The Land Cover Proportion metrics calculate the percentage and total area of any land cover type and they were used to define the distribution of the land cover classes over the park and landscape regions. (2) The Patch metrics are commonly generated for forest classes and give an indication of forest health but the ATtILA also gives an opportunity to calculate them for any other class or group of classes (Ebert et al., 2004). It includes metrics like the number of patches, the largest patch, the average patch size etc. There is an option to define the minimum patch size, which allows the user to eliminate some isolated small patches and also it gives an opportunity to extract only large patches that could be important in some landscape ecology analyses. (3) The Fragmentation metrics include categories defined according to Riitters et al. (2000) and the analysis is performed using a moving window. For this work its size was defined in 7 × 7 cells. According to this analysis the selected land cover class is divided into five categories: patch, transitional, edge, perforated, and interior. There is also a parameter, which is calculated as the average connectivity of the land cover class. These metrics are useful tools to assess the opportunity of the different land cover classes to ensure enough habitats for the species with conservation value.

4. Results

4.1. LAND COVER CLASSIFICATION

The classification of the ASTER TERRA image was satisfactory and the main vegetation classes could be differentiated from each other with a relative accuracy. The classification accuracy was estimated according to the field points, visual interpretation, and the comparison of the classification image and ASTER TERRA VNIR false color image. Especially the pine forests were classified with the best accuracy. Most difficulties were in differentiating deciduous forest types from each other and the results show that there is a particular difference in the spectral reflectance between oak and beech forests during the beginning of the growing season (April–May).

The beech and hornbeam forests have similar spectral reflectance and it is not possible to differentiate them at this stage. Two meadow types *Dry* and *Meadow with bushes* also caused some difficulties. The mountain terrain of Sinite Kamani caused some systematic misclassifications. The spectral reflectance of the same vegetation types varied in the northern and northwestern slopes compared to the southern and southeastern slopes due to the shadow and different illumination angle. These errors could have been corrected by using the DEM in classification process. The resulting classification includes 11 main classes, which are distributed irregularly around the park area (Table 1).

TABLE 1. Land cover classes in the park area and their distribution within the landscape regions.

Land cover class	Park area		Region 1		Region 2	
	Area km^2	%	Area km^2	%	Area km^2	%
Beech–hornbeam forests	41.1	36.3	17.3	28.8	23.0	43.4
Oak forests	27.0	23.9	16.4	27.4	11.2	21.1
Oak-flowering ash class	10.1	8.9	6.1	10.1	4.3	8.1
Artificial coniferous forests	3.4	3.0	2.7	4.4	0.9	1.7
Dry meadow	6.7	5.9	2.5	4.2	4.0	7.5
Meadow with bushes	21.6	19.1	13.4	22.3	8.9	16.7
Bare rocks	1.4	1.3	1.3	2.2	0.2	0.5
Water bodies	0.005	0.004	0.005	0.004	0	0
Urban	0.9	0.8	0.32	0.5	0.5	1
Roads	0.8	0.7	0.26	0.4	0.02	0.03
Sport facilities	0.08	0.07	0	0	0.02	0.05
Total	113.08	100	60.04	100	53.04	100

4.2. LAND COVER CLASSES AND THEIR CONSERVATION VALUE

1. *Beech and hornbeam forests.* This is the most widely distributed type, including the misian beech (*Fagus sylvatica L ssp. Moesiaca (K. Maly.) Hjelmquist*) and the hornbeam (*Carpinus betulus L.*) vegetation formations, according to the geobotanical classification (Stoeva et al., 2004). Most of the beech forests are mono-dominant with little or no grass coverage. The beech occupies wet habitats at higher altitudes (above 800 m), while the hornbeam is typical of the transitional zone between the oak and beech forests. There are 12 different vegetation associations (the lower level of the geobotanical classification), four of which are beeches mono-dominant and the others represent mixed associations of beech, hornbeam, oaks, and sometimes maple or

flowering ash. There are 16 plant species with conservation value found in this land cover class, 11 of which grow only here. Among them are *Digitalis viridiflora*, *Orchis militaris*, *Trachystemon orientalis* etc.

2. *Oak forests*. This type includes oak forests (*Quercus dalechampii L.*, *Q. frainetto Ten*, and *Q. cerris L*). They occupy less humid habitats compared with the beech forests and have well-developed grass coverage. Most of the forests are mixed between three types of oak and other trees like beech, hornbeam, lime, maple, aspen, flowering ash, forming 23 different vegetation associations. Eleven plant species have conservation value but only one (*Anemone sylvestris*) of them has habitat only in these forests. The others are distributed also in the beech and hornbeam forests or in grasslands, which mean that oak forests represent habitats with transitional conditions and could be used by plants from both beech and meadow land cover types.

3. *Oak-flowering ash forests and shrub*. This class includes secondary forests and shrubs developed in dry habitats with a predominantly southern exposure and pure, partially eroded soils. The vegetation formations are represented by flowering ash (*Fraxinus ornus L.*), oriental hornbeam (*Carpinus orientalis L.*), and downy oak (*Quercus pubescens*). They are combined into 13 different associations including also subdominants like *Quercus cerris*, *Sirinda vulgaris*, *Cotinus coggigria* etc. and some Mediterranean plants like *Ruscus acculeatus*, *Corronila emerus*, *Philadelphus coronarius*, and *Silene nutans*. This land cover class could be evaluated as the richest, due to the existence of plant species with the conservation value – 35, which is about 40% of the registered in the park. Twelve of them (*Aethionema arabicum*, *Fibidiaclypeolata*, *Micromeriafrivaldskyana* etc.) are observed only in this class and another 15 plant species are shared with the dry meadows.

4. *Coniferous forests*. They include predominantly pine (*Pinus silvestris L.*) forests, which are afforested by people on the place of formerly deforested areas. No rare or endangered species are registered so they have no real conservation value.

5. *Dry meadows*. They are relatively dry habitats with a thin soil cover and mainly with southern exposure. The vegetation is represented by the formations of *Festuca valesiana*, *Dichantium ischaemum*, *Sesleria latifolia*, *Koeleria nitidula*. This is the second rated land cover class according to the existence of plants with conservation value. There are 29 species with conservation value, 11 of them (*Moehringia jankae*, *Erisimum comatum*, *Iris reichenbrachii*, etc.) are found only in this class and 15 are shared with the oak-flowering ash forests and shrub class. Most species with conservation value (12) were registered in the formation of *Sesleria latifolia*, which is typical of the habitats with calcareous bedrock.

6. *Meadow with bushes.* This type includes grass and bushy vegetation, occupying more humid habitats than the previous. The grassland formations are represented by Agrostis capinalis, *Festuca platensis*, *Festuca rubra*, *Poa platensis*, *Pterideta aguilini*. Most of the meadows are also covered with shrubs of dog rose (*Rosa canina*), hawthorn (*Crataegus monogina*), Christ's torn (*Paliurus spina-cristii Mill.*), juniper (*Juniperus communis L.*) etc. The number of plant species with conservation value here is 8, most of them are found in the meadows dominated by *Festuca platensis*.

7. *Bare rocks.* This type is both of natural and anthropogenic origin. The bare, mainly quartz-porphyry rocks in the southern part of the park, with a specific blue–purple color at sunset that has given the name to the park and the area. Some of these rocks (especially in Kutelka reserve) provide habitats for endangered vultures and eagles like *Neophron perenopterus*, *Gypaetus barbatus*, *Aquila chrysaetos*, and *Circaetus gallicus* as well as 20 other predatory bird species. Sand and gravel quarries, large sand and gravel surface yards have predominantly anthropogenic origin.

8. *Water bodies.* This type includes open water areas and wetlands. They are represented by a small artificial lake in the area of the tourist complex Karandila.

9. *Urban.* This class includes separate buildings with predominantly tourism function like hotels, rest-houses, huts and some facilities of the forestry enterprise.

10. *Roads.* The park area includes about 50 km of paved roads, which are situated mainly in the western part.

11. *Sports facilities.* This class covers open air sports facilities like ski tracks.

The evaluation of the land cover classes, based on the number of species; their protection status and number of unique species, gives an opportunity to divide them into three groups. The first includes *beech–hornbeam*, *oak-flowering ash forest and scrub*, and *dry meadow*, which are characterized with high conservation value. The second have middle conservation value and includes *oak forest*, *meadow with bushes*, and *bare rocks*. The other five classes were characterized with low and without conservation value.

4.3. LANDSCAPE METRICS

The landscape pattern in the park has some specific features, which can be represented by the distribution of the main land cover types and landscape metrics. Firstly, there is a significant difference between the two landscape regions. The *mid mountain temperate humid* region is dominated by beech forests land cover class (Table 1), which is defined as the matrix of this landscape. This can be illustrated with the size of the largest patch from this class, which exceeds 10 times all the others (Table 2) as well as the patch connectivity and the percentage of the patch interior (Table 3).

The matrix definition in the second landscape region is not so easy because of the larger fragmentation almost equal share between the beech–hornbeam and oak forests here and the significant representation of other classes like meadows with bushes. The patch metrics represented in Table 2 give the opportunity to evaluate the spatial distribution of the land cover classes connected to their function as habitats for species with conservation value. There are four land cover classes, which provide habitats for the biggest number of endangered and rare species (Table 2) that is why the analyses of patch metrics was concentrated on them.

TABLE 2. Patch metrics: Pnum – number of patches within the region; Pavgs – average size of the patches within the region in ha; Pnum L – number of the large patches within the region; Pavgs L – average size of the large patches within the region in ha; Pdens – patch density (number of patches/km^2); Plrdst – size of the largest patch within the region in ha; Plgr – proportion of the largest patch to total area of the corresponding class within the region.

Patch metrics	Pnum	Pavgs ha	Pnum L	Pavgs L ha	Pdens	Plrdst ha	Plgr
LC classes				Region 1			
Beech–hornbeam	619	2.8	14	59.4	35.8	479.3	27.7
Oak	1,189	1.4	26	41.7	72.3	223.3	13.6
Oak-flowering ash	785	0.8	12	19.0	129.4	35.1	5.8
Dry meadow	456	0.6	3	16.3	182.4	23.1	9.2
				Region 2			
Beech–hornb	551	5.3	27	106.2	24.0	985.3	33.8
Oak	1,686	0.8	11	27.8	150.5	95.2	6.7
Oak-fl. ash	1,038	0.5	6	23.2	242.9	59.5	11.0
Dry meadow	387	1.3	9	33.3	97.3	86.5	17.1

The oak forests have the highest fragmentation, which is bigger in Region 2, where they occupy less area. The beech–hornbeam forests are the most compact, with a relatively low number of patches. The average patch size in Region 2 is twice in the size of Region 1. This is mainly due to the different ratio between the beech and the hornbeam. The beech forests are dominant in the Region 2, where hydro-climatic conditions are the most favorable for their development. The hornbeam forests dominate in the Region 1, where they share the habitats with oak forests. The diversity of different land cover classes is naturally determined in Region 1 and the people's activity has an additional effect on it, while in Region 2 the anthropogenic influence is the main factor for the present diversity of the primarily relatively homogenous dominant beech forests. These features can be illustrated by comparing the number and size of the large patches of the two regions (Table 2). The anthropogenic influence in the primarily more homogenous Region 2 leads to the formation of more and larger patches. According to Forman (1995) "large natural vegetation patches serve many ecological roles and provide many benefits in a landscape", therefore they should be treated with special attention in the biological conservation planning. Taking into account the size of the park and its landscape pattern, an area of 15 ha was defined as a minimum size for the large patches.

TABLE 3. Fragmentation metrics: Pff – average connectivity for the land cover class within the region; Pptch – percentage of the area that is patch class; Ptran – percentage of the area that is transitional class; Pedge – percentage of the area that is edge class; Pperf – percentage of the area that is perforated class; Pintr – percentage of the area that is interior class; reg. – percentage from the region area; cl. – percentage from the area of the corresponding land cover class.

Fragm. metrics	Pff	Pptch		Ptran		Pedge		Pperf		Pintr	
		reg.	cl.	reg.	cl.	reg.	cl.	reg.	cl.	reg.	cl.
LC classes						Region 1					
Beech-hornbeam	86.6	2.2	7.8	3.7	12.7	12.4	43.0	2.8	9.7	6.7	23.3
Oak	75.9	4.1	15.0	7.2	26.3	9.3	33.9	2.3	8.5	4.6	16.8
Oak-fl. ash	72.2	2.6	25.4	2.7	27.1	3.6	35.6	0.6	5.8	0.6	6.1
Dry meadow	67.5	1.4	33.4	1.3	31.6	1.4	32.6	0.1	2.7	0.03	0.7
Bare rocks	66.8	0.8	37.0	0.6	25.5	0.6	27.8	0.1	3.0	0.1	6.9
						Region 2					
Beech-hornb	87.8	2.4	5.5	5.4	12.5	15.2	35.0	4.1	9.5	15.3	35.3
Oak	71.5	5.0	23.6	5.8	27.3	7.9	37.2	1.4	6.7	1.1	5.2
Oak-fl. ash	66.7	2.8	34.2	2.3	28.8	2.4	29.7	0.3	3.9	0.3	3.4
Dry meadow	80.2	1.0	13.2	1.7	22.3	3.2	42.3	0.6	7.8	1.0	13.8
Bare rocks	50.8	0.3	64.9	0.1	21.1	0.1	14.0	0.0	0.0	0.0	0.0

The fragmentation metrics give us important information concerning the species with conservation value. Some of these species have their habitats only in one land cover class, while the others are found in different classes. We assumed that the first would occupy predominantly the interior of the patches and the habitats for the second would be located mainly on the edge or in transitional patches. From that point of view, the land cover classes with most interior conservation value species are *beech–hornbeam forests* (with 11 species), *oak-flowering ash shrubs* (12), and dry *meadows* (12). The beech–hornbeam forests in Region 1 provide enough space with the patch interior that is 15% of the whole area (35% of this class).

The interior patch habitats of the species with conservation value in Region 1 are scarce but adequate for their normal development (Table 3 and Figure 2). *Dry meadow* and *oak-flowering ash* land cover classes have a larger fragmentation, less connectivity and the percentage of the patch interior is very low. Consequently, the interior species with conservation value, in these classes, are not ensured enough space for the habitats and their existence is under real threat. This fact very much concerns some endangered species like *Tulipa urumoffii* and *Aethionema arabicum*, which are observed in limited places. An additional threat to these species in the dry meadows is their accessibility because they grow in an open area and can be easily reached by people. The species, which are typical of the edges and the transitional areas between meadows and oak-flowering ash shrubs, are in more favorable situation with larger habitats (Table 3).

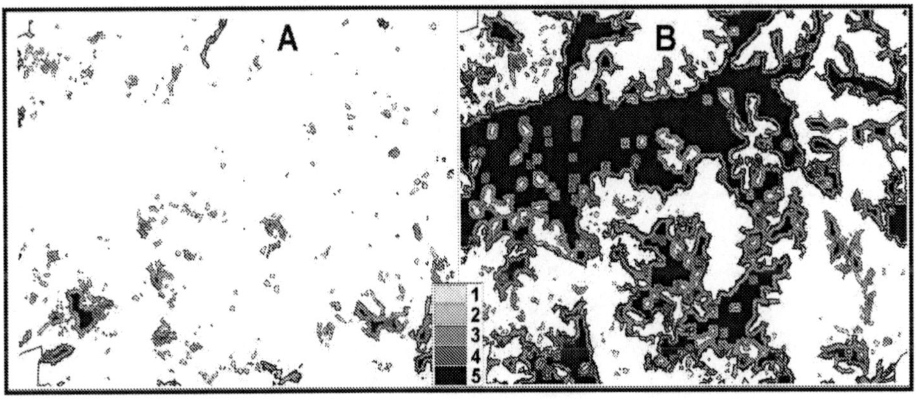

Figure 2. Patch fragmentation in the eastern part of the Sinite Kamani Nature Park for oak-flowering ash (A) and beech–hornbeam (B): 1 – patch class; 2 – transitional class; 3 – edge class; 4 – perforated class; 5 – interior class.

5. Discussion and conclusions

Land cover classification of the Sinite Kamani Nature Park using ASTER TERRA VNIR image was satisfactory. Further field work and image processing is required to be able to produce more detailed classification of the area. A further objective to improve the classification should be directed to the differentiation of beech and hornbeam forest, which requires a choice of an image from a particular time, when these kinds of trees are in different vegetation stage. We can recommend the ASTER TERRA VNIR images to be used in land use and vegetation studies in Bulgarian parks. These images could also be used when verifying the accuracy of the CORINE land cover classification, which is derived from Landsat TM with a 30 m resolution. Comparison allows the examination of the questions as how much more detailed resolution (15–30 m) enhances the classification details and what is the maximum number of classes (with 15–30 m resolution) and whether the 15 m resolution data allows a more detailed vegetation and land use pattern study.

The landscape pattern derived on the basis of the land cover classification represents correctly the real landscape heterogeneity of the area. It is an appropriate source for the extraction of different kinds of metrics useful for landscape ecology analyses. Further improvement of the land cover classification would improve its quality and usability. The land cover classes in Sinite Kamani Nature Park have various conservation values. It is higher for *oak-flowering ash forests and shrub*, *beech–hornbeam forests*, and *dry meadows*, which provide habitats for the most endangered species. A lot of them are typical only of these classes, which increases their importance. *Oak forests* and *meadows with bushes* classes provide habitats for endangered species, which inhabit predominantly the patch edges. Parts of the *bare rocks* class provide habitats for rare and endangered bird species. The landscape metrics analysis shows that *beach–hornbeam* class provides best habitat conditions for interior species, while the spatial distribution of the *oak-flowering ash* and *dry meadow* do not ensure enough space for such species, which determines additional danger for their preservation. This information is a good basis for further evaluation of the habitat ecosystem services as well as a part of the process of environmental security improvement.

The analysis from this work gives basis to recommend some measures that can be involved in the management plan of the park in order to improve the efficiency of its activity. There are particular differences between the landscape regions detected in the park, which requires differential approach in their management. The anthropogenic influence in the Region 1 is lower in general. There are no tourism bases (hotels, huts etc.) but only transport infrastructure (roads, paths, lifts), which allow access to the more "urbanized" Region 2. On the other hand its boundary touches the populated areas in the south, which requires more precise protection regime. Region 2 has well developed tourism infrastructure and management measures, which should be directed to the coordination of this activity with the objectives of the biological conservation. The land cover classes, *oak-flowering ash* and *dry meadow* include species with conservation value and need to be under special observation because of their higher vulnerability. A further improvement of the habitat function would be achieved by developing a Multiple Species Habitat Conservation Plan, which would provide a complex reserve pattern in which to hypothesize and test outcomes for different species and groups of organisms (Chen et al., 2006).

Acknowledgments

This work was carried out within the project "Environmental assessment in protected areas based on implementation of Remote Sensing and GIS" funded by Bulgarian Academy of Sciences and Finnish Academy of Sciences. We would like to thank Mr. Ivan Ivanov Director of the Sinite Kamani Nature Park for practical help and hospitality.

References

Abrams, M., 2000. ASTER: Data products for the high spatial resolution imager on NASA's Terra platform. International Journal of Remote Sensing 21, 847–853.

Chen, X., Li, B.-L., Scott, T., Tennant, T., Rottenbery, J.T., and Allen, M., 2005. Spatial characteristics of multispecies' habitats in southern California, USA. Biological Conservation, 124, 169–175.

Chen, X., Li, B.-L., Scott, T., and Allen, M.F., 2006. Tolerance analysis of habitat loss for multispecies conservation in western Riverside County, California, USA. International Journal of Biodiversity Science and Management 2, 87–96.

de Groot, R.S., Wilson, M.A., and Boumans, R.M.J., 2002. A typology for classification, description and valuation of ecosystem functions, goods and services. Ecological Economics 41(3), 393–408.

Gikov, A. and Nedkov, S., 2005. GIS-based model for spatial distribution of climatic elements in mountain areas, Scientific Conference "Space, Ecology, Safety", 10–13 June 2005, Varna, Bulgaria, 121–127 (in Bulgarian).

Ebert, D.W. and Wade, T.G., 2004. Analytical Tools Interface for Landscape Assessments (ATtILA) Version 2004 User Manual. US-EPA Report EPA/600/R-04/083. http://www.epa.gov/nerlesd1/land-sci/attila/

Farina, A. (Editor), 2000. Principles and Methods in Landscape Ecology. Kluwer, Dordrecht, pp. 225.

Forman, R.T. and Gordon, M. (Editors), 1986. Landscape Ecology. Wiley, New York, pp. 619.

Forman, R.T., 1995. Some general principles of landscape and regional ecology. Landscape Ecology 10, 133–142.

Kaab, A., 2005. Combination of SRTM3 and repeat ASTER data for deriving alpine glacier flow velocities in the Bhutan Himalaya. Remote Sensing of Environment 94, Issue 4(28), 463–474.

Klug, H. and Zeil, P., 2004. The choice and use of landscape metrics for catchment characterization, Proceedings of the Eco-Geowater conference "GI for International River Basin Management", Budapest, June 3–5, 73–82.

Knight, R. and Landres, P., 2002. Central concepts and issues of biological conservation. In: K. Guitzwiller (Editor), Applying Landscape Ecology in Biological Conservation, Springer, Berlin, Heidelberg, pp. 22–34.

Kumpula, T., Burkhard, B., and Muller, F., 2005. Can you see it all in the image? Using remote sensing techniques for landscape assessment in northern Fennoscandia. EcoSys 11, 13–23.

Riitters, K., Wickham, J., O'Neill, R., Jones, B., and Smith, E., 2000. Global-scale pattern of forest fragmentation. Conservation Ecology 4(2): 3. http://www.consecol.org/vol4/iss2/art3/

Stoeva, M., Stoichev, G., Mladenov, R., Grozeva, N., Velichkova, K., Georgieva, M., and Vulkova, M., 2004. Biological diversity of nature park Sinite Kamani, Stara Zagor, 218 (in Bulgarian).

Velchev, A., Todorov, N., Asenov, A., and Beruchashvili, N., 1992. Landscape map of Bulgaria in scale 1:500,000. Annual of the Sofia University "St. Kliment Ohridski", Faculty of Geology and Geography, Book 2 – Geography, 84, 25–37 (in Bulgarian).

Welsh, R., Jordan, T., Lang, H., and Murikami, H., 1998. ASTER as a source for topographic data in the late 1990s. I.E.E.E. Transactions on Geoscience and Remote Sensing 36, 1282–1289.

QUANTIFYING AND QUALIFYING URBAN GREEN BY INTEGRATING REMOTE SENSING, GIS, AND SOCIAL SCIENCE METHODS

STEFAN LANG [*], ELISABETH SCHÖPFER, DANIEL HÖLBLING, AND THOMAS BLASCHKE
Salzburg University, Centre for Geoinformatics (Z_GIS), Schillerstr. 30,5020 Salzburg, Austria

MATTHIAS MOELLER
Arizona State University, Global Institute of Sustainability (GIOS), P. O. Box 873211, Tempe, AZ 85287-3211, USA

THOMAS JEKEL
Austrian Academy of Sciences, GIScience, Schillerstr. 30,5020 Salzburg, Austria

ELISABETH KLOYBER
Institut für integrative Stadt- und Regionalentwicklung (IISR), Rifer Hauptstraße 39a/4, 5400 Hallein, Austria

Abstract. Sustainable urban planning in growing urban agglomerations encompasses the active development of urban green spaces. The loss of urban green not only threatens urban climate and ecosystems, but may also affect a city's image and the residential satisfaction in general. Quantifiable information about green structures and the amount and distribution of green spaces is essential for sustainable planning. Monitoring tools for outlining differences in urban green space are required, which – more than merely measuring the overall percentage of green – may reflect the different importance of green areas in specific environments. This implies both spatial explicit characterizations of green areas and the consideration of relative importance of certain green structures from a citizen's perspective. In this chapter we will present two approaches for advanced urban green mapping in the Phoenix Metropolitan Area, Arizona, USA (PHX-US), and Salzburg, Austria, Europe (SBG-AT). The

[*]Corresponding Author: Dr. Stefan Lang, Centre for Geoinformatics, Salzburg University, Schillerstr. 30, 5020 Salzburg, Austria; e-mail: stefan.lang@sbg.ac.at

I. Petrosillo et al. (eds.), Use of Landscape Sciences for the Assessment of Environmental Security, 93–105.

approaches discussed were designed for monitoring urban green development in a repeatable and transferable manner by using (1) proxies derived from remotely sensed data, (2) spatial concepts and spatially explicit measures for spatial characterization, and (3) subjective, social science data reflecting the perception of urban green and therefore, the quality of some aspects of urban life and residential satisfaction.

Keywords: Urban green structures; monitoring; fraction of surrounding vegetation; weighted green index; environmental perception.

1. Urban green in a transforming urban society

1.1. URBAN GREEN AND THE QUALITY OF URBAN LIFE

In our growing urbanized environments the maintenance and development of urban vegetated areas are among today's challenges of sustainable urban planning. Being the most densely populated regions on Earth, urban settlements are far from natural in status and much more adapted to the special needs of our economic and social lives. With respect to environmental security in urban agglomerations, the remaining green spaces may play a key role, not only in terms of their overall percentage, but also in terms of their specific arrangement and attractiveness. Three main functions of urban green have been discussed recently.

First, in these highly transformed urban settings, green vegetated areas or remnants of natural habitats play a crucial functional role as elements of ecological networks. Policies of nature conservation include both the restoration of green space in the densely built-up inner cities and an improvement of the connectivity between green spaces (Pauleit, 2004). For instance, according to metapopulation theory (Hanski, 1999) species may survive in fragmented, but functionally connected urban habitats. However, even if it is recognized that dissected habitats to a certain degree allow for maintaining viable populations, the very characteristics of fragmented urban spaces (e.g. habitat quality) may also influence the presence of species.

Second, the climate-related meaning of green areas is threefold: first, oxygen is produced by healthy vegetation; second, the overall temperature of cities is lowered at an amount which is perceptible to humans; third, open green spaces usually function as corridors for fresh air supply and air circulation in general. Ecological considerations require balancing the limits of both the increase of human population density and the maintenance of free space.

Third, green spaces enhance recreational quality for residents and the overall image of a city. Green areas which people can directly perambulate and aesthetically enjoy gain attractiveness (Lang et al., 2006). Studies on urban morphology have been performed in geography, landscape planning, and architecture, environmental psychology and sociology. Similar to the arrangements of buildings, squares, and public green spaces, the arrangements of other green areas may likewise be important. For evaluating socio-residential conditions, a quantitative analysis of the mere amount of green areas may be insufficient (Schöpfer et al., 2005). Thus the studies presented here investigate means how citizens' perceptions can be included to enhance the quality and acceptance of planning decisions.

1.2. SUSTAINABLE URBAN PLANNING, MONITORING SYSTEMS, AND THE ROLE OF CITIZEN'S PERCEPTION

Quantifiable information about green structures and the amount and distribution of green spaces is essential for sustainable planning. When considering sustainable planning as a way of guided development that meets the "*needs of the present without compromising the ability of future generations to meet their own needs*" (Brundtland, 1987), we may favor one of the following strategies of city planning: (1) expanding the city footprint (urban sprawl), but preserving existing gaps between built-up areas; (2) filling these gaps and increase urban density. Tradeoffs between expansion and densification makes planning authorities think of monitoring concepts for observing spatio-temporal changes of urban green. A monitoring system should fulfill the following requirements: (1) be transparent and objective, i.e. relying on quantifiable measurements; (2) be transferable, yet cost-effective; (3) be fully repeatable in regular intervals, yet flexible in conceptualization. Such a system is expected to be adaptable to different levels of spatial disaggregation, i.e. referable to any arbitrary spatial entity, thus allowing the results to be reported on different scales and reference units (Lang et al., 2006).

While for decades aerial flight campaigns have been used for visual interpretation and map production, digital workflows are currently taking over more of the necessary information update tasks. The use of digital data from airborne and satellite sensors in combination with existing GIS information is widespread (Moeller, 2003). Quantitative measures like percentage of impervious surface within blocks or cells or green area per resident provide – even if aggregated – aims to identify and localize areas of restructuring parts of a city (Schöpfer et al., 2005). If, however, planning processes are to involve participation of local stakeholders in the decision-making process, this information needs to be spatially disaggregated and reflect the citizen's perspective (Lang et al., 2006). In this paper we demonstrate that by using high-resolution satellite data and ancillary GIS information in combination with subjective social data we may be able to map and qualify urban green areas in a more citizen-centered approach. We discuss approaches to achieve differentiations of urban neighborhoods with respect to their "greenness" as perceived by residents.

2. Mapping urban green structures on high-resolution image data

2.1. STUDY AREA, DATA, AND STUDY DESIGNS

The Phoenix/USA (PHX-US) study area is subdivided into two parts, one located in the Tempe area, about 15 km east of the Phoenix city center. The area represents a mixture of different housing types: private residential settlements, commercial buildings in the southwest part of the study area and parts of Arizona State University in the northeast part (Figure 1, left). The color orthophotos were distributed commercially by Landiscor. Pixel resolution was 0.61 m. For the other PHX-US test site we used pan-sharpened, multispectral four band QuickBird satellite imagery (0.61 m). This site covers parts of the city of Scottsdale, about 20 km northeast of the Phoenix city center. It represents mainly residential settlements and in the north east parts of a shopping center. The size of both study areas is 4 km².

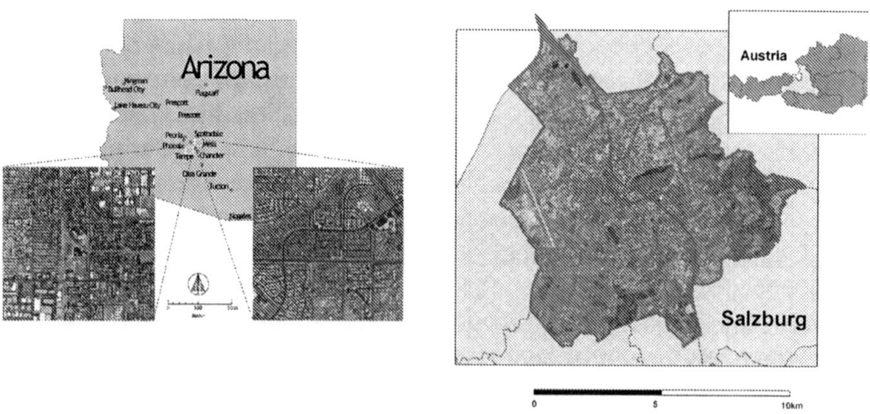

Figure 1. Study areas. Left: PHX-US; Right: and SBG-AT.

The study site of SBG-AT is defined by the city boundary of Salzburg and comprises a polygonal area of 65.7 km^2 (Figure 1, right). The city with its 150,000 inhabitants is covered by approximately 60% urban green, which currently is being considered for further residential and business space. The city itself is characterized by its core and the adjacent historic inner city, both densely built up but interspersed with forest-covered hills. Surrounding city quarters are characterized by single-family houses and some blocks of multistorey buildings. Agricultural fields and forests extend into the city. A remarkable structural element is the Salzach River crossing from South to North. The SBG-AT study has been threefold (Lang et al., 2006). In a first and experimental stage we used a 500 ha orthophoto mosaic recorded July 2002 (0.1 m pixel resolution) representing a limited area of the inner city. In the second, operational step, the study was extended to the entire city based on pan-sharpened QuickBird data acquired June 21, 2005 (0.61 m pixel resolution). In the third, integrated step, we introduced the results of a social survey evaluating the subjective importance of green structure types (see 3.2).

2.2. DATA PREPROCESSING

QuickBird data in both studies were corrected for atmospheric influences (haze, dust) first. Then, in order to merge the high spatial resolution of the panchromatic band with the multispectral information, a resolution merge (pan-sharpening) was performed, though in different ways. In PHX-US we used a texture based image fusion algorithm (Ehlers, 2005), whereas in SBG-AT we applied pan-sharpening according to Liu (2000).

The orthophoto material used in the first part of PHX-US did not provide infrared information, thus we developed a *Normalized Difference Green/Red Vegetation Index* (NDGRVI, cf. Moeller and Blaschke, 2006). In SBG-AT using the standard *Normalized Difference Vegetation Index* (NDVI) was essential generating an extended classification set and distinguishing among several green structure types. In addition, shadowed areas could be separated in those parts where shadow fell on vegetation or impervious surface. Ancillary data like cadastral information has been used to separate buildings and other sealed surfaces (e.g. roads).

2.3. CLASSIFICATION APPROACHES

Remote sensing applications increasingly use image segmentation as a primary step to derive image objects and subsequently treat these as homogenous regions for processing and classification. This technique allows for image classification with spectral, geometrical, and textural information of objects as well as their mutual relationships (For details about segmentation parameters, levels, and objects characteristics refer to Lang et al., 2006 and Moeller and Blaschke, 2006).

Dealing with highly variable structures in urban areas we relied on an object-based image analysis approach (Blaschke et al., 2000) in all stages of either study. The approach has been proved successfully for the analysis of human settlements with their very detailed, small-scale objects (cf. Hofmann and Reinhardt, 2000). The classification is based upon a flexible cognition network, which not only contains the descriptive definitions of classes in a diagnostic view, but also controls class assignment on a higher aggregated, semantic level (e.g. whether or not agricultural fields belong to urban green). Thus, it also serves as a discussion tool for user interaction and stakeholder participation. In the SBG-AT study this cognition network (Figure 2) flexibly handles the (semantic) definition of the aggregated classes "green" or "not-green" (Lang et al., 2006).

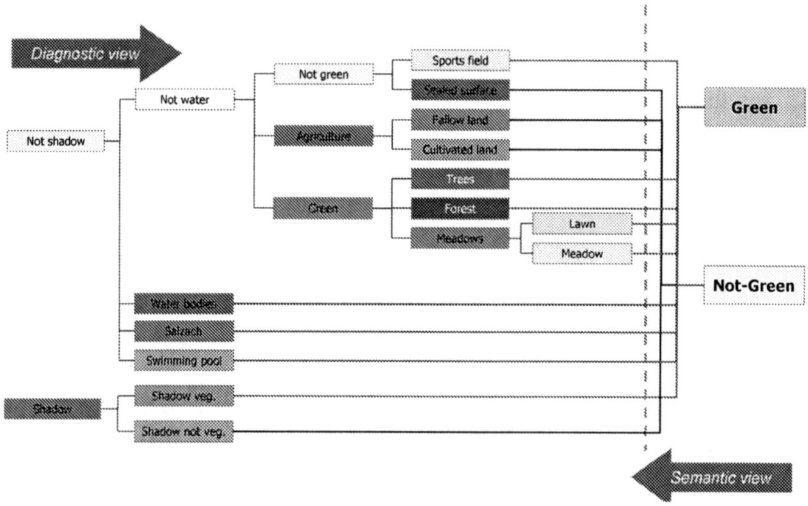

Figure 2. Cognition network as being used in the SBG-AT study. (From Lang et al., 2006.)

The PHX-US classification was performed utilizing the NDGRVI layer, and vegetated areas and building footprints were extracted. Object-based image analysis has been applied for the classification of several urban land use/land cover (LULC) classes. The class hierarchy used in SBG-AT contains 11 green structure types plus classes not considered as green. Hierarchical relations between objects were addressed in the class definitions (e.g. by the function *existence of super objects*). Larger-sized objects were detected on higher levels, for example, the class *Forest* was defined at the seventh segmentation level and then a relation to the fourth level was established, on which most of the classes were classified. Some classes were defined by a combination of different spectral features (e.g. class *Water Bodies*: (a) mean value of green band, (b) mean value of the infrared band, (c) standard deviation of blue band, and (d) NDVI.

2.4. RESULTS AND ACCURACY ASSESSMENT

In the PHX-US study area we could differentiate between four major LULC classes and nine subclasses. The overall classification accuracy for the major LULC classes was 82%, tested with 150 randomly distributed points for the Tempe test case using orthophotos. Transferring the classification rule set to the Scottsdale test case on QuickBird data, we could better differentiate between surface vegetation and taller vegetation. Accuracy was based on 100 randomly distributed points (83% for the major LULC classes).

Classification results of the SBG-AT study were proved by statistical point-based accuracy assessment (stratified random, 500 points, 30 points per class as minimum), which revealed an overall accuracy of 83.2%. By combining the results, we calculated four different green indices (i.e. percentages of green), each of them to be used for specific planning purposes: the first, GI-I (70.3%), includes all green structure types covered by the survey, the second, GI-II (68.5%), excludes *Water Bodies* and *Rivers*. A third, GI-III, was calculated similar to GI-II, but without *Pools* and *Sports Fields* (68.4%). The last, GI-IV (65.5%), the class *Fallow Land* is also exempted. The indices were calculated for different spatial units: a 50 × 50 m regular grid as well as several administrative units (e.g. enumeration blocks).

3. GIS-based modeling and indicator development

3.1. FRACTION OF SURROUNDING VEGETATION, BUILDING ARRANGEMENTS, AND STRUCTURAL ASSESSMENT

The classification result of the PHX-US study was used in a GIS environment for a detailed analysis of vegetation surrounding the extracted building footprints within a given distance. First the centroid points were calculated for each building. Around these, ten buffer fringes were computed, each with an increasing diameter of 10 m (Figure 4). Then the area percentage of vegetation was calculated for each fringe and each individual building. This calibrated measurement we named *Fraction of Surrounding Vegetation* (FSV, Moeller and Blaschke, 2006). The method was originally developed on aerial orthophotos and then successfully transferred to satellite imagery.

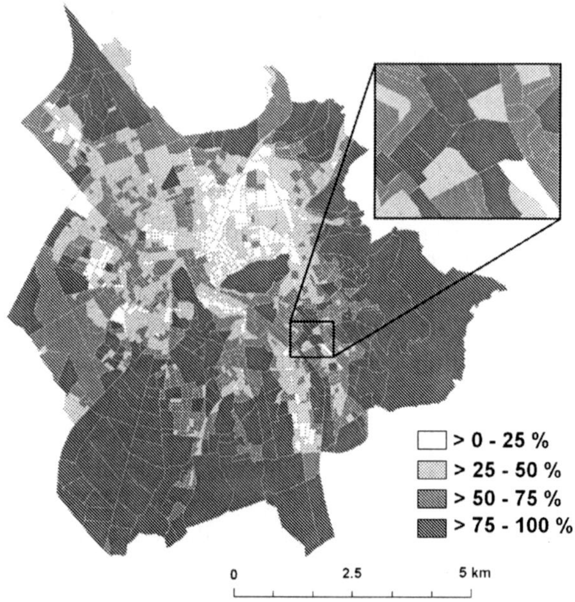

Figure 3. Green Index GI-I, aggregated on enumeration block level, divided into four classes, SBG-AT.

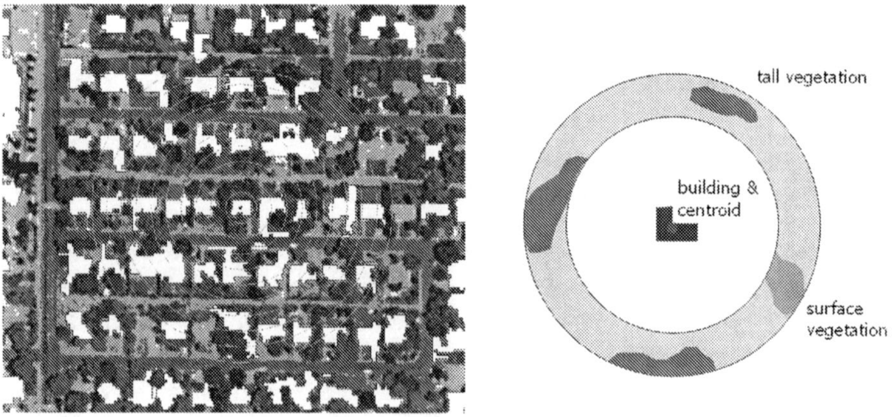

Figure 4. Extraction of building footprints and FSV. Buildings and fringes around a given centroid (left); example of vegetation extracted within one fringe (right).

Specific arrangements of buildings may influence the impression of greenness as well. A method modeling the arrangements of different building types has been discussed by Lang and Schöpfer (2005). The authors also demonstrated how structural assessment can be performed in a spatial explicit manner by investigating the specific shape and the spatial configuration of green structure types inside a given spatial unit. By this, the arrangement of green structures and their granularity could be automatically derived from remote sensing data (see Figure 5).

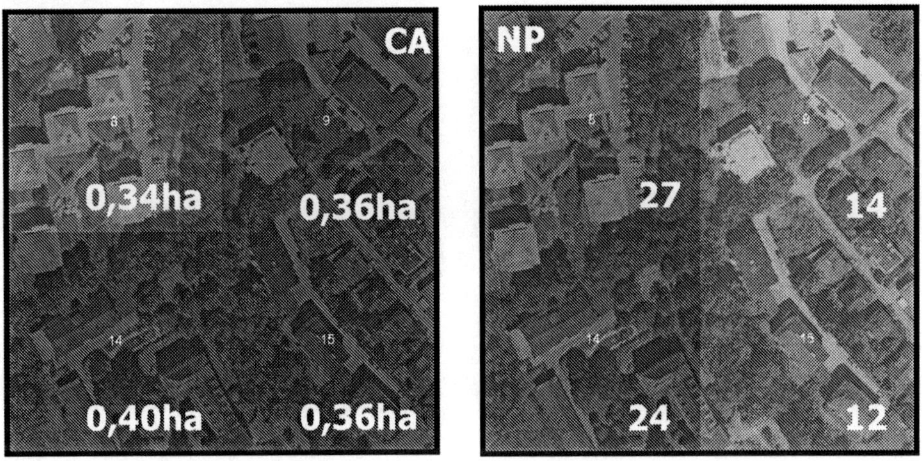

Figure 5. Automated structural assessment of green structure types. Area covered by a certain green structure type (left); number of elements belonging to this type (right).

3.2. WEIGHTED GREEN INDEX (GREEN IMPRESSION)

In SBG-AT we integrated the citizen's perception of green space in our classification results. In an interview-based survey among 128 respondents the classified green structure types were ranked with regard to their perceived importance. Respondents were asked for a ranked evaluation (on a scale from 0 to 5) of 29 different green structure types. This was done using a questionnaire containing sample images for each green structure type to enhance recognition (Lang et al., 2006). The time frame of the interviews coincided with the acquisition date of the satellite data; participants were distributed throughout the city. The ranks for specific green structure types varied markedly, ranging from values of 2.0–4.5.[†] The results were used for modeling perceived greenness (or green impression) of given spatial units. A *Weighted Green Index* (GI_w, ibid.) was used, which integrates both occurrence and distribution of relevant green structure types as well as the results of relative importance of these types in the eyes of the citizens. GI_w can be calculated for any spatial reference unit, since it is based on the percentages of green structure types and their respective importance (Figure 6). GI_w was calculated for 26,878 raster cells, each 0.25 ha in size. Statistical evaluation revealed 4,919 cells (12.3 km^2) having values of $GI_w > 3.9$ and 13,953 cells (34.9 km^2) with $GI_w < 3.0$. GI_w ranged from 0 to 4.3.

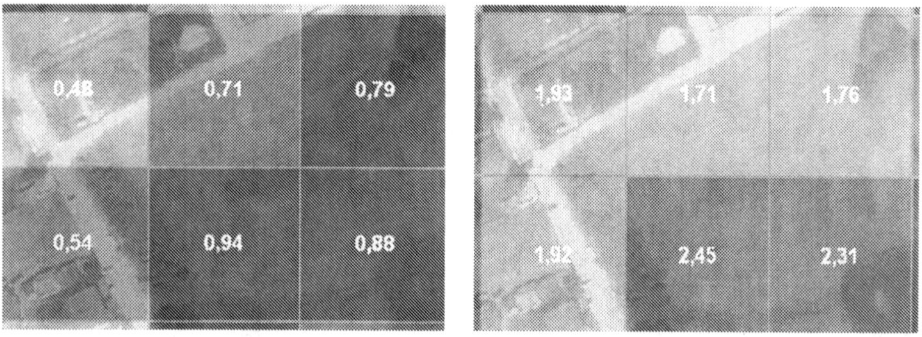

Figure 6. Comparison between GI (left) and GI_w (right) in six neighboring cells (each 50 × 50 m). GI_w considers the relative importance of the structure types, rather than the mere percentage of these.

4. Discussion

The approaches discussed above were designed for monitoring urban green development in a repeatable and transferable manner by using (1) spectral and spatial information about urban green structure types derived from remotely sensed data, (2) spatial modeling and spatially explicit measures for spatial characterization, and (3) quantified subjective, social science data reflecting the aggregated individual importance of specific green structure types. The inclusion of citizens' perception in the monitoring system should help to make expert planning decisions more acceptable among the public (Lang et al., 2006).

[†]Structure types and ranks (examples): *group of trees*, *rivers*: 4.3; *forest*: 4.1; *water bodies*: 4; *lawn/meadows*: 3.6; *sport grounds*: 2.5, *fallow land*: 2.1.

Besides the automated classification of green areas different indicators were developed: the FSV in PHX-US and GI (with its derivates) as well as GI_w in SBG-AT. In the first study, PHX-US, we assume that percentages of green in different buffer zones around individual buildings could be used as a potential measurement for urban life quality. Similar to indicators mentioned before (Lang and Schöpfer, 2005; Schöpfer et al., 2005), this also can be considered a *structural indicator*. It may be a useful tool for urban planning as parts of a city are comparable to others and planning measures can be performed based on it. While easily repeatable and therefore cost-effective, it is also widely open to criticism. First, it is based on the assumption of a direct correlation of green and urban life quality, which is not empirically tested and may show considerable regional variation. Second, this assumption is an expert's, not a citizen's assumption. Still structural indicators may be usefully integrated in spatial decision support systems if their weights are transparently argued and debated between stakeholders.

The second study, SBG-AT, again first uses GI as structural indicator which may be flexibly defined by planners and decision-makers according to public opinion. The other indicator, GI_{w}, attaches collective public opinion to spatial entities. In other words, it reveals how a certain spatial unit is publicly evaluated in terms of its greenness. This accounts for a statement made by Amerigo and Aragones (1997), that monitoring urban green should not simply be based on objective measurements from, e.g. satellite sensors, but also on subjectively perceived attributes of the residential environment. Indicators of this type may be called *collectively weighted indicators*. They are seen as a step towards participation in GI-based spatial planning that may enhance acceptance of spatial decision-making.

Returning to the requirements of the monitoring system as stated in section 1.2, we have the following conclusions: From a technical perspective, the forthcomings and advancements in remote sensing – with respect to data quality and analytical capabilities – account for considerable success in repeatability, transferability and cost-efficiency. On the other hand, aspects like transparency, flexibility, and acceptance are primarily attributed to the way in which existing analytical methods and the various sources of information are integrated and merged. This is still challenging, and so is the integration of qualitative judgments and their conversion into quantifiable spatial information. No doubt, with an increasing level of automation more continuous and time-effective monitoring may be supported. The inclusion of subjective attributes into a GI-based planning support means participation that reciprocally validates the results of automated classification. A complementary study on combining quality-of-life parameters with spatially disaggregated population data, and also considering urban green structures, has recently launched in an interdisciplinary approach with environmental psychologists.

Acknowledgments

We thank Josef Reithofer from the Salzburg City Council for funding as well as critical statements and stimulating discussions. We also highly appreciate the fruitful collaboration with Thomas Prinz from ARC Research Studio *i*Space, Salzburg. We would like to express our thanks to Arizona State University and the administration of the federal state of Salzburg for funding.

References

Amerigo, M. and Aragones, J.I., 1997. A theoretical and methodological approach to the study of residential satisfaction. Journal of Environmental Psychology 17, 47–57.

Blaschke, T., Lang, S., Lorup, E., Strobl, J., and Zeil, P., 2000. Object-oriented image processing in an integrated GIS/remote sensing environment and perspectives for environmental applications. In: A. Cremers and K. Greve (Editors), Environmental Information for Planning, Politics and the Public, Marburg, pp. 555–570.

Brundtland, G.H. (Editor), 1987. Our Common Future. Report of the World Commission on Environment and Development, Oxford University Press, Oxford.

Ehlers, M., 2005. Urban remote sensing: new developments and trends. In: M. Moeller and E. Wentz (Editors), Int. Arch. Photogramm. Remote Sensing. XXXVI, 8/W27 on CD.

Hanski, I. (Editor), 1999. Metapopulation Ecology. Oxford University Press, Oxford.

Hofmann, P. and Reinhardt, W., 2000. The extraction of GIS features from high resolution imagery using advanced methods based on additional contextual information – first experiences. Int. Arch. Photogramm. Remote Sensing. XXXIII, Amsterdam.

Lang, S. and Schöpfer, E., 2005. Measuring urban green feeling. Geoinformatics 8, 50–51.

Lang, S., Jekel, T., Hölbling, D., Schöpfer, E., Prinz, T., Kloyber, E., and Blaschke, T., 2006. Where the grass is greener – mapping of urban green structures according to relative importance in the eyes of the citizens. First Workshop of the EARSeL Special Interest Group on Urban Remote Sensing, "Challenges and Solutions" on CD-ROM.

Liu, J.G., 2000. Smoothing filter-based intensity modulation: a spectral preserve image fusion technique for improving spatial details. International Journal of Remote Sensing 21(18), 3461–3472.

Moeller, M., 2003. Urbanes Umweltmonitoring mit digitalen Flugzeugscannerdaten. Wichmann, Karlsruhe (in German).

Moeller, M. and Blaschke, T., 2006. GIS-gestützte Bildanalyse der städtischen Vegetation als Indikator urbaner Lebensqualität. *Photogrammetrie – Fernerkundung – Geoinformation*, 1/2006, 19–30 (in German).

Pauleit, S. (Editor), 2004. An Ecological Approach to Greenstructure Planning. COST Action C11, University of Manchester, Manchester.

Schöpfer, E., Lang, S., and Blaschke, T., 2005. A "Green Index" incorporating remote sensing and citizen's perception of green space. In: M. Moeller and E. Wentz (Editors), Int. Arch. Photogramm. Remote Sensing. XXXVI, 8/W27 on CD.

ALLOMETRIC SCALING AS AN INDICATOR OF ECOSYSTEM STATE: A NEW APPROACH

BAI-LIAN LI*
Ecological Complexity and Modeling Laboratory, Department of Botany and Plant Sciences, University of California, Riverside, CA 92521-0124, USA

VICTOR G. GORSHKOV AND ANASTASSIA M. MAKARIEVA
Theoretical Physics Division, Petersburg Nuclear Physics Institute, Russian Academy of Sciences, 188300, Gatchina, St. Petersburg, Russia

Abstract. It is argued that the problem of regional environmental security cannot in principle be solved without involving the regulatory environmental potential of the natural biota. Ecological allometry, i.e. the analysis of ecological and environmental phenomena based on body size and spatial scale regularities, is shown to provide clues to understanding the principles of environmental homeostasis in natural ecosystems.

Keywords: Allometric scaling; biotic regulation; energy consumption; forest; leaf area index; primary productivity; transpiration; water cycle

1. Introduction

Environmental security can be considered as a bifocal problem. On the one hand, it is important to protect the environment from being poisoned by the industrial pollution. On the other hand, irrespective of the presence/absence of direct pollution, it is vital to ensure sustainable maintenance of those

*Corresponding Author: Professor Bai-Lian Li, Ecological Complexity and Modeling Laboratory, University of California, Riverside, CA 92521-0124, USA; e-mail: bai-lian.li@ucr.edu

I. Petrosillo et al. (eds.), Use of Landscape Sciences for the Assessment of Environmental Security, 107–117.

environmental characteristics on which humans, as a biological species, are critically dependent: favorable climate conditions, high biological productivity, and availability of freshwater.

At present, according to satellite data, the civilization actively uses over 60% of the continental areas. Struggle with industrial pollution presents no conceptual difficulties and, given sufficient investments, invariably yields visible environmental improvements. In contrast, despite huge economic investments into the maintenance of exploited lands, their global degradation continues. Soils erode, biological productivity and river runoff decline.

Here we outline a new conceptual approach to the problem of environmental security. It is based on the study of the self-sustainable, self-organizing properties of the natural biota, its apparent ability to maintain all environmental characteristics in a state optimal for life in general and human life in particular for practically infinite periods of time.

Living world is characterized by a unique diversity of forms. This would preclude a quantitative scientific analysis of the biological and ecological phenomena unless one succeeded to find some common variables that could be relevant for all living organisms. Spatial scale and body size belong to the few universal biological variables of this kind. Allometric scaling represents a particular type of functional dependence between any biological, ecological, or environmental parameter x and size or spatial scale L. It can be written as $(x/x_0) = (L/L_0)^{\beta}$, where x_0 is the value of x at a characteristic size scale $L = L_0$.

Allometric scaling has been actively studied, although predominantly at the organismal level (Peters, 1983), and currently represents one of the hot spots of theoretical research in biology. Here we show that the bases of environmental stability in natural ecological systems can also be analyzed on these grounds. To illustrate the potential of the new approach, two examples are considered: (1) It is shown that the allometric distribution of energy consumption over different-sized species in the ecosystem dictates the magnitude of the fluctuations of biological productivity; and (2) It is shown that the dependence of precipitation on the distance from the ocean is coupled to the degree of disturbance of the natural vegetation cover. Both examples illustrate that allometric scaling of landscape-level ecosystem characteristics can be developed into an informative indicator of the ecosystem's state and of the degree of security of the life-support services that the ecosystem provides.

2. Theoretical platform: biotic regulation of the environment

All living organisms produce certain impact on their environment, consuming some substances from it and returning different ones. Synthesis and decomposition of organic matter in the biosphere occur at such high rates that if all the organisms in the biosphere had performed these processes chaotically and independently of each other, then the concentrations of all biogenic elements could have changed by a 100% over a time period of the order of 10 years, thus rendering the environment unsuitable for human existence.

The fact that the environment has remained suitable for macroscopic life, which has persisted during at least the last 600 million years, as testified by the available paleodata, indicates that functioning of the ecological communities of the biota does not occur chaotically. Traditionally, this nonrandom character of the biotic processes of organic synthesis and decomposition has been linked to the concept of closed biogeochemical cycles (Commoner, 1971; Schwartzman, 1999). According to this concept, the mean rates of synthesis and decomposition of organic matter coincide with each other with a practically infinite precision, so that functioning of the biota does produce any net impact ("pollution") on the state of the environment.

Indeed, if the environmental impact of the biota were nullified, the environment could only change under the impact of abiotic physicochemical processes. These processes include the emission of inorganic matter from the Earth's interior and the reverse processes of deposition of various substances in sediments. These fluxes are about 10,000 times less powerful than the fluxes of chemical elements during biological synthesis and decomposition. Since the abiotic fluxes of emission and deposition are independent physicochemical processes having different causes, they cannot compensate each other. Their net rate is of the same order of magnitude as the gross rate. This means that under the impact of abiotic processes the environment would be changing by 10^4 times more slowly than in the case of chaotic life functioning, i.e. it would change by a 100% over a time period of a 100,000 years, i.e. on a timescale 10,000 times shorter than the documented age of life.

Moreover, it is in principle impossible to equate the fluxes of synthesis and decomposition precisely. Synthesis and decomposition of organic matter represent independent biochemical processes that are generally performed by different species under different environmental conditions (temperature, humidity, etc.). While primary productivity is limited by the incoming solar radiation, there are no physical limitations on the rate of decomposition, since the latter rate is ultimately dictated by the population numbers of heterotrophic organisms. Characteristic ecosystem values of P^+ and P^- are determined by the individual design of every species, population abundance and overall numbers of autotrophic and heterotrophic species inhabiting Earth. The values of P^+ and P^- cannot coincide with an infinite precision a priori.

Indeed, the global amount of atmospheric CO_2 is of the order of $M^- \sim 10^3$ Gt C (1 Gt = 10^9 t). The mean global rates of biochemical synthesis P^+ and decomposition P^- are of the order of $P^- \sim P^+ \sim 10^2$ Gt C year^{-1}. If the rates of synthesis and decomposition were not correlated, i.e. if they coincided by the order of magnitude only, their relative difference would be of the order of unity, $|P^+ - P^-| / P^+ \sim 1$. In such a case, if synthesis exceeded decomposition, $P^+ > P^-$, the global biota would use up the entire store of atmospheric carbon on a timescale of $M^- / P^- \sim 10$ years. This would render further photosynthesis and existence of life impossible. The amount of organic carbon in the biosphere (living biomass, humus, and oceanic dissolved organic carbon) is of the same order of magnitude, $M^+ \sim 10^3$ Gt C. If the rate of decomposition exceeded the rate of synthesis, the global biota would be able to destroy itself completely in equally short periods of time. To extend the biotic life span to the documented billion years of life existence, $T \sim 10^9$ years, one has to demand that the living organisms and their ecological communities are designed such that the mean rates of synthesis and decomposition performed by them coincide to the accuracy of $M^{\pm} / |P^+ T| \sim 10^{-8}$, one millionths of a percent, which is absolutely improbable.

This unambiguously suggests that the biota should use a conceptually different principle to achieve the environmental homeostasis. Correlation of the ecological fluxes of synthesis and decomposition of the organic matter is achieved indirectly, via continuous sensing of information about the current state of the environment that is performed by the living organisms. The biota reacts to any environmental change as soon as its relative magnitude reaches some critical value, biotic sensitivity ε_b, a fundamental parameter of life organization. As long as the relative magnitude of the environmental change remains lower than biotic sensitivity, synthesis, and decomposition of organic matter by the biota may proceed in a noncorrelated manner at different rates. As soon as some environmental parameter changes by ε_b, the biota initiates compensating negative feedback processes and keeps them going until the disturbance is diminished to a level below ε_b, when the biota no longer notices it. The optimal state to which the ecosystem ultimately returns is thus defined to an accuracy of ε_b.

Environmental control based on biotic sensitivity is performed on a local scale by every individual organism of the biota. The particular forms of this control are dictated by the genetic programs of biological species. Individual organisms acting to compensate the environmental disturbances improve their local environment and enhance their competitive capacity. Local ecological communities with an overly high number of individuals with impaired genetic programs incapable of environmental regulation suffer from environmental degradation, lose competitiveness, and are replaced by normal communities free from such defective individuals. In the result, all ecological communities in the natural biota act in the same direction supporting the optimal state of the environment already on a global scale. This principle of environmental stabilization forms the basis of the biotic regulation concept (Gorshkov, 1995; Gorshkov et al., 2000, 2004; Makarieva et al., 2006).

3. Allometric scaling of energy consumption in stable ecosystems

In this section we shall see how the biotic regulation principle is implemented in the organization of energy fluxes within natural terrestrial ecosystems; how it allows to minimize fluctuations of plant biomass and to stabilize the flux of primary productivity.

3.1. STABILIZATION OF PHOTOSYNTHESIS

Plants form the basis of ecosystem's energetics. They are responsible for the synthesis of organic matter, which further fuels all life processes in the community. To make this flux stable, the photosynthesizing parts of plants dominating natural terrestrial ecosystems represent a large number of weakly correlated objects of relatively small size (e.g. leaves, needles). The fluctuations of phytomass of individual plant are thus minimized in accordance with the statistical law of large numbers (Li et al., 2004; Makarieva et al., 2004). For example, an adult fir tree has several million needles, $N \sim 10^6$, that fall off and regrows every year. Thus, the yearly relative fluctuation of individual plant's phytomass (when detrended against nonrandom processes like tree growth) does not exceed $\varepsilon \sim 1/(N)^{1/2} \sim 10^{-3}$, i.e. 0.1%. The same figure approximately characterizes the yearly fluctuation of primary productivity.

Clearly, low values of relative fluctuations ε of phytomass and primary productivity could have only evolved and been further maintained in the course of competitive interaction of individual plants with different ε values – plants demonstrating the inability to stabilize their phytomass and productivity at a needed value were losing to those that did so. It would not be advantageous for a plant to keep the process of photosynthesis stable and fluctuations ε of the phytomass minimized, unless the fluctuations of plant biomass introduced by the process of decomposition are equally low. In other words, a plant providing food to heterotrophs that consume it in an erratic fashion would be the same suffering from biomass fluctuations as a plant with unstable photosynthesis. Hence, it is justified to expect that fluctuations of plant biomass introduced by its consumption by heterotrophs do not normally exceed the fluctuations ε introduced by the process of photosynthesis.

3.2. ENERGETICS OF HETEROTROPHS

Whilst photosynthesis is performed by objects of similar size (e.g. leaves), decomposition of the organic matter is performed by organisms with linear size varying from several micrometers (bacteria) to meters (large mammals). Due to the biochemical universality of life organization energy consumption by unit live mass or volume should be the same in organisms of all sizes. This theoretical prediction agrees well with observations: within most taxa, from bacteria to mammals, most species have their mass-specific metabolic rate q in the interval of 1–10 W kg^{-1} (Makarieva et al., 2005). Energy enters the organism via its body surface, while it is spent within the body volume. The volume/area ratio is constant and size-independent in flat ("two-dimensional") organisms that can therefore occupy an arbitrary area on the ground surface having a fixed thickness L determined by the energetic demand q (W kg^{-1}) of the organism and by the incoming flux of external energy f (W m^{-2}): $L = f/q$.

Plants, fungi, and bacteria feature this type of metabolic organization. Plants absorb solar radiation which is uniformly distributed over the ground surface, so plants can form a continuous cover. The upper limit to the flux of energy that can be consumed by heterotrophs is set by the primary productivity, $f = P^+$. Thus, small organisms with linear body size $L \leq P^+/q \sim 10^{-4}$ m can afford being immobile like plants. Such organisms (bacteria, fungi) can form a continuous cover under the plant canopy and consume primary productivity P^+ without exempting the standing stock of plant biomass. In other words, the smallest heterotrophs can satisfy their energetic needs with the flux of organic matter produced on the area of the projection of their tiny bodies on the ground surface.

A very different situation is realized for large animals with linear body size $L >> P^+/q$. These organisms cannot afford being immobile and live on plant production; they have to move exempting plant biomass. Their home range must greatly exceed their projection area. This energetic property of animals has fundamental implications for ecosystem stability.

3.3. ECOLOGICAL ALLOMETRY AND STABILITY

While moving over their large home territories and feeding, animals introduce local disturbances of plant biomass, which, in their turn, destabilize the flux of photosynthesis and primary productivity. The larger the animal, the greater fluctuations that it introduces. The overall magnitude of the fluctuations introduced per unit surface area by a population of large plant-feeding heterotrophs depends on their body size, population density N (ind km^{-2}) and individual metabolic rate Q (W ind^{-1}). The value of $\beta \equiv NR/P^{+}$ characterizes the share of primary productivity allocated to heterotrophs of a given body size.

Theoretical analysis has shown that if all organisms, irrespective of their body size, consumed equal shares of ecosystem primary productivity, the relative fluctuation ε of plant biomass introduced by organisms of a given body size would increase proportionally to linear body size L. In such a case, in the course of evolutionary increase of body size the fluctuations of biomass would increase by approximately 10^4 (Makarieva et al., 2004).

To suppress the growth of plant biomass fluctuations with growing body size of heterotrophs it is necessary to decrease the share of primary productivity allocated to large heterotrophs at the expense of diminishing their population density. Then, although the per capita fluctuations remain large, as dictated by large body size, the absolute magnitude of ecosystem disturbance introduced by large heterotrophs can be kept at a low level not threatening the ecosystem's functioning. It was estimated that in stable ecosystems the share of primary productivity allocated to plant-feeding heterotrophs should scale inversely proportionally to the linear body size, Figure 1. As indicated by the analysis of empirical evidence, this pattern is indeed realized in the natural forest ecosystems of the boreal zone in Eurasia and North America, Figure 1.

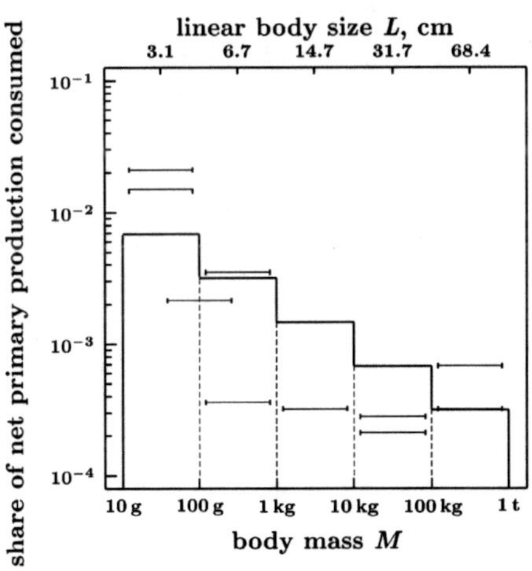

Figure 1. Relative share $\beta(L)$ of net primary production consumed by plant-feeding organisms from different logarithmic body size classes in natural forest ecosystems of the boreal zone. The histogram (thick lines) shows the theoretical distribution $\beta(L) \propto 1/L$; height of each bar is equal to the cumulative energy consumption by all community's organisms with body size falling within the respective logarithmic interval. Thin lines denote the observed $\beta(L)$ values: data from boreal ecosystems in Eurasia and North America (Makarieva et al., 2004).

3.4. IMPLICATIONS FOR ENVIRONMENTAL SECURITY

It is clear from Figure 1 that in stable ecosystems the largest heterotrophs are allowed to consume no more than several tenths of percent of ecosystem's primary productivity. As predicted by the developed approach, in unstable ecosystems, where environment is shaped by abiotic processes uncontrollable by the local biota, no ecological restrictions can be imposed on biotic environmental fluctuations. Consequently, in unstable ecosystems the community's energy flow can be distributed chaotically over differently sized animals, showing on average no dependence on body size.

Conversely, ecosystems where large animals consume large portions of primary productivity cannot be stable. The derived $\beta(L)$ distribution of energy consumption over body size is therefore important for long-term conservation practices. It allows estimating the optimum population densities of the threatened species that can be sustained on a given territory without undermining the long-term stability of the ecosystem.

At present man directly consumes about 10% of the global net primary productivity of the biosphere. (This includes food of man and cattle and consumption of wood (Gorshkov et al., 2000). As is clear from Figure 1, this figure exceeds the energy consumption quota allocated to similarly sized animals in environmentally stable ecosystems by two orders of magnitude. Expectedly, the terrestrial part of the biosphere is undergoing rapid environmental degradation. Further studies of the natural $\beta(L)$ distribution can be useful for elaborating strategies of optimization of the man–biosphere interaction.

4. Allometric scaling and terrestrial water cycle

The central idea behind the biotic regulation concept is that the natural biota creates and maintains its own favorable environment, both internal and external. While achieving this goal life had to repeatedly face the fundamental allometric problem of meeting the demand of matter or energy in an n-dimensional area by a flux of matter or energy via an $(n-1)$-dimensional area. For example, living organisms consume energy via the two-dimensional (2D) body surface and spend it within the 3D body volume. To ensure that energy is delivered to living tissues at a size-independent optimum rate, large organisms had to invent active pumps (e.g. lungs, heart), which pump matter and energy into the organism and distribute them within it. Passive diffusion fluxes of organic matter could only satisfy the needs of the smallest unicellular organisms with their linear body size L much less than the characteristic distance to the source of food.

On the ecological scale, one of the most important environmental parameters is moisture. In terrestrial ecosystems, atmospheric moisture enters a river basin via the linear coastline, while it is spent within the 2D area of the river basin. Passive geophysical fluxes of moisture can ensure sufficient soil moistening at small distances $L \sim 10^2$ km from the ocean only. To keep large territories with $L > 10^3$ km biologically productive, an active biotic pump of atmospheric moisture is necessary. That such a biotic pump does exist is confirmed by the observation that natural forests in South and North America and Eurasia are able to ensure high precipitation fluxes at any distance from the oceanic coast over several thousand kilometers inland. In contrast, in non-forested scarcely vegetated world's regions precipitation drops exponentially with distance from the ocean over a few hundred kilometers (Gorshkov and Makarieva, 2006; Makarieva et al., submitted).

The high leaf area index, a conspicuous feature of the organization of natural forests, is associated with a high rate of transpiration which can exceed evaporation from the open water surface of the ocean. The difference between the forest and ocean evaporation fluxes creates an upwelling flux of air above the forest canopy. This flux is compensated by the horizontal influx of moist air from the ocean. This regular and continuous functioning of the forest pump of atmospheric moisture prevents formation of extreme weather phenomena like hurricanes in the regions with substantial natural forest cover, which thus acts as the main guarantor of environmental security on the regional and global scale. A novel physical mechanism involving the nonequilibrium distribution of atmospheric water vapor was proposed to explain how the high transpiration fluxes developed by forests enable them to pump atmospheric moisture from the ocean to any distance inland to compensate for the runoff losses from the optimally moistened forest soil. Our results suggested that extensive forest cover plays a significantly larger role in the atmospheric circulation than previously assumed (Makarieva et al., submitted). However, more precise physical laws governing the action of the forest moisture pump demand a detailed exploration.

5. Perspectives

It is becoming progressively more evident that the problem of keeping the environment stable and secure is extremely demanding in terms of information resources. At present people attempt to meet this demand by continuously enhancing the computing facilities of the civilization in order to process the avalanche of data collected by the numerous programs of environmental and societal monitoring. No question is posed as to whether it is at all possible to collect an amount of information sufficient for environmental stabilization.

In the meantime, robust estimates indicate that in the natural biota the environmental homeostasis is maintained on the basis of huge information fluxes of the order of 10^{20} bits m^{-2} s^{-1} on a global average (Gorshkov, 1995; Gorshkov et al., 2000; Makarieva et al., 2006). In the biosphere each square micron is inhabited by several living cells whose per capita information processing capacity is comparable to that of a modern personal computer. Living cells continuously sense the environment and react to its changes in a nonrandom manner dictated by their genetic programs. Since the efficiency of this biotic regulation mechanism has been perfected during the several billion years of biological evolution, it is impossible to replicate this mechanism in technology using a smaller information flux. However, the information processing capacity of the modern civilization amounts to only 10 bits m^{-2} s^{-1}; the 19 orders of magnitude's gap dismisses all hopes for a global technology-based environmental control.

This means that the problem of environmental security on a global or regional scale cannot in principle be solved without involving the already existing regulatory potential of natural ecological systems. For this purpose, natural ecosystems will have to be conserved and restored on sufficiently large areas of continental scale. Current demographic trends in the developed countries, where the population numbers gradually decrease, create a favorable opportunity for this scenario. In this perspective the study of the bases of environmental homeostasis in the extant natural ecosystems becomes very important. Here we outlined one of possible approaches to this problem, where ecosystem's properties were analyzed within the allometric framework, i.e. based on regularities dictated by spatial scale and organismal body size. The first results are promising and clearly justifying further efforts in this direction.

Acknowledgments

This work was partially supported by the US National Science Foundation Biocomplexity Program, the University of California Agricultural Experiment Station, the Russian Science Support Foundation, and Russian Fund for Basic Research.

References

Commoner, B. (Editor), 1971. The Closing Circle. Nature, Man, Technology. Knopf, New York.

Gorshkov, V.G., 1995. Physical and Biological Bases of Life Stability. Man, Biota, Environment. Springer, Berlin.

Gorshkov, V.G., Gorshkov, V.V., and Makarieva, A.M., 2000. Biotic Regulation of the Environment: Key Issue of Global Change. Springer, London.

Gorshkov, V.G., Makarieva, A.M., and Gorshkov, V.V., 2004. Revising the fundamentals of ecological knowledge: the biota-environment interaction. Ecological Complexity 1, 17–36.

Gorshkov, V.G. and Makarieva, A.M., 2006. Biotic Pump of Atmospheric Moisture, Its Links to Global Atmospheric Circulation and Implications for Conservation of the Terrestrial Water Cycle. Preprint No. 2655, Petersburg Nuclear Physics Institute, Gatchina, Russia.

Li, B.-L., Gorshkov, V.G., and Makarieva, A.M., 2004. Energy partitioning between different-sized organisms and ecosystem stability. Ecology 85, 1811–1813.

Makarieva, A.M., Gorshkov, V.G., and Li, B.-L., 2004. Body size, energy consumption and allometric scaling: a new dimension in the diversity-stability debate. Ecological Complexity 1, 139–175.

Makarieva, A.M., Gorshkov, V.G., and Li, B.-L., 2005. Energetics of the smallest: do bacteria breathe at the same rate as whales? Proceedings of the Royal Society of London B 272, 2219–2224.

Makarieva, A.M., Gorshkov, V.G., and Li, B.-L., 2006. Conservation of water cycle on land via restoration of natural closed-canopy forests: implications for regional landscape planning. Ecological Research 21, 897–906.

Makarieva, A.M., Gorshkov, V.G., and Li, B.-L., submitted. Precipitation on land versus distance from the ocean: Evidence for a forest pump of atmospheric moisture.

Peters, R.H. (Editor), 1983. The Ecological Implications of Body Size. Cambridge University Press, Cambridge.

Schwartzman, D. (Editor), 1999. Life, Temperature, and the Earth: The Self-Organizing Biosphere. Columbia University Press, New York.

PART III INTRODUCTION

LANDSCAPE INDICATORS AND LANDSCAPE CHANGE DETECTION

THOMAS BLASCHKE
Salzburg University, Centre for Geoinformatics (Z_GIS), Schillerstr. 30, Salzburg, Austria

SERGEY VICTOROV
Research Institute of Remote Sensing Methods for Geology, St. Petersburg, Russia

The purpose of quantitative and qualitative landscape indices is twofold: (1) to be additional attributes for classification of landscape types or regions, and (2) to be indicators of landscape changes and disturbances. In practice, the potential for quantitative analyses depends fundamentally on the availability of geographical data, particularly from maps, remote sensing imagery, and in situ information. The six contributions in this section cover a significant part of this wide spectrum of landscape changes and indicators which characterize these changes and give evidence that significant progress has been achieved over the last years, both technically and methodologically.

Tiknius et al. consider the spatial arrangement of lands and the environment as a regulation mechanism. They achieve metric and topological properties of land structures with the aim to sustain development processes and environment regulation function. In applying binomial analysis procedures to study spatially expanding structures they put emphasis on the quantification, mathematical simulation, and comparison of patterns corresponding to some conception of their regular transformation considering the specific role of critical regimes.

Fjellstad and Dramstad identify driving forces of change and assess the success of environmental policies in managing change for a case study in Norway. The Norwegian Monitoring Programme for Agricultural Landscapes was designed to provide an empirical foundation on which to assess the effectiveness of agri-environmental policies. This sample-based programme aims to deliver indicators of change for four main themes: spatial structure, biodiversity, cultural heritage, and accessibility. The aim is to safeguard the environment based on precise, objective information about how that environment is changing over time. The authors present some of the first measurements of landscape change from the monitoring programme and discuss the implications of the changes in terms of landscape functions.

In the third paper Popova and Bychkova focus on capabilities of the aerial and Landsat TM data to identifying landscape types of the Natural Park "Vepssky Forest" in Leningrad's region. Multiple aerial photos satellite images and topographical maps were used to locate temporal changes in 1930–2000 in the territory of Park. This case study provides empirical insights to the changes in the reserve and its 26 vegetation types and aims to serve as a framework for other environmental security studies the same climate zone.

Victorov et al. use remote sensing methods including aerial and satellite imagery to study 30 year changes of the state of the marine and coastal environments, monitoring of water quality, study of water dynamics, change detection in coastal zone, study of drainage basins; detection, study, and prediction of environmental disasters (floods, emergency pollution, soil erosion, landslides etc). The focus is on erosion and accumulation processes. Landscape ecology aspects are reflected and historical data and recent remotely sensed data are used to analyse the dynamics of the coastal landscapes. The study of coastal zones of the Black Sea, the Azov Sea, the Volga River reservoirs, and the Ladoga Lake (Baltic Sea basin) includes environmental security aspect such as the threat of a pulp factory near the Ladoga Lake to the drinking water reservoir for 5-million inhabitants.

Haynes-Young et al. illustrate threats to landscape functions and the relation of a sense of place through exemplifying the concept of landscape character with examples from the English CQC (Countryside Quality Counts) Project. They suggest to move "beyond data" in the strategic assessment of environmental change. They conclude that supplementing data on the extent and quantity of changes with contextual information against which to judge "whether these changes matter" in a particular location is vital for the practical use of change data in policy support and environmental assessments.

DERIVING THE SPATIAL TRAITS OF ORGANIZED LAND STRUCTURES

ALGIMANTAS TIKNIUS*
Department of Geophysical Sciences, Klaipeda University, Klaipėda, Lithuania

KĘSTUTIS DUČINSKAS
Department of Statistics, Klaipeda University, Klaipėda, Lithuania

BENEDIKTAS JANKAUSKAS
Kaltinėnai Research Station of Lithuanian Agriculture Institute, Kaltinėnai, Lithuania

Abstract. Our environment is known to be under the control of the living systems. However, the natural regulation of environment is an eventual phenomenon and specific conditions are needed for its efficient manifestation. Too strong accidental disturbances can overstep the potentials of this regulation; while others slighter of biased action can disrupt the regulation structures turning the environment gradually into a state adverse for the higher forms of life (humans included). Accordingly, two main threats losing the security of the environment can be expected: one is coming from the uncontrolled hyperactivity of humans changing the environment sharply and excessively, and another – from the mass of biased factors disrupting by small degrees the "mechanisms" of environment regulation. This study is concerned with the latter hazard arising from the insufficient understanding of the significance of spatial arrangement of lands for the environment regulation. The goal of the study is to derive the metrical and topological properties of land structures essential for sustaining the development processes and herewith for the environment regulation function – that is to specify the distinctive traits and indices of organized structures. The specific analysis procedure has been applied to study the spatial transformations of developing structures. It is based on the simulation and comparison of patterns comprising before and changing after the critical regimes of development.

*Corresponding Author: Dr. Algimantas Tiknius, Department of Geophysical Sciences, Faculty of Mathematics and Natural Sciences, Klaipeda University; H. Manto st. 84, LT-92294 Klaipėda, Lithuania; Phone: + 370 46 398854; Fax: + 370 46 398802; e-mail: algis.tiknius@gmail.com

I. Petrosillo et al. (eds.), Use of Landscape Sciences for the Assessment of Environmental Security, 121–130.

Real and artificial map projections of farms, settlements, enterprises, and other expanding and developing structures resembling the organism populations (called the quasi-populations) have been examined and the specific traits and indices of their organization have been derived using this approach.

Keywords: Organized land structures; quasi-populations; organizing correlations; critical regimes

1. Introduction

This study is concerned with the spatial traits imperative or suitable for development processes and self-regulation of environment. Structures distinguished by these properties are called the organized structures. Insufficient organization of lands may affect badly the atmosphere, hydrosphere, and security of environment on the whole (Gorshkov et al., 2000, 2004). Consequently, the possible effects of spatial organization of lands should be explored thoroughly. The goal of this study is to reveal the metrical and topological features common to organized structures. Certain organism populations, as well the farms, settlements, enterprises, and other particular growing, expanding and developing structures were elaborated as being simple enough to study these features and typical enough to represent the large areas of the Earth associated with environment security problem. It is common for such structures to experience the recurrent congestion of individuals (overpopulation). A long trail of vexatious consequences follows these crises – spatial straits, scarcity of resources, rigorous competition, etc. As the mutation increases sharply, shrinkage of a population, and extinction of most individuals are the natural outcomes from such regimes. However, some of systems undergoing series of self-transformations manage to survive by overcoming crises and to prevail over extensive periods and large territories. Why do they succeed and for what reason? What is the role of structural changes in such situations? There are no clear answers to these questions despite of many serious studies with application of sophisticated models (Dieckmann et al., 2000; Epperson, 2000; Hanski and Ovaskainen, 2003; Henson et al., 2002; Ovaskainen and Cornell, 2006). The logistic model, the cellular automata and other conventional approaches also provide insufficient information on what happens with spatial structures during critical regimes and how the organized patterns emerge. This was the motive for venturing to study the processes from the very beginning. Known models have been revised and the new standpoints have been applied to simulate the simplest structures and to try to answer these questions.

1.1. METHODS

The binomial analysis procedure has been applied in this study (Tiknius, 1996). It is based on the simulation of spatial transformations of developing structures. A quasi-population structure is described using the simple algebraic functions of random variables – metrical features of individuals and their functional zones; while the spatial transformations are detected from comparison of regular stages of population change corresponding to the modified Verhulst's model (Hanski and Ovaskainen, 2003).

The habitat and the critical regime notions as the axiomatic assumptions are significant for this procedure. The population habitat is considered being a constant and indiscrete, while the critical regimes are presumed emerging repeatedly when the expanding population is filling its habitat completely. The quasi-population is unable to extend or leave its habitat of limited size while the functional stereotype of individuals remains constant. This basic assumption corresponds to the inequality (Tiknius, 2002):

$$f(\bar{x}_1, \bar{x}_2, ..., \bar{x}_k) \geq \overline{f}(x_1, x_2, ..., x_k); \quad (1)$$

where $f(x_1, ..., x_k)$ = function describing a particular spatial unit; $f(\bar{x}_1, \bar{x}_2, ..., \bar{x}_k)$ = size of uniform units occupying the habitat H at the stage of unconstrained replication: $H = nf(\bar{x}_1, \bar{x}_2, ..., \bar{x}_k)$; n = number of units filling completely the given habitat; $\bar{x}_1, \bar{x}_2, ..., \bar{x}_k$ = mean values of metrics used to describe a spatial unit (an individual with its functional zone); k = number of metrics considered in the description of a unit; $\overline{f}(x_1, ..., x_k)$ = mean size of the same units strained by the disturbances and conditions of crisis.

1.1.1. Common term of inequality satisfaction

The inequality Eq. (1) is satisfied only for *concave* functions $f(x_1, ..., x_k)$ otherwise it would be contradicting the *Jensen's* inequality (Rao, 1973). While the individuals are described initially by the *convex* functions, a contradiction arises and the structure composed of such individuals is unable to satisfy this inequality. So it is absolutely imperative that a particular structure to overcome the critical regime, must transform itself adequately. Really, the specific correlations between metrics have been found to resolve this contradiction giving a chance for a population to prolong its development (Tiknius and Jankauskas, 2005). Commonly there is no problem to detect such correlations using the inequality Eq. (1). For instance, consider the case when the spatial unit is defined by two metrics x_i and y_i simulating the measures of rectangular patches $x_i y_i$ of different size and elongation. Entering the function $f(x_i, y_i) = x_i y_i$ values into the inequality Eq. (1), it can be modified as follows:

$$(\overline{x_i}\,\overline{y_i} - \overline{x_i y_i}) = -Cov(x_i, y_i) = -(r_{x,y}\sigma_x\sigma_y) \geq 0\,; \qquad (2)$$

where $Cov(x_i, y_i)$ = covariance of metrics x_i and y_i; $r_{x,y}$ = correlation coefficient of metrics x_i and y_i; σ_x and σ_y = standard deviations of these metrics; meanings of other symbols – as in Eq. (1).

It is evident now that only the *negative* correlation $r_{x,y}$ between metrics x_i and y_i is able to satisfy the inequality Eq. (2) making the structure organized.

2. Results

Some examples of application of the given procedure with interpretation of solutions will be presented further.

2.1. A MEASURE OF METRICAL ORGANIZATION

A measure of the degree of metrical organization (*M)* naturally follows from the inequality Eq. (1):

$$M = 1 - [\overline{f}(x_1,...,x_k) / f(\overline{x}_1,...,\overline{x}_k)]\,; \qquad (3)$$

where the meanings of symbols are the same as in inequality (*1*).

This measure could be calculated either for the actual or for the regular shape of spatial units $f(\overline{x}_1,...,\overline{x}_k)$. In first case the measure would be indicating the relative metrical organization of a given pattern achieved at the later crisis; in second – the absolute organization achieved during the history of this population. Values $M_i \leq 0$ and $M_i > 0$ refer to the absence and presence of this property respectively.

2.2. RECOGNIZING THE ORGANIZED STRUCTURES

It is uneasy to know visually the ordered and the organized structures one from another (Figure 1). This task will be studied using two examples of expanding structures with application of the criterion (1).

Example 1. Consider a point pattern of an organism population distinguished by the complexity of shapes of patches depicting the functional zones of individuals. It is feasible to reckon and describe such patches (units) as the fractal objects:

$$a_i = (p_i / 2\sqrt{\pi})^{d_i}\,; \qquad (4)$$

where p_i is the perimeter; a_i is the area; and d_i is the fractal dimension ($1 \leq d_i \leq 2$).

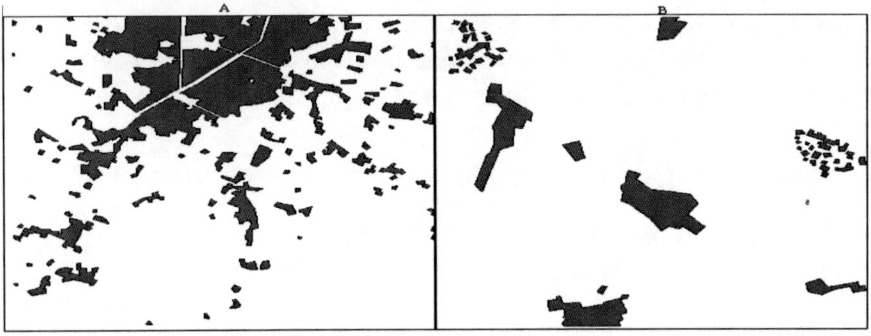

Figure 1. Metrically organized (left) and ordered though unorganized (right) settlement patterns (*Plunge* town and *Pasraučiai* village localities, each of 10 km^2 extent, Western Lithuania region).

Suppose this pattern is an organized structure.

The task is to test this assumption. In this case the criterion Eq. (1) becomes the expression:

$$\overline{(p_i / 2\sqrt{\pi})^{d_i}} \leq (\bar{p}_i / 2\sqrt{\pi})^{\bar{d}_i}; \qquad (5)$$

where hyphens above indicate the mean values of the variables.

It will be satisfied if the function $(p_i / 2\sqrt{\pi})^{d_i}$ is of concave type. A function is said to be concave over a given interval of variables if its second derivative is negative everywhere in this interval. Supposing the continuity and negative correlation of perimeters p_i and dimensions d_i, the given function can be modified to the expression $(c / d_i)^{d_i}$ (c is a positive constant here). The second derivative of the latter function has the negative sign in the segment of $1 \geq d_i \leq 2$. In this case the negative correlation of variables p_i and d_i is that making the given structure metrically organized.

Example 2. Consider the map layer of farms and their landed properties composed of units of rectangular shape:

$$a_i = w_i l_i + 2d_i(w_i + l_i) + 4d^2{}_i \qquad (6)$$

where w_i = widths of farms; l_i = lengths of farms; and d_i = mean values of widths of landed properties (indicating the distances between farms).

The task is to verify if this layer satisfies the term of organization. The inequality Eq. (1) in this case obtains the expression (Tiknius and Jankauskas, 2005):

$$4\sigma_d^2 \pm r_{w,l}\sigma_w\sigma_l \pm 2r_{d,l}\sigma_d\sigma_l \pm 2r_{w,d}\sigma_w\sigma_d \leq 0; \qquad (7)$$

where σ_i = the standard deviations of adequate metrics; $r_{i,j}$ = pair correlation coefficients of respective metrics.

The values of statistics σ_i and $r_{i,j}$ of respective metrics can be simply received using an ordinary software from the given pattern to find out if the last inequality is satisfied and the assumption of its organization approved (or not).

2.3. INDICATING THE FEATURES OF AN ORGANIZED STRUCTURE

Some tasks concerning the indication of features of an organized structure from example 2 will be studied to show the effectiveness of the derivative indices.

2.3.1. Indicating the metrical features of anomalies

The anomalies are common and observable in various structures from molecular to cosmic scale. So there is no need to prove that the land structures are not the exceptions. It is reasonable only to exploit this axiom to study the features of conspicuous localities of organized structures. *First step* in this way is to define the number of feasible solutions of inequality Eq. (7). Only some of correlations may satisfy it. For instance, the pattern distinguished in two negative correlations corresponds to the regressions $w_i - \bar{w} = -b_{w,d}(d_i - \bar{d})$ and $w_i - \bar{w} = -b_{w,l}(l_i - \bar{l})$; where $b_{w,d}$ and $b_{w,l}$ are the regression coefficients; meaning of other variables – as in previous equations. From these two the third positive regression evidently follows $d_i - \bar{d} = (b_{w,d} / b_{w,l})(l_i - \bar{l})$. Similarly other pairs of positive and negative correlations determine the negative sign of a third one. Also one negative correlation is compatible with other two zero correlations. Following this consideration only *6* combinations of *27* formally conceivable appear capable to organize this structure indicating the particular types of organized structures. *Second step* is to define the size intervals of metrics based on their standard deviations. For instance, the *low*, *medium*, and *high* values can be defined respectively by intervals: $x_i < (\bar{x} - \sigma_x)$; $(\bar{x} - \sigma_x) \le x_i \le (\bar{x} + \sigma_x)$ and $(\bar{x} + \sigma_x)$ (where $\bar{x}$ = mean value, and σ_x = standard deviation of a varying metric x_i). *Third step* is to track how the correlations organizing the structure recognized at first step may affect the metrical features of patches composing the clusters and rarefactions. The patches composing the virtual concentrations (patch swarms of high density) evidently distinguish in small distances apart. This can serve as a starting point to detect other metrical features of such swarms. For instance, entering the low values of distances $d_i \le \bar{d} - \sigma_d$ to the regression from step 1

$$w_i = \bar{w} + b_{w,d}\sigma_d = \bar{w} + (r_{w,d}\sigma_w / \sigma_d)\sigma_d = \bar{w} + r_{w,d}\sigma_w; \tag{8}$$

(where w_i, l_i, d_i = widths, lengths of patches, and distances respectively; $\bar{w}_i, \bar{l}_i, \bar{d}_i$ = mean values of these metrics; $b_{w,d}$ = regression coefficient; $r_{w,d}$ = correlation coefficient; σ_i = standard deviations of respective metrics) the medium values of patch widths composing the swarms follow. Such consideration and the given procedure can be applied to obtain the other metrical features of possible anomalies. One can find in this way that a swarm may not be composed of small patches (Figure 1); a rarefaction – of big and elongated patches, and so on.

2.3.2. Indicating the topological features

What are the typical positions of the substructures making the ternary compositions? It is of common experience that extremes (like poles) emerge at different places. So the anomalies of developing structures should be frequently separated by the typical individuals making the matrix.

What is the typical number of constituents composing an organized structure? The trivial answer to this question follows from the inequality Eq. 1:

$$n \geq F / f(\bar{x}_1, \bar{x}_2, ..., \bar{x}_k); \tag{9}$$

where F = total area covered by the given population: $F = n\bar{f}(x_1, x_2, ..., x_k)$; meanings of other symbols – as in inequality Eq. 1.

More specified answer can be obtained only for some of simplest structures. For instance, let us consider the modified farm structure from example 2 composed of patches of square shape. It can be described as follows:

$$F_i = \sum_1^n (\sqrt{a_i} + 2d)^2 \tag{10}$$

and modified to the expression (Tiknius, 2002):

$$\sigma^2{}_s = [4nd^2 F_n - (F_i - F_n - nd^2)^2] / 4n^2 d^2; \tag{11}$$

where $\sigma^2{}_s$ = variance of metrics $\sqrt{a_i}$ indicating a kind of diversity of farms; F_i = size of a given structure; a_i = sizes of patches representing farms; s_i – length of a square patch side: $s_i = \sqrt{a_i}$; d = constant distance between farms; F_n = total area covered by farms.

Now the number of components n maximizing the metrical diversity σ_s^2 of this kind of structure can be derived solving the equation $(\sigma_s{}^2)' = 0$:

$$n = (F_i - F_n)^2 / d^2 (F_i + F_n); \tag{12}$$

where $(\sigma_s{}^2)'$ – fluxion of a given function; other meanings of symbols – as in Eq. (11).

Similarly can be obtained the virtual area of farms F_n

$$F_n = n\bar{d}^2 \overline{s^2} / (\bar{d}^2 \overline{s^2}) \tag{13}$$

corresponding to the biggest value of organization of given structure (Tiknius, 2002). From this the number of constituent's maximizing the metrical organization of structure follows:

$$n = F_n(\overline{d}^2 - r_{s,d}\sigma_d^2) / \overline{d}^2\overline{s^2} ; \tag{14}$$

where hyphens above mark the mean values of respective variables; $r_{s,d}$ = correlation coefficient; σ_d – standard deviation; d_i – varying distance between farms; meanings of other symbols – as in previous Eqs. (11), (12).

2.3.3. Transformations and compositions of organized structures

The organizing correlations condition both the differentiation and aggregation of structures. For instance, the combination of correlations ($r_{l,w}$;$-r_{w,d}$;$-r_{d,l}$) can be detected applying the given procedure to induce exactly the big and compact farms composing their spatial concentrations and the small and compact farms covering the rarefied localities. While the remaining space of this landscape comes to the typical farms of medium size and common shape. From the other side, the same correlations determine the macro-constitution of given structure – the types of organization of substructures. Consequently, any particular organized structure serves as a ground for breeding of substructures of specific organization, which in their turn can branch in to the particularly organized sub-substructures, etc. The aggregation of substructures obviously should match the order of their differentiation. This conception is quite consistent with some of ecological suggestions. For instance, the anomalies are known to represent the spatial forms of the functional specialization and the matrix – the spatial forms of functional generalization.

2.3.4. Analyzing the map layers

Results of this study can be used to construct the artificial organized layers and to compare them with the real maps. Is it probable to observe the deduced signs of organization in reality? How frequent are the metrically organized map patterns? In order to obtain these signs and structures, 47 map layers of settlements, farmlands, and forest ecosystems (of 625 km^2 area each) have been studied using the CORINE data and digital maps of the Western region of Lithuania (M 1:50,000). The results have been quite unexpected. The organization criterion (3) has been satisfied only in 11 cases of 47 studied (23%). In most cases it was satisfied by the slight correlations of metrics indicating likely the inceptive, remnant or the accidental character of structural organization. Only few patterns have been organized by the correlations of the 0.05 significance.

2.3.5. Relating the spatial and functional properties

Estimation and discussion of the role of the metrical organization for the biotic regulation of environment can be found in the specific study (Tiknius et al., 2003).

3. Discussion

One thing, which is certain from this study, is that a number of distinctive traits of organized structures can be derived using the same procedure. Each description of organized structures uncovers a *set* of particularly organized structures. Now let us conceive all possible descriptions and solutions using the given procedure. The totality of such virtual (different) solutions determines a *class of organized structures* distinguished by specific heterogeneity of their metrical and topological traits. The diversity of types of spatial organization appears to be finite. Consequently, the change of organized structures should be reversible and the metrical evolution of structures is impossible. Considering this limitation, the metrically organized structures resemble the crystallized substances. It is strange yet true that such a specific knowledge can be obtained using so simple mean. Unfortunately some shortcomings are also characteristic for this technique worthy of discussing. First, there is a problem related to the quality of initial data. Criterion (1) does not require the high precision of initial data. However the biased estimations of initial data may give the deceptive results. So the direct measurements should be valued most of all. It is uneasy though to measure directly or define some of the initial data (boundaries and dimensions of functional zones in particular) and simulation or other estimations should be used here by necessity. Consequently, the additional research is needed to ascertain how the distortions of data would impact results of such studies. Second, the features of structures are quite dependent on the correlation of their metrics (Šaltytė-Benth and Dučinskas, 2005; Epperson, 2000). So it is worth to estimate further the possible impact of correlation and its significance to the results of application of the given procedure. Third, it is a complicated task to verify the deduced expectations using the map data. Really, can the given result "*11 organized patterns from 47 studied*" be treated as an assessment of the proportion between organized and random patterns of a real landscape? Surely – not. Since the distinctive features of only class of organized structures are known, the remaining 35 cases should be treated as unrecognized yet. All things taken together, a landscape is too complex an object for studying the simplest forms of organization. Probably some microstructures (like organisms expanding in vitro) would be a more suitable object for studying these problems, bearing in mind that none scale limitation appeared in this study. Meanwhile continuing the research of land structures a new approach is needed shifting the focus from the particular layers to the intercovered and interdependent structures.

References

Dieckmann, U., Law, R., and Metz, J.A.J. (Editors), 2000. The Geometry of Ecological Interactions: Simplifying Spatial Complexity. Cambridge University Press, Cambridge, pp. 207–228.

Epperson, B.K., 2000. Spatial and space–time correlations in ecological models Ecological Modelling 132(1.2), 63–76.

Gorshkov, V.G., Gorshkov, O.V., and Makarieva, A.M., 2000. Biotic Regulation of the Environment: Key Issue of Global Change. Springer, London, pp. 212–314.

Gorshkov, V.G., Makarieva, A.M., and Gorshkov, V.V., 2004. Revising the fundamentals of ecological knowledge: the biota-environment interaction. Ecological Complexity 1(1), 17–36.

Hanski, I. and Ovaskainen, O., 2003. Metapopulation theory for fragmented landscapes, short communication. Theoretical Population Biology 64(1), 119–127.

Henson, S.M., Constantino, R.F., Desharnais, R.A., Cushing, J.M., and Denis, B., 2002. Basins of attraction: population dynamics with two stable 4-cycles. Oikos 98, 17–24.

Ovaskainen, O. and Cornell, S.J., 2006. Asymptotically exact analysis of stochastic metapopulation dynamics with explicit spatial structure. Theoretical Population Biology 69, 13–33.

Rao, C.R. (Editor), 1973. Linear Statistical Inference and Applications, 2nd edition, Wiley, New York.

Šaltytė–Benth, J. and Dučinskas, K., 2005. Linear discriminant analysis of multivariate spatial-temporal regressions. Scandinavian Journal of Statistics 32, 281–294.

Tiknius, A. and Jankauskas, B., 2005. Some efects of population growth on its spatial structure. EcoSys 11, 99–107.

Tiknius, A., Jankauskas, B., and Juška, V., 2003. A method for environmental quality assessment at the landscape level. EcoSys 10, 56–70.

LANDSCAPE MONITORING AS A TOOL IN IMPROVING ENVIRONMENTAL SECURITY

WENDY FJELLSTAD* AND WENCHE DRAMSTAD
Norwegian Forest and Landscape Institute, P. O. Box 115, N-1431 Ås, Norway

Abstract. In order to safeguard our environment, we need precise, objective information about how that environment is changing over time. In particular, we need to be able to identify driving forces of change and to assess the success of environmental policies in managing change. The Norwegian Monitoring Programme for Agricultural Landscapes was designed to provide an empirical foundation on which to assess the effectiveness of agri-environmental policies. This sample-based programme aims to deliver indicators of landscape change, primarily based on measures of spatial structure derived from aerial photographs. In this paper we present some of the first indicators of change from the monitoring programme and discuss the implications of the changes in terms of landscape functions and environmental security.

Keywords: Landscape change, indicators, agri-environmental policy

1. Background

With global food demand predicted to double during the next 50 years, we face considerable challenges to ensure the sustainable development of agriculture and increase food production capacity (Tilman et al., 2002). Nevertheless, huge areas of agricultural land continue to be built upon or abandoned (Gardner, 1996). Since population centres have frequently grown up around the best agricultural land, the level of conflicts over land increase as towns expand.

In Europe and elsewhere, concern over farmland as a source of food has been matched by an increasing interest in the non-food values attached to

*Corresponding Author: Dr. Wendy Fjellstad, Norwegian Forest and Landscape Institute, P. O. Box 115, N-1431 Ås, Norway, e-mail: wef@skogoglandskap.no

I. Petrosillo et al. (eds.), Use of Landscape Sciences for the Assessment of Environmental Security, 131–141.

farming landscapes. These values include the provision of biological diversity and cultural heritage, contributions to the viability of rural areas, and the maintenance of cultural landscapes that give people a sense of identity and belonging. Increasingly, national policy objectives are being broadened and diversified to take account of these public goods, often referred to as the "multifunctionality" of agricultural landscapes. The international interest in these issues has been made particularly apparent through the work of the Organisation for Economic Co-operation and Development (OECD) in their requests for international reporting of agri-environmental indicators (OECD, 2001).

Norway is one of the OECD countries with relatively difficult conditions for agriculture, comprising 44% mountainous land and ca. 38% forest. Only about 3% of the total land area in Norway, or some 10,000 km^2, is cultivated agricultural land, and of this only a third is considered suitable for the production of grain for human consumption. In most OECD countries, cropland accounts for a much larger share, e.g. 57% in Denmark, 34% in Germany, 37% in Italy, and even other Northern countries like Finland (7%) and Sweden (6%) have a larger proportion cropland than Norway (Bruyas, 2002; OECD, 2001). Although the population of Norway is also rather small on a European scale - only 4.6 million - the amount of agricultural area per inhabitant is still lower than the world average (2.1 daa/person in Norway compared to 2.7 daa/person in the world).

This scarcity of agricultural land in Norway has lead to strong national policies to protect cultivable land. The Land Act ("Jordloven" May 12, 1995, section 1) states that "*Land resource management shall be environmentally sound and, among other things, take into consideration protection of the soil as a production factor and preservation of land and cultural landscapes as a basis for life, health and well-being for human beings, animals and plants*" (Jordloven, 1995). The Land Act explicitly states that "*All cultivated land that can provide a basis for profitable operations shall be maintained*" and if the Ministry of Agriculture and Food finds that cultivated land is poorly maintained or unused, it may issue an order regarding the measures that shall be implemented in order that the land may be cultivated profitably. The owner may even be ordered to lease out land for a period of up to ten years. "*Cultivated land must not be used for purposes that do not promote agricultural production. Cultivable land must not be disposed of in such a way as to render it unfit for agricultural production in the future*" (section 9).

In spite of this protection of agricultural land through law, municipalities can – and frequently do - allow exemptions. Nevertheless, the municipalities are all working towards the national goal that the rate of conversion of agricultural land shall be reduced by 50% by the year 2010.

As for any attempt to improve environmental security, the safeguarding of agricultural landscapes requires both quantifiable goals and a system to establish whether these goals are met. This paper describes a method to quantify landscape change in Norwegian agricultural landscapes. We give examples of observed changes over a five-year period and address a range of challenges related to the assessment of observed changes in relation to goals. Whilst different countries may have different emphasis in their environmental policies and different areas of special concern, we expect that many of the challenges associated with setting environmental goals and monitoring progress towards these goals may be similar.

2. The Norwegian monitoring programme for agricultural landscapes

Due to limitations of time and resources, any attempt to monitor landscape change is faced with a trade-off between the area to be studied and the level of detail that can be recorded. The most cost effective methods rely on some form of sampling, whereby estimates for large areas are based on the results from representative sample sites. The Norwegian monitoring programme for agricultural landscapes uses a sample of one thousand 1 × 1 km monitoring squares. The selection of sampling squares occurred in two steps: first, land use was recorded for a systematic grid of points covering the entire country, with 3 km between points. From this grid, all points falling on agricultural land were selected. Centred on each of these points, a 1 × 1 km monitoring square was established. This provided a sample of squares that was representative for the agricultural area in Norway.

True colour aerial photographs (scale 1:15,000) provide the main data source for the monitoring programme. Each square is to be photographed at five year intervals. The workload is spread over a period of five years, so that around one fifth of the sample squares are photographed each year, achieving national coverage after five years and then returning to the first squares to re-photograph. Maps are constructed by professionals with extensive experience in interpreting aerial photographs (see Dramstad et al., 2002 for methodological details) and landscape change is analysed by comparing the maps from different time periods. The monitoring programme started in 1998 in the south-eastern part of the country and data on landscape change exist for the counties of Oslo/Akershus, Østfold and Vestfold.

3. Results from the monitoring programme

Analysis of landscape change over the five-year period 1998–2003 indicates an average annual rate of conversion of agricultural land to non-agricultural land-use of 0.62% in Vestfold, 0.66% in Østfold, and 0.87% in Oslo/Akershus. In this context, "agricultural land" is used to include both arable land and in-field pastures. About 20% of this area (ca. 300 hectares) was converted to built-up land, whilst about 70% became unmanaged grassland and 10% became forest.

Landscape change accounting involves both additions and subtractions. Table 1 shows the full accounts, revealing a net negative result for agricultural land.

TABLE 1. Additions to, and subtractions from, the agricultural area in three counties in the south-eastern part of Norway. Numbers indicate the average annual change as a percent of the original agricultural area.

	Østfold %	Oslo/Akershus %	Vestfold %
Change from agriculture to…			
… unmanaged grassland	−0.43	−0.63	−0.40
… forest	−0.08	−0.06	−0.10
… built-up land	−0.14	−0.18	−0.12
… other land types	−0.01	0.00	0.00
Sum	−0.66	−0.87	−0.62
Change to agriculture from…			
… unmanaged grassland	0.15	0.19	0.22
… forest	0.07	0.08	0.11
… built-up land	0.01	0.02	0.05
… other land types	0.01	0.00	0.00
Sum	0.24	0.29	0.39
Net change	−0.42	−0.58	−0.23

Since agricultural land is a limited resource in Norway, it is worth exploring whether soil quality is taken into consideration when land is taken out of production. This analysis was carried out only for Østfold County. In Figure 1, soil quality is defined in terms of the degree to which the soil limits the type of production possible. There was no significant difference in the soil quality (Figure 1) or erosion risk (Figure 2) of land taken out of production compared with the original agricultural area.

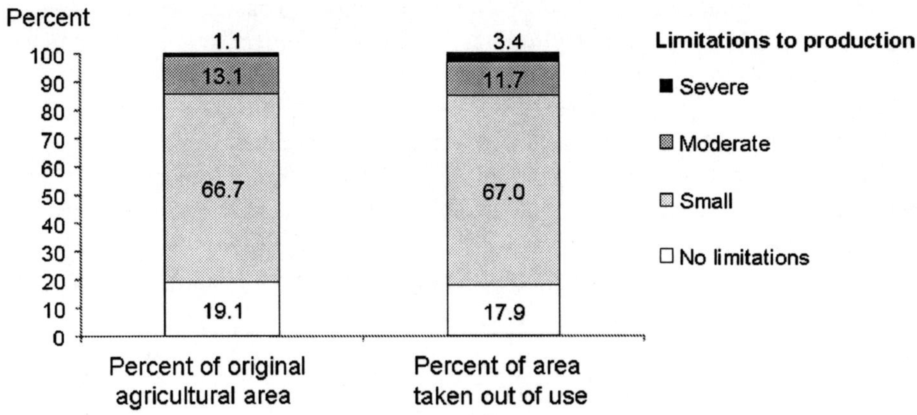

Figure 1. The proportion of agricultural land in different soil quality categories – for the original agricultural area and for the area taken out of production.

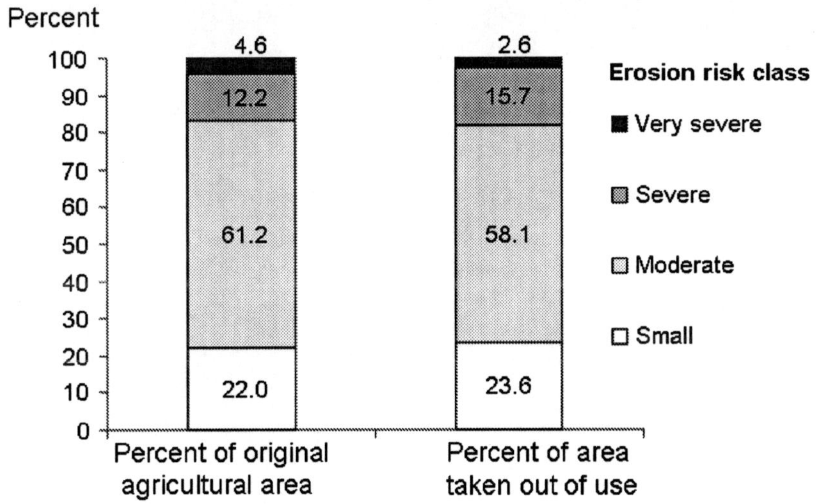

Figure 2. The proportion of agricultural land in different erosion risk categories – for the original agricultural area and for the area taken out of production.

In Norway, there is considerable interest in processes of land abandonment and the subsequent regrowth of bushes and trees on agricultural land. To provide more insight into the extent of bush regeneration, the categories "unmanaged grassland" and "in-field pasture" are both broken down into different categories according to the percentage cover of bushes and trees (see Figures 3 and 4).

Figure 3 shows that the greatest decline in the area of pasture has occurred in Oslo/Akershus and that the decline has been accompanied by a slight increase in the proportion of pasture land with bushes. Figure 4 shows that Oslo/Akershus is also the region where both the area of unmanaged grassland and the proportion of grassland with bushes have increased the most.

Environmental security in agricultural landscapes involves not only preservation of productive soils, but also preservation of a range of landscape elements that provide other goods and services. One obvious example in Østfold and Oslo/Akershus counties is the occurrence of farm ponds. Such

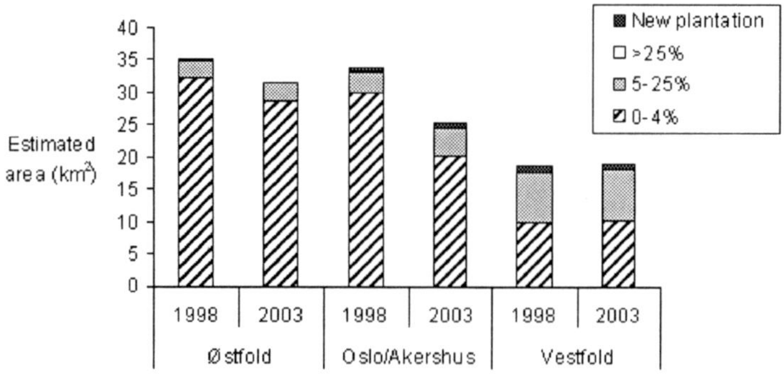

Figure 3. The area of in-field pasture in 1998 compared with 2003, showing divisions into different categories according to percent cover by bushes.

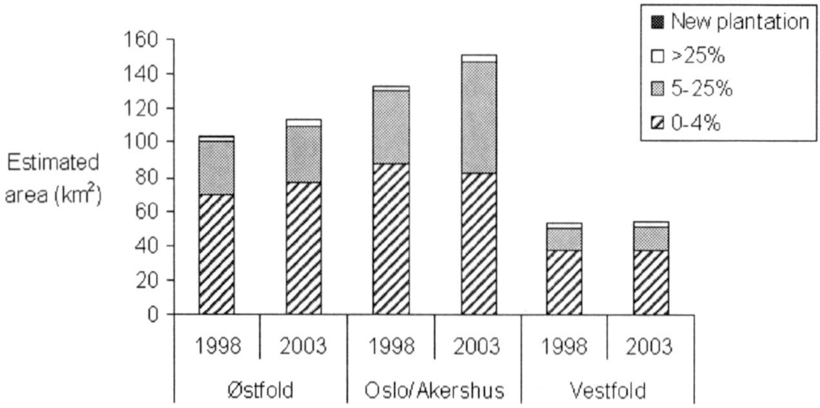

Figure 4. The area of unmanaged grassland in 1998 compared with 2003, showing divisions into different categories according to percent cover by bushes.

ponds contribute to regional landscape character, have aesthetic value, are valuable wildlife habitats, and in some places are used to improve water quality by filtering run-off from farmland before it can reach larger waterways. Results suggest that there has been an average annual decline in the number of farm ponds of 1.4% in Oslo/Akershus and 0.4% in Østfold. (Although there was also a negative trend in the number of farm ponds in Vestfold, there were too few ponds in total in the monitoring squares to be able to calculate meaningful estimates for the county as a whole).

4. Discussion

Whilst monitoring programmes like the 3Q Programme can provide us with indicators of landscape change, it is not always easy to translate these indicators into advice for policy makers. Linking the decisions of farmers and land managers to the development of specific parcels of land is hampered by issues of data availability. The same applies for linking land use changes at the farm level with driving forces at municipal, regional, national and even international scales (e.g. effects of international agreements, such as those of the World Trade Organisation). The complexity of factors influencing landscape change at multiple spatial scales, combined with time lags between cause and effect, make the identification of the driving forces of change extremely challenging and costly (Rindfuss et al., 2004). Not least, we need a better understanding of how people react within this social-ecological panarchy (Gunderson and Holling, 2002). Although more research on this subject is clearly needed, we can start by assessing whether agri-environmental goals are being met.

The clearest goal for agricultural landscapes in Norway is the preservation of agricultural land. Considering the potential strength of this protection, declines of around 2.5% of the agricultural area over the 5 year period are considerable. Securing the environment for future generations demands a long-term outlook and if the current rate of net loss continues for 170 years there will be no agriculture remaining in these counties – in the area of Norway most suited for agriculture. Rounsevell et al. (2005) refer to a similar situation for the European Union, where there was a decline in the agricultural area of 13% between 1960 and 2000, in spite of policies aiming at first to increase both production and production area and then to maintain the area.

Although the data from the monitoring programme indicate a significant loss of agricultural land, it is important to point out that about 70% of this loss was to “unmanaged grassland”. Since the mapping relies solely on the interpretation of aerial photographs, this is a category that includes rather different types of grassland, from temporary fallow land, to field margins, to abandoned pastures. Land that has been allocated to building purposes will often be classified as uncultivated grassland before building work commences and, considering that the counties studied are centrally placed in relation to population centres, this is certainly the path of change for some former agricultural land. The slight increase in the proportion of unmanaged grassland with bushes, particularly in Oslo/Akershus, may also indicate some degree of land abandonment.

When trying to assess the significance of different types of landscape changes, it is important to take into account the relative sizes of the different land categories. If two land types dominate the landscape, even random processes of change could cause large transitions between the two. To identify systematic changes in relation to the total sizes of the different land categories we used the method of Pontius et al. (2004). We found that the transition from agricultural land to built-up land occurred less than would be expected by chance, but that the most systematic transition when built-up land increased was that it replaced unmanaged grassland, supporting the suggestion that this may be the ultimate path of change for some agricultural areas. The most systematic change when unmanaged grassland increased was that it replaced pasture. In the next round of monitoring we will see whether unmanaged grasslands are again taken into use as agricultural land or whether they are built on. This illustrates some of the challenges posed by time lags between decisions affecting the landscape and the visible effects of those decisions. It also underlines the importance of long term data sets, since the interpretation of trends is easier when data is available for a long time-period, rather than at just two snapshots in time. Longer time series are also important to minimise the effects of uncertainty due to mapping inaccuracies, which is particularly relevant when the overall degree of change during each time interval is small.

As well as directly affecting food production capacity, changes in the total agricultural area influence the ecological properties of the landscape, and its appearance and recreational value for people. Indeed, even if losses and gains of agricultural land balance one another, landscape appearance will be changed by the dynamics of changing land use. In addition, it is commonly recognised that the pattern of different land uses, as well as their total area, is important in the production of many ecosystem services (Lant et al., 2005). Potschin and Haines-Young (2006) propose that there are many possible sustainable landscape options. However, the number of landscape options will be dependent on the number of functions that any particular piece of land can fulfil (Potschin and Haines-Young, 2006). Due to the marginal conditions for agriculture in Norway, the number of landscape parcels that can support agriculture is limited. Agricultural production functions can only be supplied by the best soils. Estimates suggest that about 33% of the world's cropland suffers reduced productivity due to soil erosion (Rounsevell et al., 1999), suggesting that freedom from erosion should be a valued property. Nevertheless, examination of the soil characteristics of agricultural land in Østfold County showed that neither soil quality nor erosion risk category seem to be taken into account when taking land out of production.

Whilst the best agricultural soils are located close to population centres, the recreational functions of landscapes are also of greatest value in these areas with most people. Lambin et al. (2001) point out that some of the pressures of urbanisation may be ameliorated by an increasing pressure from urban populations for recreational areas. This suggests that the strong protection of agricultural land, should be accompanied by a safeguarding of environmental qualities that make the cultural landscape an attractive recreational landscape for local populations. The process of urbanisation not only affects land use directly but also influences people's attitudes towards farming and towards the landscape. This means that the educational values associated with agricultural landscapes close to towns would also be considerable. In addition, remaining natural areas should be maintained, both for their recreational value and as habitat. Even small patches of natural habitat in the agricultural landscape are of high importance for biodiversity, since the croplands can supply relatively little in the way of regulation and habitat functions (*sensu* de Groot, 2006). The logical conclusion is that the sustainable option for agricultural landscapes in this part of Norway would be to move building projects to land that cannot support agriculture, whilst at the same time maintaining regulation and habitat functions. In practice, this implies counteracting current trends of urbanisation and centralisation, although the far-reaching consequences that this would have, for example, in terms of extra infrastructure and energy use, are beyond the scope of this paper.

As shown by Lambin et al. (2001), the clearest drivers of land use change are changes in markets and policies, and these in turn are increasingly influenced by global factors. Some land use/climate change scenarios for Europe consider that increased plant productivity due to increased carbon dioxide levels in the atmosphere, together with technological advances, will more than offset increased demand for food and fibre (Rounsevell et al., 2005). In these scenarios as much as 50% of current grassland and cropland may become surplus and can be put to other uses. Rounsevell et al. point out, however, that lower prices for agricultural products due to lower demand, may lead to agricultural expansion to compensate for lower income per unit area.

Other European scenarios describe an increase in intensive agricultural land use in Norway due to the beneficial effects of warming at higher latitudes (Berry et al., 2006). These scenarios suggest a reduction in food production capacity in lower latitudes due to increased temperatures and aridity. Other plausible futures include an increased production of bioenergy crops and policy interventions that encourage extensification for environmental reasons (Rounsevell et al., 2005).

One thing that scenario approaches make clear is that the number of plausible futures is endless. However, scenarios can make us more aware of the many different options we face and the possible consequences of today's actions – or failure to act. By taking stock of the natural resources in today's landscapes and considering the functions of the various components, we are more likely to be able to keep the sustainability choice space open for future generations.

References

Berry, P.M., Rounsevell, M.D.A., Harrison, P.A., and Audsley, E., 2006. Assessing the vulnerability of agricultural land use and species to climate change and the role of policy in facilitating adaptation. Environmental Science & Policy 9, 189–204.

Bruyas, P., 2002. Land use – land cover: LUCAS 2001 Primary Results. Statistics in focus. Theme 5–8/2002. http://www.eu-datashop.de/en/downloads/sif/nn_02_08.pdf. Eurostat, Luxembourg.

de Groot, R., 2006. Function-analysis and valuation as a tool to assess land use conflicts in planning for sustainable, multi-functional landscapes. Landscape and Urban Planning 75, 175–186.

Dramstad, W.E., Fjellstad, W.J., Strand, G.-H., Mathiesen, H.F., Engan, G., and Stokland, J.N., 2002. Development and implementation of the Norwegian monitoring programme for agricultural landscapes. Journal of Environmental Management 64, 49–63.

Gardner, G., 1996. Shrinking Fields: Cropland Loss in a World of Eight Billion. Worldwatch Paper 131. Worldwatch Institute, Washington, DC.

Gunderson, L.H. and Holling, C.S. (Editors), 2002. Panarchy: Understanding Transformations in Human and Natural Systems. Island Press, Washington DC.

Lambin, E.F., Turner, B.L., Geist, H.J., Agbola, S.B., Angelsen, A., Bruce, J.W., Coomes, O.T., Dirzo, R., Fischer, G., Folke, C., George, P.S., Homewood, K., Imbernon, J., Leemans, R., Li, X.B., Moran, E.F., Mortimore, M., Ramakrishnan, P.S., Richards, J.F., Skanes, H., Steffen, W., Stone, G.D., Svedin, U., Veldkamp, T.A., Vogel, C., and Xu, J.C., 2001. The causes of land-use and land-cover change: moving beyond the myths. Global Environmental Change-Human and Policy Dimensions 11, 261–269.

Lant, C.L., Kraft, S.E., Beaulieu, J., Bennett, D., Loftus, T., and Nicklow, J., 2005. Using GIS-based ecological-economic modeling to evaluate policies affecting agricultural watersheds. Ecological Economics 55, 467–484.

Jordloven, 1995. Norges Lover, www.lovdata.no (in Norwegian).

OECD, 2001. Environmental indicators for agriculture. Volume 3: methods and results. Organisation for Economic Co-operation and Development, Paris. 213 pp.

Pontius, R.G., Shusas, E., and McEachern, M., 2004. Detecting important categorical land changes while accounting for persistence. Agriculture Ecosystems & Environment 101, 251–268.

Potschin, M. and Haines-Young, R., 2006. "Rio+10", sustainability science and landscape ecology. Landscape and Urban Planning 75, 162–174.

Rindfuss, R.R., Walsh, S.J., Turner, B.L., Fox, J., and Mishra, V., 2004. Developing a science of land change: challenges and methodological issues. Proceedings of the National Academy of Science USA 101, 13976–13981.

Rounsevell, M.D.A., Evans, S.P., and Bullock, P., 1999. Climate change and agricultural soils: impacts and adaptation. Climatic Change 43, 683–709.

Rounsevell, M.D.A., Ewert, F., Reginster, I., Leemans, R., and Carter, T.R., 2005. Future scenarios of European agricultural land use II. Projecting changes in cropland and grassland. Agriculture Ecosystems & Environment 107, 117–135.

Tilman, D., Cassman, K.G., Matson, P.A., Naylor, R., and Polasky, S., 2002. Agricultural sustainability and intensive production practices. Nature 418(6898), 671–677.

LANDSCAPES OF THE NATURAL PARK "VEPSSKY FOREST"

TATIANA POPOVA* AND IRINA BYCHKOVA
Research Institute of Remote Sensing Methods for Geology, St. Petersburg, Russia

Abstract. This paper focuses on capabilities of the aerial and Landsat TM data in identifying landscape types of the Natural Park "The Vepssky Forest" in Leningrad's region. The studied landscape includes 3 landscape regions, 19 subregions and a variety of elements of morphological structures of the landscapes. Utilization of the aerial photos has revealed 26 types of vegetation in reserve "The Vepssky Forest". Multiple aerial photos satellite images and topographical maps are used to locate temporal changes in XX c. in the territory of Park. The natural state of the Park made them ideal targets to test possibilities to classify landscape types in the east of Leningrad's region from satellite images in a correct way.

Keywords: Landscape; aerial photos; satellite imagery; landscape change; elements of landscape; vegetation map; bog's systems

1. Introduction

Authors investigated landscape types of the Natural Park "The Vepssky Forest". Park "Vepssky forest" (Popova and Victorov, 2003) is situated in the extreme east of Leningrad's region (Figure 1). The Natural Park occupies an area of 1,900 km^2. Territory of the Natural Park is situated in the Valdai-Onega sub province of Euro-Asian taiga area. Flora of the Park has an emigrational genesis. Boreal species compose the base of it. This is an extensive tract of southern taiga forest, interspersed with numerous lakes and bogs in landscapes formed during the last glaciations. An immensely rich flora presents many

* Corresponding Author: Dr. Tatiana Popova, Research Institute of Remote Sensing Methods for Geology, 6 Birzhevoi proezd, St. Petersburg, 199026 Russia; Phone: +7 812 3282801; Fax: +7 812 3283916; e-mail: vniikam@mail.wplus.net

I. Petrosillo et al. (eds.), Use of Landscape Sciences for the Assessment of Environmental Security, 143–154.

species typical of the north, including several rare orchids. Park includes part of drainage area of *Caspian basin* (rivers Logozerka, Melozerka, Lijma) and *Baltic basin* (river Ojat with affluents: r. Verkhnaj Kurba, Nijnaj Kurba, Tjanuksa, Atshina, Kapsha with Sarka, Genuja, headwater of Pasha and others). Park is situated within the Carboniferous Plateau. Pristine landscapes shaped by glacial action feature more than 400 lakes, which, together with a network of small rivers of the purest water, have produced a territory highly conducive for wildlife watching and scientific research.

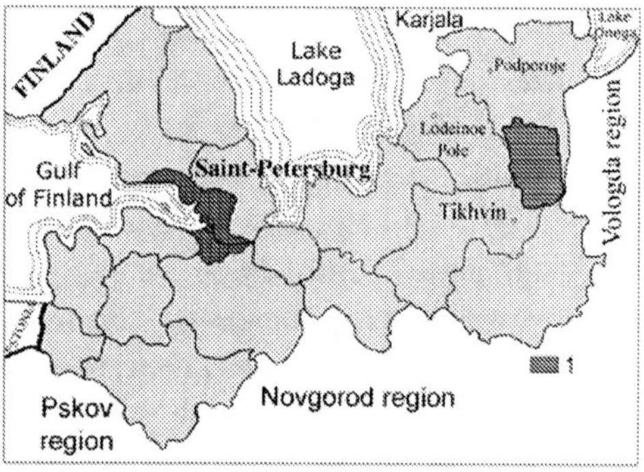

Figure 1. Park "Vepssky forest"(1) in the map of the Leningrad region.

2. Materials and methods

Authors used for landscape change studies: land surveys, topographical maps from 1847, 1940, 1980s aerial photos and satellite imagery (Popova et al., 2005). Aerial photos are particularly useful in describing land cover over the past 70 years.[1] The photos are black and white stereo images at color images. Satellite imagery presented the middle growing season and early autumn. Authors used air photos from 1937, 1977, and 1990s for identifying land cover. Satellite imagery is valuable for describing recent (last 25 years) land cover, but it is expensive to purchase images and then have them interpreted.

Studies in which Landsat imagery was the principal data base for classifying and mapping types of landscape began to appear in the literature by the mid-1980s (Nicolas et al., 1986).

The first step in studies of the landscape is the identification and mapping of the land cover types, using aerial and satellite images of the area. We used the some criteria (vegetation phase, moisture, and clouds) for identification the very best aerial and satellite image for mapping the different land cover types in the Park. The unsupervised classification of the satellite data (channels with 30 × 30 m resolution, classification with IDRISI program) set was used to locate preliminary limits for the land cover types.

The second step in the survey is the fieldwork. The accuracy of the remote sensing interpretation must be checked on site, and measurements of small features must be made. Questions of the history of land use must be ascertained. Detailed studies of the land cover types include the analysis of historical topographical maps and archives materials (Popova et al., 2003).

When the fieldwork has been completed and the preliminary map has been thoroughly checked, the final detailed land cover types map is drawn (Figure 2).

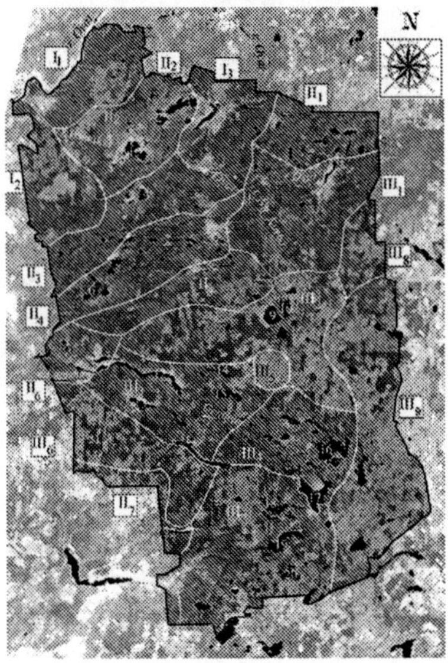

Figure 2. The map of landscapes of Natural Park "The Vepssky Forest"; I1 – number of the subregion (see in the text of Table 1), 1–20 – very big lakes of the Park (*I – 1 Hanging lakes, I – 2 Basin of the upper current of r. Sarka and left affluent of r. Tyanuksa, I – 3 lake Ozerskoe – submeridional cut of the valley of r. Oyat, II – 1. Small lakes north of lake Ladvinskoe, II – 2 Lakes Pethevskoe – Jandozero, II – 3 Sources of r. Tyanuksa – lake Sarozero, II – 4 Lakes Atchozero , II – 5 Lakes Dolgosero, II – 6 Upper current of the r. Kapscha – west point of lake Kapschozero, II – 7 Under current of the r. Uriya, III – 1 Ladvinskoe, III – 2 Lakes Kurbozero, III – 3 Lakes Kapschozero, III – 4 Lakes Lerinskoe and sources of the r. Genuya, III – 5 Carboniferous carbonate rocks, III – 6 Sources of r. Ulianitsa, III – 7 Lakes Pupozeo, III – 8 Area between the left affluents of r. Oyat, III – 9 Drenage area of the brooks of the Volga basin*).

3. Landscape division

Landscape division is the division of the Earth's surface into territorial unites. All unites are characterized by unities of natural conditions and resources. These unities are stipulated to:

- The community of historical development
- The originality of the geographical position of area

- Actual geographical factors and processes in territory
- Intimate internal territorial connections

Isolation of the landscapes submits to the zonal and azonal regularities.

Major factors in the landscape features of the Park are: geological foundation, history of the creation these territorial, actual processes. The authors divided the Park in to 3 landscape regions: Region Devonian lowland, Carbon shelf, and Carbon plateau. Within the Park we distinguished 19 subregions (Figure 2). Every subregion is characterized by peculiar relief, quaternary deposits, hydrology, vegetations, and soils (Table 1).

Region of Devonian lowland occupies small area in north-western part of Park. Lowland is transformed in sandy–clay bedrock Devonian age. Quaternary deposits presented by Lugsky Stage of Valdaisky glaciation. Region of Carboniferous shelf occupies central area of the Park. This region included the greatest number of the agricultural lands in the Park. Region of Carboniferous Plateau occupy big area in south and east of the Park. This region is characterized by the greatest number of the bogs in the Park (Figure 3). We used in this map a 2002 Landsat/TM satellite image.

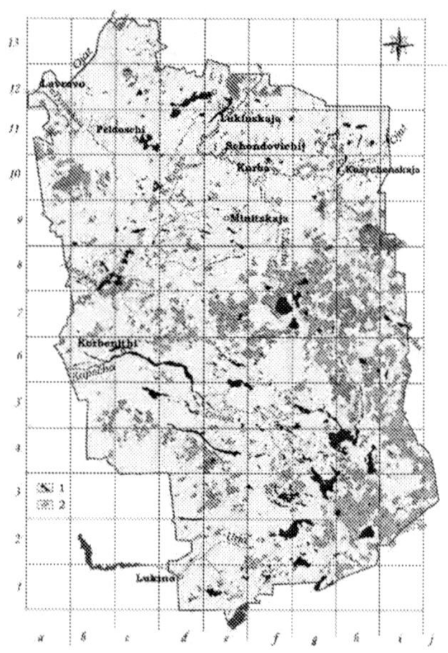

Figure 3. Spatial distribution of the bogs in the Natural Park "The Vepssky Forest" from satellite images. 1 – lakes, 2 – bogs.

TABLE 1. Landscapes regions of the Park with the description of the subregion present only.

Subregion	Relief	Dominant soils	Dominant vegetation	Cutting area	Agricultural lands
		I. Region of Devonian lowland			
I – 1	big hill-morain, h = 60–62 m (near river) and 160–180 m near east boundary	medium-podzolic soils	Spruce forest with pine	Old area, 1930s–1970s	West part, around villages Lavrovo and others
I – 3	lake-glaciation terrace, locally with hill, h = 150–160 m	superficially podzolic, slightly and medium-podzolic soils	Pine forest (by highland), spruce forest (by lowland)	Small area	Around lake Ozerskoe and village Schondovichi
		II. Region of Carboniferous shelf			
II – 1	wavy moraine terraces with hill-morain parts	medium- and slightly podzolic, peat-podzolic, humus-peat soils.	Secondary small-leaved and mixed forests	Small area in west part	Small abandoned meadows north of lake Ladvinskoe and west, north-west of lake Azmozero
II – 2	hill-morain and kames, h = 200–227 m	medium-podzolic and peat- podzolic, peat soils.	Mixed coniferous – small-leaved forests	Western part, 1980s–2000s, big area	Around lake Pethevskoe
II – 3	slightly-wavy moraine terraces, h = 180–210 m	medium-podzolic, peat-podzolic, bog-gley and peat soils	Spruce forest, birch	Central part	Around lake Sarozero Barskoe, overgrowing by birch, aspen

II – 4	lake-glaciation terrace with part of kame relief, h = 180–200 m	slightly- and superficially-podzolic, peat-podzolic and peat soils.	Pine forest, spruce forest and birch with mixed forest. Aspen present	Old logging around lake Mutnosero (early 1940)	North of Jurgozero, Abandoned lands around Mutnosero
II – 5	morain terrace, h = 180–250 m. Part of hill-morain	medium-podzolic, peat-podzolic and peat soils	Spruce forest	Central and eastern part, 1980s–1990s	Abandoned lands north of Dolgozero
II – 6	hill-morain with part of moraine loamy terraces, h = 200–250 m in hill	slightly and medium-podzolic, peat-podzolic, bog-gley and peat soils	Mixed forest with dominance of small-leaved forests	Big area, 1950s	Around lakes Kapschozero, Alekseevskoe, north of lake Novozero
II – 7	hill-morain, h = 190–240 m	medium- and slightly podzolic, peat-podzolic soils	Spruce forest	Northern area, 1980s	North-eastern of Pyalozero
III. Region of Carboniferous Plateau (properly Vepsovskaya elevation)					
III – 1	combination wavy morain terrace with slightly hill-morain lake-glaciation terrace and with fluvio-glacial cover		Pine and spruce forest	West part, around the road, 1970s–1980s	Around lake Ladvinskoe

III – 3	hill-morain and moraine terrace, h = 220–240 m.	medium-podzolic, peat-podzolic, bog-gley, and peat soils	Spruce forests, birch	East part, around the lake Sarozero, 1970s	Around lake Haraginskoe, northern coast of Kapschozero, abandoned lands north of Dolgozero
III – 4	Small hill-morain, h = 235–245 m. Great number of small lakes		Spruce forests and mixed	Old	abandoned lands north of Murmozero, around Alozero, west of Lerinskoe lake
III – 7	Small- and medium hill-morain with strong dissection	Slightly and medium-podzolic, peat-podzolic, peat and humus-peat soils	Spruce forest, aspen with spruce, aspen, birch forest. Small fragments of southern taiga forest	Big area from north to south-eastern boundary, 1970s–1990s	Near villages Syargozero, Korvala and Lukino, partly abandoned lands

4. Different landscape elements in the subregions of the Park

Every subregion of the Natural Park "The Vepssky Forest" is characterized by rather high level of a landscape variety. We will look in details subregion III_9 (Drainage area of the brooks of the Volga basin).

Bogs (raised and transitional) occupies about the third part of this subregion. In this region now are formed bogs in the lakes. Coastal–aquatic vegetation is expanded in all area of the shallow lakes (Figure 4, number 1). Region includes the bogs and lakes with pronounced boundary (due to Earth division) Figure 4, number 2). In the 1970s landscape elements in this territory have transformed due to human activities (forest exploitation). Now more of the half of forest's surface is cutting area. Detailed study of present-day vegetation (with aerial photos) has revealed 26 types of vegetation in investigated area (Figure 4) and many types of landscape morphological elements. For example, we will study only a few landscape morphological elements – bog and bog's systems. Bog is the type of hierarchical systems with three levels. Superior level is *bog's systems*, which composes from few sites. Site is an aggregation of intercorrelational micro landscapes, capable to auto-development. These *microlandscapes* are timed to one hollow bordered by terrain or lakes. Microlandscape corresponds (territorially) of the plant communities.

4.1. BOG LOUDOZERSKOE

This bog is a complex system (Figure 5). Bog is situated in area of watershed between two rivers: Vadjega and Logozerka. Bog's system consists of the two parts. Area near river is transitional mesotrophic bog. Eastern part of the bog is a raised oligotrophic bog. River Vadjega, in the north-western part of bog's system, drains this bog. Lines of drain crossed bog from north-east to south-west. Borderlands of the bogs are swampy, sphagnous, and sedge-sphagnous; locally are observed forest bogs with pine. Swampy stripe (horse-sedge-sphagnous vegetation with single shrubs of the willow) boarded this bog in northern part. Cottongrass-sphagnous bog is situated on the other side of the swampy stripe. Beyond the sedge-sphagnous stripe in southern part we look mosaic cover of cottongrass-dwarf schrub-sphagnous, crowberry-sphagnous (in hillocks), and sphagnous fragments with Scheuchzeria palustris. Sedge-sphagnous areas with cranberry are found near forest's island.

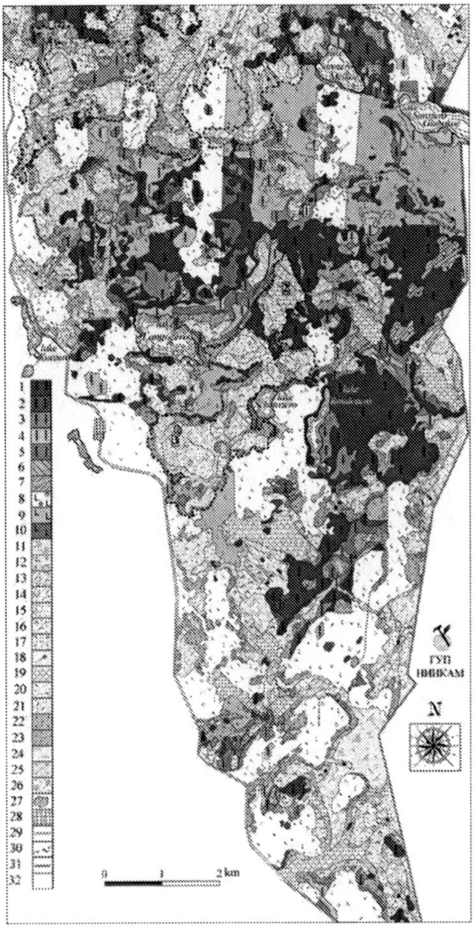

Figure 4. Vegetation map of reserve "The Vepssky Forest" (*1. Spruce green-moss forest (bilberry, grassy, short grassy, 2. Spruce green-moss – sphagnous and sphagnous forest (bilberry, sedge-horse tail), 3.Pine forests lichen – green-moss and green-moss (red bilberry – bilberry, bilberry, grassy), 4. Pine forests green-moss forest (red bilberry – bilberry, bilberry), 5. Pine sphagnous forests (cottongrass, dwarf schrub), 6. Aspen forests with pine and spruce, 7. Birch forests with pine and spruce, 8-10. Cutting area of 1965–2004, 11. Raised bogs sphagnous with short dwarf schrub, unfigured texture, 12. Sedge bog, 13. Bog with ridge-pool complex, 14. Bog with unclear ridge-pool complex, 15. Bog with ridge-small lakes complex, 16. Bogs abounding in large hillocks (with plane hillocks – hills), 17. Bogs abounding in hillocks (relatively smalls with clear boundaries), 18. Bogs swampy with pronounced line of drain, 19. Bogs swampy with unpronounced line of drain, 20. Bogs overmoistering, unfigured, 21. Bogs with windows of water, 22. Forest bogs with dense pine, 23. Forest bogs with sparse pine nanou, 24. Trembling bog, 25. Coastal–aquatic vegetation, 26. Association forest's ridge with pools, 27. Hydrographical objects (rivers, lakes), 28. Forest storage, 29. Main road for exportation of wood, 30. Country roads and paths, 31. Boundary with Vologda region, 32. Boundary of bogs and bog's systems: 1 – lake Ekchozero; 2 – bog with pronounced boundary (due to Earth division); 3 – bog Schatozero; 4 – Bog Loudozerskoe*).

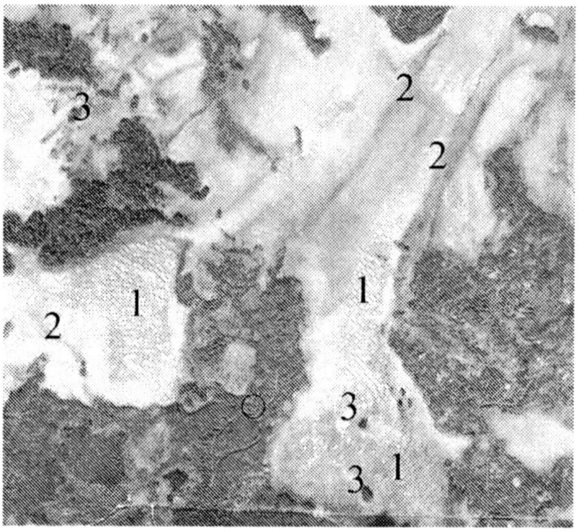

Figure 5. Bog Loudozerskoe. Aerial Image. 1 – ridge-pool complex; 2 – lines of drain; 3 – mineral islands.

Surface of this peat bog is estimated of 1,911 hectare. Average peat thickness is 1.65 m; maximal – 4.2 m. Type of the peat in the raised bog: pine-cottongrass and woody transitional sphagnous with dominant of Sphagnum magellanicum. Woody transitional and sedge-sphagnous peats occupied transitional bog. Road for exportation of wood has divided bog into two parts: northern and southern. In certain places road crossed bog by the dumps. Now here locally are observed processes of formation of secondary bogs (re-bogging): swampy with reed-mace (Typha latifolia).

5. Landscape change detection

Environment Changes in the Park depend of two groups of the factors:

1. *Natural succession processes* are gradually changes with one direction of the development. For example, age's change in vegetation covers, changes in soil's cover (accumulation of the humus), changes in micro- and mesorelief.
2. *External factors*. These changes can be in the form of the jump or gradually. Principally difference of these changes consists in velocity. External factors provoke most quick changes compared with process of the development. Changes may be sequence of the natural cataclysms (flood, storm etc.) or humanity activities (deforestation, melioration, construction, and other).

For example, Landscape changes near bog Vaguzozera in the reserve "The Vepssky Forest" from 1937s to 1993s include: deforestation in the 1980s, building of new rails road in south of this bog, change in the draining of the bog. Observations from air images indicate increased of the surface of the open water in this bog (due to changing building road).

6. Conclusion

Remote sensing data is an excellent approach to gain information over landscape division. Generalization of the result over wide ranges gives objective information which is difficult to reach and map by other means. Multiple aerial photos and satellite images are used to locate temporal changes. Satellite images are thought to provide similar views of the scene with only a few differences which are mapped. Satellite manipulation is worthless without the ground-truth information. Satellite data can be used in finding the commonness, location of each landscape morphological elements. After locating the most interesting landscape morphological elements and natural limits it is easy to make detailed fieldwork into the most appropriate natural limits to get the information about their relief, soil, hydrological objects, dominant vegetation.

In this work the changes on the landscape types were used to provide additional information for the amelioration in nature conservation in the Park. The authors recognized the need for an interdiction of the cutting the forest in subregions III-1(in the upper basin of the Ojat), III-5, III-9. The authors recommended a minimization of the anthropogenic impacts in the ecosystems in subregions III-9 for to conserve natural state of the bogs. The results could also be considered as a possible input to the environmental security/landscape issue at the global scale. In this context the case study of the registered protected area can represent the real long-term changes in the state of the environment and, thus, could be used as an important indicator of environmental security.

Acknowledgments

The authors thank for satellite images Landsat/TM Maryland University (USA) (Program Global Land Cover Facility).

References

Nicolas, M., Short, S., and Robert, W., 1986. Bleir Geomorphology from space. NASA.

Popova, T.A. and Victorov, S.V., 2003. Study of landscapes with high resolution imagery: VNIIKAM's experience. EcoSys Bd. 10, 146–151.

Popova, T.A., Berezkina, L.I., Bychkova, I.A., Leontieva, E.V., Semenova, N.N., and Schubina, M.A., 2005. Natural Park "Vepssky Forest". Vesti, SPb, 340 p.

MULTI-TEMPORAL COASTAL ZONE LANDSCAPE CHANGE DETECTION USING REMOTE SENSING IMAGERY AND IN SITU DATA

SERGEY VICTOROV*, EUGENE KILDJUSHEVSKY, LEONTINA SUKHACHEVA, AND TATIANA POPOVA
Research Institute of Remote Sensing Methods for Geology, St. Petersburg, Russia

Abstract. At present remote sensing (RS) methods are widely used for solving large amount of applied and scientific environmental problems. Among them are: assessment of the state of the marine and coastal environments, monitoring of water quality, study of water dynamics, change detection in coastal zone (CZ), study of drainage basins; detection, study and prediction of environmental disasters (floods, emergency pollution, soil erosion, landslides, etc.) through the prism of environmental security. Case studies of landscape change detection in various geographical sites located in the coastal zones of the Black Sea, the Azov Sea, the Ladoga Lake (Baltic Sea basin), and the Volga River reservoirs are presented. The focus is on erosion and accumulation processes. Both multi-temporal aerial photos and satellite imagery are used to study these phenomena during a 30 year observational period. Customized knowledge bases are widely used. They contain historical data, maps and charts. Of special interest are the examples dealing with the disputed geographical objects (Russia–Ukraine border conflict in the Azov Sea) and the dangerous situation with landslides in the Volga River reservoirs. Landscape ecology aspects are reflected when historical data and new remotely sensed data are used to analyze dynamics of coastal landscape in the vicinity of the Syas pulp factory located in the coastal zone of the largest European Lake of Ladoga – the source of drinking water for the city of St. Petersburg with its 5 million inhabitants.

Keywords: Environmental security; landscape ecology; remote sensing; coastal zone

*Corresponding Author: Professor Sergey Victorov, Research Institute of Remote Sensing Methods for Geology, 6 Birzhevoy proezd, St.Petersburg, 199034, Russia, e-mail: s1941vic@yahoo.com

I. Petrosillo et al. (eds.), Use of Landscape Sciences for the Assessment of Environmental Security, 155–164.

1. Coastal zone landscape of the Taman peninsula (Black/Azov Sea)

The Taman peninsula is washed by the waters of the Black Sea, Azov Sea, and the Kerch straight. The coastline is very much fragmented with large sand spits, islands, "limans" (local name for very shallow bays), and bays. According to historical data, there were periods when the Taman peninsula was a mere cluster of islands. These changes in landscape and coastline configuration are caused by neotectonic motions and the oscillations of the World Ocean level. According to Balabanov et al. (1988) 7,000–8,000 years ago the sea level in the Kerch straight was 9–12 m higher than now, and 6,000–7,000 ago it was 3–5 m lower. According to Porotov et al. (2004) considerable transgression started in the second half of the first millennium.

There are several peculiar features of coastal zone landscapes of the Taman peninsula.

1. The coast here is of abrasion-landslide type with wide coastal zone band.
2. There are long narrow sand spits which can change their shape and location under the influence of currents and meteorological factors. Tuzla "spit" is an example of these features (Figure 1). In historical archives at different times Tuzla was recorded as a chain of small islands, or an island, or a spit, or was not visible at all. Since 1925 Tuzla appears as an island connected with Taman peninsula with a chain of underwater shallow features. Tuzla as a geographical object recently was in the focus of a border conflict in the Azov Sea between Russia and the Ukraine.

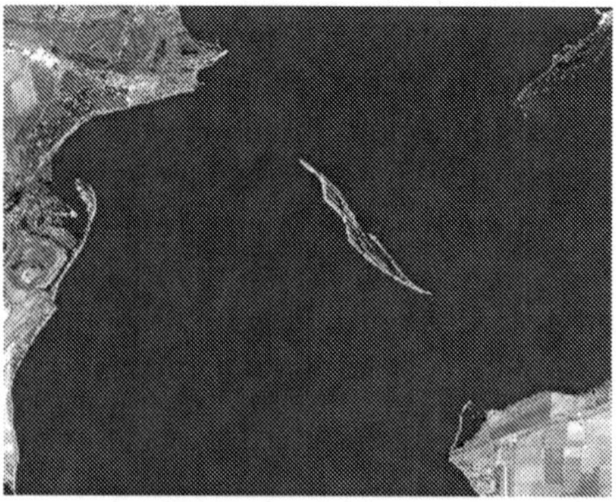

Figure 1. Tuzla spit appearing as an island in satellite Landsat TM image of 2 September 1999.

3. "Limans" (Figure 2) are the remnants of the sea bays; now many of them are isolated from the sea or connected with it by very narrow straights.
4. According to Shniukov et al. (1992) there are 42 registered mud volcanoes located within the Taman peninsula; some of them are situated on the coastal or in the coastal shallow waters (Figure 3). Usually the mud volcanoes are affiliated with the domes of anticline folds and could be used as indirect indicators for oil-bearing structures.

Figure 2. "Limans" as they appear in satellite RESURS-F/KFA (ground resolution about 10 m) image of 19 August 1985.

Figure 3. Mud volcano (low right corner) as it appeared in aerial photo of 8 August 1958.

In the context of landscape ecology and environmental security one should consider the safety of human beings living here and mention several dangerous factors affecting local population. In long term one should take into account the coastal erosion and remember the fate of ancient towns. There are historical facts dating back to the 1 century AC; according to these sources many towns were situated in the coastal area and are now at the depth of several meters. In view of the predicted World Ocean level rise (Houghton et al., 1992) many contemporary settlements here could face the same fate in the future.

Another dangerous factor is the problem of coastal landslides. And finally one should mention the eruptions of mud volcanoes as some of those are situated nearby the settlements.

2. Monitoring of landslides in the coastal zone of the Volga River artificial reservoirs

In the 1950s to 1960s a cascade of hydropower stations has been built in the valley of River Volga – the great and most well-known river of European Russia. It is a common knowledge that making large water reservoirs to accumulate water for hydropower stations is accompanied by floods of the river valleys, intensive destruction of coasts and different violations of natural conditions of adjoining territories. All these phenomena are manifested in coastal landscapes. Scientists of Research Institute of Remote Sensing Methods for Geology have been investigating reservoirs of the rivers Kama and Volga cascade (the Kuibyshevsky, the Rybinsky, the Volgogradsky, and others) on the basis of field observations and airborne remote sensing for many years. The most significant complex investigations accompanied by airborne remote sensing were carried out on the Kuibyshevsky reservoir in two phases: 1957–1966 and 1986–1990 years (Figure 4) (Bamburov et al., 1991).

As at present the hazardous landslides at this region became acute problem, we made retrospective analysis of historical experimental data and examined an archive of high-resolution satellite images in order to select images for the mentioned two periods of intensive investigations. On the basis of available ground truth and collected historical RS data and with support of relevant regionally oriented knowledge base the analytical approach for (landslides) was developed. Thus we can use this approach for detection, identification and study of landslide patterns also in contemporary satellite images, even in cases when field observations are not available (Sukhacheva and Popova, 2005).

On large reservoirs created on the basis of the river valley geological processes, like landslides, have extremely dynamic character. While filling reservoirs and during their exploitation an intensive reformation of a reservoir's

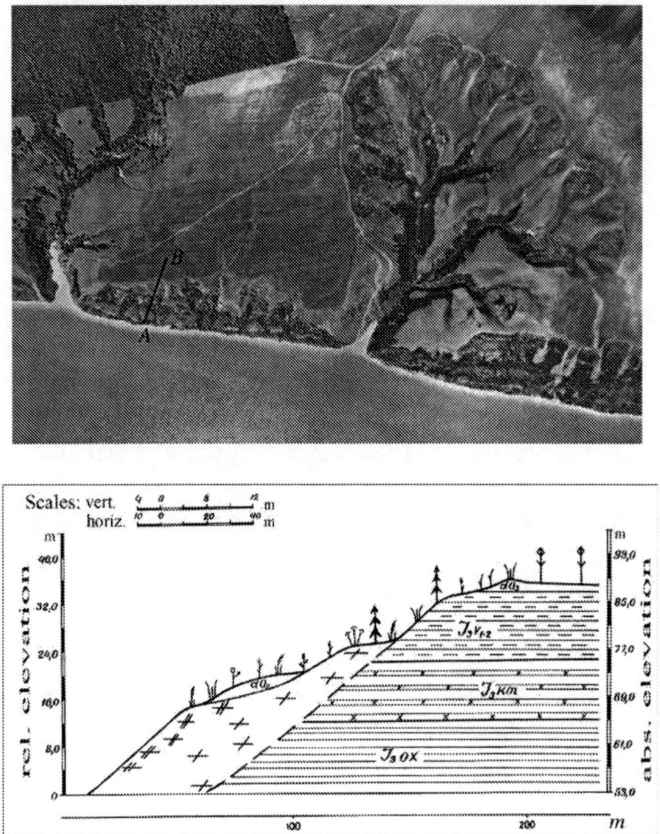

Figure 4. Fragment of right-hand coast of the Kuibyshevsky reservoir to the north from Ulianovsk: (top) airborne photo of landslide from 28 May 1963; (below) an elevation profile through the line AB.

«cup», coastal line and wide coastal band occurs. The annual seasonal oscillations of a level of water in reservoirs and a wind–wave action, prevents stabilization of processes and they are prolonged for many decades. Depending of a lithological composition and character of bedding of rock, relief type, delineation of coast, hydrogeological conditions, landscape structure, extension and type of land use, various kinds of remaking processes in coastal landscapes occur at various velocities.

Kuibyshevsky reservoir was created in October 1955 after construction of the dam nearby Zhigulevsk. The filling of the Kuibyshevsky reservoir (design topographic level of –53 m) took place in 1956. By the volume the Kuibyshevsky reservoir (58 cubic km) is the largest one among all the Volga reservoirs. Its length is 650 km, the width varies from 2 to 27 km, the mean depth is 9 m, and maximum depth is 41 m. The Volga valley here is rather asymmetric: the right coasts are steep and of high elevation, the left coasts are plane. Near Ulianovsk the height achieves 125 m. With the rising of level of water an intensive erosion of reservoir's coasts occurred, and its perimeter achieved 2,400 km. It is obvious that at such extension the geological, hydro-geological and landscape conditions essentially vary in various latitudinal sites, and also on the right and on the left coasts. So for the Kuibyshevsky reservoir different types of processes and consequently types of coasts were established. For a 30 year observational period landslide processes were not stabilized. The fresh frontal landslides, that captured coast at 200–400 m, were observed in many sites. Circus-shaped, stream-shaped, and other landslides were detected.

Analysis of the maps of landslides sites compiled on the basis of multi-temporal airborne photography and field observations allow us to make conclusion that the process of landslides formation will last at least for the next 20–30 years.

In the early 1990s our experts have made recommendations for prevention of environmental hazards, including implementation of complex monitoring (on the basis of field observations, airborne and satellite data of high resolution), protective construction, drainage and diminishing of water use for gardening at the slope of the Ulianovsk city with its two railway bridges. It is known that Ulianovsk (former Simbirsk) suffered from landslides for hundreds of years. The most disastrous year was 1785, when the landslide of Simbirsky Mountain Venets has happened. A high level of landslides activity was marked here in 1902, 1915, 1955, and 1957 (Figure 5).

Figure 5. A village after landslide event in 1957 (Rogozin, 1961).

Due to the change of sociopolitical system and economical crisis in this country the recommendations were not taken into account, and no actions were made. So the dangerous processes continue to develop. Within the urban area of Ulyanovsks City 24 km of the Volga River and 7 km of the Sviaga River slopes are in danger. On territory from settlement Sengeleia up to Undor 420 active landslides were registered. The problems related to landslides processes development became especially acute in connection with construction of the new bridge over the Volga River near Ulyanovsks City.

Our study allowed us to deduce that aerial photos were more sufficient for landslides studies in comparison to satellite imagery of LANDSAT-7 type with its "coarse" spatial resolution. However the LANDSAT-7 images (especially PAN mode with 15 m resolution) can be used for coast sites visualization and identification of relatively large landslides and extensive areas of coast erosion. And even more promising would be the use of satellite data of very high resolution of IKONOS and Quick Bird type.

For effective management of Volga valley reservoirs and coastal zone landscape protection the following measures are to be undertaken: (a) restoration of network of in situ observations in dangerous sites of reservoirs coasts, (b) Development of region- and subject-oriented database, (c) Implementation of methods of remote sensing of exogenous processes in reservoirs on the basis of modern very high resolution satellite images, (d) Integration of field and remote sensing data using GIS technology, (e) Development of measures for coasts defense, (g) Improvement of legislation (including legislation for coastal zones and hydro-technical constructions protection).

3. Landscape change detection in coastal zone in the vicinity of Syas pulp factory

Lake Ladoga is the largest freshwater lake in Europe and among the largest lakes of the world. Lake Ladoga is the unique nature object and the main source for drinking and industrial water for St. Petersburg region and for the Karelia republic. The main parameters of Lake Ladoga are: length – 220 km, width – 83 km, surface area – 17,680 km^2, mean depth – nearly 50 m (maximum – 230 m), water volume – nearly 980 km^3. Its basin has a complicated system that includes the water catchment areas of the Lakes Onega, Il'men and Saimaa. The water catchment of the Ladoga has in general the area about 260,000 km^2 and extends from the north to the south at more than 1,000 km, and from west to east at almost 600 km. The 32 rivers, with size longer that 10 km, fall into the Lake Ladoga, and only one – River Neva runs out. The largest Ladoga inflows are: River Svir, River Vuoksa, River Volkhov, and River Syas. The time required for renewal of the water in Lake Ladoga is 11 years; this indicates that the ecosystem is rather conservative. River discharge accounts for 86% of water balance input, which implies that catchment processes have a major influence on water quality. Ladoga is one of the northernmost of the world's large lakes with a cold-temperate climate type of ecosystem.

Changes in landscape of the CZ of Lake Ladoga that occurred during the last decades, influenced by anthropogenic impact on environment due to the resort infrastructure extension and due to other economic factors, were detected on the basis of satellite imagery. From the infrared satellite data sets the duration and spatial variability of thermal front and upwelling events have been examined. While the radar imagery was used mainly for Ladoga ice parameters and water dynamics features study.

In the context of landscape ecology and the state of the coastal and lake environments the Syas pulp factory has been in the focus of public awareness and severe discussions for several years as a source of water contamination. The Syas pulp factory was founded in 1923 on the coast of River Valgomki (a tribute of River Syas) at a distance of 4 km from the Ladoga coast. Industrial waste waters with suspended substances contaminated with sulfate ions, phosphorus, and nitrogen and oil products were transported to the lake through a collector. It crosses the lake plain with a series of coastal swells and bogs, the Novo-Ladoga channel and appears at the shore of the Ladoga Lake (Figure 6).

Aerial observations of the six-tube collector outlet accompanied by in situ field measurements were carried out in the period from 1977 to 1991. The landscape features around the outlet were registered in aerial imagery at the scale 1:10,000 and 1:5,000 more than 50 times from the end of May to the end of December. Coastal swells, inter-swell low areas and beaches were detected. Coastal vegetation has been also studied in detail.

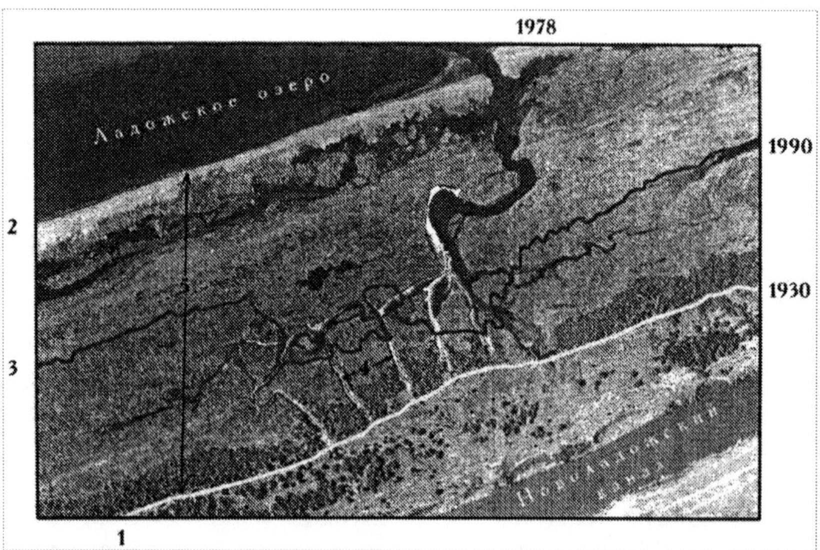

Figure 6. Aerial photo of 1978 showing the outlet of the Syas pulp factory into the Ladoga Lake (top). For explanations see the text.

Figure 6 shows the change of location of coastline in the vicinity of collector outlet. In 1930 (position of coastline is shown as white line 1-1-1930) the outlet was under the water. Photo registered the location of coastline in 1978 (marked as 2-2-1978). Curve black line (3-3-1990) in between shows coastline as it appeared in 1990. The position of coastline is affected by the annual level of the Ladoga Lake and local winds and currents. Thus the coastal band about 400 m wide was influenced by these factors during several decades. In this period the sediments from the pulp factory waste waters changed the composition of coastal soils and vegetation in this band. The plume of contaminated waters was detected in the coastal area in aerial and satellite imagery taken in visible and thermal infrared channels. It was suggested that the change of coastal landscape enhanced the migration of contaminants into the Lake and their redistribution in the coastal soils.

4. Conclusion

Case studies of landscape change detection in various geographical sites located in the coastal zones of the Black Sea, the Azov Sea, the Ladoga Lake, and the Volga River reservoirs proved that the remotely sensed satellite imagery is a powerful source of information for timely landscape change study. These data should be complemented with detailed aerial imagery and in situ archived data sets to provide reliable and nonambiguous assessments of the state of the environment in the fragile coastal zone using landscape features as indicators of natural and man-made processes. The results of our case studies can foster environmental security because they highlight situations that require attention of the proper federal, regional, and local authorities.

References

Balabanov, I.P. and Izmailov, Y.A., 1988. Change of level and hydro chemical regimes of the Black Sea and Azov Sea during the recent 20,000 years. Water Resources 6, 54–62 (In Russian).

Bamburov, I.S., Bespalyi, V.G., Viculov, A.I., Zubenco, F.S., Popova, T.A. et al., 1991. Dynamics of landscapes in a zone of influence of Kuibyshevsky reservoir. Science, St. Petersburg, 222 pp. (In Russian).

Houghton, J.T., Callander, B.A., and Varney, S.K. (Editors), 1992. Climate change 1992. The Suppementary Report to the IPCC Scientific Assessment. Cambridge University Press, Cambridge, 200 pp.

Porotov, A.V., Gorlov, Y.V., Yanina, T.A., and Fuash, E., 2004. Peculiar features of development of The Black Sea coast of Taman peninsula in the late Holocene. Geomorphology 4, 63–77 (In Russian).

Rogozin, I.S. (Editor), 1961. Landslides of Ulyanovsk and experience of struggle against them. Academy of Science of USSR. Moscow, 166 pp. (In Russian).

Shniukov, E.F., Gnashenko, G.I., Nesterovsky, V.A., and Gnatenko, O.V., 1992. Mud volcanoes of Kerch-Taman region. Naukova Dumka, Kiev, 200 pp. (In Russian).

Sukhacheva, L.L. and Popova, T.A., 2005. Study of landslides using multitemporal remotely sensed data and field observations: a case of reservoirs created on the basis of the Volga River Valley. The Abstract Volume of the 15th Stockholm Water Symposium "Drainage Basin Management – Hard and Soft Solutions in Regional Development", August 21–27, pp. 299–300.

LANDSCAPE CHARACTER AS A FRAMEWORK FOR THE ASSESSMENT OF ENVIRONMENTAL CHANGE

ROY HAINES-YOUNG*, TOBIAS LANGANKE, AND MARION POTSCHIN
Centre for Environmental Management, School of Geography, University of Nottingham

Abstract. In the context of widespread changes in European rural landscapes we underline the importance of considering threats to landscape functions relating to a sense of place, exemplified using the concept of landscape character. Illustrating our argument with examples from the English CQC (Countryside Quality Counts) Project we strongly suggest to move "beyond data" in the strategic assessment of environmental change. Supplementing data on the extent and quantity of changes with contextual information against which to judge "whether these changes matter" in a particular location is vital for the practical use of change data in policy support and environmental assessments.

Keywords: Landscape character; land use change; environmental assessment; joint character area

1. Introduction

Rural landscapes in Europe are changing due to a combination of complex cultural, economic, environmental and social drivers that act at a number of spatial and temporal scales (Palang et al., 2004; Westhoek et al., 2006). This trend reflects the rapid and extensive changes in global land cover and associated degradation of ecosystem services at global scales, highlighted by the Millennium Ecosystem Assessment (2005) and Project *Global Land*

* Corresponding Author: Professor Roy Haines-Young, School of Geography, NG7 2RD University Park, Nottingham, England; e-mail: Roy.Haines-Young@nottingham.ac.uk

I. Petrosillo et al. (eds.), Use of Landscape Sciences for the Assessment of Environmental Security, 165–174.

Project.[†] The changes faced by many rural areas in Europe often manifest themselves through agricultural intensification in favourable areas and the abandonment of traditional land use practices in more marginal situations (Verburg et al., 2006). In other areas development pressure transforms the landscape so that the traditional distinction between rural and urban is often blurred. Such trends lead not only to a loss in biodiversity (Reidsma et al., 2006), but also to a loss in the diversity and distinctiveness of the cultural landscape.

In the context of environmental security, the current rate of land cover and land use change therefore potentially threatens key ecosystem and landscape values. In this paper we consider the issue of local distinctiveness, and argue that the loss of "sense of place" that results from erosion of landscape character needs to be considered as part of our wider monitoring strategies (cf. European Landscape Convention (Council of Europe, 2000); Wascher, 2005). Local distinctiveness not only reflects the rich historical and cultural diversity of Europe, but also, with increasing globalisation of economies, constitutes a resource that can contribute to directly improve people's well-being by helping, for example, to "market" different localities and their associated products in the context of tourism or the local and regional labelling of food and other goods (Moore-Colyer and Scott, 2005).

While change-detection methods based on EO (Earth-Observation)-data are being developed and tested at a range of spatial and temporal scales, the spatial and socio-cultural context of these changes is hardly ever considered – even though we need such information if we are to judge whether such changes matter or not. For example, woodland loss or gain can have very different impacts depending on *where* that change is occurring. In this paper we therefore argue that for any strategic environmental assessment, change information *as well* as contextual data is vital for decision-making (Warren, 2002). While many of the contributions in this book explore and describe where and how much change is occurring, we assert that as a scientific community we must not loose sight of the question of how to determine whether those changes matter in a specific geographical context. In this paper we show how the analysis of landscape character can help us make progress in this important area.

2. Landscape character assessment

The potential that recent advances in data capture, storage and analysis hold for policy development and appraisal have often not been realised in terms of

[†] http://www.globallandproject.org/ and http://www.glp.colostate.edu/report_53.pdf

improving the quality of decision-making. The problem, we suggest, is not necessarily due to the timeliness or quality of the data products themselves, but rather to the lack of any systematic understanding of the contexts in which the significance of change (or lack of it) can be assessed. Although Warren (2002) has stressed the importance of understanding the spatial, temporal, economic, environmental and cultural context information for scientific studies on land degradation, the issue of context has largely been overlooked in the recent scientific literature.

In order to illustrate how questions of change detection and the understanding of context can jointly be addressed, we describe a case study based on the Countryside Quality Counts (CQC) Project, which has been undertaken for Countryside Agency in England. The project aims to identify how and where landscape or countryside character is changing, and to assess the significance of such changes in relation to the qualities that give the different landscapes of England their distinctive local identities.

Although the UK Government has long recognised the importance of understanding the nature of countryside change, the 2000 *Rural White Paper for England* (Department of the Environment, Transport and the Regions, 2000) recognised that still more needed to be done. In particular, this policy document stressed the importance of future monitoring, and made a commitment to publish an indicator of change in countryside quality that would take account of aspects such as biodiversity, heritage, and the overall character of the landscape. The case for such an indicator was based on the belief that the linkage between people and their environment needed to be more clearly identified, so that future social, economic, and environmental goals become more closely aligned.

The Countryside Agency took up the task of developing this indicator through the CQC Project. At the outset it was recognised that many different ideas were embedded in the original concept of "countryside quality" as promoted in the *Rural White Paper*, but that a key aspect was the notion of local distinctiveness. The subsequent work built on the earlier character areas initiative promoted by the Countryside Agency, which had resulted in the mapping and description of the "character areas" of England.[‡]

The CQC project started from the assumption that since the distinctive properties of the landscape character in each area could be described in terms of patterns and qualities of the elements, such as woodland, boundaries, agricultural land cover, and settlement patterns, then the change in landscape character over time ought potentially to be detectable. The advantage of combining notions of

[‡] http://www.landscapecharacter.org.uk/

landscape character with the measurement of change was that the significance of change in a landscape element could be judged against the criteria of whether local distinctiveness was being maintained or transformed. For administrative reasons the character areas of England have come to be known as "Joint Character Areas" (JCAs). The CQC project assembled a range of national data sets that could be disaggregated to at least JCA level that could be used to detect change in the elements that contribute to character. The base-line for these data was 1990, and the first period of assessment was for the period 1990–1998. Current work seeks to update the assessment for the period 1999–2003. The work showed that for the first assessment period, about 40% of the JCA were stable or showed changes that were consistent with either maintaining or strengthening their character (Figure 1). By contrast about 23% showed marked patterns of change that were transforming or weakening the characteristics that made them distinctive. The remaining JCAs showed change that was less pronounced but which were nevertheless inconsistent with our understanding of their traditional character (Haines-Young et al., 2004).

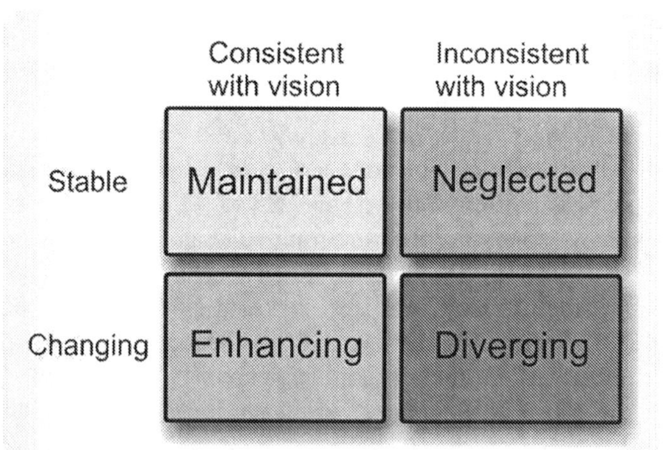

Figure 1. CQC Indicator framework which describes status and trend for each character areas of England.

The methodology used for the assessment is best illustrated by reference to some example assessments. The Cumbria High Fells, for example, is a JCA that makes up a large part of one of England's National Parks. Its distinctive qualities result from its upland character, and include the extensive tracts of unimproved rough grazing land in the higher areas, with semi-improved and improved pasture and rectilinear fields in the valleys. There are relatively few trees on the exposed higher land, but in more sheltered sites there are extensive areas of ancient, semi-natural broadleaved, mixed, and conifer woodlands. Settlement density is low throughout. Analysis of the national data-sets that were available for the area suggested that change in the elements that produced its strong sense of place were limited, and so the overall conclusion was that character was maintained.

The Mid-Somerset Hills is a distinctive more elevated area of countryside in SW England, which is predominately pastoral in character. Woodlands dominated by ash and maple are a common feature on the ridge tops and steeper side slopes. Analysis of national datasets for this JCA showed that between 1990 and 1998, significant areas of grassland appeared to have been converted to other agricultural cover types, and uptake of management agreements that would have improved the quality of woodlands or increased woodland cover had been limited. The area was therefore assessed as showing marked changes inconsistent with maintaining the existing character of the area.

For the second CQC assessment the robustness of the contextual information that is used to define character and provide the basis of the judgements made about the significance of change, were improved by an extensive consultation exercise. Drawing upon a range of sources, a set of statements describing the types of change that would help maintain and strengthen the character of each JCA or sere to weaken or transform it, was constructed, and these were tested by exposing them for comment to a range of landscape, conservation, and planning professionals using a web-based consultation tool. Consultees were asked to confirm for the JCAs that they were familiar with, the current and future relevance of each proposition, any modifications to them that were necessary to capture the "situation on the ground", and to add new statements to cover any issues that have been overlooked when the original set of statements had been constructed. Respondents also had to identify the evidence on which they based their comments. As a result, for the second CQC assessment we have an refined and better tested "vision" for the character of each area, and thus a more robust and acceptable template against which the significance of change in the elements that make up to landscape can be judged. For the update, the terminology used to describe what is happening to each JCA has been modified, and will reflect the nature of the judgement made in to first phase of the project.

Thus on the basis of the changes observed, JCAs will be described as having been either "maintained", "enhanced" if they are stable or showing changes that strengthen their overall character, or "degraded" or "diverging" if they fail to show changes that would redress previous loss of character or it they show changes that continue to erode it further (Figure 2).

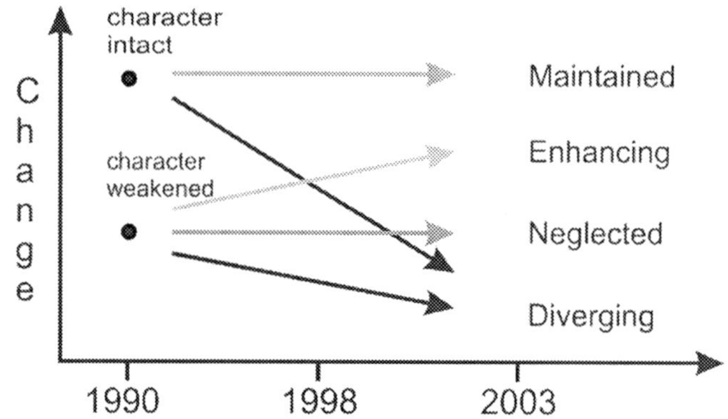

Figure 2. Concept behind assigning the CQC Indicator to JCA.

3. Contextual data for environmental assessment

Although designed to address a particular national issue, the CQC project is of interest more generally because it shows that an effective monitoring programme has to go "beyond data" to include the systematic construction of a contextual framework that can be used to make judgements about the significance of any changes identified by the monitoring process. The study shows how the notion of evidence can be expanded to include both a set of quantitative indicators and the more qualitative values or visions that people bring to the table when confronted with the question of whether a particular environmental change matters or not. The approach is, we suggest, relevant to a wider range of applications beyond the assessment of landscape character, and could be used to help monitor environmental security issues more widely. This is particularly the case where the consultation process is used to identify and quantify the limits or thresholds of acceptable or desirable change for the environmental parameters under consideration. A shortcoming of the present CQC Project was the fact that consultation was limited to "professionals". The expansion of the consultation exercise to include a wide range of publics would clearly enable a richer and more nuanced assessment to be made. The goal of such work should not be to prescribe what kinds of change are appropriate or acceptable, but to provide a more systematic body of evidence that can inform debate about the implications of environmental change for different groups within society.

The need to go "beyond data" to the "construction and identification of context" can be illustrated by reference to the recent development of a number of policy support tools. Methodologies for Strategic Environmental Assessments (SEA) are, for example, somewhat similar in character to Environmental Impact

Assessments (EIA) except that they are meant to ensure that the environmental implications of decisions are taken into account in a more strategic way. They also emphasise the need to look for alternatives, ensure early participation of stakeholders in decision-making, and subsequent monitoring and biodiversity protection activities (Sheate et al., 2005). SEAs have been required since 2004 under EU Directive 2001/42/EC (EC, 2003), and are currently being implemented in the member states. A key issue is to find ways of monitoring and assessing the significance of the environmental effects a given implementation of plan or programme. The SEA-Directive stresses the importance of using sufficient baseline information to provide the basis for predicting and monitoring environmental effects. In the UK, SEA guidelines (see ODPM, 2005) suggest a number of questions that need to be answered using such data for each selected indicator, including:

- *How good or bad is the current situation? Do trends show that it is getting better or worse?*
- *How far is the current situation from any established thresholds or targets?*
- *Are particularly sensitive or important elements of the receiving environment affected, e.g. vulnerable social groups, non-renewable resources, endangered species, rare habitats?*
- *Are the problems reversible or irreversible, permanent or temporary?*
- *How difficult would it be to offset or remedy any damage?*
- *Have there been significant cumulative or synergistic effects over time? Are there expected to be such effects in the future?*

Although quantitative data produced by monitoring programmes can help to answer such questions, "data" alone are clearly insufficient to resolve them fully. All of the questions noted above imply some understanding of the visions or values that different people or groups bring to the assessment of change, and so require that the evidence base be expanded to include such intelligence. Similar points could be made in relation to other decision support tools such as sustainability appraisals, EIA and Quality of Life Capital Assessment (Potschin and Haines-Young, 2003).

The kind of contextual information generated by CQC can also be of assistance in developing targeting strategies for policy. For example, following moves in the EU to decouple farm subsidies from production incentives, agri-environmental schemes have been implemented to deliver a range of "environmental goods". In England guidance has been published to encourage land managers to select scheme options that are appropriate to the environmental needs of their local area. Contextual information of the kind assembled by CQC can be used to further refine these guidelines, and can provide a framework for monitoring the extent to which the schemes have been successful in maintaining of restoring particular environmental qualities.

As these examples demonstrate, key component of the contextual frameworks that need to be developed to fully utilise SIA and other policy targeting and monitoring tools, is a better understanding of the concept of the limits of acceptable or desirable change. This assertion is supported by the fact that questions about environmental limits, and their implications for policies related to natural or environmental resource protection, have emerged as an important focus in recent discussions of how the goals of sustainable development might be achieved (Haines-Young et al., 2006) Following publication of the Millennium Ecosystem Assessment (2005), for example, it now recognised that not only do we need to view ecosystems in terms of the range of benefits to people, but also to better understand how pressures, such as pollution or overuse, may impact upon them and diminish the level or quality of the benefits that they provide. A review of the recent literature on limits and thresholds suggests that while the definition of an environmental limit may be based on the biophysical properties of a natural resource system, its identification also depends on the way people value the outputs from it. Increasingly an important aspect of the contextual information that we require for effective policy development and appraisal, is to understand how people and groups in society make the judgment that that given the scale of actual or potential environmental change, a critical point has been reached and that the reduction in benefit derived from natural resource systems is no longer acceptable or tolerable. It is likely that future work on constructing the kinds of conceptual framework we need to implement SIA and other assessment tools will grow out of a better understanding of notions of ecological integrity, and resilience, and the capacity of ecosystems and indeed whole landscapes to absorb change.

4. Implications for environmental assessment

While the concept of environmental security covers a very wide variety of issues, as this publication emphasises, the threat to ecosystem goods and services and the support they provide for the well-being of people is an important focus for current work. In order to take this research agenda forward, an understanding of the link between land cover and land use change and their impacts on ecosystem goods and services is now pressing.

In general the literature suggests that we have a good conceptual framework for describing and understanding the processes of land cover and land use change. A number of the chapters in this book show that in the scientific realm we are able, for example, to measure, document, monitor and model many aspects of these changes. By contrast, in the policy arena it seems that much less progress has been made. It is apparent from the literature that studies often overlook or lack the kinds of information that would allow us to go beyond merely monitoring change to making judgements about its wider significance.

The CQC Project illustrates the kinds of additional information that we require to undertake effective policy development and appraisal more generally. The description of landscape or countryside character in England has been used in this work as a body of contextual information that helps us understand what gives different landscapes their "sense of place". As a result, we can take the outcomes of monitoring activities that focus on the individual elements of landscape and make more holistic judgments about the changes they exhibit based on the implications for maintaining local distinctiveness. Following Hamilton and Selman (2005) the study illustrates how it is possible to blend national datasets with local knowledge and thus help achieve an approach to rural policy delivery that is less strictly linked to administrative boundaries and is more sensitive to countryside character, natural dynamics and time depth of landscapes and communities. The approach can be applied generally, and could also be relevant, for example, in understanding the implications of land cover change at European scales using the dominant land classes suggested by the European Environment Agency (EEA, 2006) as part of its environmental accounting initiative.

We acknowledge, however, that landscape or countryside character is but one environmental issue amongst many that we need to consider in relation to discussions about environmental security. Nevertheless, the experience this case study provides is important because it illustrates that if systematic contextual information is available, it is possible to go beyond reporting that environmental change has occurred to explain to people why such changes matter. This work suggests that the development of concepts and tools to help us understand how people view and value environmental or ecological integrity at different geographical and temporal scales is an essential focus for future research.

Acknowledgements

The CQC project (http://www.cqc.org.uk/) is funded by the Countryside Agency Landscape, Access and Recreation division, in partnership with English Nature, Defra RDS, English Heritage and Defra.

References

Council of Europe, 2000. The European Landscape Convention. http://www.coe.int/t/e/Cultural_Co-operation/Environment/Landscape/

Department of the Environment, Transport and the Regions, 2000. Our countryside: the future. A Fair Deal for Rural England, 176 pp.

European Commission (EU), 2003. Implementation of Directive 2001/42 on the Assessment of the Effects of Certain Plans and Programmes on the Environment. http://ec.europa.eu/comm/environment/eia/030923_sea_guidance.pdf

European Environment Agency, 2006. Land Accounts for Europe, 1990–2000. http://reports.eea.europa.eu/eea_report_2006_11/en

Haines-Young, R., Potschin, M., and Cheshire, D., 2006. Defining and identifying environmental limits for sustainable development. A Scoping Study. Final report to Defra, project code NR0102, overview report, 49 pp.

Haines-Young, R.H., Swanwick, C., Martin, J., and Tantram, D., 2004. Countryside Quality Counts. Final report, countryside agency.

Hamilton, K. and Selman, P., 2005. The "Landscape Scale" in planning: recent experience of bio-geographic planning units in Britain. Landscape Research 30, 549–558.

Millennium Ecosystem Assessment, 2005. Ecosystems and Human Well-being: Synthesis. Island Press, Washington, DC.

Moore-Colyer, R. and Scott, A., 2005. What kind of landscape do we want? Past, present and future perspectives. Landscape Research 30, 501–523.

Office of the Deputy Prime Minister (ODPM), 2005. A Practical Guide to the Strategic Environmental Assessment Directive, 110 pp.

Palang, H., Sooväli, H., Antrop, M., and Stetten, G. (Editors), 2004. European Rural Landscapes: Peristence and Change in a Globalising Environment. Kluwer, Dordrecht, 482 pp.

Potschin, M. and Haines-Young, R., 2003. Improving the quality of environmental assessments using the concept of natural capital: a case study from southern Germany. Landscape and Urban Planning 63, 93–108.

Reidsma, P., Tekelenburg, T., Van den Berg, M., and Alkemade, R., 2006. Impacts of land-use change on biodiversity: an assessment of agricultural biodiversity in the European Union. Agriculture, Ecosystems and Environment 114, 86–102.

Sheate, W., Byron, H., Dagg, S., and Cooper, L., 2005. The Relationship between the EIA and SEA Directives. Final Report to the European Commission. Imperial College London Consultants.

Verburg, P.H., Schulp, C.J.E., Witte, N., and Veldkamp, A., 2006. Downscaling of land use scenarios to assess the dynamics of European landscapes. Agriculture, Ecosystems and Environment 114, 39–56.

Westhoek, H.J., Van den Berg, M., and Bakkes, J.A., 2006. Scenario development to explore the future of Europe's rural areas. Agriculture, Ecosystems and Environment 114, 7–20.

Warren, A., 2002. Land degradation is contextual. Land Degradation & Development 13, 449–459.

Wascher, D.M. (Editor), 2005. European Landscape Character Areas – Typologies, Cartography and Indicators for the Assessment of Sustainable Landscapes. Final Project Report as deliverable from the EU's Accompanying Measure project European Landscape Character Assessment Initiative (ELCAI), funded under the 5th Framework Programme on Energy, Environment and Sustainable Development (4.2.2), p. 150.

PART IV INTRODUCTION

INTEGRATED STUDIES OF CATCHMENTS AND BASINS

K. BRUCE JONES
US Geological Survey, Geography Discipline, Reston, Virginia, USA

Understanding key functions of catchments (watersheds) and basins, as well as their status and trends, are critical in understanding environmental security for any given area or region. Catchments and basins provide a number of valuable environmental goods and services, including clean water for consumptive use and recreation, and habitat for a wide range of commercial important species. Moreover, healthy catchments and basins help mitigate environmental variability and extremes. For example, healthy catchments help increase environmental security by reducing flooding associated with broad-scale precipitation events. Healthy catchments reduce energy of incoming precipitation (through interception by vegetation) and through infiltration (pervious surfaces and soils). They also provide filtering capacity via wetlands and riparian zones of sediment and potential pollutants. As such, catchments integrate several important aspects of the atmosphere, land, and patterns of human-uses of the land. A comprehensive understanding of these interrelations is important in protecting and enhancing water-related goods and services, as well as to identify how to reduce risks associated with environmental extremes and hazards, including those associated with global climate change.

Another important aspect of integrated studies of catchments and basins, and their relative level of environmental security, is the degree to which a catchments or basins are interrelated. For example, to what degree is the condition of a particular catchment dependent upon the conditions of adjacent or upstream catchments? Answers to these questions are important in determining the kinds of management and policies that need to be developed and implemented. This is why it is often necessary to bring together management authorities and stakeholders from across entire basins.

In this section, many of the important issues related to catchment and basin assessments are addressed. The first paper discusses important linkages between catchment hydrology, important landscape features, and stream habitat and biota. The second paper links landscape features and patterns with water quality and potential flood hazards, and discusses how landscape change can influence water quality and flooding risks. The third paper links land cover characteristics and pattern to water quality through a spatially distributed hydrologic model. This approach is critical in identifying sources of potential impairments to water quality, which in turn, offer potential for mitigation through comprehensive land management strategies. The fourth paper uses an alternative landscape futures approach to integrate biophysical data into a decision support framework. The result is a set of spatially explicit, landscape change options that will improve environmental security of water resources. The last paper demonstrates a multi-scaled landscape assessment approach for Europe and identifies key core data gaps and needs. Such an approach is needed to identify and prioritize potential environmental risks over broad areas.

AN ECOHYDROLOGICAL APPROACH FOR THE PROTECTION AND ENHANCEMENT OF ECOSYSTEM SERVICES

KINGA KRAUZE*
International Centre for Ecology, Polish Academy of Sciences, Poland

IWONA WAGNER
Department of Applied Ecology, University of Łódź, Poland

Abstract. Water is the only factor linking all ecosystem services – provisional, supporting, regulatory, and cultural. For this reason water resources are highly vulnerable to human pressures and become a cause of environmental insecurity in many regions of the world. The methods of coping with water-related problems were built on the conviction that the risk of losing water driven services can be, to a great extent, anticipated and diminished by policy and technical and technological measures. However, considering the number of factors regulating accessibility and quality of water, the risks associated with water scarcity should be considered as having higher damage potential, persistency, ubiquity and irreversibility, especially under increasing climate variability, than previously assumed. Handling risk is problematic also due to high uncertainty and severe outcomes associated with rising water needs, conflicting demands, and low awareness among citizens. Many important issues affecting water management are often rooted in the past, e.g. poverty, dramatic population growth, industrialization, land transformations, inefficient policy and overengineering. Unknown risk, damages difficult to assess, and low capacity for worldwide introduction of top, but expensive technologies led back to questions concerning a potential for using natural mechanisms to maintain ecosystem resilience and to mitigate human activities, which are handicapping. A proposed ecohydrology approach scrutinizes the interrelation between components of river catchments, focusing especially on those between biota

* Corresponding Author: Dr. Kinga Krauze, International Centre for Ecology, Polish Academy of Sciences, 3 Tylna Str. 90-364 Łódź, Poland, e-mail: kingak@biol.uni.lodz.pl

I. Petrosillo et al. (eds.), Use of Landscape Sciences for the Assessment of Environmental Security, 177–207.

and hydrology. It suggests using these relations for increasing an adaptive capacity of ecosystems and thus the stability and security of ecosystem services.

Keywords: Environmental risk; ecosystem services; ecohydrology; ecosystem functions; biotic structure

1. Introduction

Freshwater abundance, distribution, and characteristics are a result of climatic and geological history, and morphology of the region. Recent decades, however, provided numerous examples of increasing human related perturbations, driven by local, regional, and global socio-economic factors. These perturbations have reshaped ecosystem functions and structure, enabling invasions of exotic species and eventually leading to emerge new, alien species dominated ecosystems (Hobbs et al., 2006). Human intervention is so strong, ubiquitous, and multi-dimensional that it is presently capable of modifying global environments with consequences comparable to those resulting from the natural Earth forces. Once the process began, we entered a new era – Anthropocene (Meybeck, 2003). Human goals, challenges, and community development are reflected by a land-scape structure. Economic, demographic, and social factors transform land cover, change intensity of geological, biochemical, and hydrological processes and affect biodiversity. The pressure is then transferred form terrestrial systems, the donors of matter and chemicals, to freshwater systems, the recipients. The consequence of a highly transformed landscape is therefore degradation of freshwater ecosystems (including peat bogs and marshes) and often, an uncon-trolled water surplus (floods) or scarcity (a result of droughts and poor water quality). Deprivation of this basic resource threatens millions of people and has triggered a number of conflicts (Gleick, 2006) and people migrations (Salehyan, 2005; World Water Council, 2006) all over the world. For decades, the commonly applied solutions for increasing water security have come from engineering, technology and finally, policy. The common belief became that many extremes (rainfalls, droughts, floods) can be anticipated, and their consequences can be prevented or diminished, with a technical approach. Those measures proved their effectiveness under normal conditions, with a high naturalness of landscape and steady landscape and ecosystem processes. However, under increasing climate variability, demographic growth, land use changes and remnants of past improper resource management and exploitation, these measures became questionable. They appeared to be unable to deal with increasing uncertainty of appearance, scale, reversibility, and damage potential of water-related risks.

Variety factors defining water quality and quantity, and an unpredictable behaviour of ecosystems, which have been experiencing for centuries, different forms of impact, do not favor the implementation of routine procedures (based on existing at national level standards, laws and regulations). Referring to the German Advisory Council on Climate Change's classification of risks and risk mitigation strategies (German Advisory Council on Global Change, 1998), we shifted from an area of "normal" risk to ecosystem function to a "transitional" one, characterized with limited ability to anticipate or counteract environmental extremes. Consequently the new standards should impose a reduction of damage potential carried by human activities, prevention of damage/impact occurrence (if possible) – including existing best practices, and strengthening of system (ecosystem, catchment) resilience against possible disaster. As long as the first two tasks remain linked to policy and technical measures, the last one refers to using and reinforcing ecosystem supporting and regulating services. The first group includes soil formation, nutrient and water cycling, and primary production, and plays a major role in defining potential for other services, including regulatory ones: climate and flow regulation, disease resistance and self-purification (MEA, 2005). The interlink between soil formation, nutrient and water flow and primary production, as well as catchment hydrology as a driving force for plant cover and primary production, were considered under the ecohydrology approach. It proposed the synergism between biota and hydrology for regulation of ecosystem functions and landscape processes, thus integrating these mechanisms for efficient risk management (Zalewski, 2000).

2. Resilience from catchment perspective

Resilience refers to the ability of a system to retain its oscillations within the boundaries (defined as steady state), despite disturbances. When disturbances force the system beyond its boundaries, it responds with changes of structure, functions, and behaviour affecting the ability and rate of delivering goods and services to society. Hence, the main problem of environmental management and security is recognizing the threshold to ecosystem resilience (adaptive capacity) before it is exceeded, and preventing its collapse, in order to secure services delivery. Although it is still a challenge at an ecosystem level, much progress has been made at catchment scale. Since the catchment can be seen as a system of interlinked ecosystems, information about catchment thresholds can be derived from observations of resilience decline of particular ecosystems. Furthermore, on a large scale, the rate of changes is low, thus information about the current state of the system and its transformations can be relatively easily derived from the analyses of the extent, tendency and intensity of change. Stress at a landscape level is often expressed as increased release of matter to freshwaters, which also in many cases shows pressure-specific responses. Therefore, analyses of a catchment adaptive capacity and assessment of environmental risks (security) are often based upon a number of existing freshwater indicators linked to, e.g. water chemistry, biodiversity, siltation, integrity, etc. The challenge for science and management, however, does not end with assessment when catchment resilience is threatened. An even more important question is how resilience can be maintained under increasing human pressure and recovered or enhanced, when necessary. Although living systems are complex, they do not emerge from a great number of interacting factors, but rather from a small number of controlling variables operating at particular spatio-temporal scales. These variables maintain a system's self-organization and provide a template for the others, of more subsidiary character (Holling, 1992, 2000). The crucial feature of such complex systems is their anarchy – hierarchy in which natural and man-made systems are interlinked in adaptive cycles including growth, accumulation, restructuring and renewal (Gunderson and Holling, 2001; Holling, 2001). Craig et al. (2005) argued that from such internal dynamics, the predictable pattern of structural discontinuities and aggregations emerges, e.g. regions of similar vegetation structure are broken by forest patches, urban areas, agricultural lands, etc. The pattern of discontinuities and resulting functions are persistent, despite the normal dynamics related to the turnover of elements, e.g. increase or decline of a species, invasions, etc. The loss of functions linked with elements obviously decreases resilience. The highest resilience does characterize systems with a variety of functions within scale/level (high diversity of species distinct, in terms of ecological niche) and high redundancy across scales/levels (broad functional groups) (Craig et al., 2005; Peterson et al., 1998). Interestingly, low cross-scale redundancy with high within-scale diversity characterizes living systems resistant to turning over elements and random disturbances, but vulnerable to, what Craig and others called, "targeted attack". Thus they are considered rigid systems. Translating those findings to catchment scale, seeking strategies enhancing resilience and increasing environmental security should be focused on distinguishing key regulating processes. In terms of structuring the landscape, such strategy must favor macro-redundancy and preserve micro-diversity. This means maintaining mosaic of patches and habitats, overlapping in functions and the services they deliver, and high biodiversity, contributing to supporting and regulating services by serving as a vector between processes and functions.

3. Sustainable resource management – ecohydrology approach

The main controlling variables at catchment level are: temperature, light, nutrient availability, and water mass dynamics (Zalewski, 2002a). Nevertheless, pattern of vegetation distribution, as well as its structure and productivity reflect, to a great extent, the hydrological conditions and their spatio-temporal variation (Neilson, 1986; Allen and Breshears, 1998; Xiao and Moody, 2004). Water is also a medium for chemicals, determining availability to biota. Thus hydrological patterns and conditions provide a framework for both – supporting and regulatory services. On the other hand, biota modifies the water cycle, and therefore nutrient circulation. Plant cover is able to significantly decrease surface runoff and river outflows (Sahin and Hall, 1996; Costa et al., 2003.) and through a modification of evapotranspiration and a surface radiation balance, it is engaged in climate regulation (Snyder et al., 2004; Zalewski et al., 2003). Finally, plant cover participates in carbon and nitrogen sequestration and enhances soil formation (e.g. islands of fertility in arid zones). Hence, vegetation is a causative factor working within a hydrological range and contributing to a reduction of environmental hazards. Consequently, the lowest organizational level, which should be considered when studying resilience and security of ecosystem services, is a catchment level, as it comprises coupled water and nutrient cycles. Synergistic regulation of a system complexity by hydrology and biota interaction became a basis for the formulation of the ecohydrology concept and a development of a scientific approach for an integrated watershed management (Zalewski, 2000). Ecohydrology postulates use of ecosystem properties as a management tool: (i) hydrology to reshape a vegetation structure, support habitats/patches functioning as nutrient traps and water storages, and to enhance optimal paths of matter and energy flows and (ii) biota to prevent nutrient leak and matter release (erosion), increase a phosphorous and nitrogen retention, regulate water outflow from a catchment and to flatten peak discharges (Zalewski, 2002b). Focusing only on two controlling variables and a few ecosystem self-regulation processes, provides a chance for long-term improvement of water and terrestrial system quality at relatively low costs. It provides also "insurance" against impacts, which are unpredictable with regard to existing data and scenarios, or which are neglected due to low decision-makers and society awareness. The system of securing ecosystem services proposed by ecohydrology incorporates the number of processes operating within, or at the borders (ecotones) of ecosystems constituting a catchment (Table 1). One could argue that most of considered processes have been already well known and described, e.g. regulatory function of land/water ecotones. However the innovation of the approach emerges from the recognizing, enhancement and integration of key processes in one coherent system, capable of maintaining catchment assimilative capacity for existing and potential pressures. The examples of an implementation of ecohydrology approach are provided and discussed in the following section of the chapter.

4. Ecohydrology for assuring water resource security: case studies

The concept of "environmental security" has arisen to define a new class of environmental problems, such as shortages of water, energy and other resources, pollution, disasters of different origin and health risk, and the loss of essential services provided by the world's ecosystems. The loss of safety is caused by natural processes or human activities triggered by ignorance, accident, mismanagement, or improper design. Therefore, desirable actions aimed at restoring security, should include: prevention of the foreseeable risks, restoration of damages, amelioration of resource scarcities, and halting environmental degradation when possible. This can be achieved only through the creation of conditions for reconcilement of social and ecological systems and development of mechanisms to address environmental crises and conflicts.

One of the major sources of conflicts is water. Therefore, several international programmes focus on this issue. Providing water security in the 21st century is the major goal of the Hague Ministerial Declaration, signed in March 2000. The key challenges identified by the Declaration provide the context for the UN World Water Assessment Programme (WWAP). Also UNESCO's International Hydrological Programme (IHP) launched in 2001, in cooperation with the Green Cross International (GCI) programme "From Potential Conflict to Cooperation Potential" (PCCP), helps peaceful management of shared water resources. Water scarcity results not only from limited resources, but is often caused by low quality. The main factor affecting both – quality and quantity is land use transformation, and especially agriculture. It imposes a threat to aquatic ecosystems in both developing and developed countries, and independently of a type of pressures, influences biodiversity and hence the delivery of ecosystem goods and services. According to MEA (2005), the development and dissemination of technologies that could provide a sustainable increase to the production of food per unit area, without harmful trade-offs related to excessive consumption of water or use of nutrients or pesticides, would significantly lessen this pressure. In some cases, appropriate technologies exist, but there is lack of financial resources and institutional capabilities for gaining and implementing them. In other cases, technologies and technical solutions are not sufficient. In such situations, ecohydrology measures can be considered as complementary or alternative ways addressing environmental security issues. The approach has been applied in several national and international projects, including demonstration projects of UNESCO IHP. Three of them – the Grabia and Lubrzanka River basin studies and the UNESCO/UNEP Pilica demonstration project – are discussed in following sections of the paper.

4.1. WATER FOR BIODIVERSITY AND RIVER ECOSYSTEM SERVICES: THE GRABIA AND LUBRZANKA CASE STUDIES: RAZIONALE

Worldwide statistics say that the water withdrawals from rivers and lakes doubled since 1960. Most water is used for agriculture (70% worldwide), as cultivated systems cover already one quarter of Earth's terrestrial surface. Agriculture is also responsible for water degradation. Since 1960, flows of biologically active nitrogen in terrestrial ecosystems have doubled, and flows of phosphorus have tripled (MEA, 2005). The problem is vital in developing as well as developed countries. According to the EEA/UNEP (1997) report on water resources in Europe, both water availability and quality are severely threatened in many parts of the continent. In one third of countries, availability of water appeared to be lower than 5,000 m^3 per head per year, but water demand is still increasing, which may eventually lead to over exploitation of ground waters (already in 60% of large European cities). Also norms for nitrate, phosphorus, and pesticides are exceeded. These facts bring into question future health, economic development, ecosystem functions, services and their security. The accentuated water scarcity problem (due to low quantity and poor quality) increase a system's vulnerability to pressures/disturbances, by impacting biodiversity and changes in vegetation cover and landscape structure (often due to ad hoc adjustment of land use practices to environmental conditions). Considering biodiversity changes, the MEA (2005) alarms us that only since 1970, populations of freshwater species included in the Living Planet Index, declined on average by 50% (compared with 30% for marine and other terrestrial species). Considering fish, approximately 20% of the world's 10,000 described freshwater species have been listed as threatened, endangered, or extinct in the past few decades. In Europe (including the former Soviet Union), there are 67 threatened species of freshwater fish, including sturgeons, barbs, and other cyprinids. Fish decline has been noted as a loss of genetic variability and hence population/community resilience, a cause/effect of energy flow disturbance, and as an indicator of drivers and pressures leading to and resulting from landscape transformations, which hit the vital functions of freshwater and terrestrial systems. Finally changes in fish communities have been viewed as a serious diminution of goods, holding a certain value for economies. The focus of the Grabia's and Lubrzanka's river basins study was to identify threats to water resources on the basis of 30 year research study and to provide ecohydrologically sounded solutions for management. Analysis was conducted on combined data of catchment transformation, water chemistry, and fish community dynamics.

TABLE 1. Ecohydrological measures are targeted at natural, inherit properties of catchment. Their aim is to secure delivery of landscape services through process tuning at designated catchment component and hence increase catchment assimilative capacity for pressures.

Tool	**Threats resulting from current man activities**	**Ecohydrological enhancement**	**Efficiency (literature examples)**
Water and matter retention in landscape	Landscape unification, fragmentation, deforestation, agriculture, urbanization, drainage	Compensating of impacts with "green spots", phytotechnology, capacity building through creation of socio-economic drivers	Annual mean water output [mm], NO_3–N and NH_4–N loss [g m^{-2}] in mosaic watershed decreases of 31.2%, 94%, and 70.7% respectively, comparing to uniform watershed (Bartoszewicz, 1994)
Natural and constructed wetlands	Wetlands reclamation, river regulation, drainage, lack of people acceptance	Hydrological (hydroperiod, hydraulics) and biological (species composition and distribution) optimization of nutrient trapping, adjustment of wetland setting to pressures	Efficiency of NO_3–N and TP retention in a constructed wetland reaching 55% and 57% respectively, may decrease to 31% and 13% under pulsing flows (Mitsch et al., 2005)
Transformation in land/water ecotones	River straightening, canalization, dredging, drainage of valleys, cutting, and grazing	Ecotones preservation and restoration based on native species adapted to hydrological conditions	Natural ecotones, e.g. wet meadows are more effective (85%) in nitrogen and phosphorus trapping then cultivated meadows (68%) (Mander et al., 1995)
Denitrification	Land drainage, stream canalization	Regulation of soil and water conditions in terrestrial habitats, hydrological control of instream carbon accumulation	10% increase of reservoir bottom area with 10% of dry weight organic carbon increases removal of TN load through denitrification by 4.5% (Bednarek, 2006).

INCREASE OF ASSIMILATIVE CAPACITY

Self-purification	River/valley connectedness, and degradation physical and biotic structure	Maintenance of major parameter values (light access, flow, trophic structure) within abiotic–biotic river continuum boundaries	Maintenance of riffles/pools sequence increases efficiency of self-purification – water restoration occurs 2–4 km downstream the source (Fauvet et al., 2001).
Biofiltration	Matter and nutrient overload, trophic structure modifications	Modification of reservoir trophic structure with water level regulation	Control of reservoir water level may increase 3–4 times zooplankton biomass and prevent cyanobacterial blooms (Zalewski et al., 1990; Tarczyńska et al., 2002)
Sedimentation	Erosion, hydraulics modifications	Reduction of land and bank erosion by enhancement of biotic structure; hydraulic control of sedimentation in backwaters and floodplains	Introduction of willow plots on can decreases flow velocity by 35%–77% and sediment transport up to 67% (Magnuszewski et al., 2005)
Regulation of hydrological regime	Improper damming, canalization, drainage	Adjustment of operational procedures of hydrotechnical infrastructure to maintain ecosystem services; tuning of water relations by impoundment	TP loads during floods risings are higher (Pilica: 206%, Luciaza: 211%) than in stable flows; Retaining rising flood waters in floodplains may enhance rivers self-purification (Zalewski et al., 2000).

5. Study area

The Grabia River is a small tributary in the watershed of the Oder River. Its catchment encompass 819.5 km^2 and was chosen due to diversified land use patterns within its component drainages – from heavily impacted agricultural to relatively undisturbed areas. The quaternary geology of the catchment is dominated by glacial remnants (sands, gravel, morainal clay, and chalk rocks) thus permeability is very high, especially in middle, chalk rock-dominated reaches. The catchment elevation varies from sea level to 270 m, with an average slope of the catchment 5.07% and the river gradient ranging from 0.8% to 3.3%. More detailed description of the river basin was given by Zalewski et al. (1990).

The Grabia is a typical lowland river with a broad floodplain present almost along the whole water course. The valley is covered with natural and semi-natural vegetation including: wet meadows, wetlands and willow patches. In the catchment, natural and semi-natural vegetation is represented by woodlands (mostly pine forests), which despite intense agriculture, still constitute over 12% of the area. Since the 1990s however, they have been exposed to the exaggerating process of fragmentation due to reclassification of land and development of recreational districts.

The study sites along the Grabia River course were established in 1983. Initially six were chosen, with the aim to collect information on distribution and diversity of fish populations against a habitat background. For the purpose of further and detailed, long-term comparisons, three of them – reflecting river character in upper (second-order stream), middle (fourth-order stream) and lower courses (sixth-order stream) – were chosen (Figure 1).

The Lubrzanka River is a tributary of the Nida, situated in the Southern part of the Vistula watershed. Its catchment covers an area of 252 km^2 and is interesting due to its function of a buffer zone between the spreading industrial agglomeration of Kielce and the natural, precious landscape of the protected areas in the central part of the Swietokrzyskie Mountains.

The river is 32.3 km long and cuts its valley through dolomitic limestone, sandstone, and clay. While gravel and sand, a major component of the valley deposits, are highly permeable, clay building the hills constrains infiltration and enhances surface flow. The process is exaggerated by the catchment morphology – elevations in the catchment vary from sea level to 360 m, the slope of the catchment reaches 23.6% and river gradients range from 0.8% to 12.5% (Figure 1). What is more, the valley lacks stabile bank vegetation cover and a floodplain, as its V-shape is not favourable for development of land/water ecotones. Thus river is deprived of a buffer against catchment born disturbances.

The Lubrzanka river basin can be divided into four parts: (1) the upper one (first- and second-order streams) of natural character, with the river passing through strictly protected peatbogs and mixed spruce and fir forest; (2) the upper one of agricultural character where streams have been regulated; (3) a middle part (third- and fourth-order streams), which is covered by a mosaic of forests (at tops of hills) and fields (on the slopes); and (4) lower part (fourth- and fifth-order streams), being heavily agricultural.

Since 1972, when the Cedzyna reservoir was constructed on the Lubrzanka River, the catchment landscape started to transform under a pressure of tourism. Although the main purpose of the reservoir was to stabilize a water budget during the growing season, currently it attracts over 5,000 people a day during summer periods.

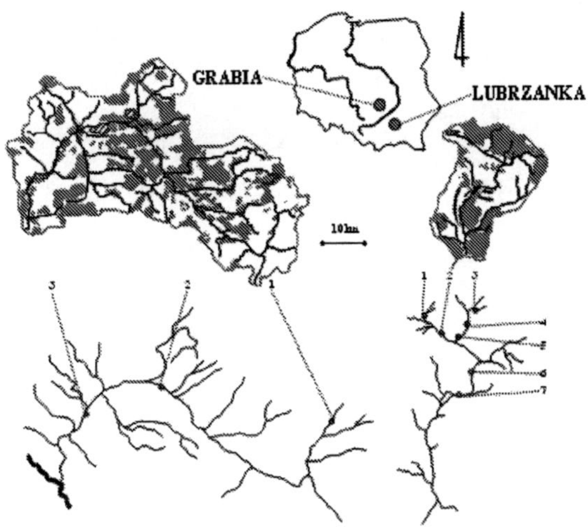

	GRABIA	LUBRZANKA
catchment area (km^2)	819,5	252
catchment slope	5,1	23,6
river length	81	32,5
gradient	0,8-3,3	0,7-12,5
catchment geology	sands, gravel, morainal clay, chalk rocks	sands, gravel – valleys, clay – slopes and hills
average temperatures [C°] (Jan/July)	-2 to -3 / 16-18	-5 to -2/ 16-17
annual precipitation [mm]	550-600	600-800
forest cover	12,5	29,4

Figure 1. Study area – location of the Grabia and Lubrzanka rivers and their characteristics (grey areas reflect forest distribution, patterned areas – distribution of urban sites). Numbers indicate location of study sections.

In order to reflect river and catchment conditions, 12 sites were established in 1982 along the river course. For the purpose of this study, seven sites situated above the reservoir were chosen (Figure 1).

6. Goal of the study

The two river systems were chosen according to their similarity in size, naturalness, prevailing catchment use, and source of pressures – agriculture, forestry, and recreation. The second criterion was related to the ecohydrological assumption that natural properties of ecosystems should be a tool for maintaining or increasing ecosystem assimilative capacity. Hence, it was desirable to look at systems differing in resilience (inherit vulnerability). The chosen rivers represent two different modes of regulation. The Lubrzanka River is shaped by harsh abiotic conditions, e.g. high flow velocities, low temperatures, rapid runoffs, seasonal and/or spatial nutrient shortages. In case of the Grabia River, relatively stable environmental conditions allowed for the development of more complex communities and increase of the role of a biotic regulation over an abiotic one.

There are also a number of similarities with regard to socio-economic settings. Both rivers play an important role for local communities. They stabilize the water budget of neighbouring areas, and maintain a farm yield by supplying crops with water in summer time; they provide water for fish farms, support natural and semi-natural plant communities, e.g. peatbogs and water meadows, crucial for preserving of terrestrial and aquatic biodiversity; they also serve as major recreational areas for two regional capital cities – Kielce (the Lubrzanka River) and Lodz (the second biggest city in Poland; the Grabia River). Such a complex role mobilizes interest of different users and therefore, also increases tensions over land and water use.

The aims of the study included: (1) identification of key pressures on water ecosystems linked with landscape transformations, (2) assessment of long-term fish community dynamics against the data on catchment transformation, and (3) linking biotic indicators (fish) with threats in order to formulate key management recommendations, focused on biodiversity protection and environmental security.

7. Methodology and approach

Step 1: Identification of key pressures on water ecosystems linked with landscape transformations

Analysis of landscape transformations was conducted using two aerial photography sets of river catchments taken in 1970 and 1997.

Trends in landscape transformation appear to be similar in both catchments – lowland one of the Grabia River and the upland of the Lubrzanka. Present build-up areas were extended and new ones started to develop. The process was the most dynamic in middle parts of catchments – along the third- and fourth-order streams of the Grabia and Lubrzanka, and their tributaries. The major difference between the two catchments was the course of settlements development: in upland catchments, development was linear, along roads and linked to existing villages, while in the lowlands, these were usually new settlements located in forests or close proximity to the rivers (Figure 1). Other observed changes included:

- Afforestation of low value agricultural land due to planting and/or succession
- Progressing land drainage (Figure 2a)
- Transformation of wetlands into meadows and pastures (mostly the Grabia River)
- Degradation of land/water ecotones due to land drainage and river regulation – the first- and second-order streams of the Grabia, and the fifth ones of the Lubrzanka.
- Development of new wetland areas along the fifth-order streams of the Grabia as a result of discharge decrease, river shallowing and an overgrowing of river bed by macrophytes and willow.
- Development of fisheries and hence, a potential increase of water uptake

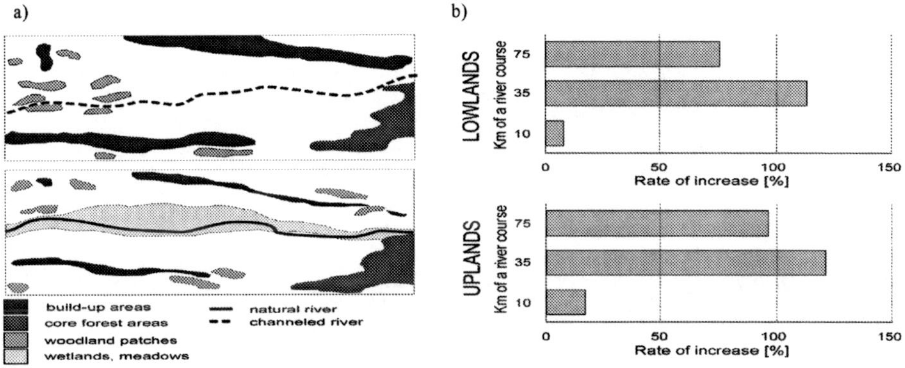

Figure 2. Changes of river valley and catchment structure – the Grabia example: (a) Large areas of wetlands and natural and semi-natural meadows were replaced by fields, especially along the first- and second-order streams, and river regulation in many of those sections; forest core areas remained almost unchanged or they increased; the same trend was observed for woodland patches: the build up areas extended and became more compact; (b) the fastest development of built-up areas (estimated with a percentage increase of number of farms between 1977 and 1997) occurred along the third- and fourth-order streams – the most natural and attractive parts of the catchment.

Landscape transformations were analysed against water quality data. The most important observation was a dramatic decline of discharges (Figure 2b) in the studied period, resulting to a great extent from climatic variability and low precipitation. However, the process was exaggerated by an increase of water outflow from both catchments due to river regulation and a decrease of water storing capacity caused by land drainage, water uptake and expansion of intensely cultivated land and built-up areas.

In order to prove the impact of land use intensification on water resources, an analysis of 15 different parameters was conducted, including heavy metals and other toxic substances and *E. coli* presence. According to our data, the quality of both rivers was lowered by contamination with *E. coli* and phosphorus and nitrogen compounds.

The expectations were not confirmed in the Grabia case. There was no deterioration of water quality, which generally meets requirements of second class, with only some sections being classified to the third one and lower due to *E. coli* contamination. On the contrary, a trophy of the Lubrzanka water increased since the 1980s. The concentrations of total nitrogen almost quadrupled, phosphate doubled and total phosphorus concentrations increased by about 30%. Simultaneously, concentrations of magnesium, potassium, and calcium increased by about 50%. Increase of phosphorus and cations concentration could be viewed as a sign of rising erosion threat.

Step 2: Assessment of long-term fish community dynamics for indicating the sources of threats to river ecosystem in contrasting catchment conditions

The fish community is sensitive to conditions of both an aquatic ecosystem and its surrounding watershed and can therefore be used to assess environmental degradation (Karr, 1987) and to identify threats to environmental security at a catchment level. The framework for using fish communities to describe river naturalness/degradation was provided by the multi-level concept for fish-based, river-type specific assessment of ecological integrity (Schmutz et al., 2000). It aroused from the hierarchical organization of biota (Odum, 1971) and the relations of the organizational levels to temporal and spatial scales (Frissel et al., 1986). Thus, assets of assessment criteria, representing different hierarchical levels of biocenosis, enabled detection of a variety of human alternations, especially when combined with an assumption, that a fish community responds to a continuum of abiotic and biotic factors (Zalewski and Naiman, 1985); hence, its organization should reflect a decrease in environmental harshness from springs to river mouths. The assessment procedure is finalized with comparing of the evaluated reaches with reference conditions.

In the described case study, the results of three surveys 1982–1985, 1990–1993, and 1998–2001, were compared and the first one was assumed as a reference. As samples in each case were taken every month, the base line constitutes 48 of them.

Results indicated a decline in fish biomass (Table 2) and density and species diversity in both rivers. The relatively smallest changes were observed in the most natural sections of both rivers, separated from agricultural land by broad floodplain meadows and riparian forest (fourth-order stream of the Grabia River and the fifth-order stream in the Lubrzanka). The major biomass changes were observed in parts of the catchments intensely used for crop production. Eventually it resulted in a rapid biomass increase in the third-order stream of the Lubrzanka basin, which was usually nutrient limited. Considering a structure of fish communities (Figure 3), a "potamalization effect", usually occurring due to river impoundment, appeared in both rivers. An indicator of the process was a decline of rheofils – a species preferring fast flows (riffles), and lithofils – fish spawning on stony or gravel substrate (also occurring in riffles). In case of mountain and upland rivers like the Lubrzanka, potamalization means a loss of many rare and valuable species in favour of common ones (e.g. trout can be replaced by roach or perch).

TABLE 2. Change in total fish biomass since the first survey – between-site comparison. Biomass was measured in kg per 100 m^2 of river bed. The only increase of fish biomass was observed in the third order Lubrzanka stream. It was caused by increase of water trophy and mass appearance of only three species.

Stream order	The Grabia River			The Lubrzanka River		
	Rate*	Trend	Tukey's test results (<0.05)**	Rate*	Trend	Tukey's test results (<0.05)**
II	16.5	–	44.67	25.6	–	6.5
III				64.68	+	15.5
IV	14.6	–	7.67	23.1	–	3.5
V	30.9	–	7.17	13.6	–	3.5

* Rate of change was estimated relatively on a basis of a regression curve slope, calculated for each site separately on a basis of data obtained in certain time spans.

** Results concern comparison between the first and last survey.

An interesting phenomenon was a decline of benthic species, which can be explained with high sensitivity of some of them (e.g. bullhead) and classification of some rheofils to this group (e.g. barbel). In the Lubrzanka, it can be considered as confirmation of the escalating erosion problem and at least a periodical increase of matter transport. The explanation for the Grabia is more problematic, possibly considering effects of indirect factors, e.g. food depletion.

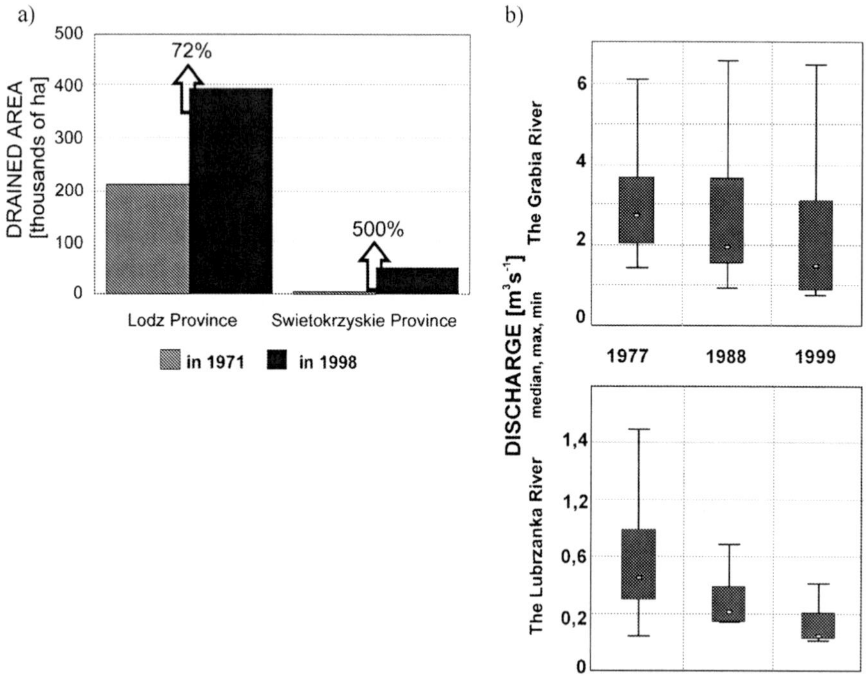

Figure 3. In order to increase area of cultivated lands, many wetlands, wet meadows, and alder forests were drained. Regulation of water balance was especially important in mountain and upland areas, as it enabled use of the most fertile soils located in river valleys (a). Land drainage addressing past hydrological conditions, thus not adjusted to present climatic variability and increased water outflow from the catchments resulted in dramatic decline of discharges (b). The upland rivers appear to be more affected due to their lower natural resistance.

The next detected change was related to a group of general insectivores. They usually occupy low order streams with low primary productivity. Hence, they are supposed to be sensitive to an increase of water trophy (Schlosser, 1982a, b).

An interesting phenomenon was also decrease of diversity of phytofils. Many of them require spring floods to enter their spawning grounds, only some of them can reproduce successfully using bank vegetation and submersed macrophytes (e.g. roach). In this case declining diversity could be considered as indicator of disturbed river hydrological dynamics – changed timing of expansion/contradiction events, connectedness of river and floodplain, critically low water level and discharges, etc.

Step 3: Interpretation of results and recommendations

The decline of fish biomass (of about 30% on average in the Grabia and 20% in the Lubrzanka) and species diversity within fish functional groups reflect a decrease of environmental carrying capacity, related to limited access to feeding and spawning areas and appropriate habitats. Although to some extent, it results from river regulation, the major factor is probably a decline of discharges. The problem of water stress applies to both case studies, but the more affected river system is a mountain one. The reason is its lower natural resistance due to weakly developed riparian zones, floodplains, and scarcity of wetlands and small retention in the basin. On the other hand, a better performance of lowland river, and especially its middle part – meandering, with numerous ponds and wet meadows in the valley – indicates a need for continuous identification and more efficient protection of such areas. Especially important are mountain catchments, first because such areas are more scarce, but also more attractive for agriculture; second because discharge alternation affects other habitat parameters, e.g. flow velocity, sedimentation, and substrate coarseness, distinctive for these particular river types.

Considering current trends in catchment development, it is unlikely that the pressures to freshwater systems will be mitigated. To secure resources (water availability, and aquatic and terrestrial biodiversity) in both studied areas, the urgent actions have to include:

- Protection and development of small retention.
- Restoring of river – valley connectivity.
- Adjustment of water uptake to current capacity of river systems.
- Development of buffering zones in valleys of lowland rivers.
- Creation of biogeochemical barriers in upland catchments – improper space planning and relying on gradual afforestation of abandoned land causes imbalance with rapidly developing rural area.
- Periodical river damming can be considered in case of lowland river for improving the water balance of adjacent systems.

8. Conclusions

The results obtained for the case studies unveiled a need for development of new, catchment level strategy of water resource management. Such a strategy should attract more attention to river biodiversity hot spots for maintaining river ecosystem functions, and the high recreational value of both areas. Despite threats mitigation, it should use existing environmental potential and consider enhancement of catchment processes. In order to implement the recommendations quantification of the processes through small scale testing was required. One of important issues was the effectiveness of buffering zones for water and nutrient storage. The question concerned possibilities of increasing their efficiency by management practices, improvement of structure (e.g. increase of land patchiness, changes of plant community composition) and regulation of hydrology. To ensure sustainability of solutions, it was also important to provide examples of alternative use of riparian areas, to create positive socio-ecological feedback, and mitigate current and future risks emerging from landscape transformations and climatic instability. In order to address all these problems, the demonstration project on the Pilica River and the Sulejowski Reservoir was launched.

9. Securing water resources for human well-being – the Pilica River case study: Razionale

Results of the research on the Grabia and Lubrzanka Rivers provided evidence for relationships between modification of the water cycle in the catchment due to climate, and land use changes and a structure and functioning of aquatic ecosystems. Addressing such processes in catchment management plans may help secure water resources for biodiversity. The question that rises is whether the knowledge about abiotic–biotic feedbacks and their regulation can be used for securing desirable water resources with the quality for human well-being. The UNESCO-IHP/UNEP-IETC Pilica River Demonstration Project in Poland (Wagner-Lotkowska et al., 2004) aimed to answer this question. Key issues addressed ecological and health hazards due to progressing eutrophication and toxic cyanobacterial blooms in a drinking water reservoir.

10. Study area

The Pilica River (Figure 4), the longest left-side tributary of the largest Polish river – Vistula, is a river of total length 342 km, a catchment area of 9,258 km^2, and mean discharge in its middle reach of 21.2 m^3 s^{-1}. The river valley has mostly semi-natural character – floodplains and bank slopes are principally forested or covered by shrubs, and some of them are used for pastures or cultivation.

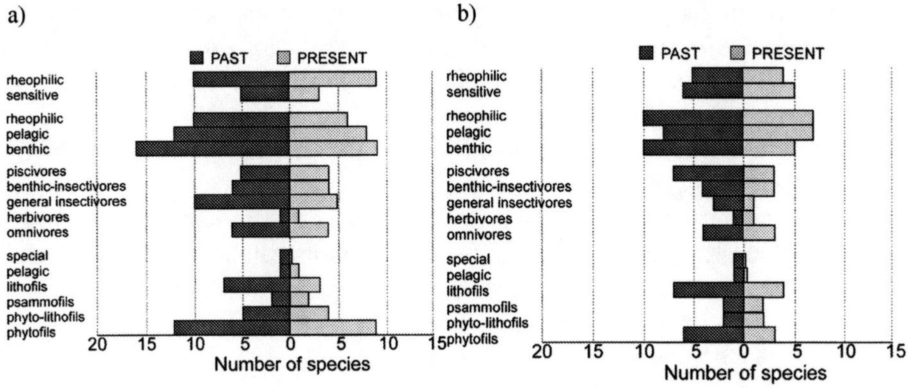

Figure 4. Over 20 year change in fish community composition, species were analysed according to their sensitivity, breeding and trophic groups they belong to, and habitat preferences: (a) The Grabia, (b) The Lubrzanka.

In 1973, the river was dammed with the purpose of creation a drinking water supply for the City of Lodz (ca. 800,000 inhabitants). Thanks to surrounding forests and the historical character of the region, the reservoir provides also a large recreational area for nearby cities. The reservoir (mean depth 3.3 m), at its maximum capacity (75×10^6 m^3), covers an area of 22 km^2, and reaches mean water retention time of about 30 days. These morphological characteristics, together with prevailing agricultural catchment use (Figure 4) generate export of nutrients and humic substances and increase the reservoir vulnerability for intensive eutrophication and escalate its symptoms. The most hazardous one is occurrence of intensive cyanobacterial blooms (*Microcystis aeruginosa*) during summers. Their maximum biomass observed in 1995, reached 60 mg l^{-1} (Tarczynska, 1998). Several studies revealed cancerous and toxic effects of the toxins to humans (e.g. Mankiewicz et al., 2003).

11. Aims of the study

The presence of toxic cyanobacteria and high concentrations of humic substances after rainfalls, induced problems with purification of potable water, lowered the quality of the final product and affected environmental security. After 30 years of the functioning of the reservoir as a drinking water source, the abstraction has been restricted. The reservoir is still maintained as an alternative water source, however a regular withdrawal takes place from a number of wells that were constructed along the banks. Moreover, progressing deterioration of water quality and the risk posed for human heath and raised economic constrains for the region. A diminished reservoir's appeal for recreation lowered income for the local society, which to a certain extent was based on agrotourism and recreation.

This complex ecological and socio-economic setting created a framework for the objectives of the demonstration project, which aimed in elaboration of a scientific basis for water quality improvement, using cost-efficient ecohydrology measures. Long-term research in the reservoir and its catchment (e.g. Zalewski et al., 1990; Frankiewicz et al., 1999; Tarczynska et al., 2001; Wagner and Zalewski, 2000; Izydorczyk, 2002; Jurczak et al., 2005) provided data that allowed for developing a comprehensive strategy.

12. Methodology and approach

Step 1: Recognizing the pattern of an external reservoir supply with nutrients

Export of nutrients from catchments with high contribution of an agricultural land, usually increases during rainy periods. An analysis of the Pilica River's hydrochemical data, superimposed on its hydrological pattern, allowed identification of the episodes of the highest nutrients loads (Figure 5a). The generalized conclusions say that:

- The highest concentrations were observed during the rising hydrograph stage ("nutrient-condensing stage") of medium floods.
- The concentrations started to decrease before the maximum discharge, and continue decreasing during the hydrograph fall ("nutrient-dilution stage").
- Nutrients' loads were significantly higher in the first stage, than in the second one.
- The regularity of this pattern can be disturbed in the case of flash floods of short duration time and high amplitude, which follow the pattern described above only in the very initial stage of the flood formation.

Step 2: Evaluation of the effect of the reservoir supply pattern on reservoir's eutrophication symptoms

The obtained results showed that the pattern of the reservoir supply with nutrients is highly dependent on the river's hydrology. Consequently, the dynamics of the reservoir's ecosystem is determined by timing and a scale of a flood event. If a supply occurs during or shortly before the growing season, intensification of growth of phytoplankton dominated by toxic cyanobacteria can be almost always expected, soon after stabilization of the hydrological parameters of the reservoir. Figure 5b presents one of the most evident example of this process – the mean phytoplankton biomass after the passage of flood through the reservoir in the summer of 1999 reached over 43 mg dm^{-3} (Izydorczyk, 2002), and was significantly higher than before the flood period. These results indicate that control of high river discharges could be potentially one of the elements of a management strategy to improve security of water resources for human use (Wagner and Zalewski, 2000).

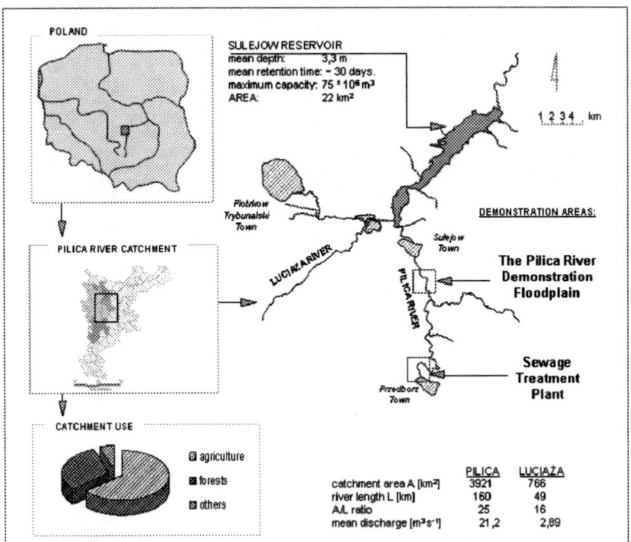

Figure 5. Location and characteristics of the Pilica River and Sulejow Reservoir – area of the UNESCO/UNEP demonstration project "Application of Ecohydrology and Phytotechnology for Water Resources Management and Sustainable Development" in Poland (Wagner-Lotkowska, 2004).

Step 3: Assessment of the potential for enhancement of nutrient retention on floodplains

The clear difference between nutrient concentrations and loads transported during the particular phases of the flood, as well as the relationship between concentrations and river discharge suggested, that retaining only the "nutrient-condensing stage" could significantly contribute to the lowering of the reservoir loading (Wagner-Lotkowska, 2001). Considering that floodplains are fundamental in the two-way exchange of water and matter masses (e.g. Junk et al., 1989), the semi-natural character of the Pilica valleys could be potentially used for this purpose. Improving the hydraulic connectedness between the river and its valley could increase water retentiveness and provide a tank for the matter transported by high waters. In the case of systems, in which human intervention changed this natural process, there is a need for appropriate management of still available floodplains, so that their capacity for nutrient retention can compensate for the lost properties of the entire system. The question raised was, how this could be possible by ecohydrological regulation. Research carried out on the experimental floodplain (26.6 ± 2.7 ha), located upstream the reservoir, was to answer the question how the relationships between the floodplain hydraulics and vegetation cover could enhance a capacity for nutrient trapping. The analyses were developed based on a Digital Terrain Model (DTM), location altitude, inundation and vegetation cover maps of the floodplain elaborated in the project (Koch, 2001). Additionally, a CCHE2D hydraulic model was chosen among various other two-dimensional hydraulic models available, due to its state-of-the-art computational module for calculating a sediment transport and river bed morphology changes. Comparisons of numerical simulation for two scenarios – natural conditions (scenario 0) and land cover altered by introducing of some willow patches (scenario 1) showed, that adjustment of vegetation cover can be used for optimization of sedimentation of matter transported by a river in designated areas. Changing the roughness of even a relatively small area reduces considerably the velocity field. Willow patches are found to be particularly useful in creating dense and compact vegetation cover with high values of Manning coefficients (Magnuszewski et al., 2005). Simultaneously, biological assimilation of dissolved fraction of phosphorus, enhanced by adjusting of floodplain vegetation to groundwater level and hydroperiod, and enriching it with willow patches, increases an ability of the system to assimilate nutrients (Figure 6), and their temporal storage in an unavailable pool of a vegetation biomass (Koch, 2001). A possible added benefit of using willow for water purification is its suitability as an alternative crop for agriculturally cultivated floodplains (Figure 7). The biomass can be utilized for bioenergy and to increase revenue for the region.

Step 4: Diminishing eutrophication symptoms in the reservoir

As far as measures described in the previous sections aimed in diminishing an external nutrients loading, the ecohydrological research in the reservoir was mostly focused on diminishing eutrophication symptoms. A long-term monitoring and research data allowed for profound recognition of the hierarchy

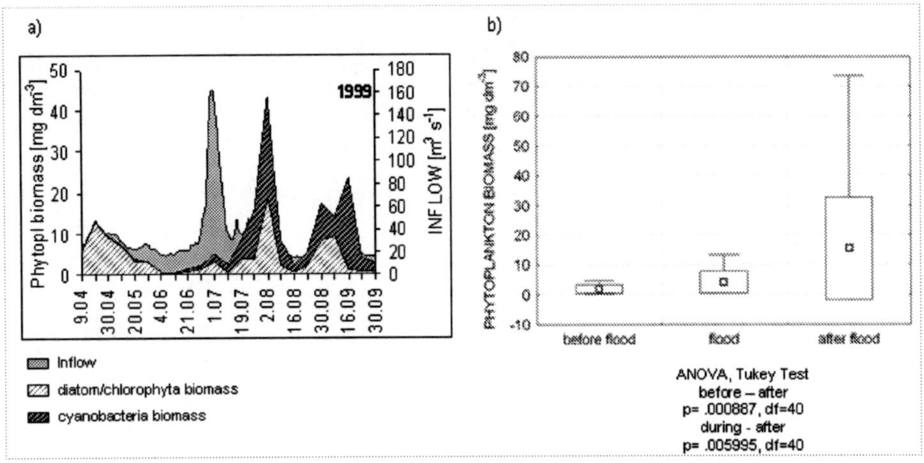

Figure 6. Effect of the hydrological pattern of the river on the rate of external nutrient supply to the reservoir and biota dynamics: (a) High inflow of water to the reservoir (max discharge 141.0 $m^3\ s^{-1}$) fed the reservoir with the load of 55.7 t of total phosphorus and 2,110.0 t of total suspended matter. This generated high a biomass of phytoplankton, dominated by toxic cyanobacterial bloom (max biomass 43.3 mg dm^{-3}; Izydorczyk, 2002). (b) The biomass was significantly higher than before the flood (Wagner-Lotkowska, 2001).

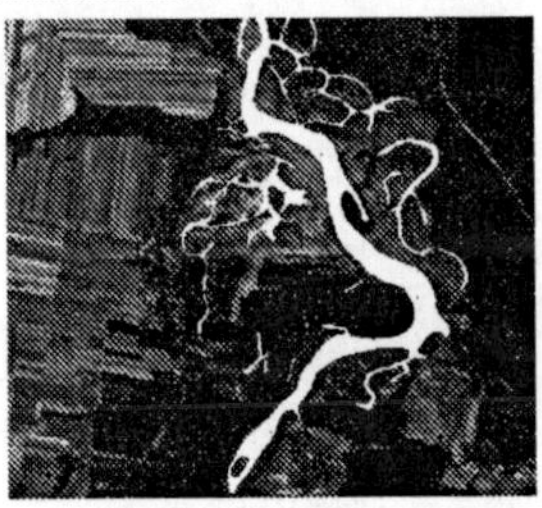

Floodplain (27ha) coverage with fast-growing willow patches [%]	0%	24%	48%
Summer TP accumulation [kg] Accumulaion [%]	255,0 0%	332,4 30.6%	399,4 56.6%

Figure 7. The efficiency of phosphorus accumulation in biotic structure of a floodplain depends, among others, on vegetation management, e.g. a percentage of the area coverage with fast-growing willow patches (Zalewski, 2006).

of factors driving the reservoirs ecosystem dynamics (e.g. Zalewski et al., 1990; Frankiewicz et al., 1999; Wojtal et al., 1999; Tarczynska et al., 2001; Wagner and Zalewski, 2000; Izydorczyk, 2002), and some of which were applied in practice. One of the examples involves withdrawing of the implementation plan, based on the wrong assumption, that an increase in the storage capacity of the reservoir may dilute polluted waters and increase water quality. Results of research by Tarczynska (1998), providing statistically significant correlations between water retention time and intensity of cyanobacterial blooms, showed that such a strategy combined with any realistic plan for catchment management reducing of the external load, would only further intensify the problem. Some possible ecohydrological measures proposed on the basis of the research carried out by Zalewski et al. (1990), use seasonal regulation of the water level by adapting dam's operational procedures. Lowering the water level and disclosing shore vegetation during spawning periods of planktivorous fish, reduces their spawning grounds and consequently recruitment. Release of zooplankton from fish pressure enhances efficient filtration of algae and lessens high summer biomass of phytoplankton. Some other methods based on regulation of the reservoir hydraulics as a tool can be used for control of sedimentation in the backwaters of the reservoir (Wagner-Lotkowska et al., 2004) and denitrification enhancement (Bednarek et al., 2002).

13. Conclusions

In the case of the Pilica River project, it was proposed to utilize nutrients converted into floodplain vegetation (biomass) as an energy source (bioenergy). Such an approach reduces expenditure of the region to fossil fuels and investments in environment protection, enhances implementation of modern technologies and reduces risk of pollution. Through analysis of the role of particular components in reservoir/river/floodplain system, the bottom-up initiative of development of willow plantations industry was directed toward creation of additional ecosystem value. It demonstrated that maintaining the mosaic structure of autochthonous vegetation and fast-growing willow species preserves biodiversity in the river corridor. Introducing willow, as an alternative crop for cultivated sections of floodplains, increases water and environment quality, contributes to reduction of toxic algal blooms in the reservoir, and improves appeal of the region for tourism. The anticipated approach combines several elements presented in Table 1, and is basically concerned with cost-efficient dislocation of nutrients to unavailable pools (e.g. dragged sediments) or designated trophic levels of the ecosystems (e.g. floodplain vegetation, zooplankton). The efficiency of the measures towards water security for human well-being (e.g. safe drinking water, health, recreation) may be increased by using natural regulatory feedbacks between hydrological conditions and dynamics of biotic components of the man-disturbed ecosystems, which allows for an increase in potential to absorb human pressure, thus preserving resilience.

14. Summary – stabilizing ecohydrological measures by socio-economic benefits

Resilience, although not priced on current markets, has a tangible value for society. In an economic context, it is interpreted as a natural insurance capital against risk of ecosystems" malfunctioning, and the consequent damages due to potential interruption of the ecosystems' ability to provide goods and services (Vergano and Nunes, 2006). Recognizing the complex nature of ecosystems is therefore crucial not only for the successful application of environmental measures, but also for proper evaluation of potential risks, limitations and opportunities that the ecosystems can provide to societies. A socio-economic framework in turn, including: legislation, mobilization of potential (e.g. human power and awareness), finances, cultural issues and others, defines the social acceptance and the strength of socio-economic drivers in successful implementation of novel approaches. The success in implementation of the ecohydrological solutions depends, to a great extent, on acceptance by local communities, decision-makers and legislative authorities. Therefore ecohydrology implementation should not only be focused on reduction of the risk to water related services, based on the adequate management of available resources. Meeting other emerging regional issues, relevancy to programmes of regional development, and creating positive socio-economic feedbacks to enforce the implementation process (Zalewski, 2002b) (Table 3) are also of crucial importance. Securing of ecosystem services is an important justification for implementing ecohydrological measures. Increasing quality and capacity of ecosystems against disturbances enhances their ability for self-purification, flood and microclimate regulation, reduction of disease/health risk (e.g. those associated with toxic cyanobacterial blooms) and others. Additionally, the use of the regulatory potential of interplay between hydrological specifics and biota dynamics of catchment assures provisioning and cultural benefits for societies, such as production of bioenergy, natural fertilizers (sediments), increasing biodiversity, and therefore guarantees water-based income (e.g. recreation, aquaculture, fisheries) (Table 3). Ecohydrological measures may encourage a shift between steady states or maintain desirable, while the thresholds are known. If thresholds are unknown, but threats are identified and ordered according to impact (scale, persistence, and harm), they may help in operating the buffering capacity of elements and processes within catchment, to keep them within boundaries of known consequences.

TABLE 3. Socio-economic feedbacks promoting ecohydrology implementation by providing provisional and cultural services for local societies.

Ecohydrological tool	**Socio-economic benefit**	**Literature examples**
Water and matter retention in landscape	Diminishing of floods and droughts risk, microclimate improvement increase of yield, esthetic, and recreational value	Native willows adapted to hydroperiod, retain nutrients improving water quality and produce up to 10 t dm ha^{-1}; biomas that is economic energy source (Production of 1GJ energy from biomass: 19PLN, coal and natural gas: 32PLN, fuel: 50PLN (estimation for 2002; Michalak, 2003)
Pollutants trapping in natural and constructed wetlands	Reduced investments for water quality improvement, use of biomass as bioenergy, development of new markets, job opportunities	Adapting of native willow patches to hydroperiod, and harvesting in the average age below 7 years allows withdrawing 43–173 g of P with every 100 kg of willow. Withdrawing 1 kg of P reduce by 2–3 t phytoplankton bloom, disabling recreational use and water uptake (Wagner-Lotkowska et al., 2004)
Nutrient transformation in land/water ecotones	Increase of fish yield and biodiversity, recreation and tourism development	Riparian vegetation keeping medium intensity (350–650 μEinst $cm^{-2}s^{-1}$) light access within: 40–50% of surface (upland river channel), 60–70% (lowland, 1st and 2nd orders) and above 80% (3rd and higher orders) maintains highest fish production and biodiversity (Lapińska et al., 2002)
Denitrification	Reduction of health risk, decrease of agriculture impact on waters	Enhancement of denitrification rate allowing to lower concentration of nitrate–nitrogen in drinking water below 10 mg/l lowers risk of – methemoglobinemia, birth defects, cancer and spontaneous abortions (Almeida et al., 1997).

Self-purification	Increase of fish yield and biodiversity, recreation and tourism development	Maintenance of self-purification potential results in high river productivity–even up to two order of magnitude higher fish density and biomass higher of about 80% (Zalewski, 1998)
Biofiltration	Reduction of health risk, fish community enrichment, recreational values	Occurrence of toxic cyanobacteria excludes reservoir from recreational use for 5–12 weeks in June–September period reducing income from tourism and posing a health hazard for humans (Tarczyńska et al., 2001)
Sedimentation	Use of sediments for forestry and agriculture	Sediments dredged from lake improves. e.g. soil water-retention (18–24.3%) and cations exchange capacity (2.8–5.6 meq per 100 g), of sandy soils and some crops yields (e.g. lettuce: 110–194 g) (Canet et al., 2003)
Regulation of hydrological regime	Diminishing of flood/drought and eutrophication risk, fisheries improvement	Recruitment of a migratory fish in the Parana River varies from 89 to 3,481 kg depending on flood duration ($r = 0.8$, $p = 0.11$); accurate dam operation can maintain river-floodplain integrity and fisheries (Agostinho et al., 2004)

References

Agostinho, A.A., Gomes, L.C., Verıssimo, S., and Okada, E.K., 2004. Flood regime, dam regulation and fish in the Upper Parana′ River: effects on assemblage attributes, reproduction and recruitment. Reviews in Fish Biology and Fisheries 14, 11–19.

Allen, C.D. and Breshears, D.D., 1998. Drought-induced shift of a forest-woodland ecotone: rapid landscape response to climate variation. Proceedings of the National Academy of Sciences of the United States of America 95, 14839–14842.

Almeida, J.S., Reis, M.A.M., and Carrondo, M.J.T., 1997. Nitrite inhibition of denitrification by *Pseudomonas fluorescens*. Journal of Theoretical Biology 186, 241–249.

Bartoszewicz, A., 1994. The chemical compounds in surface waters of agricultural catchments (under the soil, weather conditions of the Kóscian lowland). Roczniki Akademii Rolniczej, Poznán 250, 5–68. (In Polish).

Bednarek, A., Zalewski, M., Błaszczyk, M., Dąbrowska, E., Tomaszek, J., and Czerwieniec, E., 2002. Estimation of denitrification rate in diversified bottom sediments in Sulejow Reservoir. In: M. Zalewski (Editor), Proceedings of the Final Conference of the First Phase of the IHP – V Project 2.3, 2.4 on Ecohydrology, Vol. 2 (1–4), pp. 297–304.

Bednarek, A., 2006. Importance of denitrification process in restoration of the Sulejow Reservoir. Ph.D thesis, University of Łódź.

Canet, R., Chaves, C., Pomares, F., and Albiach, R., 2003. Agricultural use of sediments from the Albufera Lake (eastern Spain). Agriculture, Ecosystems and Environment 95, 29–36.

Costa, M.H., Botta, A., and Cardille, J.A., 2003. Effects of large-scale changes in land cover on the discharge of the Tocantins River, Amazonia. Journal of Hydrology 283, 206–217.

Craig, R.A., Gunderson, L., and Johnson, A.R., 2005. The use of discontinuities and functional groups to assess relative resilience in complex systems. Ecosystems 8, 958–966.

EEA/UNEP, 1997. Water stress in Europe – can the challenge be met? European Environment Agency (EEA) and United Nations Environment Programme (UNEP) Report, http://www.unep.ch/roe/publications/assessment/01_eurowater.pdf, p. 16.

Fauvet, G., Claret, C., and Marmonier, P., 2001. Influence of benthic and interstitial processes on nutrient changes along a regulated reach of a large river (Rhône River, France). Hydrobiologia 445, 121–131.

Frankiewicz, P., Dąbrowski, K., Martyniak, A., and Zalewski, M., 1999. Cannibalism as a regulatory force of pikeperch, Stizostedion lucioperca (L.), population dynamics in the lowland Sulejǿw Reservoir (Central Poland). Hydrobiologia 408/409, 47–55.

Frissell, C.A., Liss, W.J., Warren, C.E., and Hurley, M.D., 1986. A hierarchical approach to classifying stream habitat features: viewing streams in a watershed context. Environment Management 10, 199–214.

German Advisory Council on Global Change (WBGU), 1998. World in transition – strategies for managing global environmental risks. Annual Report, Springer, Berlin, p. 359.

Gleick, P.H. (Editor), 2006. Water conflict chronology. Pacific Institute for Studies in Development, Environment, and Security, www.worldwater.org/conflictchronology.pdf.

Gunderson, L. and Holling. C.S. (Editors), 2001. Panarchy: understanding transformations in human and natural systems. Island Press, Washington, DC.

Hobbs, R.J., Arico, S., Aronson, J., Baron, J.S., Bridgewater, P., Cramer, V., Epstein, P., Ewel, J., Klink, C., Lugo, A., Norton, D., Ojima, D., Richardson, D., Sanderson, E., Valladares, F., Vila, Zamora, R., and Zobel, M., 2006. Novel ecosystems: theoretical and management aspects of the new ecological world order. Global Ecology and Biogeography 15, 1–7.

Holling, C.S., 1992. Cross-scale morphology, geometry and dynamics of ecosystems. Ecological Monographs 62, 447–502.

Holling, C.S., 2000. Theories for sustainable futures. Conservation of Ecology 4, 7.

Holling, C.S., 2001. Understanding the complexity of economic, ecological, and social systems. Ecosystems 4, 390–405.

Izydorczyk, K., 2002. Influence of abiotic and biotic factors on the space and spatial succession of phytoplankton community of the Sulejow Reservoir. Ph.D thesis, University of Lodz.

Junk, W.J., Bayley, P.B., and Sparks, R.E., 1989. The flood pulse concept in river-floodplain systems. In: D.P. Dodge (Editor), Proceedings of the international large river symposium. Can. Spec., Publications from Fisheries Aquatic Science 106, 110–127.

Jurczak, T., Tarczynska, M., Izydorczyk, K., Mankiewicz, J., Zalewski, M., and Meriluoto, J., 2005. Elimination of microcystins by water treatment processes – example from Sulejow Reservoir, Poland. Water Research 39, 2394–2406.

Karr, J.R., 1987. Biological monitoring and environmental assessment: a conceptual framework. Environmental Management 11, 249–255.

Koch, E., 2001. Nutrient de-position rate in the pilica river floodplain of different typology. M.Sc. thesis. Department of Applied Ecology, University of Lodz.

Łapińska, M., Kaczkowski, Z., and Zalewski, M., 2002. Restoration of streams for water quality improvement and fishery enhancement. In: M. Zalewski (Editor), Guidelines for Integrated Management of Watershed, UNEP, Osaka, p. 155.

Magnuszewski, A., Kiedrzyńska, E., Wagner-Lotkowska, I., and Zalewski, M., 2005. Immobilizing of sediments in a lowland river floodplain. Publications from Institute of Geophysics, Polish Academy of Sciences E-5 (387), 239–259.

Mander, Ü., Kuusemets, V., and Ivask, M., 1995. Nutrient dynamics of riparian ecotones: a case study from the Porijogi River catchment, Estonia. Landscape Urban Plan 31, 333–348.

Mankiewicz, J., Tarczyńska, M., Jurczak, T., Wojtysiak-Staniaszczyk, M., and Zalewski, M., 2003. Test with luminescent bacteria for the toxicity assessment of cyanobacterial bloom samples. FEB 12(8), 861–864.

Meybeck, M., 2003. Global analysis of river systems: from earth system controls to anthropocene syndromes. Phylosophical Transactions of Royal Society (B) 358(1440), 1935–1955.

Michalak, Ł., 2003. Efficiency of Salix species to phosphorous accumulation in natural and experimental conditions. M.Sc. thesis. Department of Applied Ecology, University of Lodz.

Millennium Ecosystem Assessment (MEA), 2005. Ecosystems and human well being – synthesis. World Resources Institute, Washington, DC.

Mitsch, W.J., Li, Zhang, Anderson, C.J., Altor, A., and Hernandez, M., 2005. Creating riverine wetlands: Ecological succession, nutrient retention and pulsing effect. Ecological Engineering 25, 510–527.

Neilson, R.B., 1986. High resolution climatic analysis and southwest biogeography. Science 232, 27–34.

Odum, H.T., 1971. Environment power and society. Wiley Interscience, New York. 331 p.

Peterson, G.D., Allen, C.R., and Holling, C.S., 1998. Ecological resilience, biodiversity and scale. Ecosystems 1, 6–18.

Sahin, V. and Hall, M.J., 1996. The effects of afforestation and deforestation on water yields. Journal of Hydrology 178, 293–309.

Salehyan, I., 2005. Refugees, climate change, and instability. Human Security and Climate Change, an International Workshop, Oslo, 21–23 June 2005, http://www.cicero.uio.no/humsec/papers/Salehyan.pdf.

Schlosser, I.J., 1982a. Trophic structure, reproductive success and growth rate of fishes in a natural and modified headwater stream. Can. J. Fish. Aquat. Sci. 39, 968–978.

Schlosser, I.J., 1982b. Fish community structure and function along two habitat gradients in headwater stream. Ecological Monographs 52, 395–414.

Schmutz, S., Kaufmann, M., Vogel, B., Jungwirth, M., and Mohar, S., 2000. A multi-level concept for fish-based, river-type-specific assessment of ecological integrity. Hydrobiologia 422/423, 279–289.

Snyder, P.K., Delire, Ć.C., and Foley, Ć.J., 2004. Evaluating the influence of different vegetation biomes on the global climate. Climate Dynamics 23, 279.

Tarczynska, M., 1998. Reasons of toxic algal bloom forming and their influence on representatives of water ecosystems in the Sulejow Reservior. Ph.D. theses, University of Lodz, Poland.

Tarczyńska, M., Frankiewicz, P., and Zalewski, M., 2002. The regulation and control of hydrological and biotic processes within reservoir for water quality improvement. In: M. Zalewski (Editor), Guidelines for Integrated Management of Watershed – Phytotechnotogy and Ecohydrology, UNEP/UNESCO. UNEP IETC Freshwater Management Series No 5, p. 155.

Tarczyńska, M., Romanowska-Duda, Z., Jurczak, T., and Zalewski, M., 2001. Toxic cyanobacterial blooms in a drinking water reservoir – causes, consequences and management strategy. Water Science and Technology: Water Supply 1(2), 237–246.

Vergano L. and Nunes, P.A.L.D., 2006. Analysis and evaluation of ecosystem resilience: an economic perspective. FEEM Working Paper No. 25.06.

Wagner, I. and Zalewski, M., 2000. Effect of hydrobiological patterns of tributaries on biotic processes in lowland reservoir – consequences for restoration. Ecological Engineering 16, 79–90.

Wagner-Lotkowska, I., Izydorczyk, K., Jurczak, T., Tarczynska, M., Frankiewicz, P., and Jorgensen, S.E., 2004. Ecohydrological methods of algal bloom control. In: M. Zalewski, I. Wagner-Lotkowska, and R.D. Robarts (Editors), UNESCO/UNEP Integrated Watershed Management-Ecohydrology & Phytotechnology – Manual, UNESCO-IHP, UNESCO-ROSTE, UNEP-IETC, International Centre for Ecology PAS, Department of Applied Ecology, University of Lodz, Venice Osaka, Shiga, Warsaw, Lodz, p. 210.

Wagner-Lotkowska, I., 2001. Influence of the selected climatic, hydrological and biological factors on eutrophication processes and symptoms in the Sulejow Reservoir. Ph.D. thesis. University of Lodz, Poland.

Wagner-Lotkowska, I., 2004. Application of ecohydrology and phytotechnology for water resources management and sustainable development. UNESCO/UNEP demonstration project. In: M. Zalewski, I. Wagner-Lotkowska, and R.D. Robarts (Editors), UNESCO/UNEP Integrated Watershed Management-Ecohydrology & Phytotechnology – Manual, UNESCO-IHP, UNESCO-ROSTE, UNEP-IETC, International Centre for Ecology PAS, Department of Applied Ecology University of Lodz, Venice Osaka, Shiga, Warsaw, Lodz, p. 210.

Wagner-Lotkowska, I., Bocian, J., Pypaert, P., Santiago-Fandino, V., and Zalewski, M., 2004. Environment and economy – dual benefit of ecohydrology and phytotechnology in water resources management: Pilica River demonstration project under the auspices of UNESCO and UNEP. Ecohydrology & Hydrobiology. Special Issue: Ecohydrology From Theory To Action 4(3), 345–352.

Wojtal, A., Frankiewicz, P., and Zalewski, M., 1999. The role of the invertebrate predator Leptodora kindti in the trophic cascade of a lowland reservoir. Hydrobiologia 416, 215–223.

World Water Council, 2006. Thematic document framework theme 1: water for growth and development. 4th World Water Forum Mexico City, March 2006.

Xiao, J. and Moody, A., 2004. Photosynthetic activity of U.S. biomes: response to spatial and temporal variability in temperature and precipitation. Global Change Biology, Online Publication: 02/20/04.

Zalewski, M., 1998. Ecohydrology and fisheries management. Italian Journal of Zoology 65(Suppl.), 504.

Zalewski, M., 1999. Minimizing the risk and amplifying the opportunities for restoration of shallow reservoirs. Hydrobiologia 395/396, 107–114.

Zalewski, M., 2000. Ecohydrology. The scientific background to use ecosystem properties as management tools toward sustainability of water resources. Guest editorial, Ecological Engineering 16, 1–8.

Zalewski, M., 2002a. Ecohydrology – the use of ecological and hydrological processes for sustainable management of water resources. Hydrological Sciences Journal 47(5), 825–834.

Zalewski, M. (Editor), 2002b. Guidelines for the integrated management of the watershed – phytotechnotogy and ecohydrology. UNEP/UNESCO, UNEP IETC Freshwater Management Series No. 5.

Zalewski, M., 2006. Flood pulses and river ecosystem robustness. Frontiers in Flood Research/ Le Point de la Recherche sur Les Crues. International Association of Hydrological Sciences (IAHS) Publication No. 305, 143–154.

Zalewski, M., Brewinska-Zaras, B., Frankiewicz, P., and Kalinowski, S., 1990. The potential for biomanipulation using fry communities in a lowland reservoir. Concordance between water quality and optimal recruitment. Hydrobiologia 200/201, 549–556.

Zalewski, M. and Naiman, R.J., 1985. The regulation of riverine fish communities by a continuum of abiotic–biotic factors. In: J.S. Alabaster (Editor), Habitat Modification and Freshwater Fisheries, FAO UN, Butterworths, London.

Zalewski, M., Tarczynska, M., and Wagner-Lotkowska, I., 2000. Ecohydrological approach for elimination of toxic algal blooms in lowland reservoir. Verh. Internat. Verein. Limnol. 27, 1–8.

Zalewski, M., Santiago-Fandino, V., and Neate, J., 2003. Energy, water, plant interactions: "green feedback" as a mechanism for environmental management and control through the application of phytotechnology and ecohydrology. Hydrological Processes 17, 2753–2767.

ENVIRONMENTAL QUALITY AND LANDSCAPE-HAZARD ASSESSMENT IN THE YANTRA RIVER BASIN, BULGARIA

MARYIANA NIKOLOVA*, STOYAN NEDKOV,
AND STOYAN IANKOV
Bulgarian Academy of Sciences, Institute of Geography, Sofia, Bulgaria

DARIUS SEMMENS
U S Environmental Protection Agency Office of Research and Development, Las Vegas, Nevada, USA

Abstract. The objective of the present work is to analyze the role of landscape for environmental security in the Yantra River Basin, exploring its relationships with river-water quality and flood hazard. The relationship between landscape and river-water quality is analyzed on the basis of landscape indicators and assessment tools like Automated Geospatial Watershed Assessment (AGWA) and Analytical Tools Interface for Landscape Assessment (ATtILA). The relationship between landscape and flood hazard is explored using set of flood-hazard indicators and the Soil and Water Assessment Tool (SWAT). The results from ATtILA implementation show that the main sources of nitrogen loading are the agricultural landscapes and the urban areas in the river basin. The SWAT simulation is done for three scenarios in which land cover (forest lands) changes are related to flood hazard. For the most unfavorable scenario, decreasing forest lands, a significant increase of the river discharge is predicted. The degree of environmental security depends strongly on the specific spatial patterns of landscape change in the river basin.

Keywords: Landscape modeling; water quality; hazard assessment

* Corresponding Author: Dr. Nikolova Maryiana, Bulgarian Academy of Sciences, Institute of Geography, Akad. G. Bonchev Str., Bl. 3, 1113 Sofia, Bulgaria, Phone: +3592 870 02 53, e-mail: mn@bas.bg

I. Petrosillo et al. (eds.), Use of Landscape Sciences for the Assessment of Environmental Security, 209–224.

1. Introduction

The Yantra Basin occupies parts of the southern slopes of the Central Stara Planina Mountain, its foothills being called the Pre-Balkan, and the Danube plain. The main river length is 285 km, its basin covers an area of 7,869 km^2, and the density of the river network ranges from 0.7 to 1.5 km/km^2 (Figure 1). Within the boundary of the basin there are more than 800 settlements distributed among 26 municipalities.

The Yantra River Basin is clearly divided into five landscape regions with common energy and matter cycles. The main sources of anthropogenic loading are the landscapes with predominantly agricultural lands and urban areas (Veliko Tarnovo, Gabrovo, and Gorna Oryahovitsa). Water quality is an integrated indicator of the environmental state and can thus be used to broadly assess environmental condition, as well as vulnerability to landscape change and associated implications for environmental security. The impact of the river water contamination is well represented by the results of Hydrobiological Monitoring (HBM) and Physico-Chemical Monitoring (PSM). For example, the assessment of data from HBM show that for a period of 20 years (1967–1987) the number of invertebrate taxa at Cholakovtci gauge (Veliko Tarnovo) decreased from 62 to 12 (Nedkov et al., 2005). Integrated analyses of the two kinds of monitoring can be used to assess the threats to important ecosystem services provided by river water.

Natural hazards, especially floods, are another source of environmental stress in the basin. Floods exert relevant impact on the environmental state in the basin and landscape patterns play a significant role in both amplification and mitigation of the environmental assimilative capacity. The flood hazard is characterized by a high level of complexity; the specific conditions in some parts of the basin, such as subbasin morphology, forest management, or land-use planning, can alter flood hazard significantly.

The aims of this investigation are:

1. To provide an assessment of the environmental security with respect to water quality, floods and flood-related hazards in the Yantra River Basin using digital elevation model, land cover, land use, landscape types, surface-water quality and water-quantity extremes
2. To use scenario analysis to assess possible changes in the system stability and ecosystem services with tools and models like AGWA, ATtILA, and SWAT

2. Materials and methods

2.1. DATA ACQUISITION

2.1.1. Water quality

The assessment of water quality is based on the results of physicochemical and hydrobiological monitoring carried out by the Regional Inspectorate for Environment and Water (RIEW) in Veliko Tarnovo. The system of HBM includes 214 sampling points. Bulgarian RIEWs have adopted the Irish Q-rating biotic index, which was adapted for local conditions and implemented in 1997. The biotic index has a five-level scale for water-quality assessment. The highest value, BI-5, is given for clear water with high quality and no anthropogenic influence; the lowest value, BI-1, is for extremely polluted water (State of the Environment, 2004). The data used in the investigation include results from 158 sampling points for 2000–2003, of which 37 were sampled twice per visit with collection of samples twice in a year (Figure 1). The short period of observation does not allow investigations on water-quality dynamics for longer time periods. For this reason we also used the information available from bioindication assessment based on the saprobiological method for the period from the 1960s to the beginning of the 1990s (Rusev, 1994; Janeva and Rusev, 1997). It is based on different biotic indices like the Pantle-Buk index, Zelinka-Marvan-Rothshtein index and others. The water quality is measured in five-level saprobity scale, which corresponds to particular values of every index. The water-quality categories of the biotic index and saprobity scale have similar characteristics and can be easily converted to the five-level biological classification of the rivers adopted in Bulgaria (Table 1).

The system for PCM includes 16 points with monthly sampling and 24 observed parameters. We used data for the period 1999–2004 from 10 sampling points for 13 parameters (dissolved oxygen, dissolved solids, suspended solids, manganese oxydability, biological oxygen demands (BOD_5), chlorides, sulfates, phosphates (PO_4), ammonia nitrogen (NH_4–N), nitrite (NO_2–N), nitrate (NO_3–N), iron, and manganese content) (Figure 1). The results from the HBM and PCM systems complement each other because the hydrobiological sampling points cover more precisely the river system, while the physicochemical ones ensure more information for selected points.

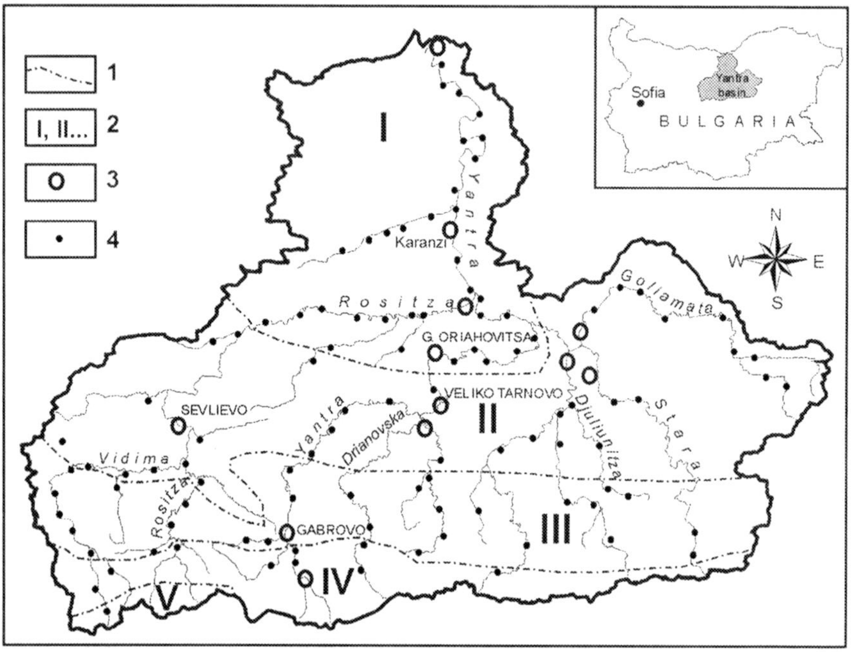

Figure 1. Yantra River Basin: 1 – Border between landscape regions; 2 – Landscape regions: I – Plain, temperate, dry subhumid; II – Hilly and plateau, temperate, moist subhumid; III–Low mountain, temperate, moist subhumid; IV – Mid mountain, temperate humid; V – High mountain, cold humid; 3 – Physicochemical monitoring sampling points; 4 – Hydrobiological monitoring sampling points.

2.1.2. Natural hazards (floods)

The objective of the investigation requires implementing indicators for the assessment of floods as a natural hazard, specifically in terms of their direct and indirect pressure on the landscape. Floods in the Yantra River Basin, which have caused significant material losses and can be the subject of flood-risk analysis, are classified as devastating when runoff is higher than 100 $m^3 s^{-1}$ (Gerasimov, 1992). During the period 1990–2005 52 floods were registered in the Yantra Basin, of which 25 were in the Yantra River, 21 in the Rositsa River, and 6 in the Vidima River.

The flood dimensions depend on the basin morphology; soils, land cover, flood-inducing rains, etc., and can be manifested as flood-related hazards like landslides and erosion. It is a complex pressure with impacts that are different for each location in the basin. To estimate this pressure, we need to know first, what are the hazardous phenomena displayed within the framework of a given territorial unit (municipality, basin, subbasin, etc.) and what are the probability and intensity with which they occur, and second, to what extent their total effect would influence the landscape security or/and vulnerability of the socioeconomic system. For this reason we propose the use of a complex parameter (R), (Nikolova, 1998, 2001), obtained as follows:

- Frequency of occurrence of each hazardous phenomenon at a given place is assigned to the territory of the municipality in which it is contained.
- Class interval (I) is calculated for the data set ($X_{1....n}$) about the frequency of the phenomenon in each municipality:

$$I = (X\ max - X\ min)/k. \tag{1}$$

- The number of hazard classes depends on the sample size and is determined according to the formula:

$$k = 1 + 3{,}3lg\ N \tag{2}$$

where: k – number of classes, and N – corresponds to the grouping of the real frequencies, to which the corresponding hazard classes are assigned for each single phenomenon.

- The assessment of the total loading of the municipalities is made on the basis of the formula:

$$R = \sum r/n \tag{3}$$

where: r – hazard class assigned to each phenomenon in a given municipality; n – number of the hazardous phenomena within the assigned risk class.

The assignment of hazard classes to the phenomena makes them comparable; at the same time it makes it possible to show clearly the relative weight of the pressure of each of them on the territory.

The following data were used to define the potential hazard: frequency of intensive rains (May–August); frequency of river raising and floods; zoning of erosion on the basis of the sediment yield; and number of active landslides, according to Nikolova et al. (1998).

2.2. LANDSCAPE CHARACTERIZATION

The construction of the landscape classification scheme has been carried out using elements from other schemes, developed for the territory of Bulgaria and adapted for the needs of the investigation. The characterization of landscapes in the Yantra River Basin has been made on the basis of a three-level hierarchical classification scheme.

The first level of differentiation corresponds to the landscape types according to Velchev et al. (1992). The territorial units are distinguished on the basis of differences determined by the effect of hydro-climatic factors. The degree of generalization conforms to a middle scale mapping. The so-formed territorial units correspond to the concept of a landscape region in the context of Forman (1995) and Farina (2000). Five landscape regions are distinguished in the Yantra River Basin (Figure 1; Iankov et al., 2004).

The second hierarchical level differentiates the Yantra River Basin into subbasins on the basis of specific features of relief, which determines the direction of water flows by the form of the topographic surface. The basin has been divided in 12 subbasins, 6 of them being on the main river, 2 on the main tributary – the Rositsa River, and 4 on the eastern tributaries – the Drianovska, Golyamata, Stara, and Djuliunitsa Rivers (Figure 1).

The third level of landscape differentiation is based on the European CORINE Land Cover classification. Of the 44 CORINE land cover classes, 37 are present in the Bulgarian territory (Vatseva, 2005). The application of this scheme, which has been adopted by almost all European countries, provides the possibility of comparing the results with other regions of the continent.

2.3. CHANGE DETECTION AND SCENARIOS DEVELOPMENT

The Executive Environmental Agency (EEA) provided the CORINE Land Cover for the years 1990 and 2000. It is structured in the form of a database in GIS format. This provides the possibility of establishing the land cover changes that took place within a 10-year period. For this purpose the land cover data for the two years have been compared, separating the detected transformation in a single layer (shape file). The obtained areas were grouped according to their land cover classes and a number of indicators were derived for analyzing their spatial position within the framework of the basin as a whole and within the subbasins.

Another source for change detection was the information from the Bulgarian Inquiry for Observation of Agriculture and Economic Conjuncture (BIOAEC). It covers the period from 1998 to 2003 and was developed on the basis of satellite images and field investigations (Nedkov et al., 2005). It was provided by the Ministry of Agriculture and includes land use data in the tables distributed according to the municipal boundaries. This information is transformed and conformed to the territorial units used in the investigation – landscape regions and subbasins.

These two sources show particular discrepancies and represent different trends in the land cover changes, which are due to differences in the methods and criteria used. This gives an opportunity to use CORINE and BOIAEC data to work out two future scenarios for the basin. The changes observed in the CORINE land cover could be interpreted as more representative of the current trend. These data were therefore used to create a future scenario characterized by relatively slight changes for the basin as a whole, with some significant differences between the subbasins. A second scenario was based on the data from BIOAEC, and represents future conditions that could be realized under the condition that plans for the development of this territory on local and regional level are realized. It is characterized by a significant increase of forest areas and reduction of agriculture lands. A third scenario was adopted to represent a pessimistic view of future changes in the basin, and is characterized by more significant replacement of forested areas with agricultural and barren lands.

2.4. INDICATORS

The assessment of the relationship between environmental security and landscape pattern was realized using landscape indicators (land cover, land use, and water quality) and flood-hazard indicators (catchment morphometry, rainfall intensity, river raising, floods, land cover, land use, landslides, and erosion). Each landscape region was characterized according to its indicator attributes (shape and size of the basins, land cover/land use changes as a percentage of the area, water-quality categories of the rivers, rains, floods, landslides and erosion intensity and potential, total loading of flood-related hazards, density of the population, and the exposure of settlements to hazards. Settlement vulnerability, or sensitivity to processes taking place in the river basin, was assessed with AGWA tool and its component SWAT model.

2.5. TOOLS AND MODELS

The AGWA tool is a GIS-based multipurpose hydrologic analysis system designed to: provide a simple, direct and repeatable method for hydrological modeling; use basic, attainable GIS data; be compatible with other geospatial basin-based environmental analysis software; and be useful for scenario development and alternative future simulation work at multiple scales (Miller et al., 2002, 2007). AGWA provides the functionality to conduct the processes of modeling and assessment for two component hydrologic models: SWAT and the Kinematic Runoff and Erosion Model (KINEROS2, Smith et al., 1995).

SWAT is a distributed lumped-parameter model developed at the US Department of Agriculture, Agricultural Research Service (USDA-ARS) to predict the impact of land management practices on water, sediment, and agriculture chemical yields in large complex basins with varying soils, land use and management conditions over long periods of time. It is a continuous-time model using daily average input values and has some major components including: hydrology, weather generator, sedimentation, soil temperature, crop growth, groundwater, and lateral flow (Arnold et al., 1994). The use of SWAT in the investigation is directed to evaluate the effect of changes in the landscape on the river flow and especially during the periods with floods. The output data can be represented as average yearly or monthly values as well as daily output, which enable the user to receive detailed information about the investigated characteristics.

ATtILA is an ArcView-based tool, which facilitates the process of deriving different kinds of landscape metrics (Ebert and Wade, 2004). It gives the opportunity to calculate four groups of metrics: landscape characteristics, riparian characteristics, human stressors, and physical characteristics. It was used for the water-quality analyses at subbasin level.

2.6. ASSESSMENT

The subbasins divided on the second level of the landscape differentiation were used to determine the effect of landscapes on water quality, and to differentiate the original sources of pollution. To evaluate water quality and define measures for its improvement, it is very important to differentiate the point and nonpoint sources. It has been established by experience that the individual land uses discharge different amounts of pollutants. For example, from one hectare of urban territory (land cover) 1.2 kg of phosphorus and 5.5 kg of nitrogen are discharged annually in water. The application of the Human Stressors option of ATtILA shows potential for exerting impact on river-water quality by nitrogen loading. A direct comparison of these data with the results from the physiochemical monitoring is difficult to accomplish because of differences in the applied methodology for obtaining them and a discrepancy between the measuring units. For this purpose we use a simple coefficient defined as the ratio between the two values: $K = m/As$, where m is the average amount of nitrogen or other water-quality constituent in the water of the respective point and As is the calculated quantity of the nitrogen incoming to the respective subbasin from the landscapes. It gives the opportunity to differentiate point and nonpoint sources of nitrogen load.

3. Results

3.1. LAND COVER CHANGES

In accordance with the performed analysis of the CORINE Land Cover data, it has been established that a total of 4,619 ha in the Yantra River Basin experienced some type of land cover change during the period from 1990 to

2000. It includes 162 areas with various types of land use (land cover) with an average area of 22 ha (min 5 ha, max 103 ha).

A typical feature of most of the changes for the period 1990–2000 is that they are observed in both directions: from transitional woodland-shrub vegetation to deciduous forests and the opposite. The largest share of changes belongs to the transformation from transitional woodland-shrub vegetation to deciduous forests (33% of the area and 54 changed areas); the reverse transformation (deciduous forests to transitional woodland-shrub vegetation) is the second largest change. Thus the real change for deciduous forests is reduced to 808.5 ha, while transitional woodland-scrub decreases with 716 ha (Table 1). This was due mostly to forestry activities, including felling of the forest during the considered period and to a greater extent the recovery of forest cut earlier. A comparison of the results shows that the areas with recovered forests exceed by about 50% the areas with felling.

TABLE 1. Land cover changes in the Yantra River Basin determined from the CORINE land cover data for the period 1990–2000. Total change includes all transformation in particular land cover type, while in real change column, the reverse transformations were subtracted.

Land cover type	Total change ha	%	Real change	%
Forests	1,829	40%	808.5	18%
Transitional woodland shrub	1,010	22%	–716	–16%
Pastures	393	9%	–84	–2%
Arable lands	1,181	26%	729	16%
Agriculture with natural vegetation	112	2.4%	–114	–2.5%
Urban areas	10.3	0.2%	10.3	0.2%
Industrial or commercial units	5.2	0.1%	5.2	0.1%

The largest portion of the changes for agricultural land is in the northern part of the basin, 95% of them being in landscape regions II and I. The areas with forests are increased in the subbasins of the rivers Dzhulyunitsa, Stara Reka, and Golyamata Reka at the expense of the areas with tree-shrub vegetation, while these areas have decreased in the subbasins of the Rositsa River upstream of its mouth, the Dryanovska and Yantra Rivers before the Cholakovtsi village.

According to the data of BIOAEC the agricultural lands on the territory of the Yantra River Basin have been decreased by about 2% during the period 1998–2003. This reduction is distributed across the whole basin with the exception of region I, where the most fertile land is concentrated (Nedkov et al., 2005). A trend of increasing forested lands is also observed, which is better expressed in landscape regions II (4%) and landscape region I (3%). The growth trend of forest areas coincides for both sources, but it is rather less expressed in CORINE data.

3.2. WATER QUALITY

According to the results of HBM, a clearly expressed tendency towards aggravation of water quality was observed in the period from the 1960s to the 1980s in accordance to social and economical changes in the region (Nedkov et al., 2005). After the late 1980s there is particular improvement of the water quality in all streams except G. Oriahovitza. Spatial analysis of the HBM data shows that during the period 1995–2004 the river-water quality has dropped down to the level observed during the 1980s and in the lower river course some quality improvement is observed relative to the 1990s (Table 2).

TABLE 2. Water quality of the Yantra River according to the biomonitoring results in 1967, 1987, 1995, and 2004. Water-quality classes: Water-quality classes according to BDS EN ISO 8689-1:2001: I – worst condition; II – bad condition; III – moderate condition; IV – good condition; V – very good condition. Locations of the sites are shown on Figure 1.

Year	Jabalka	Gabrovo	V. Turnovo	Samovodene	G. Oriahovitza	Karanzi	Mouth
1967	V	IV	IV	IV	IV	IV	IV
1987	V	I	I	II	I	II	II
1995	V	II	III	III	I	IV	III
2004	V	I	I	II	I	III	III

Analysis of the physicochemical data for the Yantra River showed biological oxygen demand (BOD_5), ammonium nitrogen (NH_4–N), nitrite nitrogen (NO_2–N), phosphates (PO_4), iron and insoluble substances exceeded the admissible limits. The points along the Yantra River after Gabrovo and Samovodene (after Veliko Tarnovo) and along the Rositza River after Sevlievo exhibit the greatest number of parameters (BOD_5, NH_4–N, NO_2–N, PO_4, Fe and insoluble substances) exceeding the admissible limits. In these places are situated the biggest urbanized and industrial centers and point sources of the river-water pollution.

To obtain more complete picture, the average values of water quality obtained from the HBM and the number of "hot spots" in each subbasin have been introduced as additional indicators. The results obtained from the application of the Human Stressors option of ATtILA show that the landscapes in the lower course of the Yantra River (Table 3) have the highest potential for exerting impact on river-water quality by nitrogen emissions. To differentiate the point and nonpoint sources of contamination with ammonium nitrogen, a coefficient (K) was calculated. Analysis of the results shows that for the subbasins where nitrogen has mainly nonpoint sources, the value of this coefficient is between 0.30 and 0.45.These data confirm some of the established results, for example the existence of a pollution source at the Gabrovo station. For other subbasins, however, the additional indicators establish the presence of point sources that were not identified previously. In particular, the subbasin of the Yantra River before the Karantsi point, where the coefficient is only 0.33 but the water quality measured is rather low (2.6). Further investigation, by means of additional differentiation for the subbasin, revealed the source of this discrepancy: It is the strongly urbanized zone around the Gorna Oryahovitsa town. An analogous procedure was applied for the subbasins of the Yantra River before Gabrovo and of the Rositsa River before Sevlievo (Table 3).

TABLE 3. Water quality indicators and characteristics in some subbasins of the Yantra River.

Subbasins	Sum N mg/l	N load kg/ha/year	K	Mean HBM	No. of hot spots
Yantra Jabalka	0.8	2.5	0.32	4.0	0
Yantra Etara		2.8		3.6	0
Yantra Gabrovo	7.0	4.1	1.73	2.3	1
Yantra Samovodene	2.2	3.8	0.57	1.9	1
Yantra G. Oriahovitza		6.2		2.1	2
Yantra Karanzi	2.2	6.6	0.33	3.0	
Yantra mouth	2.2	7.2	0.31	2.9	0
Rositza before Sevlievo		4.0		3.8	
Rositza Sevlievo	2.2	5.6	0.39	2.5	1
Rositza mouth	2.9	5.8	0.50	2.9	1

3.3. NATURAL HAZARDS

The Yantra River Basin is characterized by high natural hazards pressure. The main hazards, floods and intensive rains, are also important factors for the activation of landslides and erosion. To evaluate the pressure of the floods and flood-induced hazards on the territory and its impact on the landscape, we first estimated the complex pressure of these hazards on the river basin, and then estimated how the changes in land cover, especially in the forest lands, enhance or mitigate the flood hazard in the catchment. Four hazard classes (1 – low, 2 – medium, 3 – high, and 4 – very high) were estimated for intensive rains, floods, landslides and erosion on the basis of frequency of occurrence of each phenomenon within each of the municipal territories in the Yantra Basin (Nedkov and Nikolova, 2006). The total loading of all investigated hazards is represented by a complex parameter (R), (Figure 2). The results show that in all landscape regions the intensive-rains hazard is estimated as medium to high, flood hazard is medium only in region I and high in regions II, III, and IV. Landslide hazard is estimated as low for region I, high for region II and very high for regions III and IV. The erosion impact is low in landscape regions I, II, and IV and medium for region III. The complex pressure of the investigated hazards according to R parameter is highest for landscape region III (R = 12), followed by region IV (R = 10), III (R = 9), and I (R = 6). The highest values of R are observed in the southern part of the basin around the municipalities Veliko Tarnovo, Sevlievo, and Elena, which occupy landscape regions III and IV. The lowest values of R are for the municipalities of Polski Trambesh, Svishtov and Biala, situated in the northern part of the basin in landscape region I.

Forests have a significant regulatory effect on surface runoff, river flow, and the whole water balance in the basin, but they also have regulatory functions for some geomorphic hazards induced from heavy precipitation and floods, such as landslides and soil erosion. SWAT modeling was implemented to analyze how some changes in the natural system, such as detected changes in forest areas, enhance or mitigate the expected treat of floods.

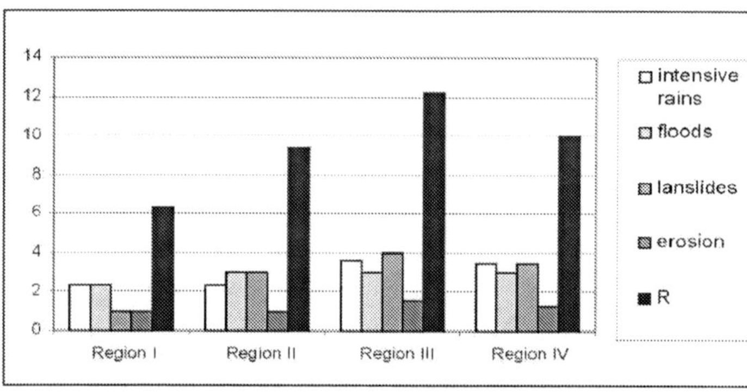

Figure 2. Distribution of the hazard classes for intensive rains, floods, landslides, erosion, and the complex hazard class R, by landscape regions in Yantra River Basin.

Two stream-flow stations were used for the process of hazard modeling: Veliko Tarnovo and Gabrovo. The subbasins of the river upstream of these stations were delineated using AGWA. Five flood events were chosen for the process of modeling. Two of them (26 May 2005 and 06 July 1991) are characterized as very high hazard. The other three (10 May 1993, 13 April and 20 June 1992) are characterized as high level of hazard. The model simulation representing the first land cover scenario shows a slight increase (between 0.5% and 1.5%) in water quantity at the Veliko Tarnovo station, and an even smaller change in Gabrovo. The second scenario is characterized by a significant increase of the forested areas at the expense of arable land, transitional woodland-scrub and some urban areas. The simulation results for this scenario show a decrease of the peak flow varying from 3% to 26% for the different events. The third and most unfavorable scenario resulted in an increase of the river discharge varying from 3% to 18% (Figure 3).

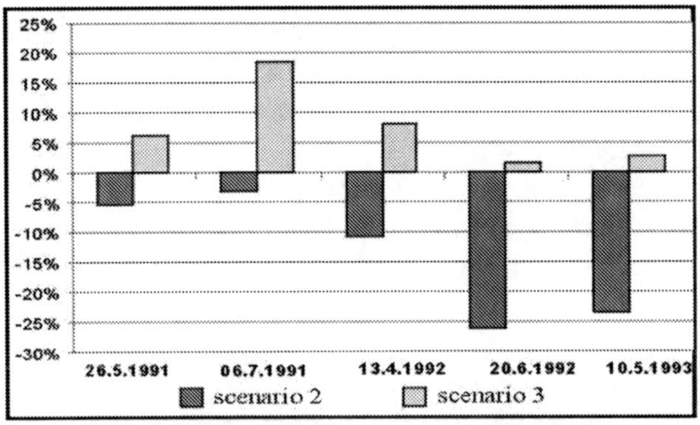

Figure 3. Change of the peak river discharge at Veliko Tarnovo station in percent for five flood events according to model simulations for two scenarios (scenario 1 was not included because of the insignificant changes).

Further investigation was directed to evaluate the change of flood hazard for the three scenarios. According to the measured water quantities for the period 1987–1998, there were 18 cases with river risings that could be characterized as a flood hazard. The modeling results from second scenario reduce the water quantity for three of them below the hazardous rating cutoff of 100 $m^3 s^{-1}$ at Veliko Tarnovo station. Of the two observed events classified as very high hazard (above 350 $m^3 s^{-1}$), one of them was reduced to the high hazard level according to the model. This translates into an overall 17% reduction of the flood hazard for that area in scenario 2. Results from the third scenario show that there are two additional cases exceeding the 100 $m^3 s^{-1}$ threshold, and another five coming very close. This increases the overall flood hazard by 11% or 28%, respectively.

4. Discussion and conclusions

The obtained results demonstrate that the level of environmental security with respect to both the river-water quality and flood hazard varies in the boundaries of the Yantra River Basin, and depends on the landscape patterns. The tendency towards aggravation of water quality observed in the period from the 1960s to the 1980s is due mainly to the processes of industrialization and urbanization during this period, which lead to increasing the share of both industrial pollutants and household wastewater. It is characterized with intensive use of ecosystem values provided by river water. Political changes in the country during the 1990s led to an economic crisis, industrial collapse and population decline. This in turn reduced anthropogenic pressure on the environment and as a result, definite improvement of water quality and the ecological state of aquatic landscapes was observed. These changes occurred spontaneously and not as result of any purposeful measures for improving the quality of environment and hence the environmental security. Economic development towards the end of the 1990s and beginning of the 21st century led to new growth of anthropogenic pressures and to aggravation of the ecological conditions under qualitatively new circumstances, which are illustrated by the substantial territorial differences in water quality. The use of AGWA and ATtILA provides the opportunity to differentiate the pressure from point and non-point sources of contamination and in this way to analyze the relationship between water quality and land-use changes. The obtained results show that the main sources of nitrogen load in the low part of the Yantra Basin are agriculture landscapes, while in the middle part the impacts of urban area (point sources) is dominant. The impact of the river-water contamination is long term, and the impact of some of the most devastating floods, like those observed in the Yantra Basin in July 2005, has a short-term impact that usually lasts until the natural restoration of the ecosystem. However, some flood-related hazards, such as erosion or landslides, lead to land degradation that could also have a long-term impact.

The implementation of SWAT enabled the evaluation of the role of land cover changes on flood dimensions. The increase of natural vegetation and especially forest-land cover reduces the hazard, while more agricultural, urban, or barren lands increase the flood threat. This influence varies around the different subbasins according to the landscape patterns.

All investigated indicators show that the anthropogenic and natural hazard pressure is highest in the low mountain landscape region III as a result of land cover change, poor water quality, and abundant natural hazards. The biggest settlements, Veliko Turnovo and Gabrovo, with high densities of population, transportation, and industrial infrastructure, are situated in the mountainous landscape regions (III and IV). It makes these regions more vulnerable and increases the ecological and socioeconomic risk. The implementation of the complex parameter R gives good results for risk management on both landscape and administrative levels.

Acknowledgments

The authors are grateful to Professor Gregory Knight who was kind to provide for this investigation hydrologic, climatologic and DEM data from the REKA project, sponsored by US National Science Foundation.

This work was supported from the Regional Environmental Inspectorate at Veliko Turnovo and we would like to express our gratitude to its Director, Ms. Elena Grigorova.

References

Arnold, J.G., Williams, J.R., Srinavasan, R., King, K.W., and Griggs, R.H., 1994. SWAT-Soil Water Assessment Tool. USDA, Agricultural Research Service, Grassland, Soil and Water Research Laboratory, Temple, Texas, http://www.brc.tamus.edu/swat/

Ebert, D.W. and Wade, T.G., 2004. Analytical Tools Interface for Landscape Assessments (ATtILA) Version 2004 User Manual. US-EPA Report EPA/600/R-04/083. http://www.epa.gov/nerlesd1/land-sci/attila/

Farina, A., 2000. Principles and Methods in Landscape Ecology. Kluwer, Dordrecht, 256 pp.

Forman, R.T., 1995. Some general principles of landscape and regional ecology. Landscape Ecology 10, 133–142.

Gerasimov, S., 1992. Chronology and classification of the big floods in Bulgaria. Proceedings from Scientific Conference on Protection of the Population from Disasters and Accidents, Book IV, Sofia, 119–126.

Iankov, S., Nikolova, M., and Nedkov, S., 2004. Use of bioindicators for landscape assessment in the Yantra river basin, Central North Bulgaria. EcoSys 42, 35–49.

Janeva, J. and Rusev, B., 1997. Veranderungen der Artenzusammensetzung und Guteklasse des Bulgarishen Donauzuflusses Yantra nach dem Makrozoobenton. Lauterbornia 31, 1–6.

Miller, S.N., Semmens, D.J., Miller, R.C., Hernandez, H., Goodrich, D.C., Miller, W.P., Kepner, W.G., and Ebert, D., 2002. Automated Geospatial Watershed Assessment (AGWA): A GIS-based hydrologic modeling tool. Proceedings of the Second Federal Interagency Hydrologic Modeling Conference, Las Vegas, NV, 7/28–8/1.

Miller, S.N., Semmens, D.J., Goodrich, D.C., Hernandez, M., Miller, R.C., Kepner, W.G., and Guertin, D.P., 2007. The automated geospatial watershed assessment tool. Environmental Modeling and Software 22(3), 365–377.

Nedkov, S., Nikolova, M., and Iankov, S., 2005. Demographical changes and their influence on the landscapes in the area of Yantra river basin. EcoSys 11, 85–98.

Nedkov, S. and Nikolova, M., 2006. Modeling flood hazard in Yantra river basin. CD Proceedings of Papers from International Conference on Water Observation and System for Decision Support, 2326 May, Ohrid, Macedonia.

Nikolova, M., Aleksiev, G., Zyapkov, L., Tishkov, H., Kroumova, J., Velev, S., and Vlaskov, V., 1998. Scientific research project “Risk Analysis of Disasters and Catastrophes of Natural and Technogenic Origin in Bulgaria”. Ministry of Education and Science, Sofia (in Bulgarian).

Nikolova, M., 2001. Natural hazard pressure on the territory. Proceedings of the International Scientific Session "Management of Natural and Technogenic Risks", 4–8 June, Sofia, 173–176.

Rusev, B. (Editor), 1994. Limnology of the Bulgarian Tributaries of the Danube, Sofia, 255 pp. (in Bulgarian).

Smith, R.E., Goodrich, D.C., Woolhiser, D.A., and Unkrich C.L., 1995. KINEROS – A kinematic runoff and erosion model. In: V.J. Singh (Editor), Computer Models of Watershed Hydrology, Water Resources Publications, Highlands Ranch, CO, pp. 697–732.

State of the Environment, 2004. Regional Inspectorate of Environment and Water – Velico Turnovo. http://www.riosv-vt.hit.bg.

Vatseva, R., 2005. Land cover and relief elements mapping and updating based on satellite imagery. Ph.D. thesis, 195 pp.

Velchev, A., Todorov, N., Asenov, A., and Beruchashvili, N., 1992. Landscape map of Bulgaria in scale 1:500000. Annual of the Sofia University "St. Kliment Ohridski", Faculty of Geology and Geography, Book 2 – Geography, Vol. 84: 25–37 (in Bulgarian).

THE INFLUENCE OF CATCHMENT LAND COVER ON PHOSPHORUS BALANCE FOR LARGE FRESHWATER SYSTEMS

SERGEY A. KONDRATYEV*
Institute of Limnology of the Russian Academy of Sciences, Sevastyanova St., 9, 196105 St. Petersburg, Russia

Abstract. A load assessment method applicable to catchments with multiple land cover types was studied using a mathematical model of phosphorus balance for the Lake Ladoga and the Neva Bay freshwater systems (an area of about 280,000 km^2). The calculation of phosphorus load included both point and nonpoint sources of loading, as well as phosphorus retention in receiving waters. The decreasing agricultural land use area for the Russian part of the studied area was documented and its effects on water quality demonstrated. The importance of Lake Ladoga in retaining total phosphorus (P_{tot}) from upstream reaches and preventing phosphorus export into the already anthropogenically eutrophied Neva Bay was demonstrated. For the largest urban area in the studied region (St. Petersburg), municipal wastewater inputs in the Neva Bay were assessed using historic data and planned future wastewater treatment facilities.

Keywords: Phosphorus balance; point and nonpoint sources; phosphorus retention; wastewater treatment

1. Introduction

The water quality of lakes and rivers and their ecological condition are strongly influenced by the nutrient loads that enter these water bodies from the catchments they drain. Not surprisingly, different types of catchment land cover have different influences on pollution load formation. For example, if runoff arises from agricultural and urban areas, rather than natural land cover, nutrient loads

* Corresponding Author: Dr. Sergey A. Kondratyev, Institute of Limnology RAS, Sevastyanova 9, 196105 St. Petersburg, Russia, Phone: +7 812 3870276, Fax: +7 812 3887327, e-mail: kondratyev@limno.org.ru

I. Petrosillo et al. (eds.), Use of Landscape Sciences for the Assessment of Environmental Security, 225–235.

tend to increase while nutrient retention tends to decrease. The role of point source discharges in load formation in urban areas and their influence on water bodies depend on different economic factors and management plans.

For large-scale aquatic systems that consist of a complex network of lakes and rivers in a complex multiuse landscape, ecological simulations derived from mathematical models provide effective practical means for assessing the influence of catchment land cover on surface water quality and ecological condition. Such tools can be used to assess the response of various water bodies to different pollution loads.

The study area is Lake Ladoga, Neva Bay of the Gulf of Finland and their catchment – the largest European transboundary freshwater system. Phosphorus is an element which defines the eutrophication of studied freshwater system (Petrova, 1982) therefore a special model of total phosphorus (Ptot) balance was developed. It is the first successful experience of collection of information sufficient for modeling of phosphorus regime which covers the whole studied water system. The goal of the present study is to demonstrate a method of load assessment for different catchment land covers using a simple mathematical model of Ptot balance in a large-scale aquatic system.

2. Studied area and description of the model

Lake Ladoga is the largest lake in Europe, with a water surface of about 18,000 km^2 (Petrova, 1982; Kondratyev et al., 2002). It is connected with the Gulf of Finland and Baltic Sea by the Neva River. The lake volume is about 840 km^3, and its average and maximum depths are 46.5 m and 230 m, respectively. Its drainage basin (about 280,000 km^2) covers parts of seven administrative regions of Russia (76% of the territory), four provinces of Finland (24%) and a small part of Belarus. This area is highly industrialized with about 600 industrial complexes and 680 agricultural enterprises present in the Russian portion of the catchment alone. The Lake Ladoga drainage includes the watersheds of lakes Saimaa (Finland), Onega and Ilmen (northwest Russia), which are connected with Lake Ladoga by the Vuoksa, Svir, and Volkhov rivers. At present time the lake is mesotrophic.

Neva Bay (400 km^2) is a freshwater portion of the Gulf of Finland and has a shallow eutrophic littoral zone and mesotrophic center. The catchment area of Neva Bay is 67% of the total Gulf of Finland catchment and it receives runoff from the most heavily loaded and eutrophied portions of the Gulf of Finland. The water quality and ecological condition of Neva Bay are most strongly influenced by the city of St. Petersburg located in the Neva River delta and by the outflow from Lake Ladoga.

Five main water bodies (lakes Saimaa, Onega, Ladoga, Ilmen, and the Neva Bay) and five sub-catchments (shown in Figure 1), were selected as the main reporting units for the modeling (Kondratyev, 2005). Areas of studied water bodies, their catchments, and water surface in catchments are presented in Table 1.

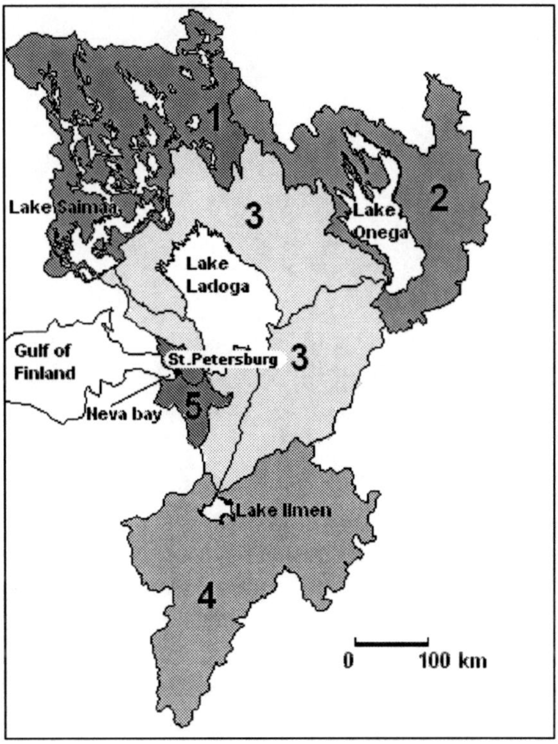

Figure 1. Scheme of the aquatic system of Lake Ladoga and the Neva Bay used in modeling: selected sub-catchments of lakes Saimaa (1), Onega (2), Ladoga (3), Ilmen (4), and the Neva River and Neva Bay (5).

Each of five selected water bodies and their catchments acts as a phosphorus sink that accepts Ptot load and retains some portion thereof, and that also serves as a source of Ptot loading to the downstream receiving waters. The mathematical background of the conceptual unit-area-load modeling with time step of 1 year is presented below.

TABLE 1. Studied water bodies and their catchment and water surface areas (Kondratyev, 2005).

Name	Area (km^2)
Saimaa catchment/including water surface	56,130/8,419
Onega catchment/including water surface	41,770/3,721
Ladoga immediate catchment/including water surface	93,058/2,857
Ilmen catchment/including water surface	66,190/1,215
Neva river and Neva Bay catchment/including water surface	6,660/36
Lake Saimaa	4,460
Lake Onega	9,720
Lake Ladoga	17,329
Lake Ilmen	1,200
Neva Bay of the Gulf of Finland	400

Annual P_{tot} load on the water body from the catchment area f_l (kg yr^{-1}) is:

$$f_1 = (\sum_{i=1}^{n} k_i F_i + L_p + L_a + l_a F_{cw})(1 - R_c), \tag{1}$$

where k_i – coefficient of P_{tot} emission in water bodies for i – type of land cover (kg km^{-2} yr^{-1}), F_i – area of i-type land cover (km^2), L_p – load from point sources (kg yr^{-1}), L_p – load from animal husbandry (kg yr^{-1}), l_a – atmospheric specific load (kg km^{-2} yr^{-1}), F_{cw} – area of surface waters in catchment (km^2), R_c – coefficient of P_{tot} retention in surface waters (dimensionless). The following empirical relationship was used for R_c assessment (Ostrofsky, 1978):

$$R_c = \frac{24}{(30 + \frac{wF_c}{F_{cw}})}, \tag{2}$$

where w is annual runoff (m yr^{-1}). The relationship is best suited to the assessment of P_{tot} retention in Lake Ladoga (Petrova, 1982). The equation of P_{tot} annual balance in water body is:

$$k\frac{d(C_p V)}{dt} = f_1 + f_2 + f_3 + f_3 + f_4 + f_5 - f_6 - f_7, \tag{3}$$

where C_P – annual P_{tot} concentration in water body (mg m^{-3}), V – water volume (km^3), k – coefficient of dimension changes, f_2 – P_{tot} inflow from upper parts of water system (for Lake Ladoga it is a sum of inflows from lakes Saimaa, Onega and Ilmen) (kg yr^{-1}), f_3 – P_{tot} input from bottom sediments (internal load) (kg yr^{-1}), f_4 – atmospheric deposition (kg yr^{-1}), f_5 – inputs of P_{tot} from direct point sources (kg yr^{-1}), $f_6 - P_{tot}$ retention in water body (kg yr^{vl}), $f_7 - P_{tot}$ outflow (kg yr^{-1}). Annual phosphorus retention in water body is:

$$f_6 = R_w \sum_{i=1}^{4} f_i \,, \tag{4}$$

where R_w is coefficient of P_{tot} retention in water body (dimensionless), calculated by using the equation similar to Eq. (2).

Annual runoff w is needed for calculating P_{tot} retention in surface waters (see Eq. 2). These values were derived from the hydrological model (Kondratyev and Shmakova, 2005). The model consists of sub-models of snow cover formation and snow melting, soil moisture dynamics and evaporation, and rain and meltwater runoff (divided into slow and fast components). The model is based on a set of semiempirical equations that account for variability in the timing of the phenomena under study without regard for their spatial distribution. The model based on hydrometerological and land surface parameters works for daily, monthly and annual time steps. As was previously shown (Kondratyev and Shmakova, 2005), the specific feature of hydrological model is dependence of the coefficient of hydrograph abatement ($k_a = Q(t+\Delta t)/Q(t)$, where Q is water discharge) on percentage of lakes in catchment area. The relationship between mentioned parameter of hydrological model and percentage of lakes in catchment area based on results of hydrological model verification with Δt = 1 month on different sub-catchments in Lake Ladoga basin is presented in Figure 2. It demonstrates the principle that the smaller an area of lakes in the catchment, the faster the runoff hydrograph abatement takes place.

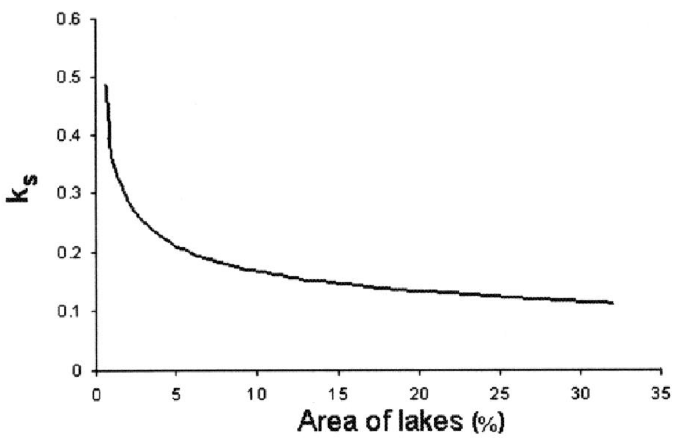

Figure 2. Dependence of hydrological model parameter on percentage of lakes in catchment (Kondratyev and Shmakova, 2005).

3. Catchment land covers – impact and results of calculation

Water bodies surrounding a catchment is one of the most important sources of its external nutrient loading. Various types of land cover contribute various types and amounts of chemical nonpoint runoff to surface waters. Land cover changes can, therefore, result in changes in catchment loading and, thus, changes in water quality and ecological condition. Annual P_{tot} runoff in a catchment's surface waters can be calculated using Eq. (1). Long-term field studies of P_{tot} concentration in runoff in selected sub-catchments were used for assessment of P_{tot} nonpoint emission in different land covers (Kondratyev, 2005; Kondratyev et al., 2003). The values of k_i for various types of land cover in the studied area used in modeling are presented in Table 2.

TABLE 2. Coefficients of nonpoint emission (k_i) of P_{tot} in surface waters for various land covers in the Gulf of Finland catchment (Kondratyev, 2005).

Land cover	k_i, kg $km^{-2}yr^{-1}$
Marsh	29
Forest	35
Agriculture	110
Urban	90
Others	30

Data on changes of various land cover areas F_i, P_{tot} load from point sources L_p such as industrial and municipal wastewaters dischargers and P_{tot} load from animal husbandry L_a were taken from materials of previous international studies (Kondratyev et al., 2000, 2002). P_{tot} atmospheric deposition and P_{tot} internal load were taken from the results of the Institute of Limnology studies (Ignatyeva, 1997; Kondratyev et al., 1996, 1997). The trend in decreasing agricultural land use in the Russian part of the study area is shown in Figure 3. It is one of the important sources of declining external P_{tot} loading into Lake Ladoga and the Neva Bay and is the primary reason for the declining P_{tot} concentration in these water bodies.

Lakes, rivers, and reservoirs strongly influence catchment phosphorus dynamics and P_{tot} retention. The values of the coefficient of retention for the studied water bodies which were calculated using mean annual runoff as w in Eq. (2) are presented in Table 3. Lake Onega had the highest retention coefficient P_{tot}, with an average water residence time of approximately 18 years, while Lake Ladoga had a retention coefficient of 0.70, and a retention time of about 11 years. The residence time for Neva Bay is only about one week and the P_{tot} retention of Neva Bay is, therefore, only approximately 8% of the total load.

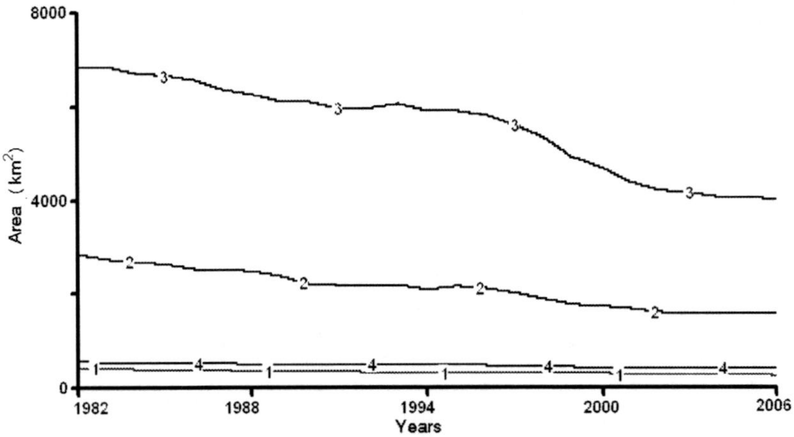

Figure 3. Dynamics of agricultural areas in the Lake Onega catchment (1), the Lake Ilmen catchment (2), the Lake Ladoga immediate catchment (3), and the Neva River and Neva Bay catchment (4).

TABLE 3. Coefficients of P_{tot} retention in water bodies and catchment surface waters.

Name	Coefficient of P_{tot} retention
Lake Saimaa	0.71
Lake Onega	0.76
Lake Ladoga	0.70
Lake Ilmen	0.53
Neva Bay	0.08

It should be noted that calculations of P_{tot} retention presented here are, at best, rough estimates because the runoff characteristics are not constants as assumed and because retention is related to the amount of surface waters in the catchments.

The comparison of calculated and measured P_{tot} concentrations in Lake Ladoga (Figure 4) shows that the model performs satisfactorily. The decrease of P_{tot} concentration in Lake Ladoga can be explained by decreased loadings from external components. The primary reasons for this decrease were effective environmental protection measures in the 1980s and the economic crisis in Russia in the 1990s.

A simulation modeling with the aim of assessment of natural an anthropogenic parts of P_{tot} catchment load on Lake Ladoga and Neva Bay was developed. Different model inputs are presented in Table 4 Atmospheric P_{tot}

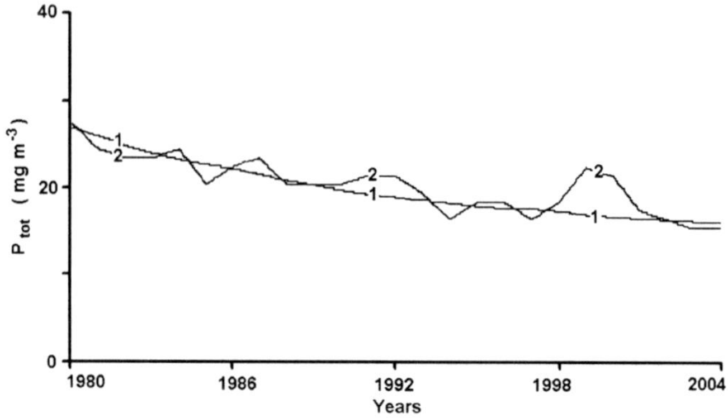

Figure 4. Calculated (1) and measured (2) P_{tot} concentration in Lake Ladoga.

deposition on Lake Ladoga and Neva Bay were estimated as 34.5 and 3.5 t year^{-1}, respectively. Internal P_{tot} load on Lake Ladoga and Neva Bay were estimated as 875 and 60 t year^{-1}, respectively.

The results of simulation modeling showed (see Table 4) that currently nonpoint runoff from natural land cover (forests and marshes) constitutes approximately 71% of P_{tot} entering Lake Ladoga, while approximately 29% of P_{tot} comes from anthropogenic sources. The values for Neva Bay are 28 and 72%, respectively.

TABLE 4. Calculated P_{tot} catchment load (t year^{-1}) in 2005 for different initial condition of simulation modeling

Model input	P_{tot} catchment load on Lake Ladoga	P_{tot} catchment load on Neva Bay
Natural load (all catchment is covered by forest and marshes)	2,412	1,235
Natural load + emission from anthropogenic lands (fields and urban areas)	2,559	1,305
Natural load + emission from anthropogenic lands + load from animal husbandry	2,974	1,606
Real load (natural load + emission from anthropogenic lands + load from animal husbandry + point load)	3,386	4,343

St. Petersburg, located in the Neva River delta (Figure 1), is the largest urban area in the study region. P_{tot} loads from the wastewaters of St. Petersburg into Neva Bay are 25–30% higher than the calculated load from Lake Ladoga with Neva River runoff (Kondratyev, 2005). The next simulation calculation was produced with the aim of assessment of P_{tot} content in Neva Bay depending on different scenarios of wastewater treatment in treatment plants.

The point source wastewater inputs into Neva Bay from St. Petersburg used in the developed model were taken from plans for future municipal wastewater treatment facilities (Karmazinov, 2002). The assessment of municipal wastewater inputs into Neva Bay is based on historic data and planned future wastewater treatment facilities is presented in Figure 5. The following values of P_{tot} concentration in treated waters were used in the calculation: 1.3 mg l^{-1} for the Central aeration station (CAS), 0.9 mg l^{-1} for the Northern aeration station (NAS), 1.5 mg l^{-1} for the South-West treatment plant (SWTP), 2.2 mg l^{-1} for the Krasnoselskaya aeration station (KAS), and 1.9 mg l^{-1} for the other municipal treatment plants. The value of P_{tot} concentration in untreated waters was assumed to be 3.2 mg l^{-1}.

An assessment of expected P_{tot} concentration in Neva Bay in 2012 under various scenarios of wastewater treatment in St. Petersburg is presented in Table 5. Wastewater treatment under HELCOM recommendations (1.5 mg P_{tot} l^{-1} for all treatment plants) will lead to a P_{tot} decrease in Neva Bay of only about 5% compared to 2004 levels, which is worse than results calculated using the present level of treatment. Additional wastewater treatment of all treatment plants according to EU recommendations (1.0 mg P_{tot} l^{-1} in treated waters) will lead to a decrease of P_{tot} content in Neva Bay of about 22%. Additional treatment up to 0.8 mg P_{tot} l^{-1} in treated waters will lead to a decrease of P_{tot} content in Neva Bay waters by 29% of the level of 2004. These results can be used by decision-makers for evaluation of perspectives of practical application of new and very expensive technologies of municipal wastewater treatment by taking into account the possible response of Neva Bay water quality.

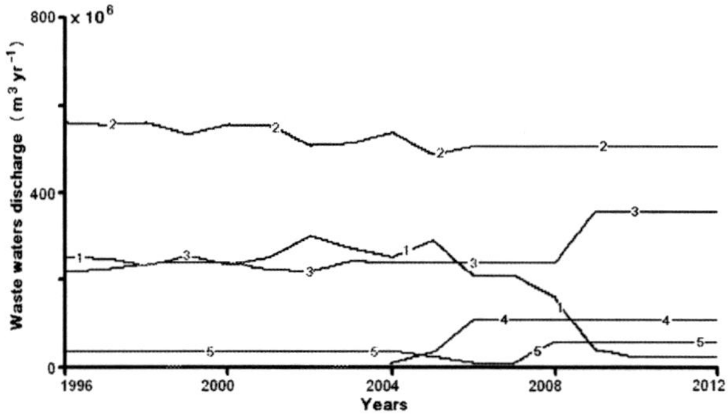

Figure 5. Dynamics of municipal wastewater inputs in the Neva Bay based on historic data and planned future municipal wastewater treatment facilities: 1 – untreated wastewaters, 2 – CAS, 3 – NAS, 4 – SWTP, 5 – KAS.

TABLE 5. Assessment of future changes of P_{tot} concentration in the Neva Bay depending on different scenarios of wastewater treatment in St. Petersburg.

Scenario	P_{tot} concentration in 2012 μg l^{-1}	% of the value of 2004 (39.8 μg l^{-1})
Without treatment	58.2	153
Plan of wastewater treatment (Karmazinov, 2002)	32.3	85
1.5 mg P_{tot} l^{-1} in treated waters	36.2	95
1.0 mg P_{tot} l^{-1} in treated waters	29.7	78
0.8 mg P_{tot} l^{-1} in treated waters	27.1	71
Without municipal wastewaters	15.7	41
Without municipal wastewaters and without retention in Lake Ladoga	48.8	128

At the same time a set of hypothetical scenarios was calculated. If the wastewaters of St. Petersburg are discharged into Neva Bay without treatment, the annual P_{tot} concentration will increase up to 58.2 μg l^{-1} (that is 53% higher than the calculated level of 2004). If P_{tot} retention in Lake Ladoga is excluded from the calculation, its concentration in the Neva River will reach 48.8 μg l^{-1} or 128% of the level of 2004.

4. Conclusions

The mathematical model presented is an effective tool for assessing the effects of loadings arising from various land cover types on water quality. The decreasing agricultural land use area for the Russian part of the study area was documented and it is one of the primary reasons for declining external P_{tot}, loading on surface waters. The results of simulation modeling shown that currently nonpoint runoff from natural land cover (forests and marshes) constitutes approximately 71% of P_{tot} entering Lake Ladoga, while approximately 29% of P_{tot} comes from anthropogenic sources. The values for Neva Bay are 28% and 72%, respectively.

Lake Ladoga is of great importance in retaining phosphorus loadings from the upstream reaches. About 70% of incoming P_{tot} is retained in Lake Ladoga, which reduces loading into Neva Bay and helps reduce anthropogenic eutrophication.

Model results show that P_{tot} concentration in Neva Bay is governed by the inputs from St. Petersburg and the outflow from Lake Ladoga. Implementation of a scenario that decreases wastewater discharges will lead to a decrease of P_{tot} content in Neva Bay in 2012 of about 15% compared to 2004 levels.

The important result for future studies is that a balance of P_{tot} for large scale aquatic systems was developed. All the main sources of P_{tot} loading (point and nonpoint) were included in this balance. It means that the developed model can be used for environmental security needs as a tool for simulation modeling and forecasting of the response of water bodies to different ansthropogenic impacts (including landscape changes).

References

Ignatyeva, N.V., 1997. Phosphorus in sediments and phosphorus exchange across the sediment-water interface in Lake Ladoga. Ph.D. thesis, Institute of Limnology, St. Petersburg, Russia, 24 pp. (in Russian).

Karmazinov, F.V. (Editor), 2002. Environmental report 1997–2001. Vodokanal of St. Petersburg, Russia, 68 pp.

Kondratyev, S., 2005. Land cover assessment and phosphorus balance modeling for large freshwater systems. Ecosystems 11, 6–12.

Kondratyev, S.A. and Shmakova, M.V., 2005. Study of river runoff using methods of mathematical modeling, watershed of Lake Ladoga as a case study. Proceedings of XII Assembly of Russian Geographical Society 6, 99–104 (in Russian).

Kondratyev, S.A., Efremova, L.V., Sorokin, I.N., Egorov, A.N., Kulibaba, V.V., and Rodionov, V.Z., 1996. Estimation of the external load on the Gulf of Finland. Ecological Chemistry 5(4), 240–249 (in Russian).

Kondratyev, S.A., Efremova, L.V., Raspletina, G.F., Tchernich, O.A., Veselova, M.F., Kulibaba, V.V., and Svistov P.F., 1997. Estimation of the external load on Lake Ladoga. Ecological Chemistry 6(2), 73–84 (in Russian).

Kondratyev, S., Gronskaya, T., Wirkkala, R.-S., Yefremova, L., Gayenko, M., Markova, E., and Aksenchuk, I., 2000. The geographical information system "Water resources of Lake Ladoga and its drainage basin". Publications of Karelian Institute, University of Joensuu, 129, Finland, 338–343.

Kondratyev, S., Gronskaya, T., Ignatieva, N., Blinova, I., Telesh, I., and Yefremova L., 2002. Assessment of present state of water resources of Lake Ladoga and its drainage basin using sustainable development indicators. Ecological Indicators 2, 79–92.

Kondratyev, S.A., Alyabina, G.A., Sorokin, I.N., Moiseenkov, A.I., and Izmailova, A.V., 2003. Assessment of external load on aquatic systems depending on the land cover of watersheds (Northwest Russia as a case study). Ecosystems 10, 107–115.

Ostrofsky, M.L., 1978. Modification of phosphorus retention models for use with lakes with low water loading. Journal of the Fisheries Research Board of Canada 35(12), 1532–1536.

Petrova, N.A. (Editor), 1982. Anthropogenic eutrophication of Lake Ladoga. Nauka, Leningrad, 304 pp. (in Russian).

THE USE OF SCENARIO ANALYSIS TO ASSESS FUTURE LANDSCAPE CHANGE ON WATERSHED CONDITION IN THE PACIFIC NORTHWEST (USA)

WILLIAM G. KEPNER*
US Environmental Protection Agency, Office of Research and Development, Las Vegas, Nevada, USA

MARIANO HERNANDEZ
US Environmental Protection Agency, Office of Research and Development, Las Vegas, Nevada, USA

DARIUS J. SEMMENS
US Environmental Protection Agency, Office of Research and Development, P. O. Box 93478

DAVID C. GOODRICH
USDA Agricultural Research Service, Southwest Watershed Research Center, 2000 E. Allen Road, Tucson, AZ 85719

Abstract. The ability to assess, report, and forecast the life support functions of ecosystems is absolutely critical to our capacity to make informed decisions which will maintain the sustainable nature of our environmental services and secure these resources into the future. Scenario analysis combined with landscape sciences can be used to characterize uncertainties, test possible impacts and evaluate responses, assist strategic planning and policy formulation, and structure current knowledge to scope the range of potential future conditions. In this study, potential impacts from three wide-ranging scenarios in a large regional area in the northwest United States are compared to current conditions (ca. 1990) of the region in terms of a set of processes that are modeled in a geographic information system (GIS). This study presents an integrated approach to identify areas with potential water quality problems as a result of land cover change projected by stakeholders within the basin. Landscape metrics in conjunction with hydrological process models were used to examine the contribution

* Corresponding Author: William G. Kepner, US Environmental Protection Agency, Office of Research and Development, P. O. Box 93478, Las Vegas, NV 89193 USA; e-mail: kepner.william@epa.gov

I. Petrosillo et al. (eds.), Use of Landscape Sciences for the Assessment of Environmental Security, 237–261.

of land use/land cover to water and sediment yield and identify subwatersheds within the Willamette River Basin (Oregon, USA) that would be most affected in the year 2050 relative to three possible future scenarios which include inherent differences related to conservation, planning, and open development. Specifically, this study provides one example of the use of landscape sciences for environmental assessment that examines the impact of both urban and agricultural development in a large river basin. In particular, it attempts to (1) answer questions that relate to future scenarios that describe contrary positions related to urban development, (2) provide information which can be used to assess the potential changes of the landscape relative to human use, and (3) provide options that could be useful for sustainable management of natural resources and thus minimize future hydrologic and environmental impacts.

Keywords: Alternative futures analysis; hydrologic modeling; watershed assessment; environmental security; geographic information systems; landscape indicators; landscape characterization; Oregon; Willamette River

1. Introduction

In a traditional sense, security has always represented individual or collective safety achieved through protection of border and political sovereignty and cast in state, regional, or national interests. Concern over the environment has now expanded the need to include it into security policy, especially as it affects strategic renewable resources. Environmental problems related to depletion and scarcity are now seen as threats to human livelihood and thus can contribute to social and economic inequality which can lead to tension or in a worse case can result in open conflict. Land degradation, water quality and quantity, climate change, and the management and distribution of natural resources (e.g. water, forests, oil, and minerals) are factors which can either directly contribute to conflict or can be intrinsically linked to social outcomes which result in poverty, migration, infectious diseases, and declining economic productivity.

The ability to assess, report, and forecast the life support functions of ecosystems is absolutely critical to our capacity to make informed decisions which will maintain the sustainable nature of our environmental services and secure these resources into the future (SCEP, 1970). It is from this perspective that we have used the state of Oregon and particularly the Willamette River basin to demonstrate the integration of advanced technologies (such as satellite remote sensing and land cover characterization, large digital primary datasets, spatial statistics, and process models) with landscape ecological theory to examine impacts in environmental processes. Specifically, this study provides one example of the use of landscape sciences for environmental assessment that examines the impact of both urban and agricultural development in a large regional area in the northwest USA. In particular, it attempts to (1) answer questions that relate to future scenarios that describe contrary positions related to urban development, (2) provide information which can be used to assess the potential changes of the landscape relative to human use, and (3) provide options that could be useful for sustainable management of natural resources and thus minimize future hydrologic and environmental impacts.

In this study, potential impacts from three wide-ranging scenarios are compared to current conditions of a region in terms of a set of processes that are modeled in a geographic information system (GIS). Alternative futures landscape analysis involves (1) describing development patterns and significant human and natural processes that affect a particular geographic area of concern; (2) constructing GIS models to simulate these processes and patterns; (3) creating changes in the landscape by forecasting and by design; (4) and evaluating how the changes affect pattern and process using models (US Environmental Protection Agency, 2000). This study presents an integrated approach to identify areas with potential water quality problems as a result of land cover change projected by stakeholders within the basin. Landscape metrics describing spatial composition and spatial configuration were computed using the Analytical Tools Interface for Landscape Assessments (ATtILA) tool (Ebert and Wade, 2004). The landscape metrics generated from ATtILA were used in conjunction with the Automated Geospatial Watershed Assessment (AGWA) tool (Hernandez et al., 2005) to examine the contribution of land use/land cover to water and sediment yield and identify subwatersheds within the Willamette River Basin that would be most affected in the year 2050 relative to three possible future scenarios which include inherent differences related to conservation, planning, and open development.

2. Materials and methods

2.1. WILLAMETTE RIVER BASIN

Oregon is the 10th largest of the US states (Figure 1). It comprises an area of 251,418 km^2 and is bounded by the Pacific Ocean on the west and the states of Nevada and California on the south, Washington on the north, and Idaho on the east. The major landforms in Oregon include the Coast Ranges and the Cascade Range, a region which includes the snow-capped and glaciated peaks of Mount Hood and Mount Jefferson (Omernik and Gallant, 1986; Thorson et al., 2003). The Willamette River Valley is a substantial basin which empties into the

Columbia River near the city of Portland and is situated between the Coast and Cascade ranges. According to US Census records, nearly 70% of Oregon's population lived in the Willamette Valley in 1990. Most of the Willamette Valley is urban, privately owned, and used heavily for agriculture. Much of Oregon's economy is directed by activities that occur in the counties that comprise this river valley and thus the economy of the study area is based most notably on the forest products, tourism, agriculture/livestock, and more recently, the electronics industries. The mountainous areas in the eastern half of Oregon are primarily managed by the federal government (e.g. US Forest Service) while most of the lower-elevation plateau areas are agriculturally based private land which support dryland wheat farming, fruit orchards, and livestock ranching operations. Annual precipitation in the study area varies greatly, from less than 25–460 cm and elevations range from sea level on the coast to as high as 3,426 m at Mount Hood. Approximately 3.5 million people reside within the state of Oregon, most of whom live in the cities and suburbs along or adjacent to Interstate 5, including Portland, Salem, Eugene, Albany, Corvallis, and Medford. There are also many smaller towns along the Pacific Coast on US Route 101, such as Florence, Lincoln City, Newport, and Coos Bay as well many state recreation areas and parks.

By the year 2050, the number of people who inhabit the Willamette River Basin is expected to double which undoubtedly will place huge demands on existing environmental services and has the potential to threaten the security and sustainable management of natural resources. Digital data were collected from a variety of public sources, e.g. Oregon State University and US Environmental Protection Agency (Heggem et al., 2003). The baseline condition of ca. 1990 was selected and a set of land cover/land use maps were developed by Baker et al. (2004) for the year 2050 based on current land management and projected census growth. For the purpose of this study, the 2050 maps were selected for three scenarios which reflected important contradictions in desired future policy based on stakeholder input. The scenarios are listed in Table 1 and basically reflect changes in population within the watershed, patterns of growth, and development practices and constraints. The Conservation Scenario is the most ecosystem protection and restoration oriented, the Plan Trend Scenario reflects the most likely census predictions with zoning options designed to accommodate growth, and the Development Scenario is the least conservation and most market-economy positioned option. All three scenarios assume the same level of population increase, i.e. 2–3.9 million by 2050.

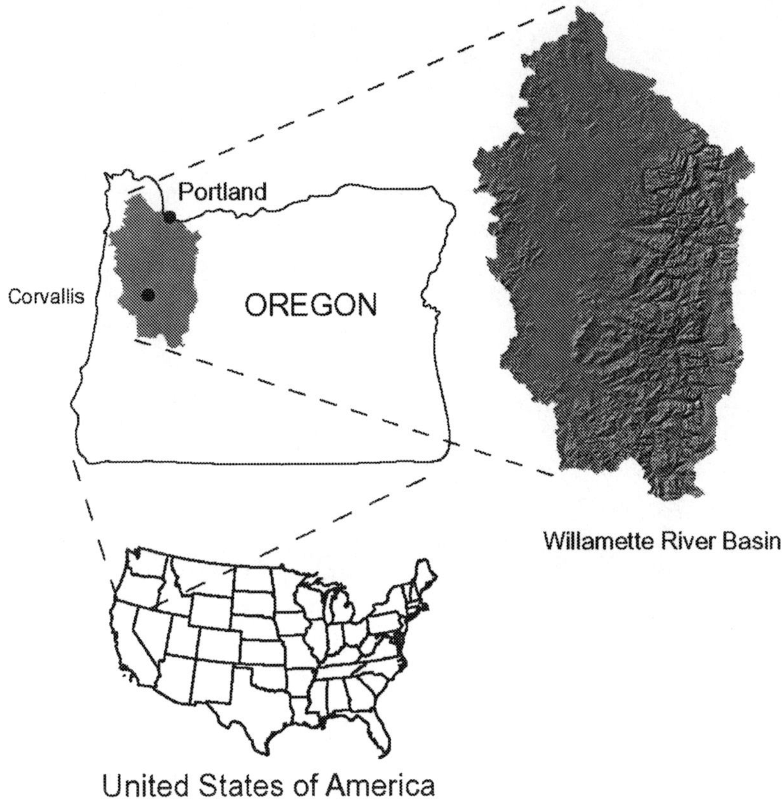

Figure 1. Location of the study area.

TABLE 1. Alternative-future scenarios in the Willamette River Basin, Oregon in the year 2050. (Adapted from Baker et al., 2004.)

Conservation 2050	Places greater priority on ecosystem protection and restoration, although still reflecting a plausible balance between ecological, social, and economic considerations as defined by citizen stakeholders
Plan Trend 2050	Assumes existing comprehensive land use plans are implemented as written, with few exceptions, and recent trends continue
Development 2050	Assumes current land use policies are relaxed and greater reliance on market-oriented approaches to land and water use

2.2. APPROACH

The general approach used in this study was carried out in three steps. The first step consisted of evaluating the entire state of Oregon using descriptive landscape metrics in 3 km grids. Landsat Thematic Mapper (TM) data were the basis in production of the statewide land cover grids (Baker et al., 2004). The data set contained 65 classes of land use/land cover in 30 m resolution pixels and was aggregated into a National Land Cover Data 21-class system (Anderson et al., 1976; Vogelmann et al., 2001). The land cover grids were also reprojected from NAD27 datum to NAD83 to be consistent with the other data utilized for the ATtILA landscape metric analysis and AGWA watershed assessment. ATtILA was used to compute the landscape metrics across the state and from the initial analysis of 65 metrics it was determined that 11 metrics representing three families of measurement, i.e. landscape characteristics, riparian characteristics, and human stressors, were the most diagnostic (Table 2). Landscape characteristics are related to land cover proportions (e.g. percentage of

TABLE 2. Eleven most diagnostic landscape metrics applied to Oregon. (Adapted from Ebert and Wade, 2004.)

I. Landscape characteristics	
A. Land cover proportions	
U-Index	Percentage that is all human land use
N-Index	Percentage that is all natural land use
Pagt	Percentage that is all agricultural use
Purb	Percentage that is all urban use
B. Slope metrics	
AgtSL3	Percentage that has any agricultural use with slopes that are equal to or exceed 3%
II. Riparian characteristics	
Ragt0	Percentage of stream length adjacent to all agricultural use with zero buffer
Rhum0	Percentage of stream length adjacent to all human land use with zero buffer
Rnat0	Percentage of stream length adjacent to all natural land use with zero buffer
III. Human stressors	
P_Load	Phosphorous loading (kg/ha/year)
N_Load	Nitrogen loading (kg.ha/year)
Popdens	Population density count/area in km^2

anthropogenic use) and slope metrics (e.g. percentage of agricultural use with slopes that are equal to or exceed 3%). Riparian Characteristics describe land cover adjacent to and near streams (e.g. percentage of stream length adjacent to all human land use with zero buffer). Human stressors are concerned with population, roads, and land use practices (e.g. population density count/area in km^2). Three types of output display were provided once metrics were generated by ATtILA. The first displayed areas ranked by individual metric value, the second ranked areas by a weighted index made up of two or more metrics, and the third displayed a bar chart of selected areas and metrics.

The 11 ATtILA metrics identified the Willamette River Basin as the most anthropogenically modified feature in the entire state of Oregon. In a preceding study, the future Willamette River scenarios were determined with input from local stakeholders and the land cover grids were modified accordingly to reflect their desired future conditions (Hulse et al., 2004). These land cover grids for the three alternative futures were then subject to evaluation relative to surface hydrologic endpoints of concern using AGWA. The AGWA tool (Miller et al., 2007; Hernandez et al., 2005) is a multipurpose hydrologic analysis system for use by watershed, natural resource, and land use managers and scientists in performing watershed- and basin-scale studies. It was used in this study to evaluate the relative hydrologic consequences of anticipated future urban and suburban development. AGWA, as with ATtILA, is an extension for the Environmental Systems Research Institute's ArcView versions 3.X (ESRI, 2001), a widely used and relatively inexpensive PC-based GIS software package. The GIS framework is ideally suited for watershed-based analysis, which relies heavily on landscape information for both deriving model inputs and presenting model results. Both the ATtILA and AGWA tools are designed to support landscape assessment at multiple spatial and temporal scales and provide the functionality to conduct all phases of a watershed assessment. Both tools also use widely available standardized spatial data to develop the input parameter files and provide tabular statistics and graphic data displays. AGWA utilizes two widely used watershed hydrologic models: the Soil Water Assessment Tool (SWAT; Arnold et al., 1994); and a customized version of the KINematic runoff and EROSion model (KINEROS2; Smith et al., 1995). SWAT is a continuous simulation model for use in large (river-basin scale) watersheds such as the Willamette River Basin whereas KINEROS2 is an event-driven model designed for small arid, semiarid, and urban watersheds. Data requirements for both models include elevation, land cover, soils, and precipitation data. Our modeling approach involved running SWAT using the 1990 baseline land cover to parameterize the model to determine reference condition. The Willamette River Basin boundary was delineated using an outlet point located at the Portland USGS streamflow gauge. The Willamette River is fully regulated by seven water reservoirs in the larger tributaries in the Coast Range and Cascade Mountains; therefore, it was necessary to incorporate all of them in the analysis to have a better understanding of the regimes along the river. In addition, because SWAT requires that water reservoirs be placed on the main channel network, it was necessary to use a small Contributing Source Area (CSA) value. Several attempts using different CSA values were necessary to generate the drainage network that would be dense enough to place all seven water reservoirs on the main channel network. At the end of the iteration process, a CSA of 1% of the total area of the Willamette Basin was used to generate the stream channel network and 111 subwatersheds. The hydrologic model was run for a 25-year period (1977–2001) following a 1-year warm up period that used data from 1976 to 1977. The warm up period establishes appropriate initial conditions for soil water storage. The 25-year period was divided into two parts to perform calibration (1977–1991) and validation (1992–2001) of the hydrological model. Calibration of the model was carried out against USGS data from four stations along the Willamette River (station # 14211720 at Portland; station #14191000 at Salem; station # 14174000 at Albany; and station # 14152000 at Jasper near Eugene, see Figure 2).

Two criteria were used to assess yearly runoff simulated by SWAT. The first criterion is the model efficiency of Nash and Sutcliffe (1970), which Sevat and Dezetter (1991) found to be the best objective function for reflecting the overall fit of a hydrograph. Model efficiency expresses the fraction of the measured streamflow variance that is reproduced by the model. For annual streamflow, model efficiency may be expressed as

$$E = 1 - \frac{\sum (Q_{obs} - Q_{sim})^2}{\sum (Q_{obs} - Q_{mean})^2} \tag{1}$$

where E is the coefficient of efficiency, Q_{obs} is the observed annual streamflow (mm); Q_{sim} is the simulated annual streamflow (mm); and Q_{mean} is the mean observed annual streamflow during the evaluation period. The value of E in Eq. (1) may range from minus infinity to one, with one representing a perfect fit data. According to common practice (Van Liew and Garbrecht, 2003), simulation results are considered to be good for values of E greater than 0.75, while for values of E between 0.75 and 0.36 the simulation results are considered to be satisfactory. For this study, E values less than 0.36 were considered to be unsatisfactory. The second criterion is regression analyses. The regression analyses of the calibration data show how well the calculations of the model matched the data used to calibrate it. The regression of the validation data shows how accurately the predictions of the calibrated model match the subsequent measurements.

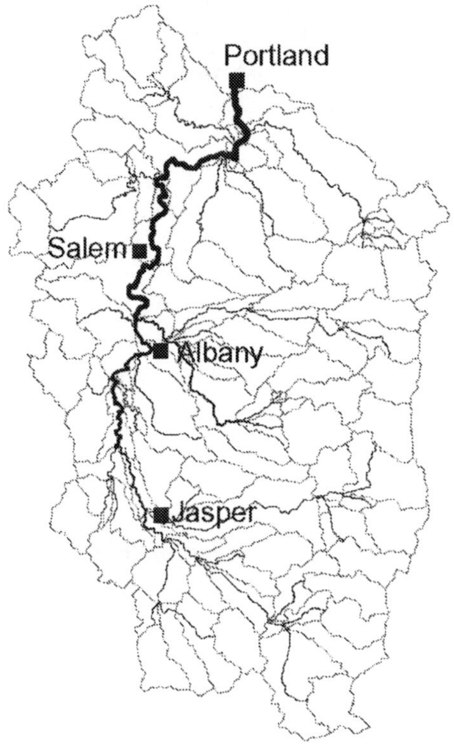

Figure 2. Location of USGS streamflow gauges.

To determine the impact of projected land cover future scenarios on the hydrologic regime of the Willamette River Basin, the 25-year period hydrologic simulation was performed using each of the three 2050 land cover scenarios to develop parameter inputs. Simulations for the future conditions were carried out using optimized parameters from the baseline conditions. Climate, and in particular precipitation, is a major source of uncertainty in hydrologic modeling. Changing climate and its associated impacts on basin hydrology, however, are a significant concern associated with future predictions. The future scenario simulations were run using precipitation and temperature from the 25-year simulation period. This treatment of climate is the most practical means of deriving climatic inputs for future simulations, and has the added benefit of eliminating climatic variations from assessments of hydrologic response to landscape change. Average annual outputs from the three alternative futures were differenced from the baseline values to compute percent change in average daily values in surface runoff, percolation, sediment yield, streamflow discharge, and sediment concentration over the 60-year period. The results focus on the relative magnitude and spatial distribution of the computed changes and the ability to provide contemporary environmental assessment with plausible predictions related to the benefits (or subsequently the consequences) of visualizing future environments and their likely impacts.

3. Results

3.1. ANALYSIS OF LAND COVER CHANGE

3.1.1. Baseline scenario

The percent of land cover distribution within the Willamette Basin for ca. 1990 is depicted in Figure 3. Forest land cover accounts for 69% of the land cover in the basin, with evergreen forest the predominant forest type. Wetland land cover types are found primarily in the northern part of the basin near Portland accounting for less than 1% of the total land cover, most of it in shrub wetlands. Agricultural land cover types comprise 19% of the total land cover. Pasture/hay and row crops are the dominant agricultural land uses in the central valley of the Willamette Basin. Urban land cover types account for 5% of the total land use. Shrub land cover comprises 5%, barren land cover types 1% and natural grasslands less than 1% of the total land cover, respectively.

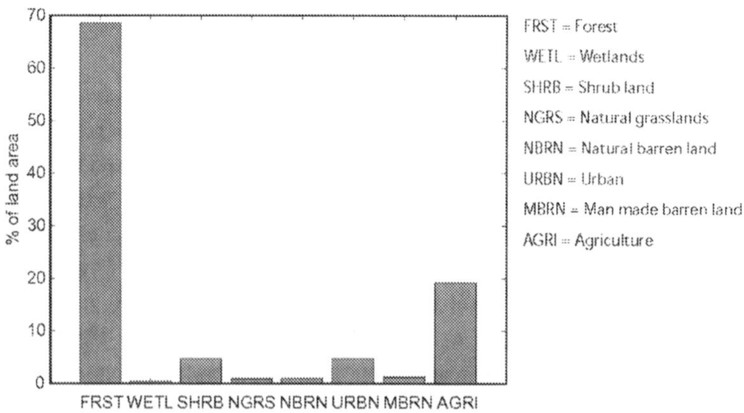

Figure 3. Percent land cover distribution for ca. 1990.

One of the most informative indicators of environmental impact is the extent to which humans have changed the natural vegetation to crops or urban land cover. The human use intensity index (U-Index) comprises agriculture and urban areas; the spatial distribution of it is depicted in Figure 4a. Urban areas are relatively minor (5%), in terms of total area, and agricultural areas are more extensive (19%). Most of the human land use has occurred in the central valley of the basin. The regional pattern of human use is reflected in the watershed rankings over the region. In Figure 4a, the accompanying bar chart shows that the highest U-Index value for a watershed is about 94%, which means that 94% of that watershed has agriculture or urban land cover. The lowest value is 0.03%, and the median value is about 11%. The proportion of area with urban or agriculture land cover exceeds 50% in about 29 watersheds, and about 53 watersheds have U-Index less than 10%. Higher elevations appear to have been less impacted by conversions to agricultural or urban land cover. The bar chart gives some details about the distribution of human use intensity among watersheds. About 3% of the watersheds have been almost completely converted to agricultural land. These are located in the fertile central valley of the Willamette Basin. About 52% of the watersheds have only small amounts of agriculture. These watersheds are primarily located in the mountainous areas.

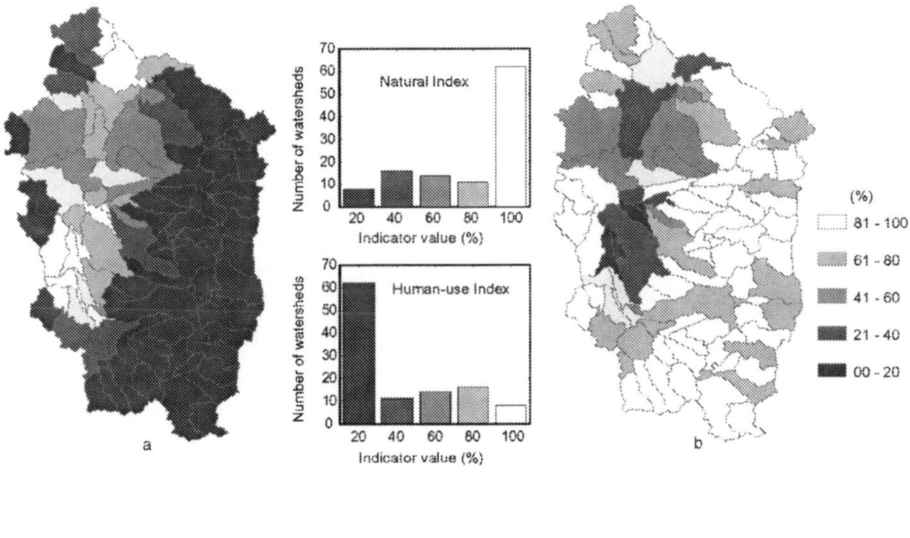

Figure 4. (a) Human-use Intensity Index and (b) Natural Land Use Index.

The spatial patterns are evident in the map of all natural land use (anything except urban and agriculture) (Figure 4b). The largest areas of natural land cover are found in the Cascade Mountains and along watersheds in coastal areas. About 48% of the subwatersheds are characterized as having all natural land use (Figure 4b). As expected, the Natural Index (N-Index) is less than 20% in subwatersheds in the Willamette valley close to urban areas.

3.1.2. Scenario analysis

This section on trajectories of land use change for the three scenarios presents data on composition of land cover expressed as percent of land base composed of natural grassland, shrubs, forest, wetland, agriculture, natural barren, man-made barren, and urban. The three alternative scenarios – Conservation 2050, Development 2050, and Plan Trend 2050 – create different patterns of land cover/ land use in the Willamette River Basin. The percent of total area and percent change relative to the baseline (ca. 1990) for each alternative scenario and for the eight main land cover/land use classes are presented in Table 3.

TABLE 3. Percent of total area and percent relative change land cover for eight classes.

Land cover	Baseline % of total area	Conservation % of total area and % change	Development % of total area and % change	Plan trend % of total area and % change
Forest	68.29	68.82	67.78	67.96
		+0.79	−0.75	−0.48
Wetland	0.22	0.56	0.21	0.22
		+158.26	−3.45	+0.0043
Shrubs	4.45	4.57	5.13	4.23
		+2.83	+15.27	−4.87
Natural	0.63	2.28	0.60	0.57
grassland		+260.92	−4.73	−9.82
Natural barren	0.65	0.53	0.57	0.54
		−19.22	−13.26	−17.90
Urban	4.44	5.59	6.74	5.67
		+25.87	+51.81	+27.75
Man-made	1.01	0.90	1.31	1.01
barren		−10.69	+29.44	−0.11
Agriculture	18.96	15.34	16.32	18.45
		−19.07	−13.91	−2.67

+ Values indicate an increment from baseline conditions

Patterns of natural vegetation across the Willamette River Basin reflect the properties of both the lowlands and upland regions. Future scenarios depict relatively minor changes in natural vegetation for the basin as a whole. Distributions of major land uses are unlikely to change greatly; for instance, changes vary from 1% to 4% of total area. As a result, the magnitudes of change by areal proportion will be relative minor, with greatest differences being the composition of agricultural land use.

3.2. WATERSHED HYDROLOGY

3.2.1. Calibration

The observed water year and predicted runoff obtained from the calibration run at the main watershed outlet and at three USGS flow gauging stations along the Willamette River were compared and are shown in Figure 5.

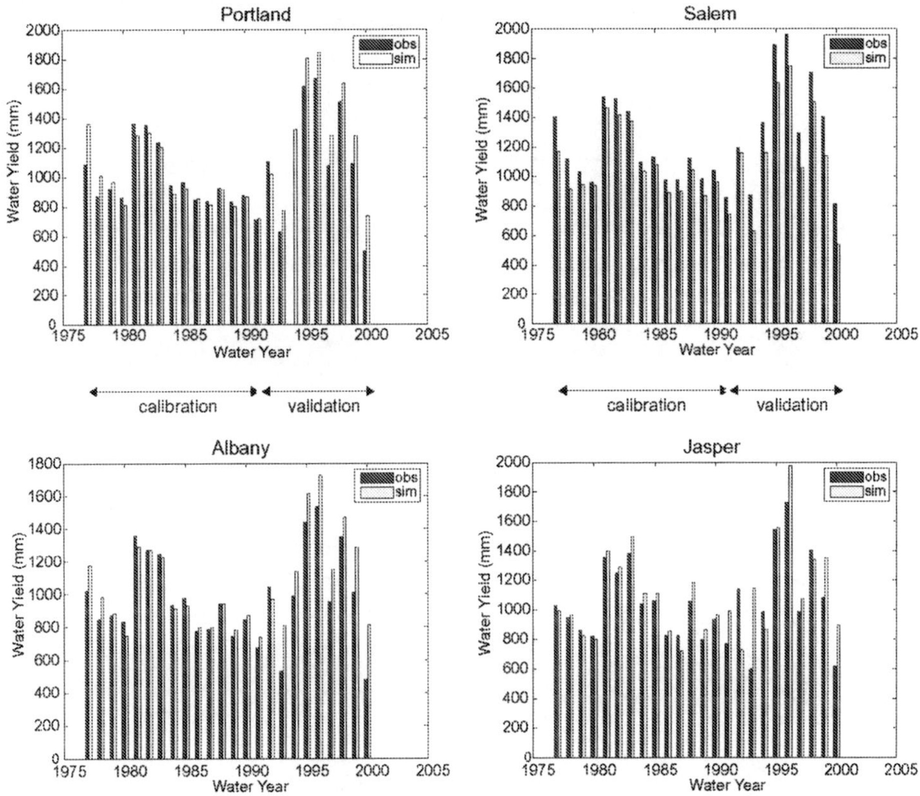

Figure 5. Simulated and observed water yield for the calibration and validation periods at four USGS streamflow stations.

The results of model efficiency and all regression analyses are presented in Table 4. The slope of the regression lines for the calibrations indicates that there is some bias in the model; i.e. some processes and/or phenomena are not accounted for adequately in the model. Notice that the largest bias is at Jasper, which is indicated by the largest deviation from unity of the slope of regression lines. The reasonably constant value of the square root of mean square residual (RMSE) indicates that there is a similar data scatter about each of these four regression lines. This could be interpreted as no degradation of uncertainty among the four USGS stations.

TABLE 4. Model efficiency and statistics from Ordinary Least Square regression analyses for the calibrations and validation periods.

Location	Period	E	Slope	Intercept	R^2	RMSE
Portland	1977–1991	0.75	0.94	68	0.89	98
	1992–2001	0.83	1.02	116	0.97	112
Salem	1977–1991	0.72	0.93	−10	0.96	57
	1992–2001	0.70	0.97	−167	0.98	80
Albany	1977–1991	0.88	0.90	104	0.94	68
	1992–2001	0.65	0.91	266	0.95	119
Jasper	1977–1991	0.88	1.16	−134	0.97	56
	1992–2001	0.42	0.74	390	0.75	262

A high value of the average of the Nash and Sutcliffe model efficiency (0.81) and a high average coefficient of determination (0.88) suggest that the model was accurately calibrated for predicting runoff from the Willamette Basin.

3.2.2. Validation

The model validation was carried out by comparing the observed surface runoff with the simulated runoff over the 1992–2001 water year period (Figure 5). A fairly good agreement was obtained throughout the validation period. However, the distribution of the predicted runoff values about the 1:1 line revealed that the data scatter about the regression lines at Portland and Jasper has been reversed. Regression analysis between the observed and predicted runoff values resulted in a high average value (0.91) of the coefficient of determination (r^2) and a slope of near unity, which indicated a close relationship between measured and predicted runoff. The average efficiency coefficient for the validation period (0.65) decreased 20% from the calibration period (0.81). In addition, the average RMSE of the validation period for the four USGS stations is twice as much as the calibration period. This could be interpreted as an increment in the uncertainty from all sources.

Based on the calibration and validation analyses, we assume that the model is adequate for effective planning and evaluation of future alternative scenarios.

3.2.3. Spatial distribution of hydrological processes

Among the various hydrological processes taking place in the Willamette River Basin, surface runoff, percolation, streamflow discharge, sediment yield and sediment concentration are of particular interests. To determine the spatial distribution of these water balance components, calculations were performed on a water year basis and results for surface runoff, percolation, and sediment yield are shown as average annual values averaged over the 25-year period and for streamflow discharge and sediment concentration results are shown as average daily values averaged over the 25-year period of simulation. Land use and plant cover structure are important factors in formation of runoff. The time lapse of watershed reaction on rainfall is greater in the presence of forest or rich plant cover, and the maximum runoff is reduced in comparison with barren soil or urban areas. Based on Figure 6a the spatial distribution of overland surface runoff follows very closely the spatial distribution of the Human-use Index depicted in Figure 4a; i.e. highly populated areas are associated with areas producing high overland flow volumes. Except that natural barren areas such as those in high elevation areas of the Cascade Mountains produced on average380 mm/year. Overland flow volumes ranged in the average from 10 mm/year in forested areas to 449 mm/year in some urban areas.

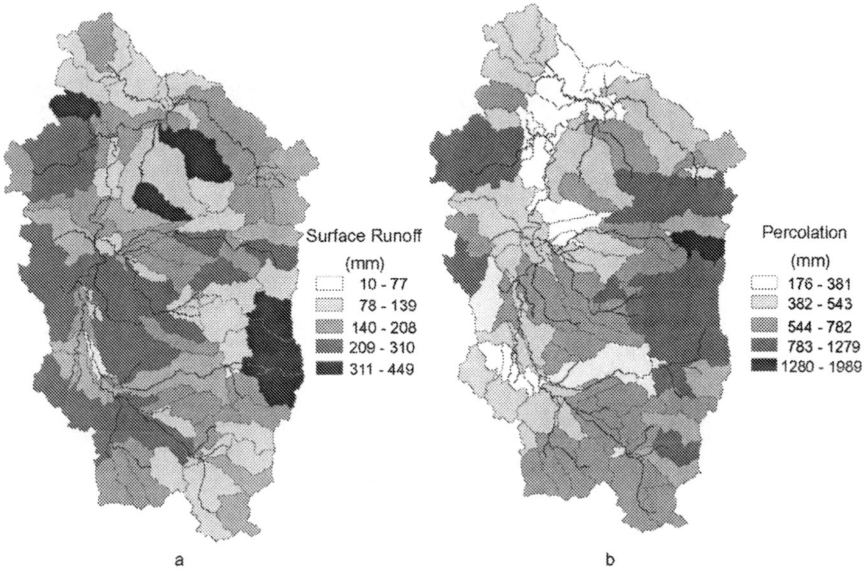

Figure 6. (a) Spatial distribution of average annual surface runoff and (b) average annual percolation.

Mean annual percolation volume estimates range from 176 mm/year to 1989 mm/year, the spatial pattern of percolation resembles the Natural Index spatial pattern shown in Figure 4b; i.e. areas with high percolation volumes are associated with forested areas and low percolation volumes with urban and barren areas (Figure 6b).

The mean daily discharge averaged over the 25-year simulation period is shown in Figure 7. The discharge along the Willamette River and its tributaries varies from 3.5 m^3/s to 1,400 m^3/s at the Portland USGS streamflow station. Furthermore, to illustrate the flow regime and goodness of fit between observed and simulated daily discharge at Portland and Salem, the hydrographs for 1981 water year are shown in Figure 7. Notice that the model over predicts the peaks during the winter season but in general the model simulates the overall trend.

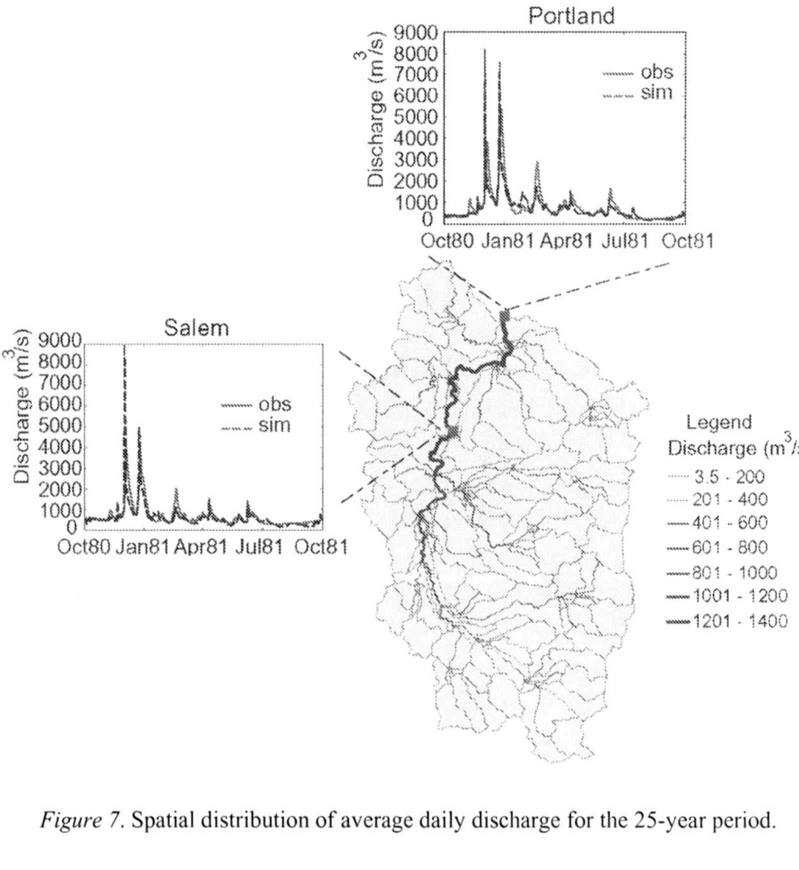

Figure 7. Spatial distribution of average daily discharge for the 25-year period.

Mean discharge for the Willamette River at its mouth averages 917 m^3/s, but varies seasonally from 233 m^3/s in August to over 5,000 m^3/s in December, reflecting the seasonal pattern of precipitation. Much of the summer runoff depends on the high-elevation snowpack because more than 35% of the precipitation occurs as snow above 1,200 m in elevation. As a result of the rain-dominated hydrograph at low elevations and snowmelt-dominated hydrograph at high elevations, streamflows overall are highest in winter and lowest in late summer. The main stem of the Willamette River is not dammed, but there are seven major flood control reservoirs in the larger tributaries in the Coast Range and Cascade Mountains and many small dams for irrigation or power generation. Reservoirs control approximately 27% of the flow of the Willamette River.

The mean spatial daily suspended sediment concentration averaged over the 25-year simulation period is shown in Figure 8. The largest concentrations occur along the Willamette River near Portland, the mean daily concentrations averaged over the 25-year simulation period vary from 16 gm/m^3 to 25 gm/m^3. However, observed daily concentration over 100 gm/m^3 was recorded on 2000 water year. Moreover, the 1981- and 2000- water year sedigraphs are shown in Figure 8; notice that the model predicted the overall trend of daily suspended sediment concentration at Portland. Suspended sediment is important to characterize water quality because nutrients, such as phosphorous and nitrogen, and toxic constituents, such as dioxin and chlorinated pesticides can be transported in association with fine sediments.

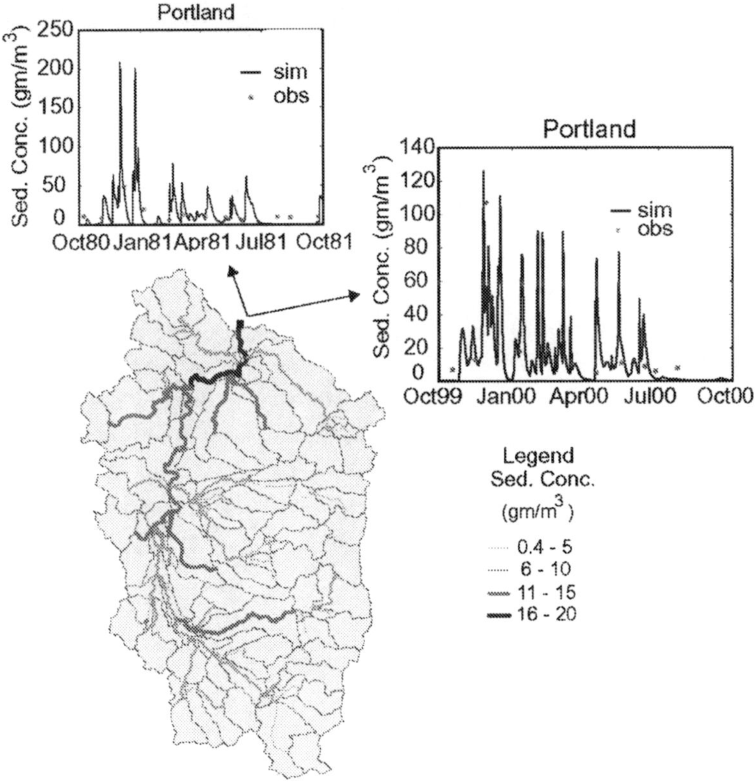

Figure 8. Spatial distribution of average daily sediment concentration for the 25-year period.

3.3. ALTERNATIVE FUTURE ANALYSES

3.3.1. Temporal and spatial variation

To evaluate each alternative future, the calibrated hydrologic model was run using each of the three 2050 land cover scenarios to develop parameter inputs. Average annual outputs for surface runoff, percolation, sediment yield, and average daily estimates for flow discharge and sediment concentration from the three alternative futures were differenced from the baseline values to compute percent change over the 60-year period. Results from the simulation runs are given in Table 5 and Figures 9–13. These figures show the relative departure from the 1990 baseline year and illustrate the spatial variability of changes to the surface hydrology. In general, the simulation results indicate that land cover changes associated with future development will significantly alter the hydrologic response of the basin. Changes are primarily associated with increasing urbanization and the associated replacement of vegetated surfaces with impervious ones.

In the case of surface runoff the simulations show average increases commensurate with increases in urbanization. There is considerable spatial variability of simulated hydrologic response. Although most subwatershed elements exhibited an increase in surface runoff, other areas showed improvement or decreasing runoff (Figure 9).

The greatest change was simulated for the Development scenario with an average increase of 1.16% over the 1990 baseline Table 5. Simulated increases in surface runoff predominately occur within subwatersheds distributed in the northern reaches of the watershed and along the Willamette Valley near Portland, Oregon City, and Eugene. The root mean square error (RMSE) of the regression analysis for the validation period shown in Table 4, which is an explicit measure of the prediction uncertainty, is larger than the surface runoff predicted relative change for each of the three development scenarios. Thus the hydrologic response of the basin as a consequence of land cover changes associated with future development can be also attributed to model uncertainty.

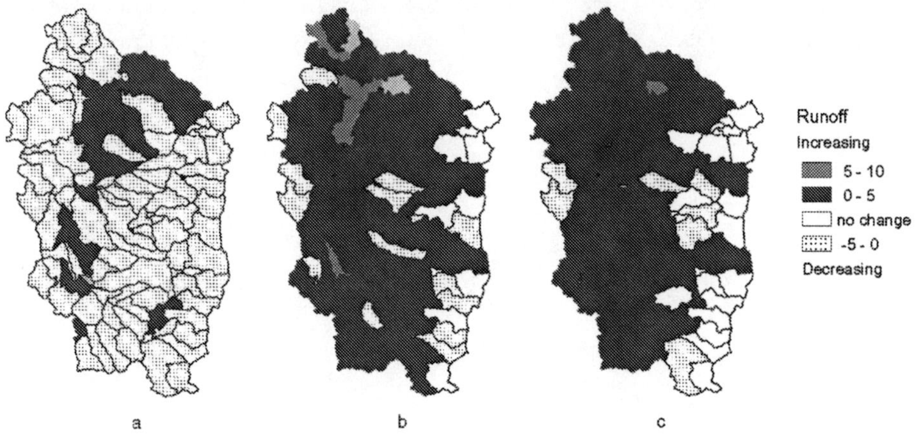

Figure 9. Percent change in surface runoff for future scenarios. (a) Conservation, (b) Development, (c) Plan Trend.

TABLE 5. Simulated average annual surface runoff, percolation, and sediment yield for the 1990 baseline and future conditions and predicted relative change for each of the three development scenarios.

Water balance component	Baseline 1990	Simulated percent relative change 1990–2050		
		Conservation	Development	Plan trend
Surface runoff (mm)	330.98	327.59 −1.02	334.81 +1.16	334.18 +0.97
Percolation (mm)	655.12	656.27 +0.18	650.28 −0.74	653.04 −0.32
Sediment yield (t/ha)	36.69	32.22 −12.16	33.70 −8.12	36.42 −0.72

Percent change in simulated channel discharge agrees closely with results for surface runoff. Figure 10 shows change in simulated mean daily channel discharge relative to the 1990 baseline for each of the three future scenarios. By mapping this model output for each stream segment in the model area it is possible to visually identify reaches that are anticipated to experience the greatest changes in their hydrologic regime as a result of the land cover/land use change. As such, the simulated changes to the hydrologic regime mapped in Figure 10 can also be viewed as an index of riparian vulnerability to the unmitigated future development. As in the previous example, channel discharge increased most under the Development 2050 scenario and although the results are spatially variable, the greatest impact seems to be concentrated in the subwatersheds in the northern portion of the basin and along the Willamette Valley where most new development is forecast.

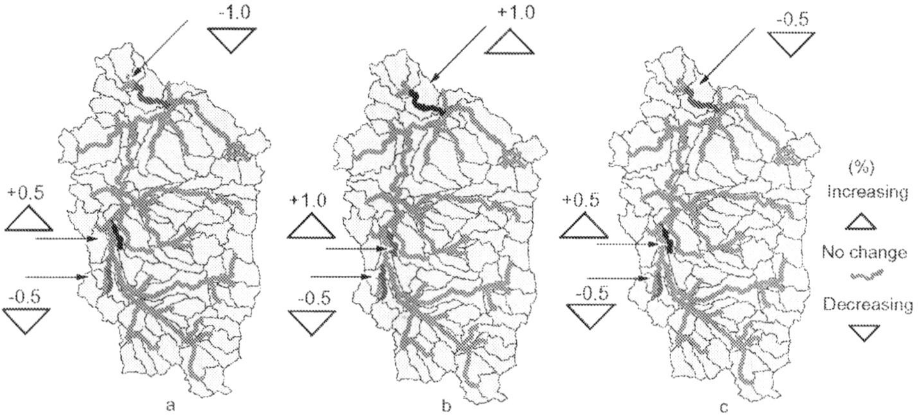

Figure 10. Percent change in discharge for future scenarios. (a) Conservation, (b) Development, and (c) Plan Trend.

Sediment yield and erosion are directly related to runoff volume and velocity. The percent change in sediment yield and sediment concentration simulated with SWAT also displayed a high degree of spatial variability across the basin and between the three scenarios (Figures 11 and 12). Subwatersheds with the greatest increase in sediment yield do not necessarily correspond with those exhibiting the greatest change in surface runoff, however those model elements in the northern headwaters and on the valley generally showed the greatest increase in sediment yield. Sediment yield decreases on the average 12% under Conservation 2050, whereas in both Development 2050 and Plan Trend 2050 the reduction is 8% and 0.72%, respectively (Table 5). In addition, sediment concentration in the Willamette River is greatest in both Development 2050 and Plan Trend 2050; notice that the greatest reduction in sediment concentration is achieved under the Conservation 2050 scenario (Figure 12). The sediment concentration map may provide some understanding about the sediment dynamics occurring between land cover and channels. For instance, forest management practices that change forest conditions will inevitably change channel conditions and must therefore be carefully tailored to mitigate adverse impacts on riverine habitat. It remains apparent, however, that under the Plan Trend 2050 scenario more sediment is expected to erode and be transported offsite than for the other two options (Table 5).

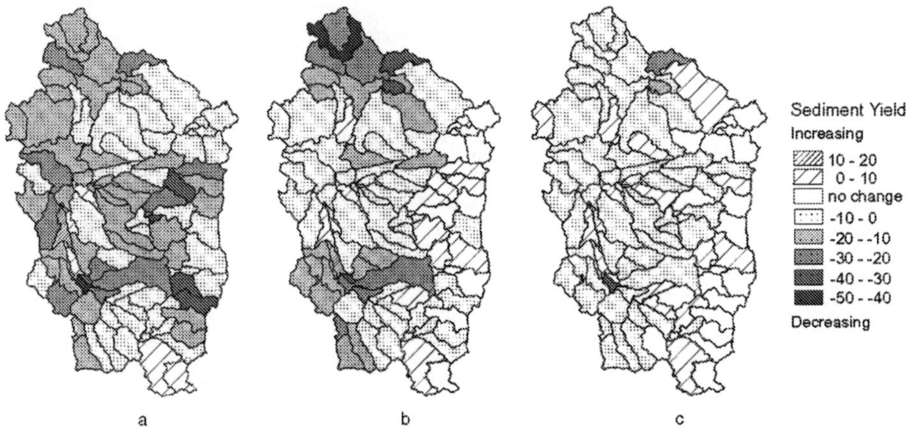

Figure 11. Percent change in sediment yield for future scenarios. (a) Conservation, (b) Development, and (c) Plan Trend.

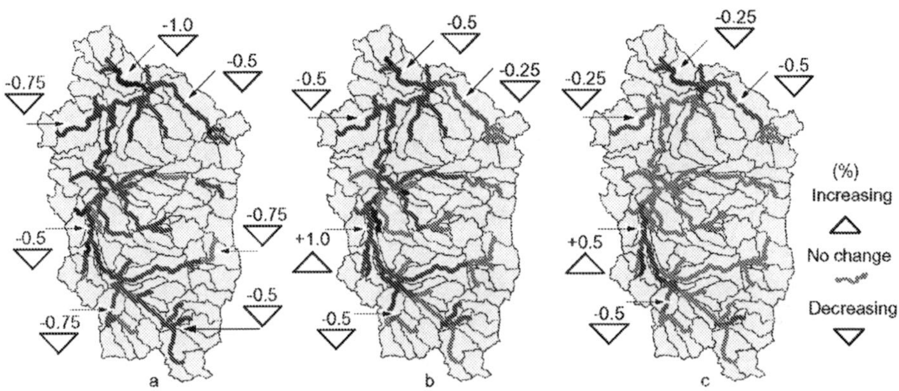

Figure 12. Percent change in sediment concentration for future scenarios. (a) Conservation, (b) Development, and (c) Plan Trend.

Percolation is a hydrologic measure of the water volume that is able to infiltrate into the soil past the root zone to recharge the shallow and/or deep aquifers. Figure 13 displays the simulated change in percolation for the three future scenarios. Although the model predicts some improvement in the watershed headwaters where human inhabitation is most dispersed, overall percolation is expected to decrease in all options as urban impervious surfaces are expanded, especially under the Development 2050 scenario.

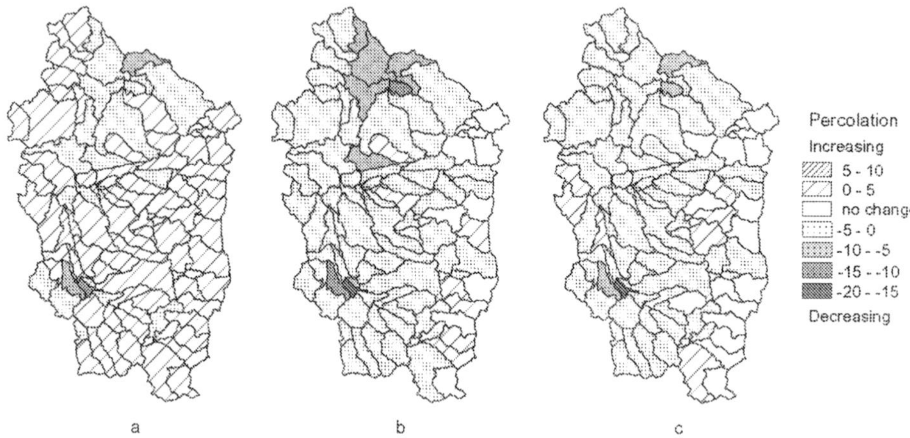

Figure 13. Percent change in percolation for future scenarios. (a) Conservation, (b) Development, and (c) Plan Trend.

However, notice that under the Conservation 2050 scenario percolation increased up to 5% in some areas along the Coast Range and along the Willamette River; percolation decreased between 4% and 13% under the other two scenarios, particularly, in the southern region of the basin comprising the city of Eugene. On average, the baseline scenario estimates percolation at 655 mm/year and that this amount will decrease 0.74% under Development 2050 and 0.32% under Plan Trend 2050. In contrast, percolation will increase 0.18% under Conservation 2050 (Table 5).

4. Summary and conclusions

Landscape pattern analysis was conducted on a watershed basis across the entire state of Oregon to characterize the heterogeneity of land cover and land use. ATtILA was used to compute the metrics associated with statewide landscape characteristics for the period of ca. 1990. From this analysis it was determined that the Willamette River Basin is the most modified hydrological feature relative to human land use within the state of Oregon. Because spatial variability of land cover alters the hydrological structure within a watershed, we then invoked AGWA to examine the watershed response relevant to surface runoff, sediment yield (i.e. soil erosion), and percolation for each subwatershed. The hydrological model was calibrated against total water yield, surface runoff and base flow using measured streamflow records at four permanent flow gauging stations along the Willamette River over a water year period from 1977 to 1991. Model validation was carried out by comparing observed surface runoff with the simulated runoff over a water year period from 1992 to 2001.

In general, the simulation results for the 2050 alternative future scenarios indicate that land cover changes associated with potential future development will significantly alter the hydrologic response of the basin. Changes are primarily associated with increasing urbanization and the subsequent replacement of vegetated surfaces with impervious ones. Under a future urbanizing environment the model simulation results appear to indicate that changes in stream condition within the Willamette Basin are subtle compared to those from ca. 1990 for most of the stream condition indicators. However, there are some subwatersheds that did show some important impacts to the subwatershed hydrology. The most notable changes are likely to be increases in the amount of surface runoff and sediment discharge, especially in the northern portion of the basin and along the Willamette Valley where most new development is forecast.

For the purpose of this study, negative impacts are considered to be increases in surface runoff, streamflow discharge, sediment concentration, and decline of percolation volume. The impacts are summarized graphically by percent change relative to the 1990 reference condition for each of the alterative futures using subwatersheds as the comparative unit. Urbanization and agriculture are presumed to be the major environmental stressors affecting watershed condition of the Willamette River Basin.

The hydrologic modeling results indicate that negative impacts are likely under all three of the future scenarios as a result of predicted urbanization; however, there is remarkable variation in their specific hydrologic responses, particularly between Development 2050 and Conservation 2050.

In general, the Development 2050 scenario has the greatest negative impact on surface hydrology and water quality and results in greater simulated surface runoff, flow discharge, and sediment concentration, especially in the downstream reaches near Portland and reaches along the Willamette Valley. Additionally, percolation and thus groundwater recharge is most reduced under this scenario. This scenario favors development and allows for the largest future population increase within the watershed.

The Conservation and Plan Trend alternative futures have the least negative impacts to the surface water hydrology. However, it is important to highlight that under the Conservation 2050 scenario, surface runoff and sediment yield are reduced considerably along the Coast Range Mountains. Consequently, percolation is increased within the same subwatersheds. In regard to water quality, the spatial pattern depicted under the Conservation 2050 scenario suggests an important improvement on the environment given that sediment concentration decreased along the Willamette Valley and Coastal Range.

There are of course many ecological and hydrological endpoints in addition to those that were measured in the present study. The endpoints reflect fundamental environmental services and the benefits that people receive from them, e.g. food supply, water purification and supply, wildlife habitat, climate moderation, flood control, forest products, recreation, and carbon and nutrient storage. Those benefits are central to human sustainability, quality of life, and ultimately to security. Conventional definitions of security have now been expanded to include environmental threats thus evoking the term environmental security. Subsequently, environmental security includes the notion that population growth and the inequitable distribution of resources can directly cause competition over resources and that any outcomes resulting in environmental degradation or depletion can therefore become a source of tension at any political or geographical scale (local, national, regional, or global).

This study was initiated to demonstrate the potential of applying landscape sciences to environmental assessment. Specifically, it sought to integrate landscape spatial analysis with hydrological modeling to evaluate the consequences (or benefits) of choice for policy decisions related to the sustainable management of natural resources. Although the Willamette findings are largely intuitive, the authors believe that (1) the process of scenario analysis coupled with the use of spatial analysis tools can develop a better understanding of alternative options for future management of strategic natural resources and (2) the collective technologies provide an ability to quantify and forecast relative impacts to ecosystem services and therefore provides important tools to help both stakeholders and decision-makers make better decisions for improved future options. Lastly, the alternative futures approach and the application of the ATtILA and AGWA tools can be deployed on any variety of landscapes, watersheds, and regions throughout the world and thus provides an important process for watershed assessment of landscape change and impacts to water resources.

References

Anderson, J.F., Hardy, E.E., Roach, J.T., and Witmer, R.E., 1976. A land use and land cover classification for use with remote sensor data. U.S. Geological Society Professional Paper 964, 28.

Arnold, J.G., Williams, J.R., Srinivasan, R., King, K.W., and Griggs, R.H., 1994. SWAT: soil water assessment tool. U.S. Department of Agriculture, Agricultural Research Service, Grassland, Soil and Water Research Laboratory, Temple, TX.

Baker, J.P., Hulse, D.W., Gregory, S.V., White, D., Van Sickle, J., Berger, P.A., Dole, D., and Schumaker, N.H., 2004. Alternative futures for the Willamette River Basin, Oregon. Ecological Applications 14(2), 313–324.

Ebert, D.W. and Wade, T.G., 2004. Analytical Tools Interface for Landscape Assessments (ATtILA). USEPA, Office of Research and Development, National Exposure Research Laboratory, Environmental Sciences Division, Landscape Ecology Branch, Las Vegas, NV (EPA/600/R-04/083), 39 pp.

ESRI, 2001. ArcView version 3.2a software and user manual. Environmental Systems Research Institute, Redlands, CA.

Heggem, D.T., Ebert, D.W., Wade, T.G., Augustine, S., Weiss, A.D., Evenson, E., and Bice D.L., 2003. EMAP western United States landscape characterization Oregon data and product browser. U.S. Environmental Protection Agency, EPA/600/R-02/025, 2003 (http://www.epa.gov/nerlesd1/land-sci/oregon/).

Hernandez, M., Semmens, D.J., Miller, S.N., Goodrich, D.C., and Kepner, W.G., 2005. Development and application of the automated geospatial watershed assessment tool. In: C.W. Richardson, A.D. Baez-Gonzalez, and M. Tiscareno-Lopez (Editors), Modeling and Remote Sensing Applied to Agriculture, USDA-INIFAP, U.S. and Mexico, pp. 127–158.

Hulse, D.W., Branscomb, A., and Payne, S.G., 2004. Envisioning alternatives: using citizen guidance to map future land and water use. Ecological Applications 14(2), 325–341.

Miller, S.N., Semmens, D.J., Goodrich, D.C., Hernandez, M., Miller, R.C., Kepner, W.G., and Guertin, D.P., 2007. The Automated Geospatial Watershed Assessment tool (AGWA); Environmental Modeling and Software 22, 365–377.

Nash, J.E. and Sutcliffe, J.V., 1970. River flow forecasting through conceptual models, I, A discussion of principles, Journal of Hydrology 10, 282–290.

Omernik, J.M. and Gallant, A.L., 1986. Ecoregions of the Pacific northwest. U.S. Environmental Protection Agency (EPA/600/3-86/033), 39 pp.

SCEP (Study of Critical Problems), 1970. Man's impact on the global environment. MIT Press, Cambridge, MA.

Sevat, E. and Dezzetter, A., 1991. Selection of calibration of objective functionsin the context of rainfall-runoff modeling in a Sudanese Savannah area. Hydrological Sciences Journal 36, 307–330.

Smith, R.E., Goodrich, D.C., Woolhiser, D.A., and Unkrich, C.L., 1995. KINEROS – a kinematic runoff and erosion model. In: V.P. Singh (Editor), Computer Models of Watershed Hydrology, Water Resources Publications, Highlands Ranch, CO, pp. 697–732.

Thorson, T.D., Bryce, S.A., Lammers, D.A., Woods, A.J., Omernik, J.M., Kagan, J., Pater, D.E., and Comstock, J.A. 2003. Ecoregions of Oregon. U.S. Geological Survey, Reston, VA.

U.S. Environmental Protection Agency, 2000. Environmental planning for communities. A Guide to the Environmental Visioning Process Utilizing a Geographic Information System (GIS). EPA/625/R-98/003, 49 pp.

Van Liew, M.W. and Garbrecht, J., 2003. Hydrologic simulation of the Little Washita River experimental watershed using SWAT. American Water Resources Association 39, 413–426.

Vogelmann, J.E., Howard, S.M., Yang, L., Larson, C.R., Wylie, B.K., and Van Driel, N., 2001. Completion of the 1990s national land cover data set for the conterminous United States from Landsat Thematic Mapper Data and Ancillary Data Sources. Photogrammetric Engineering and Remote Sensing 67, 650–662.

CROSS-EUROPEAN LANDSCAPE ANALYSES: ILLUSTRATIVE EXAMPLES USING EXISTING SPATIAL DATA

K. BRUCE JONES* AND SHARON HAMANN
US Geological Survey, Geography Discipline, Reston, Virginia, USA

MALIHA S. NASH, ANNIE C. NEALE, AND WILLIAM G. KEPNER
US Environmental Protection Agency, Las Vegas, Nevada, USA

TIMOTHY G. WADE
US Environmental Protection Agency, Research Triangle Park, North Carolina, USA

JOE WALKER
CSIRO Land and Water, Canberra, ACT Australia

FELIX MÜLLER
University of Kiel, Kiel, Germany

GIOVANNI ZURLINI AND NICOLA ZACCARELLI
University of Salento, Lecce, Italy

ROB JONGMAN
Alterra, Wageningen UR, Wageningen, The Netherlands

STOYAN NEDKOV
Institute of Geography, Bulgarian Academy of Sciences, Sofia, Bulgaria

C. GREGORY KNIGHT
Department of Geography, Pennsylvania State University, University Park, Pennsylvania, USA

Abstract. Thirty-nine landscape metrics related to (1) conditions of terrestrial habitat, water quality, and ecosystem productivity, (2) potential pressures on or stresses to environmental resources, and (3) changes in conditions, were generated

* Corresponding Author: Dr. K. Bruce Jones, U.S. Geological Survey, Geography Discipline, Reston, Virginia USA, 20192, Phone: +1 703.648.4762; Fax: +1 703.648.5792; e-mail: kbjones@u sgs.gov

I. Petrosillo et al. (eds.), Use of Landscape Sciences for the Assessment of Environmental Security, 263–316.

for Europe from existing spatial data. The core land cover data used were available at resolution scales of 1 km (International Geosphere Biosphere Program or IGBP) and 100 m (Corine). These core data were used to calculate landscape metrics on water catchments (average area of approximately 2,500 km^2 for 1,888 catchments) and on 64 km^2 areas to capture finer-scale patterns. We also calculated the same metrics using finer-scale landscape data on the Yantra River Basin of north-central Bulgaria, permitting a comparison to broader-scale results from across Europe. We found that data to calculate all of the metrics did not exist for all of Europe and this resulted in analysis of two different spatial extents of Europe and different mixes of metrics in the landscape analyses. The Corine data set did not cover all of Europe but where it existed it was available for the approximate periods of 1990 and 2000. Greenness (Normalized Difference Vegetation Index or NDVI) estimates were available for all of Europe for approximately the same time period (1992–2003), but at a resolution scale of 64 km^2. These data sets in total offered an opportunity to compare results for the different metric sets used, and different spatial scales and changes in values between sample times. The results showed some differences in several key metrics between the different data sets but that it was possible to map areas with regards to relative condition with reasonable agreement. As expected the 64 km^2 analysis units showed greater detail and variation in landscape conditions and change than did catchments. However, the relatively course-scaled nature of the stream and river database for Europe precluded an analysis of riparian habitat conditions on the 64 km^2 areas. Overall changes in the landscape metric values between the 1990 and 2000 sample times were small, but there was considerable spatial variation in the amount of gain or loss. For example, relatively large percent gains in forests were observed in Spain, southern France, and in east-central Europe, whereas relatively large percent losses were observed in southwest France and western Spain. Forest change was inversely associated (from most to least important) with changes in shrubland, total agriculture, grassland, and urban land cover ($p < 0.05$). Agriculture lands were inversely (in decreasing order of importance) associated with changes in grassland, forest, shrubland, and urban land cover. However, because the Corine 1990 and 2000 databases were created from different methodologies, the change results must be interpreted with caution. On average, Europe became significantly greener between 1992 and 2003. Significant ($p < 0.05$) positive trends in greenness were observed across Europe, but in larger patches in eastern Spain, Wales and Scotland (UK), and in Romania. Significant negative trends were observed in southern Spain and southwestern Russia along the Caspian Sea. Trends in greenness and land cover change were uncorrelated. Results from the European-wide analyses of the Yantra River Basin compared favorably to the more detailed analyses that were based partially on finer resolution biophysical data. However, estimates for riparian land cover metrics were much higher for the more detailed analyses than the broader-scale analyses, a result of a denser stream network used in the former. Additionally, because of differences in the scale of Digital Elevation Model data used in the two analyses (90 m and 25 m), estimates for agriculture on greater than 3% slopes differed as well. A principal components analysis (PCA) was used to combine multiple landscape metrics to evaluate the relative environmental condition of and change in catchments and the 64 km^2 areas. Additionally, a simple index of relative vulnerability was calculated and mapped by combining PCA results for landscape condition and change. We discuss results and limitations of this analysis. We also discuss the value of this preliminary assessment for broadscale analyses to identify geographic areas where environmental security may become an issue. We note limitations in the analytical techniques used, data gaps and issues regarding interpretation of these results and make suggestions for future landscape analyses.

Keywords: Landscape metrics; European landscape analysis; integrated analysis; catchments; goods and services

1. Introduction

Evaluation of broadscale landscape conditions and change is critical in understanding potential threats to and vulnerabilities of a wide range of ecological goods and services, including clean water, habitat for species, and ecosystem productivity (Zurlini et al., 1999, 2004; Bradley and Smith, 2004). Moreover, degraded environmental conditions can lead to conflict among local indigenous people, communities, regions, and countries (Tuchel, 2004; Herrero, 2006; Liotta, 2006). Loss of environmental security has resulted in population shifts, reduced human health, increased exposure to floods, fires, and other hazards, and reduced quality of life (Brauch, 2006). Additionally, changes in human population demographics can dramatically affect environmental condition and security (Swiaczny, 2005). In recognition of these issues, the European Union (EU) has established a number of directives and strategic actions related to environmental resources and their sustainability. These include protection of biodiversity and habitats (CEC, 1992), water resources (CEC, 2000), and soils (CEC, 2002). Moreover, the EU has established a comprehensive strategy for sustaining human well-being and the environment (CEC, 2001). All of these directives and strategies call for development of indicators and comprehensive environmental assessments. Additionally, NATO views environmental security as a factor that may influence political stability and potential conflict among countries (Kepner et al., 2006). Spatially continuous, digital databases on land cover and other important biophysical attributes (soils, elevation, topography, etc.) have become increasingly available via web sites and data portals (Jones et al., 2005). This availability coupled with advances in computer technology, including processing speed, the amount of data that can be stored and processed, and software development (e.g. geographic information systems (GIS)), now make it possible to conduct landscape assessments at multiple scales over relatively large geographic areas. Spatially continuous, digital data, as well as in situ data, have been used to conduct landscape analyses at many scales relative to several important environmental issues, including assessments of forest fragmentation and urbanization (McCollin, 1993; Riitters et al., 2000; Galleo et al., 2004), impervious surfaces (Slonecker et al., 2001), landscape change and consequences of change to ecological resources (Vogelmann, 1995; Wickham et al., 1999; Jones et al., 2001a), ecosystem productivity (Minor et al., 1999; Young and Harris, 2005; Nash et al., 2006), catchment condition (Walker et al., 2002; Jones et al., 2006), biological diversity (Saunders et al., 1991; Ojima et al., 1994; Kattan et al., 1994; Koopowitz, et al., 1994; O'Connor et al., 1996; Zurlini et al., 1999; Scott et al., 2003), water quality and quantity (Behrendt, 1996; Mattikalli and Richards, 1996; Wickham et al., 2000; Jones et al., 2001b; Jennings and Jarnagin, 2002; Kondratyev, 2007, this volume), surface water biological condition (Donohue et al., 2006), future water quality risks (Wickham et al., 2002), soil loss (Van Rompaey and Govers, 2002), management options related to individual species or groups of species (White et al., 1997; Burkhard et al., 2004; Mucher et al., 2004), and ecological forecasting (Reynolds et al., 2000; Domenkiotis et al., 2004; Peters et al., 2006). These and other studies form the foundation for application of landscape metrics to assess multiple environmental themes over broad geographic areas. A major advantage of spatially explicit biophysical data is the ability to relate and compare multiple landscape metrics and model results to each other for many specific places over relatively large geographic areas. Moreover, spatial intersection of biophysical data allow for the assessment of important processes that affect environmental quality. For example, intersection of elevation (slope) and land cover (cropland) data results in a metric of potential soil loss and nutrient export (cropland on steep slopes, Jones et al., 1996, 1997). Spatially integrated, multiple indicator and model results offer the potential for comparative and weight-of-evidence conclusions about environmental conditions, potential causes of environmental conditions, and threats to and vulnerabilities of specific geographic areas (Wickham et al., 1999; Bradley and Smith, 2004). This differs from more traditional assessments that report on individual indicator and model results in the form of graphs that often represent many different spatial scales and extents (e.g. see Australia State of the Environment Committee, 2001; Heinz Center, 2002; EPA 2003).

The primary aims of this project were to (1) determine the spatial extent, spatial resolution, and level of landscape analyses possible given currently available data and statistical approaches; (2) conduct landscape analyses over different spatial extents from existing data; and (3) identify limitations and issues in conducting European-wide, landscape analyses. As such, we used readily available, digital biophysical data, geographic information system (GIS) analysis tools, and statistical analyses to calculate and interpret 39 landscape metrics that related to water resources, terrestrial habitat, and ecosystem productivity, as well as potential threats or stresses to these environmental resources. We report the results of these assessments and discuss issues related to data quality, scales of applications, and landscape metric generation, synthesis, and interpretation. The assessments conducted in this study are thus illustrative rather than final and definitive.

2. Materials and methods

2.1. ANALYSIS UNITS AND EXTENT OF ANALYSIS

We generated landscape metrics on two types of analysis units; catchments represented by polygons with on average an area of approximately 2,500 km^2, and 64 km^2 areas (grid cells). This resolution of grid cells was chosen because it represented the finest scale of spatial resolution upon which multiple landscape metrics could be generated from available data. We downloaded a spatial coverage of catchments for Europe from the European Environmental Agency (EEA) web site. The 64 km^2 grid cells were those generated from the Normalized Difference Vegetation Index (NDVI) grid (see discussion below). However, because certain core biophysical data were not available for all of Europe, the extent of the metric analysis varied (Table 1; Figure 1A–D). As such, we calculated the available landscape metrics for the areas shown in C and D (Figures 1C, D). Additionally, several very small catchments were removed from the analysis. Analysis of grid cells permits a finer-scale analysis of landscape metrics, and potentially, the ability to identify environmental correlates of landscape condition and change.

2.2. LANDSCAPE METRICS

We generated 39 landscape metrics known to have qualitative and quantitative relationships to environmental themes or values, including water (quality, flooding), terrestrial ecosystem productivity, and wildlife habitat.

TABLE 1. Corine land cover reclassification used to generate landscape metrics.

Corine land cover type	Reclassifiction
Urban, Industrial, Airport	Urban
Mine, Construction, Dump	Man-made Barren
Urban Grass and Recreation	Urban other
Arable	Cropland
Permanent Irrigated	Cropland
Rice, Vineyards, Fruit Trees, Olive Groves	Ag Other
Pastures	Ag Pasture
Annual Crops, Complex Cultivation	Cropland
Ag/Forest Mix, Deciduous Forest, Coniferous Forest, Mixed Forest	Forest
Grassland	Grassland
Moors and Heath, Fires	Other Natural
Sclerophyllus, Woodland/Shrubland Transition, Sparse Veg.	Shrubland
Beaches, Outcrop/Barren, Glaciers/Snow	Natural Barren
Inland Marsh, Peat Bogs	Wetland
Salt Marsh, Salines, Intertidal Flats, Lagoons, and Sea, Water Courses, Water Bodies	Water Estuaries, Ocean
No Data	No Data

Additionally, some of the metrics calculated in this study represent potential impairment of or stress to environmental resources, including marginal land use, potential for soil loss and export of nitrogen and phosphorus to streams and rivers, and intensive types of land use that can be harmful to environmental resources, including croplands and urban environments (Table 1). We used the Analytical Tools Interface for Landscape Analysis (ATtILA, Ebert and Wade, 2004) ArcView extension (ESRI, 1996) to calculate all of the landscape metrics on catchments and grid cells, with the exception of NDVI slopes (see below). Six separate analyses of landscape metrics (Table 1) representing two different geographic extents (Figure 1C, D) were conducted based on data availability: (1) catchments (total of 1,888) and (2) 64 km^2 grid cells (total of 112,770) where International Geosphere Biosphere Program (IGBP) land cover, digital elevation model (south of 60° north latitude), and agriculturally limited area databases were available (Figure 1C); (3) catchments (total of 908) and (4) grid cells (total of 54,815) where Corine 2000 land cover data were available (Figure 1D); and (5) catchments (total of 908) and (6) grid cells (total of 54,815) where Corine 1990 and Corine 2000 land data were available (for change analysis, Figure 1D). Two types of land cover databases were used to

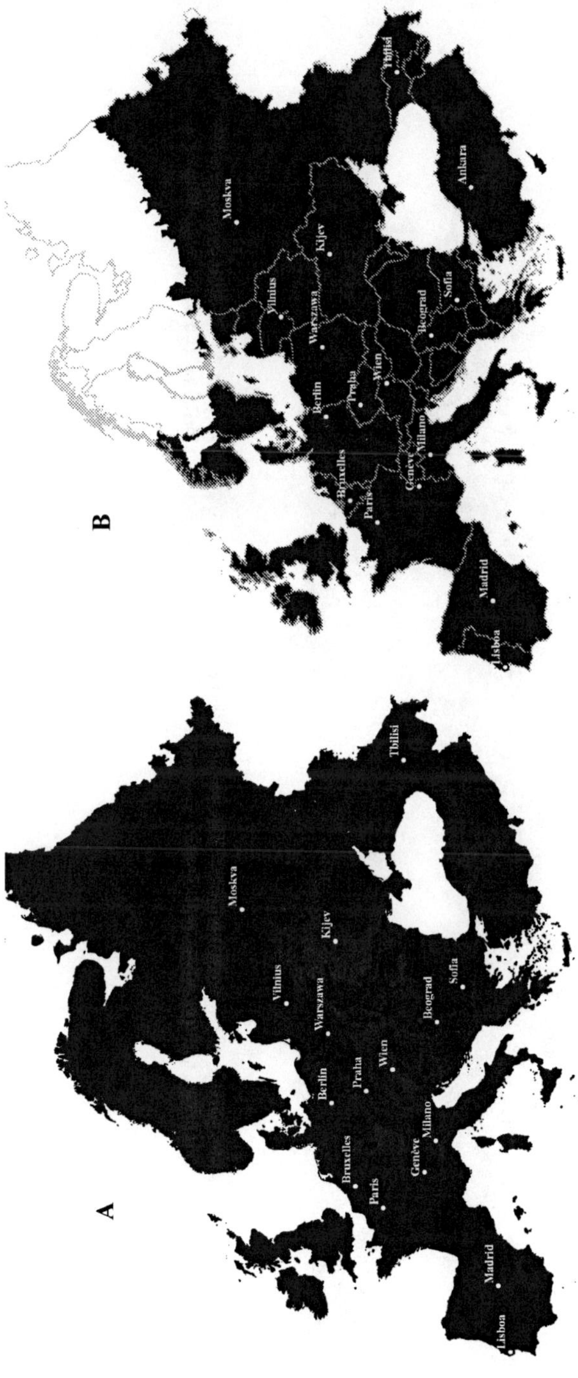

Figure 1A and 1B. Spatial extents of landscape analyses that are possible based on existing spatial data. (A) All of , (B) all of Europe south of 60°, north latitude.

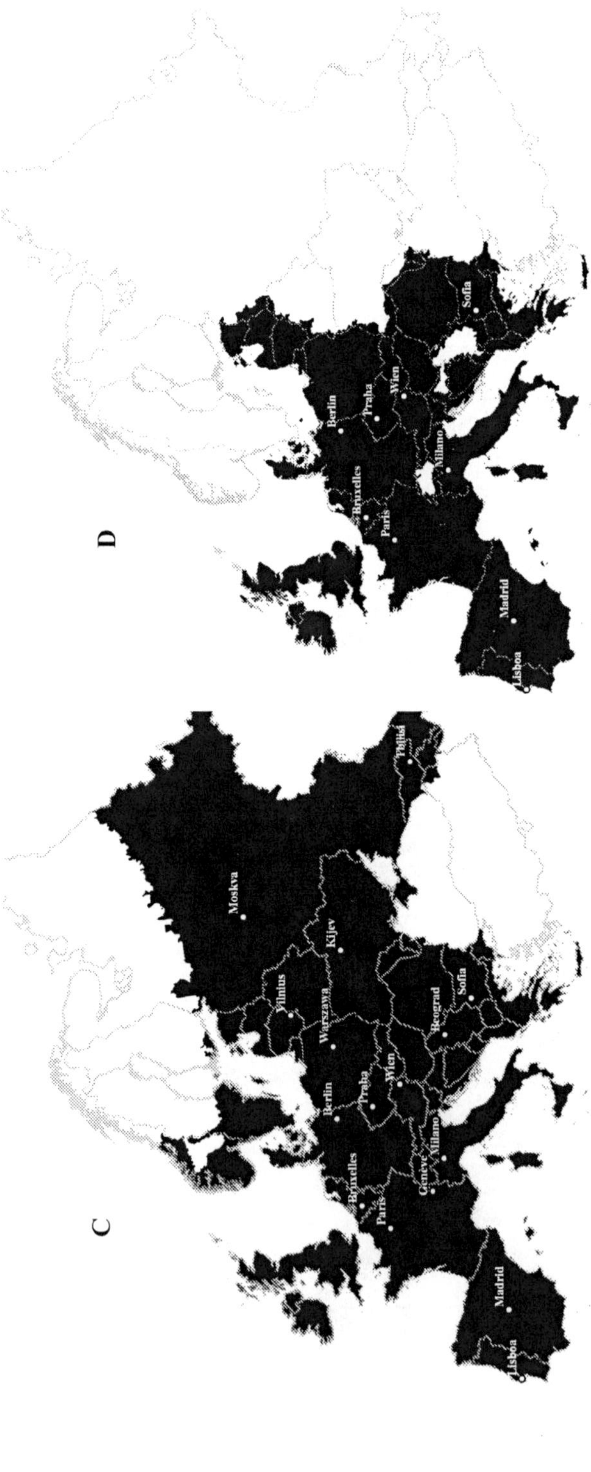

Figure 1C and 1D. Spatial extents of landscape analyses that are possible based on existing spatial data. (C) All of Europe south of 60° north latitude, excluding Turkey, and (D) areas with Corine land cover data. Landscape analyses reported in this study were limited to spatial extents C and D.

calculate percentages of land cover types in each analysis unit; the IGBP land cover database, which was available for all of Europe, and the Corine land cover databases (the 1990s and early 2000). The IGBP database was obtained from the US Geological Survey's EROS Data Center (EDC) web site and the Corine dataases from the EEA. To generate land cover-based metrics in ATtILA, we had to reclassify each of the land cover databases into fewer classes (see Tables 2 and 3).

TABLE 2. International Geosphere Biosphere Program (IGBP) land cover reclassification used to generate landscape metrics.

IGBP Land Cover	Reclassification
Evergreen Needleleaf Forest, Evergreen Broadleaf Forest, Deciduous Needleleaf Forest, Deciduous Broadleaf Forest, Mixed forest	Forest
Closed Shrubland, Open Shrubland	Shrubland
Woody Savannas, Permanent Snow/Ice	Other Natural
Savannas, Grassland	Grassland
Permanent Wetland	Wetland
Cropland	Cropland
Urban	Urban
Cropland/Natural Mix Agriculture	Other
Barren/Sparse	Natural Barren
Unknown	No Data

We obtained a 90 m digital elevation model (DEM) data set from the EDC (Shuttle Radar mission) and created a digital mosaic of these data for all of Europe south of 60° north latitude; data were not available for areas north of this latitude. A digital slope database was created from the DEM in ArcGIS (ESRI, 1996). These data were used to generate a metric of cropland on greater than 3% steep slopes, an indicator of potential soil and nutrient loss (Jones et al., 1996).

We obtained a digital database (100 km^2 grid cell resolution) that identified agriculturally limited areas from the European Commission's Joint Research Center (JRC) in Ispra, Italy. We resampled the data to 64 km^2 grid cells for comparison with other metric results. This database characterizes areas that have limited ability to support agricultural land uses. This designation was based on soil type (e.g. hydric and highly erodible soils), topography, and geology not conducive to agricultural production. These data were used in combination with land cover to determine where cropland existed on areas that were marginal for agriculture (hence, an indicator of marginal land use).

TABLE 3. Landscape metrics calculated on catchments and 64 km^2 grid cells. Environmental themes: EP = ecosystem productivity, TH = terrestrial habitat, BD = biological diversity, W = water, AH = aquatic habitat. Map extent numbers correspond to those given in Figure 1. Metrics with an * relate to stress or potential impairment to environmental themes or values. # = modeled from look-up tables.

Metric type	Themes	Source	Scale	Map
Land cover %	All	IGBP Land Cover	1 km	C
Land cover %	All	Corine Land Cover	100 m	D
% area cropland on >3% slope*	W, EP	IGBP LC/90 m DEM	1 km	C
% area cropland on >3% slope*	W, EP	Corine LC/90 m DEM	100 m	D
% area marginal land use*	W, EP	IGBP LC/Ag Lim. database	1 km	C
% area marginal land use*	W, EP	Corine LC/Ag Lim. database	100 m	D
% river miles with different land cover types (riparian land cover)	W, BD, AH	IGBP LC/River Network	1 km	C
% river miles with different land cover types (riparian land cover)	W, BD, AH	Corine LC/River Network	100 m	D
Population density (km^2)*	All	Landscan	1 km	C, D
Nitrogen and phosphorus export*#	W, EP	IGBP LC/ATtILA look-up tables	1 km	C
Nitrogen and phosphorus export*#	W, EP	Corine LC/ ATtILA look-up tables	100 m	D
% area w/ NDVI sign and nonsign slopes of negative/positive change (catchments only)	EP, TH	AVHRR, bimonthly samples 1992–2003	8 km	C, D
NDVI characterization of significant and nonsignificant slopes of NDVI (grid cells only)	EP, TH	AVHRR, bimonthly samples 1992–2003	8 km	C, D
NDVI slope (mean for catchments; single value for grid cells)	All	AVHRR, bimonthly samples 1992–2003	8 km	C, D
% Land cover change	All	1990/2000 Corine LC	100 m	D
% Land cover change near rivers (riparian land cover change)	W, BD	1990/2000 Corine LC/River Network	100 m	D
% Change in cropland on >3% slopes*	W, EP	1990/2000 Corine LC/ 90 m DEM	100 m	D
% Change in marginal land use*	W, EP	1990/2000 Corine LC/Ag Lim. database	100 m	D

We obtained a digital database of human population estimates for all of Europe from the Oak Ridge National Laboratory Landscan program (Dodson et al., 2000). This database provided an estimate of population density on each of the analysis units. A digital line graph of rivers covering all of Europe was obtained from the EEA web site. These data were used in conjunction with land cover databases to calculate land cover percentages within riparian zones (areas immediately adjacent to rivers). Because of computational limitations, and the relatively course-scale nature of the river data, riparian land cover metrics were limited to catchments. This metric was the proportion of river miles with certain types of land cover adjacent to the river. The type and amount of land cover adjacent to rivers is an indicator of riparian habitat and water quality (Peterjohn and Corell, 1984; Borin et al., 2005).

We quantified changes in vegetation greenness and cover using the slope value obtained from a regression model of NDVI from 1992 to 2003 ($n = 312$ for a complete time series) for each 64 km^2 pixel. These data were obtained from the University of Maryland Global Change web site. NDVI was calculated from the following equation:

$$NDVI = IR(Band\ 4) - Red(Band\ 2)/Red\ (Band2) + IR(Band\ 4)$$

Prior to the regression, we used NDVI values that were more than –9,999. An error in NDVI values was defined as an extreme change in value from one date to the next. Groten (1993) developed a method to correct for dust storm and cloud effects on NDVI for a drought study in Burkina Faso. He found that if an NDVI value was less than that of the preceding day by more than 10% it was due to a dust storm. His algorithm replaced those values by interpolating from prior to subsequent dates. We used this algorithm when consecutive values differed by more than 200 (Table 1, Nash et al., 2006). When an extreme value was the first or last date in a year, it was averaged with the neighboring observation for the same year only. We used a time series regression to estimate significant trends in NDVI at the single pixel level. Time series regression was used because errors in temporal data may be dependent. If dependency exists, then the standard error of the estimate (e.g. slope) will be inflated, and the significant level value of the slope will not be correct. We used SAS to conduct the regression analysis (SAS/ETS, 1999). The significant level of the slope was also calculated using 0.05 as the significance threshold. The significance slopes for the NDVI were mapped for the study area. Positive slopes in NDVI represented a trend of increasing vegetation cover; negative slopes in NDVI represented decreasing vegetation cover. NDVI for catchments was split into four separate metrics (Table 4). NDVI for grid cells was a ranking from 1 to 4, where 1 = significant positive trend, 2 = nonsignificant positive trend, 3 = non-significant negative trend, and 4 = significant negative trend. This was done because the grid cell size equaled the area of the NDVI analysis (64 km^2).

TABLE 4. Landscape metrics used in Principal Components Analysis.

Landscape Metric	Indication	PCA
% Forest	+	1,2,3,5
% Shrubland	+	1,2,3,5
% Natural Grassland	+	1,2,3,5
% Urban	+	1,2,3,5
% Agriculture – Total	–	1,2,3,5
% Cropland	–	1,2,3,5
% Pasture	–	1,2,3,5
% Cropland on >3% Slopes	–	1,2,3,5
% Marginal Land Use	–	1,2,3,5
Nitrogen Export (kg/ha/year)	–	1,3
Phosphorus Export (kg/ha/year)	–	1,3
Population Density (people/km^2)	–	1,2,3,4,5,6
% River km with Riparian Forest	+	1,3
% River km with Riparian Natural Cover	+	1,3
% River km with Riparian Cropland	–	1,3
% River km with Riparian Agriculture – Total	–	1
% River km with Riparian Urban	–	1,3
% River km with Riparian Anthropogenic Cover	–	1
% Forest Change	Gain(+);loss(–)	4,6
% Shrubland Change	Gain(+);loss(–)	4,6
% Natural Grassland Change	Gain(+);loss(–)	4,6
% Urban Change	loss(+);gain(–)	4,6
% Agriculture – Total Change	loss(+);gain(–)	4,6
% Cropland Change	loss(+);gain(–)	4,6
% Pasture Change	loss(+);gain(–)	4,6
% Cropland on >3% Slopes Change	loss(+);gain(–)	4,6
% Marginal Land Use Change	loss(+);gain(–)	4,6
Nitrogen Export (kg/ha/year) Change	loss(+);gain(–)	4
Phosphorus Export (kg/ha/year) Change	loss(+);gain(–)	4
% River km with Riparian Forest Change	Gain(+);loss(–)	4
% River km with Riparian Natural Cover Change	Gain(+);loss(–)	4
% River km with Riparian Cropland Change	loss(+);gain(–)	4
% River km with Riparian Urban Change	loss(+); gain(–)	4
Regression Slope (trend)	pos(+); neg(–)	2,6
Mean Regression Slope (trend)	pos(+); neg(–)	1,4
NDVI Regression Slope Classification	pos(+); neg(–)	2,6
% area w/NDVI Significant Positive Slopes	+	1,4
% area w/NDVI nonsignificant Positive Slopes	+	1,4
% area w/NDVI Significant Negative Slopes	–	1,4
Mean NDVI nonsignificant Negative Slopes	–	1,4

We also used finer-scale spatial data from the Yantra River Basin in north-central Bulgaria to calculate landscape metrics similar to those from the broader European analysis (Nikolova et al., 2007, this volume). This included enhanced land cover, digital elevation (25 m resolution), and digital line data for streams (generated from 1:100,000 topographic maps, Knight et al., 2002). The purpose of this analysis was to see how well the broader assessment captured landscape patterns generated from more detailed data.

2.3. STATISTICAL ANALYSES

We performed Principal Components Analyses (PCA) on six different data sets using SAS: (1) catchments with IGBP land cover, (2) grid cells with IGBP land cover, (3) catchments with Corine 2000 land cover (status), (4) catchments with Corine 1990 and 2000 land cover (change), (5) grid cells with Corine 2000 land cover (status), and (6) grid cells with Corine 1990 and 2000 land cover (change) (Table 4). We evaluated the first five principle components (PCs) based on their tendency to indicate less desirable or stressed conditions versus more desirable or less stressed conditions (Table 4). PCs with high positive values of stress-related metrics (e.g. marginal land use), and negative values of desired conditions (e.g. large amounts of forest), were considered to be negative. PCs with high positive values of metrics that indicate desired conditions, and negative values of stress-related metrics, were considered to be positive. We then ranked each of the first five PCs for each catchment and grid cell using the quartile rank function in SAS (equal interval ranks of 1–4, with a rank of 1 being the poorest relative condition and 4 being the best relative condition). We then added the rankings of each of the first five PCs to obtain an overall score for each catchment and grid cell (possible scores from 5–20). Lower scores indicated less desirable conditions, higher scores more desirable conditions. The groups of landscape metrics used in the PCAs are given in Table 4. We also calculated a simple index of the relative vulnerability of catchments and grid cells to potential future environmental degradation (or loss of environmental security). The analysis was limited to areas with Corine 1990 and 2000 data. The most vulnerable catchments and grid cells were those that were in the lowest 25% of the PCA results for landscape condition and change (see above), whereas the least vulnerable were in the top 25% of the PCA results. Therefore, areas with the least desirable landscape conditions that were changing in a negative direction were more vulnerable to future environmental degradation than areas that had more desirable and improving environmental conditions.

We used the nearest-neighbor algorithm in the "Animal Movement Analysis" ArcView extension (Hooge and Eichenlaub, 1997) to evaluate the degree to which the spatial pattern of changes in forest, grassland, cropland, and

NDVI were randomly versus nonrandomly distributed. To conduct this analysis, we converted grid cell polygons (Corine areas only) to point shapefiles and then ran the Nearest Neighbor Distance Test. We ran stepwise multiple regression analyses in SAS to evaluate the relationships between forest change, cropland change, and NDVI change (dependent variables), and other metrics of landscape change.

3. Results

Because of the large number of maps needed to display the spatial distribution of each individual metric, we have included only a few maps of metric results to illustrate how these results can be mapped and displayed.

3.1. LANDSCAPE STATUS

Means, standard deviations, and minimum and maximum values for landscape metrics are presented in Table 5. The 64 km^2 grid cells captured finer-scale variation in the landscape metrics (Figures 2, 3, and 4), and results for many grid cell metrics were more variable than results for catchments. This reflects the finer-scale nature of grid cells, as opposed to catchments, which give an average for a much larger area. The amount of forest across IGBP land cover areas tended to be less than those estimated for Corine areas, whereas cropland was more abundant in the former than in the later (Table 5). These differences may reflect differences in the grain size of the land cover data, or different patterns of land cover in eastern Europe. Urban land cover was nearly three times more common in Corine areas than IGBP areas, and population density was also greater in the former (Table 5). This likely reflects the addition of rural, less populated areas in east and northeast Europe in the IGBP analyses. Values for riparian forests were higher in Corine analysis areas than across the IGBP analysis area (Table 5). This too may reflect the finer scale nature of Corine land cover data (100 meters) versus the IGBP land cover data (1 km).

3.2. LANDSCAPE CHANGE

Two types of landscape change metrics were generated: (1) those related to differences between the 1990s and 2000 Corine land cover data and (2) those related to trends (slope) in the Normalized Difference Vegetation Index (NDVI). Although there was considerable spatial variation in the amount of change, overall, changes in landscape metric values between the early 1990s and 2000 were small (Table 6). Additionally, the mean values for catchments

TABLE 5. Summary statistics for European-wide landscape metrics.

Metric	Mean	SE	Min	Max
IGBP catchments				
% Marginal Land Use	20.7	0.54	0.0	100.0
% Cropland on >3% Slopes	16.0	0.37	0.0	100.0
% Agriculture – Total	70.7	0.66	0.0	100.0
% Cropland	44.1	0.62	0.0	100.0
% Pasture	26.0	0.63	0.0	99.1
% Urban	1.9	0.17	0.0	100.0
% Riparian Urban Land Cover	2.5	0.16	0.0	100.0
% Riparian Cropland Land Cover	33.0	0.64	0.0	100.0
% Forest	17.2	0.50	0.0	100.0
% Grassland	4.6	0.28	0.0	100.0
% Shrubland	2.1	0.12	0.0	83.1
% Riparian Natural Land Cover	14.9	0.49	0.0	100.0
% Riparian Forest	9.5	0.40	0.0	100.0
Phosphorus Export (kg/ha/year)	1.3	0.01	0.1	2.3
Nitrogen Export (kg/ha/year)	5.6	0.04	0.2	8.5
Population Density (people/km^2)	57.9	10.33	0.0	2,709.7
% area Sign. + NDVI Change	9.7	0.35	0.0	100.0
% area nonsign. + NDVI Change	58.3	0.54	0.0	100.0
% area Sign. – NDVI Change	1.3	0.13	0.0	100.0
% area nonsign – NDVI Change	30.7	0.57	0.0	100.0
Mean NDVI Slope	0.1	0.01	–2.7	5.3
IGBP grid cells				
% Marginal Land Use	13.3	0.08	0.0	100.0
% Cropland on >3% Slopes	20.4	0.07	0.0	100.0
% Agriculture – Total	71.0	0.10	0.0	100.0
% Urban	1.1	0.01	0.0	100.0
% Pasture	26.3	0.09	0.0	100.0
% Cropland	45.0	0.10	0.0	100.0
Population Density (people/km^2)	50.0	42.2	0.0	14,135.2
% Forest	19.7	0.09	0.0	100.0
% Shrubland	2.6	0.04	0.0	100.0
% Grassland	2.5	0.03	0.0	100.0
Phosphorus Export (kg/ha/year)	1.2	0.01	0.1	2.3
Nitrogen Export (kg/ha/year)	6.1	0.01	0.2	8.5
NDVI Regression Slopes	0.3	0.01	–4.4	9.5
NDVI Regression Slope Classification	2.3	0.01	1.0	4.0

TABLE 5. Continued

Metric	Mean	SE	Min	Max
Landscape metric statistics (Corine 2000) – catchments				
% Marginal Land Use	12.9	0.50	0.0	85.6
% Phosphorus Export	1.1	0.02	0.1	2.2
% Nitrogen Export	4.8	0.05	0.2	8.3
% Cropland on >3% Slopes	13.8	0.34	0.0	54.6
% Urban	4.9	0.19	0.0	75.1
% Pasture	9.9	0.42	0.0	78.2
% Cropland	39.2	0.71	0.0	94.0
Population Density (People/km^2)	86.0	4.89	0.0	2,709.7
% Riparian Urban Land Cover	9.1	0.42	0.0	100.0
% Riparian Cropland Land Cover	36.1	0.87	0.0	100.0
% Forest	26.7	0.54	0.0	96.1
% Shrubland	6.1	0.32	0.0	71.01
% Grassland	2.5	0.17	0.0	52.8
% Agriculture – Total	56.9	0.69	0.0	99.3
% Riparian Natural Land Cover	19.4	0.68	0.0	100.0
% Riparian Forest Land Cover	13.8	0.54	0.0	100.0
Landscape metric statistics (Corine 2000) – grid cells				
% Marginal Land Use	16.5	0.12	0.0	100.0
% Cropland on >3% Slopes	14.8	0.12	0.0	91.3
% Urban	4.3	0.03	0.0	97.9
% Pasture	10.3	0.14	0.0	100.0
% Cropland	37.0	0.13	0.0	100.0
Population Density (People/km^2)	74.0	73.00	0.2	14,135.2
% Forest	26.7	0.10	0.0	100.0
% Shrubland	7.3	0.06	0.0	100.0
% Grassland	3.1	0.04	0.0	98.9
% Agriculture – Total	55.7	0.13	0.0	100.0

versus grid cells were quite similar, although grid cells were always more variable (Table 6). Again, this likely reflects the finer-scale nature of grid cells. In general, metric results tended to be highly variable, which points to the importance of analyzing spatial variability. Figure 3 illustrates percent changesin forest on catchments and grid cells. Although there were similarities in the spatial distribution of forest change across Corine areas, there were some noticeable differences (Figure 3). Both analyses picked up relatively high percentages of forest losses in southeastern and southestern France, and in areas of

A

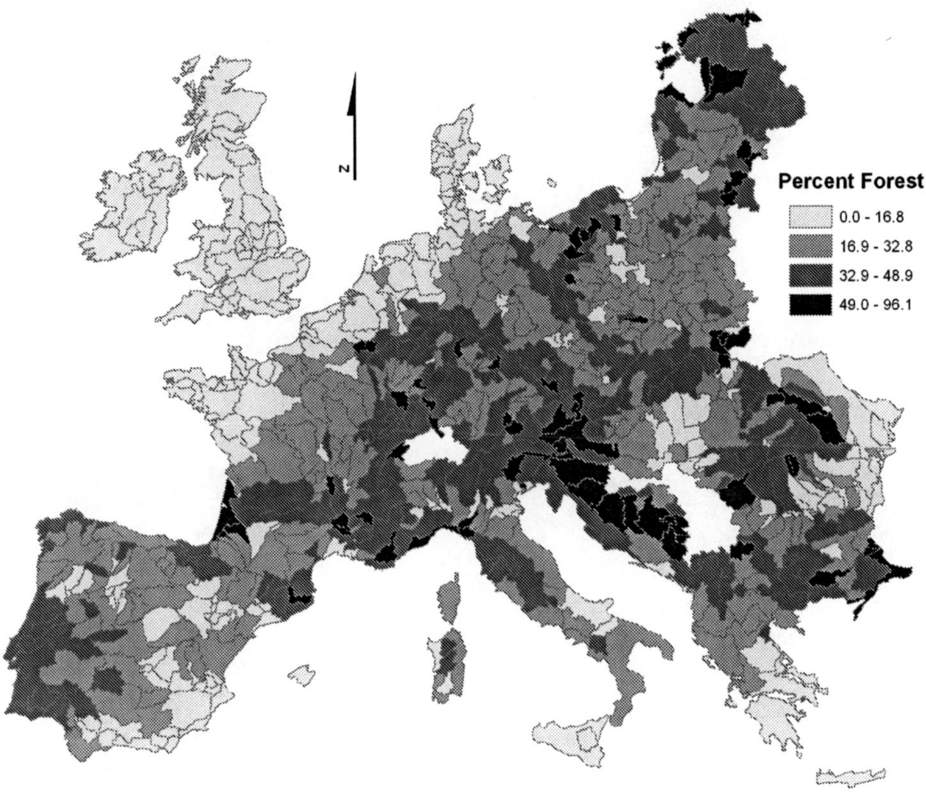

Figure 2A. Percent forest by catchment. Percent forest is based on an analysis of the Corine 2000 land cover data.

southeastern Spain, but grid cells picked up addiional losses in western Spain whereas catchment analyses showed greater relative forest decline in Eastern Europe (Figure 3). There seemed to be better agreement on the pattern of forest gain between catchments and grid cells (Figure 3). These differences may result from the averaging of values over larger areas in the catchment analysis than in the grid cell analysis. Overall, Europe exhibited a pattern of increased greenness over the period from 1992 to 2003 (Tables 5 and 6; Figure 4). However, most positive changes in greenness were not significant. Significant positive changes in greenness were clustered in northeast Spain and southeastern Europe,

B

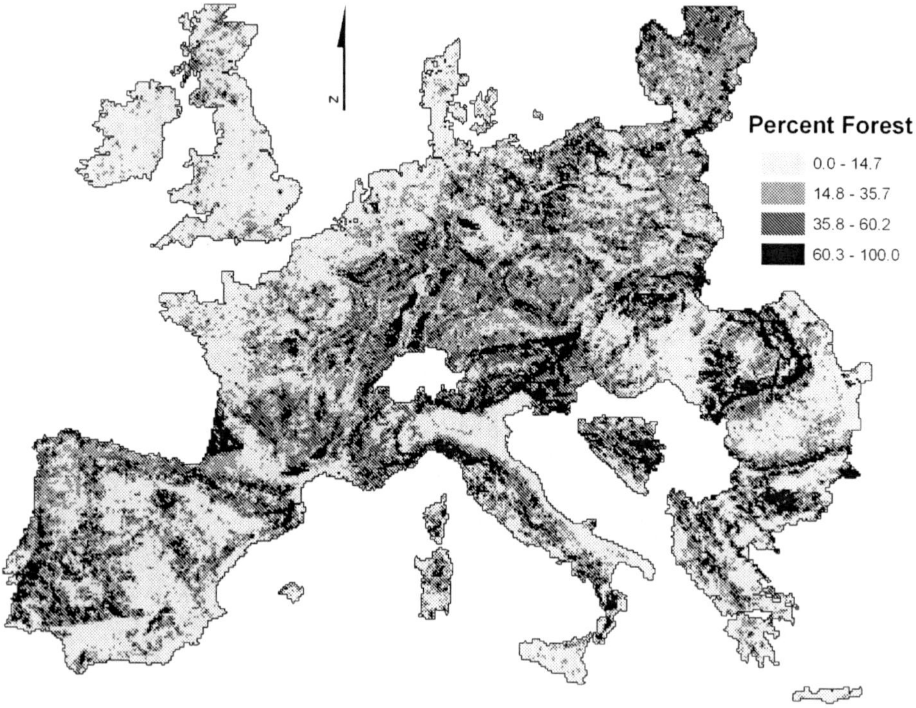

Figure 2B. Percent forest by grid cells. Percent forest is based on an analysis of the Corine 2000 land cover data.

and significant negative changes in southeastern Spain and extreme Eastern Europe near the Caspian Sea (Figure 4). Tests for nonrandom patterns of landscape metric change (based on analysis of grid cells) found that all spatial patterns of change had a tendency to be clumped and nonrandom. In all cases the null hypothesis of randomness was rejected. Some of these results may reflect the fact that certain land cover types were clumped in their distributions. For example, forests tended to occur in mountainous regions and, therefore, were nonrandom in nature. However, others tended to be more ubiquitous and less clumped in distribution.

A

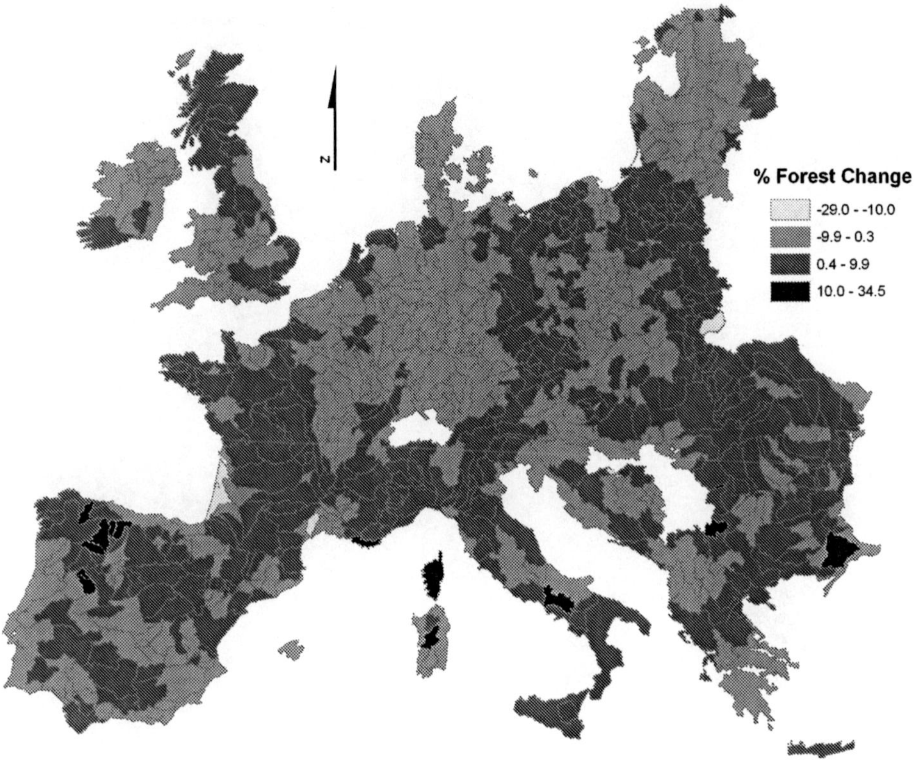

Figure 3A. Percent forest change by catchments. Percent forest change is based on an analysis of differences in Corine land cover from the 1990s and 2000 databases.

B

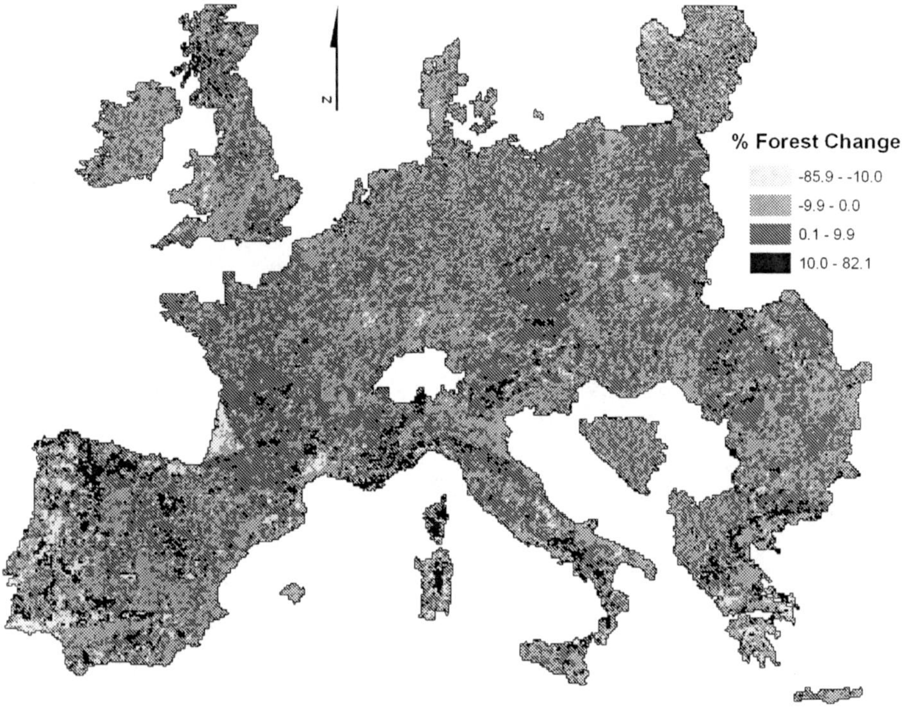

Figure 3B. Percent forest change by grid cells. Percent forest change is based on an analysis of differences in Corine land cover from the 1990s and 2000 databases.

A

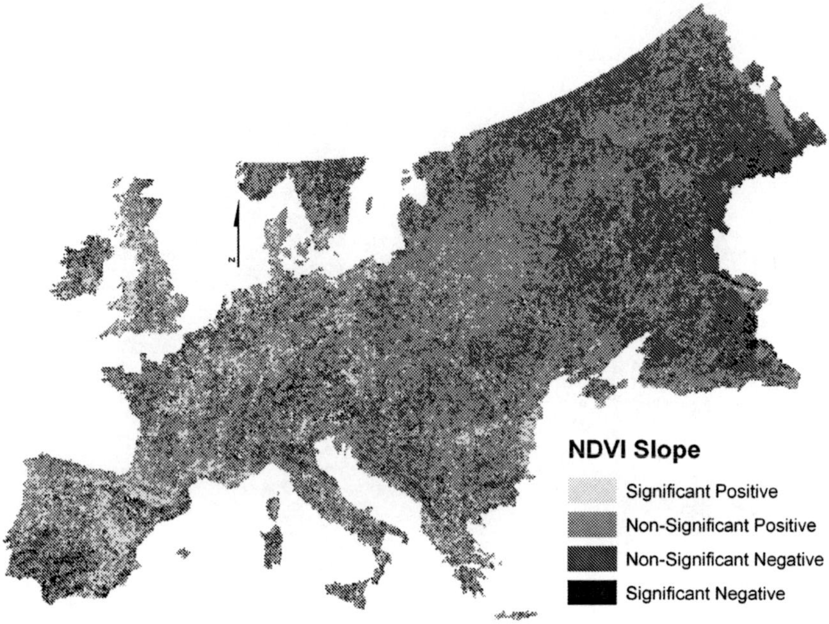

Figure 4A. Slopes of changes in the Normalized Difference Vegetation Index (NDVI) between 1992 and 2003 for grid cells.

B

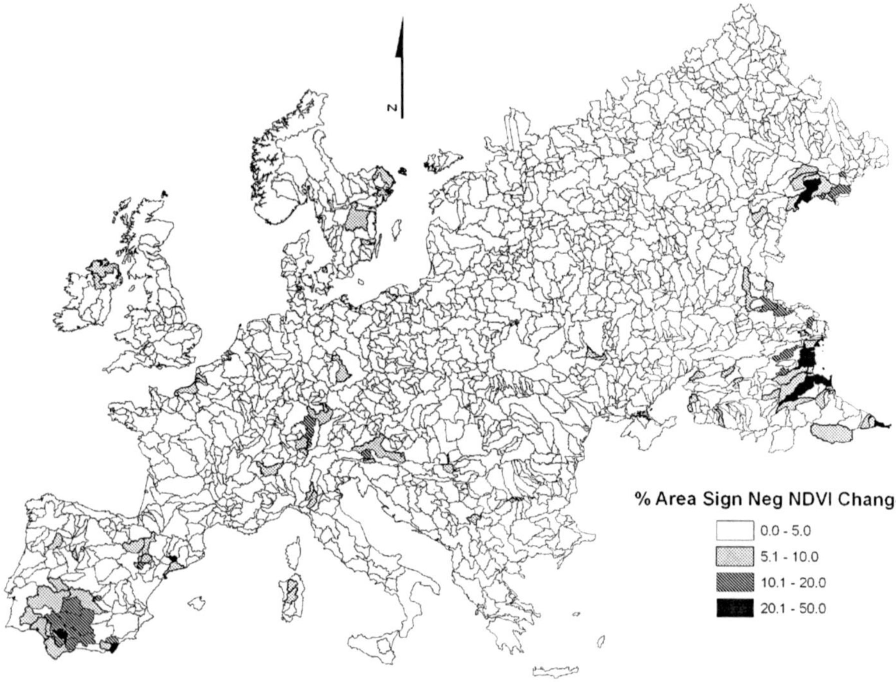

Figure 4B. Percent area of significant negative Normalized Difference Vegetation Index (NDVI) change between 1992 and 2003 for catchments.

TABLE 6. Landscape change statistics.

Metric	Mean	SE	Min	Max
Landscape change statistics (Corine 1990s/2000) – Catchments				
% Marginal Land Use Change	–0.3	0.09	–24.1	38.5
Nitrogen Yield (kg/ha/year) Change	0.1	0.01	–2.3	2.6
Phosphorus Yield (kg/ha/year) Change	0.1	0.01	–0.9	0.7
% Urban Change	0.5	0.04	–18.4	12.4
% Pasture Change	0.3	0.17	–36.3	35.7
% Cropland Change	–0.7	0.18	–57.9	30.3
% Agriculture – Total Change	–0.8	0.12	–27.7	25.5
% Cropland on >3% Slopes Change	–10.5	0.30	–58.2	6.0
% Area with NDVI Sign. Neg. Slopes	1.0	0.10	0.0	34.7
% Area with NDVI non-sign. Pos. Slopes	65.4	0.64	0.0	100.0
% Area with NDVI nonsign. Neg. Slopes	20.1	0.56	0.0	100.0
% Riparian Urban Change	0.8	0.15	–79.2	45.8
% Riparian Cropland Change	–2.5	0.37	–97.0	35.9
% Forest Change	0.7	0.12	–29.0	34.5
% Shrubland Change	0.8	0.10	–19.3	22.9
% Grassland Change	–0.6	0.11	–23.1	36.1
% Area with NDVI Sign. Pos. Slopes	13.5	0.51	0.0	100.0
% Riparian Forest Change	–0.1	0.23	–70.5	90.6
% Riparian Natural Cover Change	–0.8	0.29	–91.2	93.6
Mean NDVI Regression Slope	0.55	0.01	–2.2	2.1
Landscape metric change statistics (Corine 1990s/2000) – grid cells				
% Marginal Land Use Change	–0.7	0.03	–95.3	73.4
% Urban Change	0.6	0.01	–59.5	50.4
% Pasture Change	0.4	0.03	–89.4	85.8
% Cropland Change	–0.8	0.03	–99.0	100.0
% Agriculture – Total Change	–0.7	0.03	–93.9	100.0
% Cropland on >3% Slopes Change	–0.7	0.02	–98.3	82.1
% Forest Change	0.8	0.03	–85.9	82.1
% Shrubland Change	1.0	0.03	–76.2	100.0
% Grassland Change	–0.8	0.02	–87.8	89.5
NDVI Regression Slope Classification	2.1	0.01	1.0	4.0
Mean NDVI Regression Slope	0.5	0.01	–4.3	9.8

3.3. YANTRA RIVER COMPARISION

Estimates of some land cover percentages derived from the broader European-wide and Corine areas compared favorably to estimates derived from finer-scale data. Estimates of the amount of forest, cropland, grassland, and urban land cover for the Yantra River Basin in north-central Bulgaria derived from the Corine 2000 database were similar to those derived from finer-scale data (Table 7). Estimates from the IGBP land cover data were less similar (Table 7). Closer alignment with the Corine databases likely reflects the use of similar data (e.g. Landsat Thematic Mapper) to derive land cover in the Yantra River Basin. However, estimates for riparian metrics differed markedly from those of the broader assessment, but especially for forest and cropland riparian metrics (Table 7).

TABLE 7. Comparison of landscape metric values generated from the European-wide analyses versus those generated from finer-scale data for the Yantra River Basin.

Metric	IGBP	Corine 2000	Corine change	Yantra
% Forest	27.5	34.6	–	34.1
% Cropland	39.4	38.7	–	35.2
% Grassland	0.7	2.2	–	1.2
% Urban	0.7	5.4	–	5.6
% Ag >3% Slopes	28.1	26.3	–	20.9
% Riparian Forest	6.3	7.3	–	41.2
% Riparian Natural	7.2	12.8	–	47.5
% Riparian Urban	0.9	12.1	–	6.4
% Riparian Cropland	52.3	58.2	–	20.4
% Forest Change	–	–	1.1	0.1
% Cropland Change	–	–	–0.4	0.1
% Urban Change	–	–	–0.3	0.9
% Ag >3% Slope Change	–	–	–1.4	0.1
% Natural LC Change	–	–	1.2	–0.1
% Riparian Forest Change	–	–	–0.1	0.1
% Riparian Cropland Change	–	–	8.5	–0.1
% Riparian Urban	–	–	–1.4	0.1

The markedly greater amounts of riparian forest, and lower amounts of cropland from the finer-scale analysis, as compared to the broader-scale analysis, reflect the far greater resolution of rivers and streams in the finer-scale analysis (Figure 5). The stream coverage used in the broader analysis only included two main rivers (the Yantra and Rositza) whereas the finer-scale analysis includes numerous rivers and streams in heavily forested headwater areas (Figure 5). Additionally, the DEM used in the finer-scale analysis may also explain lower values for agriculture on steep slopes (Table 7). The broader-scale analysis tended to have higher values of metric change than did the finer-scale Yantra analysis, although change values for both assessments were very low (Table 7). Given the potential for errors in land cover classification (see EEA, 2006), it is unlikely that these differences are statistically significant at the Yantra River Basin level.

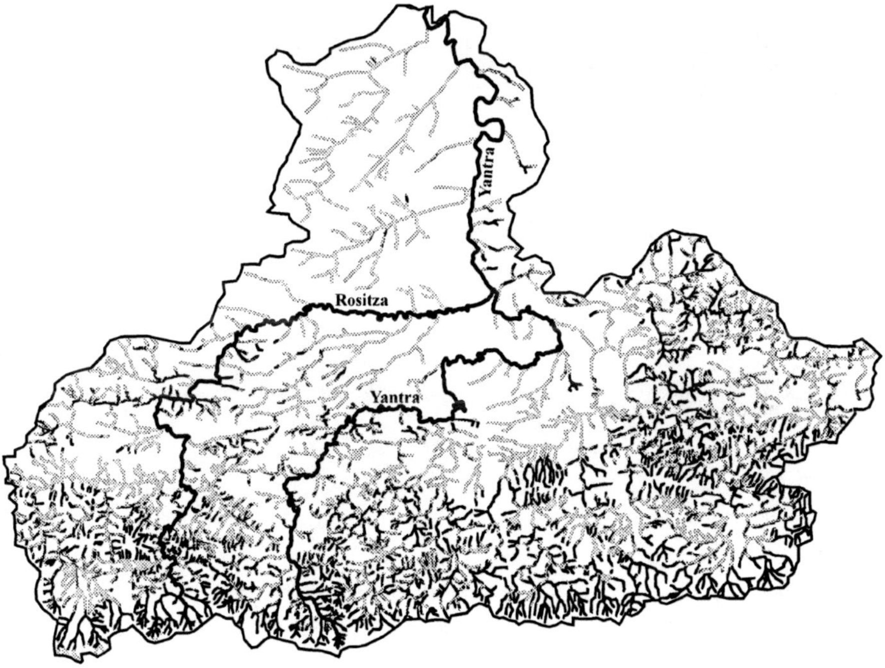

Figure 5. Riparian forests along finer-scale rivers and streams in the Yantra River Basin. Small black lines indicate rivers and streams with adjacent forest. Gray lines indicate rivers and streams with adjacent urban or agriculture. Large black lines indicate the extent of rivers used in the European-wide analyses (from the European digital line graph for rivers).

3.4. PCA ANALYSIS

Although there was considerable similarity in the principle components (PCs) of IGBP catchments and grid cells (Table 8), there were differences in the spatial pattern of cumulative PCA scores (e.g. overall condition, Figure 6). These differences may result from differences in the total number and types of metrics used in the two analyses (Table 4), as well as differences in the quartile values. The IGBP catchment analysis included riparian metrics and the proportion of the catchment with different types of NDVI slopes; the IGBP grid cell analysis lacked riparian measures and only had characterizations of NDVI change (Table 4).

TABLE 8. Principal Components Analysis results. Metrics used in the analysis are given in Table 4. Higher values are indicated with a (+); lower values with a (–). Each PC is rated as being an indicator of positive or negative landscape condition or vulnerability. Population density = = status only (no change)*.

PC	Metrics	Eigenvalue	Indication
(1) IGBP Catchments-status (n = 1888). Cumulative Proportion of variance explained = 0.75			
1.	+ Total Agriculture; – Forests, Grassland, Pos. NDVI	0.39	Neg
2.	+ Pos. NDVI Slope, Forests, Riparian Forest; – Neg. NDVI Slope	0.13	Pos
3.	+ Urban, Pop. Density*; – Total Agriculture	0.09	Neg
4.	+ Pasture, Pop. Density*, Urban; – Grassland	0.07	Neg
5.	+ Marginal Use, Neg. NDVI Slope; – Forest, Riparian Forest	0.07	Neg
(2) IGBP Grid cells-status (n = 112,770). Cumulative Proportion of Variance Explained = 0.75			
1	+ Total Agriculture; – Forests, Grassland, Pos. NDVI Slope	0.24	Neg
2	+ Pos. NDVI Slope, Forests, Riparian Forest; – Neg. NDVI Slope	0.17	Pos.
3	+ Urban, Pop. Density*; – Total Agriculture, Pasture	0.13	Neg
4	+ Pasture, Pop. Density*, Urban; – Grassland, NDVI Slope	0.12	Neg
5	+ Shrubland; – Forest	0.09	Neg
(3) Corine2000 Catchments-status (n = 908) Cumulative Proportion of Variance Explained = 0.76			
1	+ N and P Export, Cropland, Cropland >3% Slope, Marginal Use; – Grassland, Forest	0.33	Neg
2	+ Urban, Pop. Density*; – Total Agriculture, Shrublands	0.15	Neg

PC	Metrics	Eigenvalue	Indication
3	+ Forests, Riparian Forests, Natural Cover; – Pasture, Total Agriculture	0.12	Pos
4	+ Marginal Use, Cropland; – Forest	0.09	Neg
5	+ Marginal Use, Urban; – Forest, Cropland	0.07	Neg
(4) Corine Catchments-change (n = 908). Cumulative Proportion of Variance Explained = 0.63			
1	+ Marginal Use, N and P Export, Cropland change; – Forest, Grassland change	0.18	Neg
2	+ Neg. NDVI Slope, P Export change; – Pos. NDVI Slope, Forest changes	0.12	Neg
3	+ Urban, Riparian Urban change; – Cropland and Riparian Forest change	0.10	Neg
4	+ Riparian Forest, Pos. NDVI Slope, Grassland change; – Pasture and Riparian Urban change	0.09	Pos
5	+ Forest, Pos. NDVI Slope.; – Neg. NDVI Slope, Cropland on >3% Slope change	0.07	Pos
(5) Corine 2000 Grid Cells-Status (n = 54,815). Cumulative Proportion of Variance Explained = 0.81			
1	+ Total Agriculture, Marginal use; – Forests, Grassland	0.31	Neg
2	+ Urban, Pop. Density*; – Cropland, Shrubland	0.17	Neg
3	+ Marginal Use, Pop. Density*, Cropland >3% Slope; – Pasture, Cropland	0.14	Neg
4	+ Grassland; – Forest, Marginal Use, Cropland >3% Slope	0.11	Pos
5	+ Marginal Use, Pasture; – Grassland	0.08	Neg
(6) Corine Grid Cells – Change (n = 54,815). Cumulative Proportion of Variance Explained = 0.73			
1	+ Marginal Use, Cropland, Cropland >3% Slope changes; – Forest, Pasture changes	0.21	Neg
2	+ NDVI Neg. Slope.; – Pasture changes	0.15	Neg
3	+ Total Agriculture Change; – NDVI Pos. Slope, Forest, and Grassland change	0.15	Neg
4	+ Shrubland change; – Forest change	0.12	Neg
5	+ Pop. Density*, Urban change; – Grassland change	0.10	Neg

When compared to grid cell results, the IGBP catchment analysis showed larger clusters of areas in less desirable condition than did the grid cell analysis (Figure 6). However, both analyses showed lower cumulative PC values in the UK, east and southeast Europe, southern and northern parts of Spain, and the Po River basin of Italy (Figure 6). Similar patterns were observed for results of the Corine 2000 analysis, although there appeared to be greater spatial variability in cumulative PC scores than the IGBP analyses (Figure 7). Greater spatial variability in the Corine 2000 analyses may reflect the finer-scale nature of the Corine land cover data. The Corine 2000 grid cell analysis showed larger areas of central Europe (north of the Alps) with lower cumulative scores than did the Corine 2000 catchment analysis. Conversely, some areas of the northern UK, Denmark, and Greece had relatively high scores in the grid cell analysis and lower scores in the catchment analysis (Figure 6). These differences may reflect differences in metrics used in the analysis (see Table 4), and differences in PC metric loadings (Table 8). The Corine catchment analysis included a PC with positive loadings of forests, riparian forests, and natural land cover; the grid cell analysis had no riparian metrics and lacked this PC (Table 8).

Cumulative PC scores of landscape change were variable but spatially clustered (Figure 8). Nearest neighborhood tests for spatial randomness indicated that landscape changes were spatially nonrandom and had a clumped tendency. Like other results, grid cells showed greater spatial variation than catchment analyses (Figure 8). Additionally, there were differences in the PCs, likely resulting from differences in the number and types of metrics used in each analysis. Analysis of changes in catchments included riparian metrics whereas grid cells lacked riparian metrics (Table 4). PCA analysis of catchment changes also had the lowest cumulative proportion of variance explained by the first five PCs (0.63, Table 8).

The southeastern UK, northwestern France, areas surrounding the Alps, and north-central Europe showed the greatest amount of negative change (e.g. loss of forest, gains in N and P export, and loss of greenness, Table 8), whereas a number of areas in Spain, central France and east-central Europe showed greater amounts of positive change (forest, riparian forest, and greenness gain, Table 8).

A

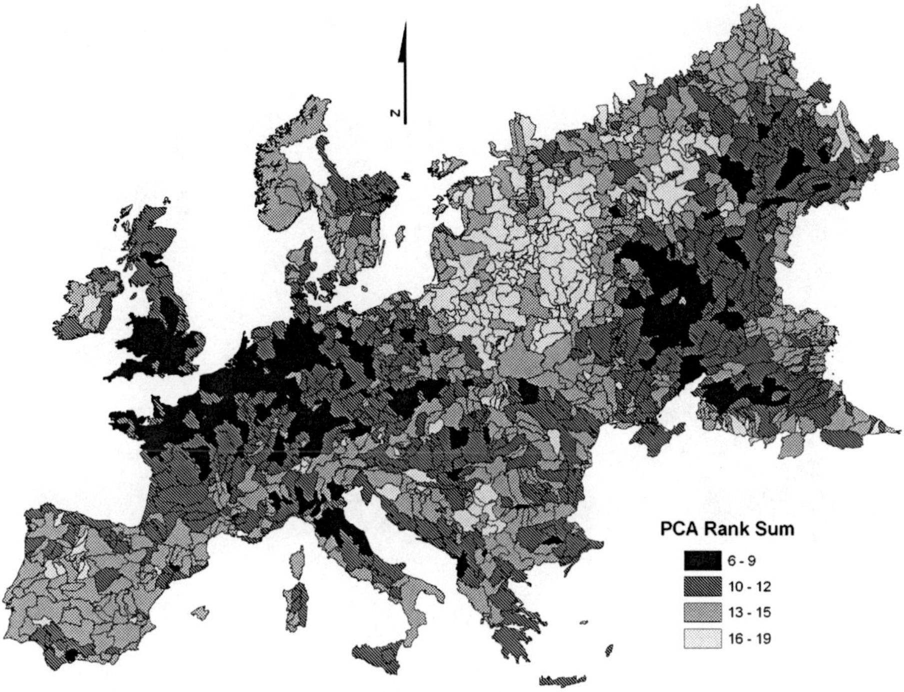

Figure 6A. Principle Components Analysis of catchments based on IGBP land cover data. See text for explanation.

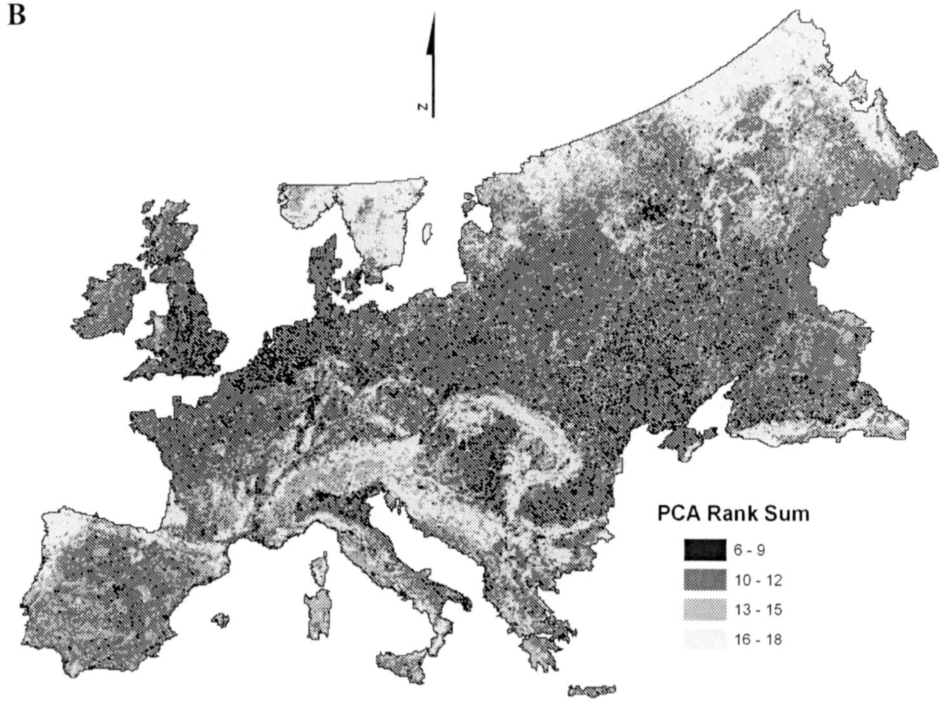

Figure 6B. Principle Components Analysis of grid cells based on IGBP land cover data. See text for explanation.

A

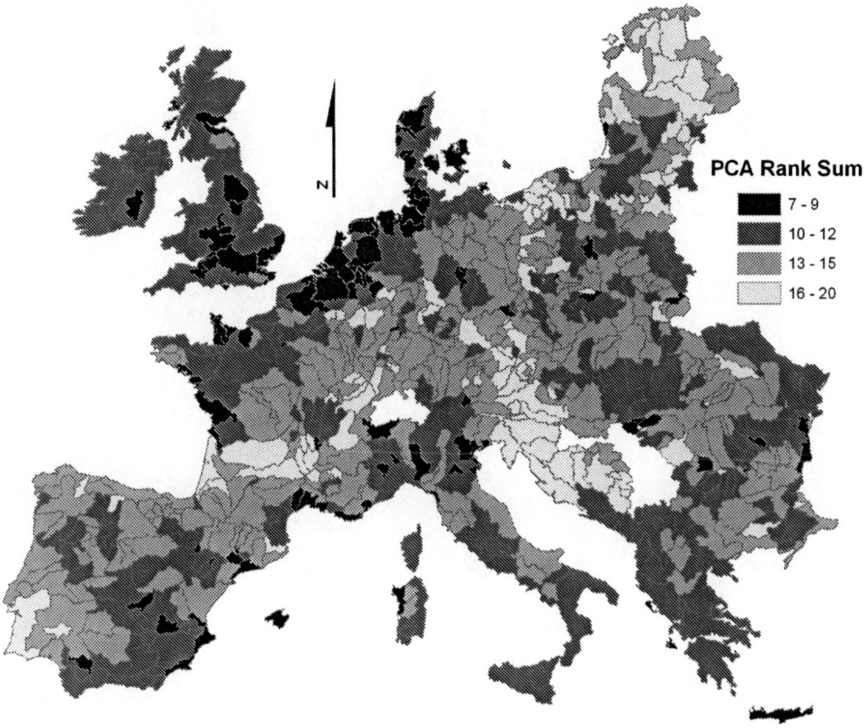

Figure 7A. Principle Components Analysis of catchments based on 100 m Corine 2000 land cover data. See text for explanation.

B

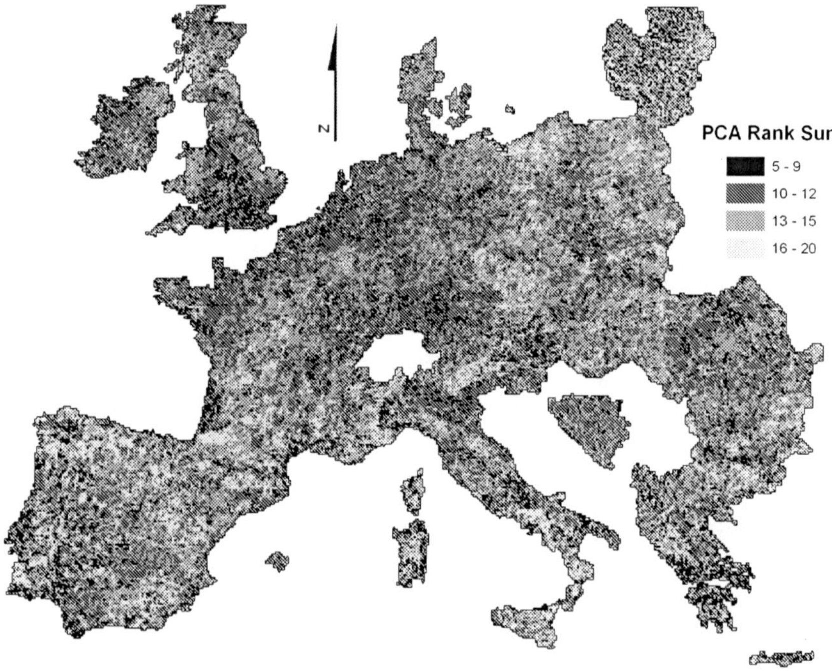

Figure 7B. Principle Components Analysis of grid cells based on 100 m Corine 2000 land cover data. See text for explanation.

A

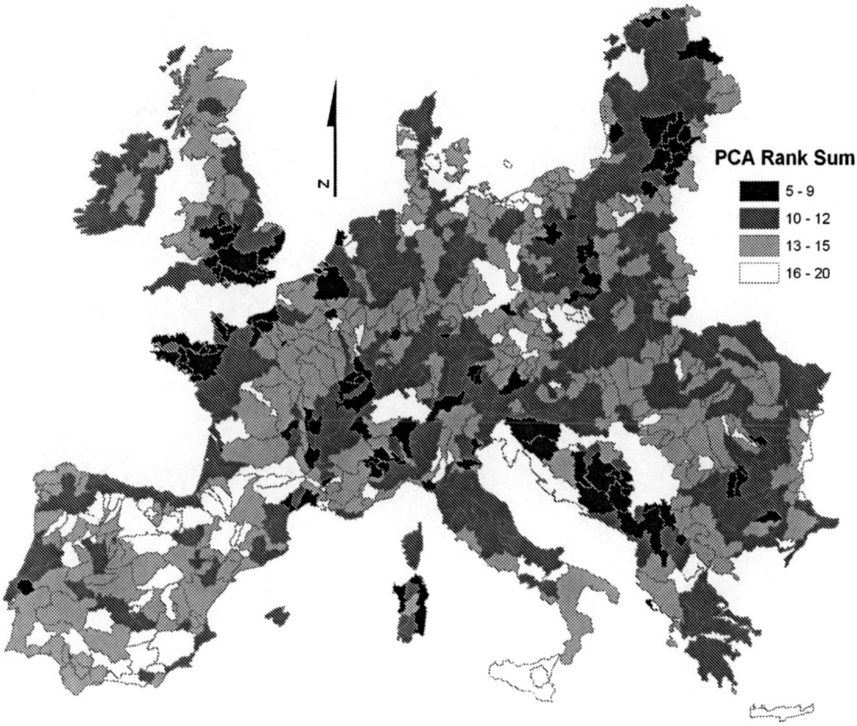

Figure 8A. Principle Components Analysis of catchments based on changes in land cover metrics (Corine 1990s and 2000). See text for explanation.

B

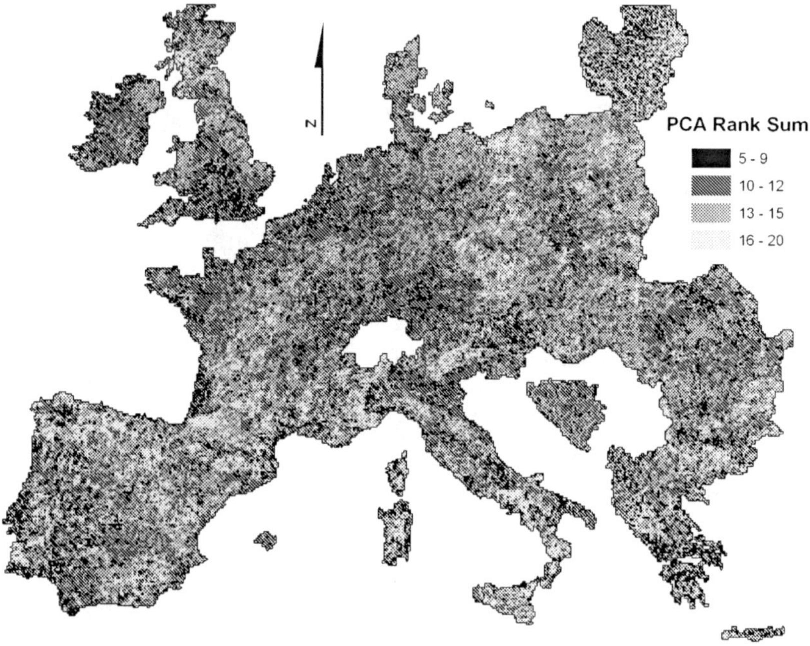

Figure 8B. Principle Components Analysis of grid cells based on changes in land cover metrics (Corine 1990s and 2000). See text for explanation.

Catchments with the lowest scores for condition (least desirable conditions) and lowest scores for change (greatest amount of negative change) occurred in the southeastern UK, parts of northern France, and isolated catchments in eastern Europe (Figure 9). Catchments with high scores for status (most desirable conditions) and change (areas with desirable changes) occurred sporadically across Europe, with a cluster of catchments in Slovenia and Croatia (Figure 9). Grid cell analysis showed a similar but finer-scale pattern of areas with relatively low values, although there appeared to be a greater number of these areas in Italy and north-central Europe (Figure 9B). Additionally, Spain appeared to have a greater number of areas with low values based on grid cell analysis than the catchment scale analysis (Figure 9).

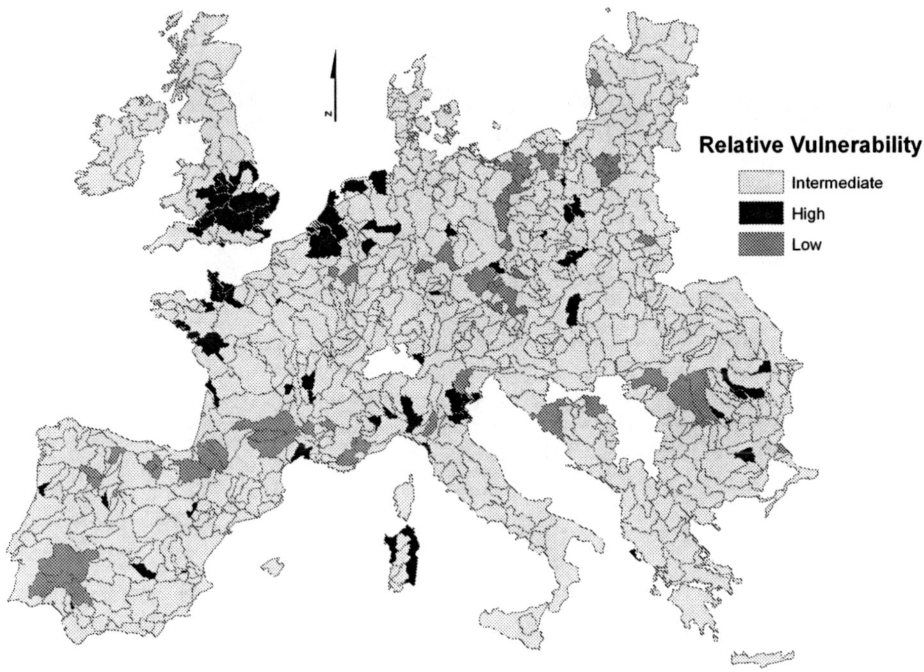

Figure 9A. Catchments indicating areas with the highest levels of vulnerability versus those with relatively low vulnerability. High levels include those catchments that were in the lowest 25% of all catchments with regards to cumulative PCA scores for condition and change; low levels of vulnerability are those catchments that were in the top 25% of all catchments with regards to cumulative PCA scores for condition and change.

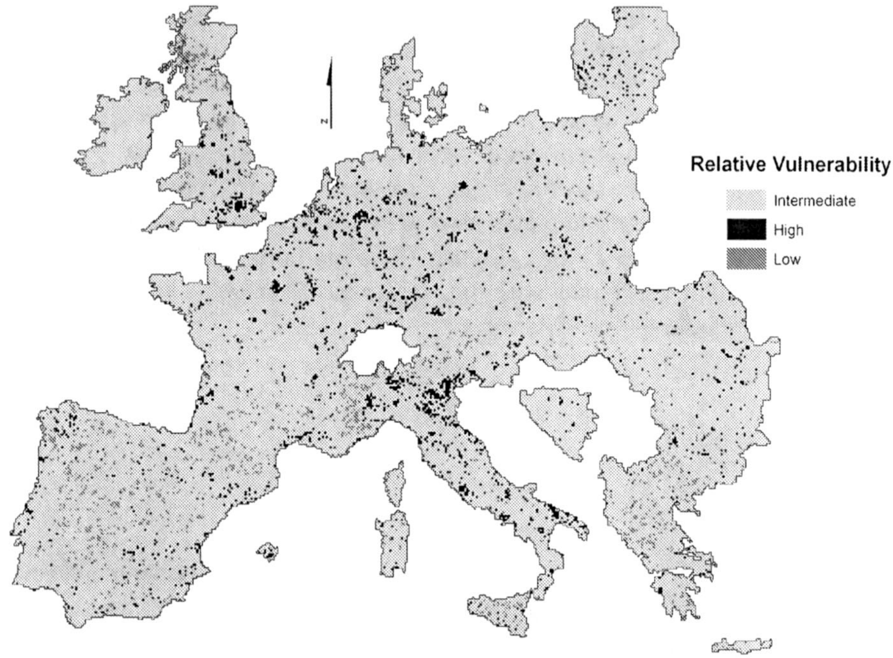

Figure 9B. Grid cells indicating areas with the highest levels of vulnerability versus those with relatively low vulnerability. High levels include those grid cells that were in the lowest 25% of all grid cells with regards to cumulative PCA scores for condition and change; low levels of vulnerability are those grid cells that were in the top 25% of all grid cells with regards to cumulative PCA scores for condition and change.

3.5. ENVIRONMENTAL CORRELATES OF CHANGE

Overall, landscape change metric values were relatively unrelated to each other. The highest levels of correlation existed between nitrogen (N) and phosphorus (P) export metrics (0.95), N export and cropland change (0.78), P export and cropland change (0.88), and change in the percentage of cropland on greater than 3% slopes and total agricultural land cover change (0.86). High correlation between N and P export was expected because they are derived from similar look-up tables based on land cover composition. Correlation coefficients for all other combinations of metrics were less than 0.60.

Sixty-seven percent of the variation in total agriculture land cover change on catchments was explained by six other variables of landscape change (Table 9).

TABLE 9. Results of stepwise multiple regression analysis for selected metrics of landscape change.

Dependent Variable	Independent Variables	Cum Var.	Sign Explained
Total Agriculture Change (Catchments)	Grassland Change	0.16	–
	Forest Change	0.32	–
	Shrubland Change	0.47	–
	Urban Change	0.56	–
	Cropland Change	0.58	+
	Pastureland Change	0.67	+
Total Agriculture Change (Grid Cells)	Grassland Change	0.18	–
	Forest Change	0.28	–
	Shrubland Change	0.45	–
	Urban Change	0.50	–
	Cropland Change	0.56	+
	Pastureland Change	0.60	+
	Marginal Land Use Change	0.64	+
Forest Change (Catchments)	Shrubland Change	0.17	–
	Agricultural Total Change	0.31	–
	Grassland Change	0.46	–
	Urban Change	0.50	–
Forest Change (Grid Cells)	Shrubland Change	0.18	–
	Agricultural Total Change	0.29	–
	Grassland Change	0.44	–
	Urban Change	0.47	–
Riparian Forest Change (Catchments)	Riparian Natural Change	0.43	+
	Agricultural Total Change	0.44	–
	Forest Change	0.46	+

Of these variables, grassland and forest change (both negative relationships) explained the greatest variation (Table 9). A similar set of variables explained total agriculture land cover change on grid cells, although the total variation explained was less than that for catchments (0.64, Table 9). Additionally, total agriculture land cover change was positively related to changes in marginal land (Table 9). Fifty and forty-seven percent of the variation in forest land cover change were explained by four landscape change variables for catchments and grid cells, respectively (Table 9). Forest change was negatively associated with shrubland, total agriculture, grassland, and urban changes (Table 9). Forest riparian change was positively associated with natural riparian change (0.43), and total agriculture and forest change at the catchment scale (Table 9). Less than 15% of the total variation in any of the NDVI metrics was explained by variation in any combination of the other change variables, and no other regression model explained more than 35% of the total variation in any other change metric.

4. Discussion

4.1. EUROPEAN LANDSCAPE CONDITIONS

Using existing data and a set of landscape metrics, we were able to conduct landscape analyses for different assessment units across all of Europe. This included calculation of basic and multi-metric statistics (e.g. Principle Components Analysis (PCA)), determining correlations and relationships among metrics, and the ability to display the results spatially for common assessment units. However, limitations in availability of certain spatial data precluded calculation of several important metrics or indicators of environmental conditions for certain geographies within Europe. For example, lack of Digital Elevation Model (DEM) data north of 60° north latitude precluded an analysis of cropland on steep slopes for areas of northern Europe; lack of spatial data on agriculturally limited lands precluded an analysis of marginal land use in Turkey; lack of two or more time series land cover data precluded analysis of change in landscape metrics in eastern Europe, Serbia–Montenegro, Slovenia, Switzerland, and Turkey. Additionally, lack of finer-scale digital line graphs for rivers and streams prevented analysis of riparian habitat conditions on relatively fine-scaled analysis units, such as the 64 km^2 areas (grid cells) used in this study. Although several landscape metrics can be generated across all of Europe, the quality and accuracy of landscape analyses will be improved by extending the finer-scale and land cover time series data to all of Europe.

The large number of landscape metrics used in our assessments precluded presentation of results for all individual metrics in this paper. Additionally, the large number of sampling units (grid cells and catchments) precluded comparisons of individual grid cells or catchments. However, radar plots can be used to compare specific metrics, or combinations of metrics (e.g. PCA) among individual catchments (Ten Brink et al., 1991; Burkhard et al., 2003). This capability can be achieved through web-based applications. For example, the Regional Vulnerability Assessment (ReVA) program has a web-based decision support tool that facilitates comparison of assessment units through radar plots and other means (http://www.epa.gov/reva).

Comparison of results from the different analyses showed some similarities and differences. The International Geosphere Biosphere Program (IGBP) level analyses showed more extensive areas of northwestern Europe and the UKin less desirable condition than did the Corine analyses. Additionally, the IGBP analyses showed extensive areas of less desirable conditions west of the Caspian Sea in areas where Corine land cover data were not available. These areas had relatively steep and significant slopes of negative NDVI trends, high amounts of cropland, and cropland on steep slopes. Differences in results between IGBP and Corine land cover may reflect the relatively course scale resolution (1 km) of the IGBP data. Course resolution land cover data may inflate the values of certain landscape metrics involving spatial intersection of land cover and other biophysical characteristics (e.g. cropland on steep slopes) and therefore produce lower principal component (PC) summary scores. However, Corine-level analyses showed large areas of the UK, north-central Europe, Italy, southern Greece, and large patches within Spain with less desirable landscape conditions. These areas possessed large amounts of cropland and marginal land use, high population density, and relatively low amounts of forest at the analysis unit and riparian habitat scales. Conversely, the Corine analyses showed large patches of more desirable landscape conditions existed in western and northeastern Spain, southern France, and central and north-central Europe. These areas tended to have higher relative amounts of forest and other natural vegetation, lower population density, and lower amounts of marginal land use.

Besides the NDVI trend analysis, analysis of landscape change was limited to areas with Corine land cover data. Many areas with less desirable landscape conditions estimated from the Corine 2000 land cover data also experienced declines in forest and increases in urban and cropland land cover between 1990 and 2000, although there were a smaller number of areas that had the worse rankings for landscape status and change. These included areas of the southeastern UK, northeastern Europe, the Po River Basin of northern Italy, and scattered areas of northern and eastern Europe. Areas with the highest positive ranks of landscape status and change occurred in Slovenia and Croatia, in catchments of north-central Europe, southwestern Spain, and in southern France. The grid cell analysis showed greater detail in the spatial variable of these conditions. In particular, Spain exhibited a salt and pepper pattern of the least and most desirable landscape conditions. Additionally, grid cells showed that some areas of the northern UK (Scotland) had desirable landscape conditions. This suggests that catchments may cancel out important differences in landscape conditions, but especially where high spatial variability in conditions exists. Therefore, the grid cell analysis may be a more effective way to capture landscape condition and change.

Loss of forest across Europe was associated with increases in agricultural and urban lands, but also with increases in shrubland and grassland. Similarly, loss of agricultural land was associated with gains in natural land cover, including grassland and forests. These change patterns reflect those that have been reported in other European assessments (EEA, 2003).

Results of the NDVI analysis suggest that Europe became greener between 1992 and 2003, although the spatial pattern of greenness change varied considerable. Increased greenness may reflect increasing trends in precipitation for most of Europe over the last 100 years (Schonwiese and Rapp, 1997). Declines in greenness in eastern Europe and along portions of the Mediterranean region may reflect land use and land cover changes (LADAMER, 2003; DISMED, 2006), and declines in precipitation in the Mediterranean region during winter months associated with the North Atlantic Oscillation (Folland et al., 2001). We found no significant relationships between metrics of land cover change and any of the NDVI trend metrics. Lack of correlation between these change metrics may reflect the small amounts of change observed in land cover between 1990 and 2000, and the relatively fine-scaled nature of land cover change estimates (100 m) as compared to NDVI change (8 km). Moreover, land use changes (e.g. changes in crop type and cropping practices) known to influence changes in greenness patterns may go undetected by land cover change analysis. Finer-scale estimates of NDVI change (e.g. 1 km) may improve correlations between greenness and land cover changes.

Results of our analysis reflect the general concern in the European community that environmental conditions across Europe continue to pose a threat to biological diversity, water quality and quantity (flooding and water availability), and productivity (EEA, 2003). Although policies and management strategies have improved environmental quality across portions of Europe (EEA, 2003), decision-makers will continue to be challenged with how to reduce the impacts of urbanization and marginal land use. However, results presented in our analyses suggest that this type of analysis can help identify those areas that are of greatest concern.

4.2. ISSUES RELATED TO INTERPRETATION OF LANDSCAPE ANALYSIS RESULTS

There are some important issues that may affect the results of the analyses and their interpretation. These include resampling and reprojection of spatial data so that data from different sources can be compared, comparison of landscape metrics across two or more dates, decisions about classification and reclassification of the original spatial data, the scale of the data, and interpretation of results as negative or less desired conditions versus positive or more desired conditions. Resampling is used to change the grain or pixel size of a particular database so that metrics can be calculated on common sample units. In this study, we resampled 100 km^2 grid cells of the agriculturally limited areas database to a scale of 64 km^2. These scales were relatively similar and therefore the reclassification likely had a small impact on the analysis of the metric. However, the wide range of pixel or grain sizes represented by the spatial data (90 m, 100 m, 1 km, 8 km, 10 km) affects the degree to which the intersection of important biophysical conditions occurs. For example, calculation of the marginal land use metric required spatial intersection of the agriculturally limited area data (8 km resolution) with 100 m Corine and 1 km IGBP land cover. The goal of the metric was to capture areas where agricultural practices occurred on lands that had been designated as agriculturally limited due to soil properties, geology, and topography. Because these biophysical characteristics are likely to vary within each of the 64 km^2 grid cells, the spatial intersection with finer resolution land cover data will also vary. Therefore, the results of the metric only indicate a general tendency for the two properties to intersect. This also is why we selected a minimum reporting unit of 64 km^2. Similarly, only catchment-scale analysis of riparian metrics was possible, primarily due to the course-scale nature of the rivers and streams database. Many of the 64 km^2 grid cells had no rivers or streams. Moreover, comparison of riparian metric results from the Corine-wide analysis versus the analysis using finer-scale data within the Yantra River Basin revealed that many riparian habitat conditions for many smaller rivers and streams were excluded from the broader analysis. Therefore, the broader-scale analysis underestimated forested and natural land cover riparian habitat conditions and their associated functions (e.g. filtering capacity). Finer-scale, river and stream network data are needed to improve the accuracy of all riparian metrics.

Other metrics used in this study that required spatial intersection of two or more sources of data had similar minimum spatial resolutions. This was especially true for areas with Corine land cover data. For example, cropland on steep slopes was estimated by intersecting a 90 m slope database with a 100 m land cover data.

Reclassification of the IGBP and Corine land cover into fewer types for the calculation of landscape metrics requires decisions about how to categorize land cover types. Because some of the classes are mixed (e.g. savannah/woodland/grassland, forest/agriculture mix), arbitrary decisions had to be made on their inclusion into more general land cover classes. These types of decisions can affect results for a wide array of metrics, but especially those that use land cover (e.g. N and P export which are based on land cover look-up tables). Therefore, documentation of decisions about land cover reclassifications is critically important.

Differences in methodologies used to construct land cover databases for different time periods can affect the accuracy of change estimates for landscape metrics. Landscape condition changes reported in this study were based on postclassification analyses of the two Corine land cover databases (1990 and 2000). This introduces the possibility that the changes observed in the landscape metrics resulted from the differences in how the two land cover databases were developed rather than real changes on the ground. Preclassification change detection approaches reduce errors associated with differences in classification methodologies since they often involve co-registration of the images and use of similar classification methods such as spectral differencing (Rogan and Chen, 2004). The Corine 1990 and 2000 databases were developed separately and only a postclassification change analysis was possible. Finally, it is important to note that the Corine 1990 and 2000 databases were constructed from Landsat satellite imagery across a range of dates before and after 1990 and 2000 (EEA, 2006). Therefore, change estimates represent time difference ranging from approximately 7 to 13 years.

Decisions about how to characterize integrated metrics, including the principle component (PC) metrics scores used in this study, can affect rankings of assessment units and their relative conditions. In the Principle Components Analyses we had to make decisions as to whether high positive values or loadings indicated more desirable conditions and change, or less desirable conditions and change. This was necessary because we summed individual PC scores to create an overall index. In a few situations there were positive values or loadings of one to a few natural land cover types (e.g. shrubland and grassland), and negative values of other natural land cover (e.g. forest). However, in each case there were loadings of stress-related metrics and we used those loadings to determine if the PC indicated more desirable or less desirable conditions and change. Moreover, these PCs tended to explain less variation in the overall model, therefore, decisions about their direction tended to have a relatively small impact on the overall PC sum value. Other statistical approaches, including simple sums and ranks (Jones et al., 1997; Walker et al., 2002), cluster analysis (Wickham et al., 1999; Smith et al., 2006), regression tree (Jones et al., 2006), and analytical hierarchy procedures (Tran et al., 2004), have been used to integrate multiple landscape metrics to assess multiple environmental themes. However, these approaches also are influenced by the scales of data and decisions about classification and the relative meaning of the results. Moreover, interpretation of results from these statistical approaches for multiple environmental themes (e.g. water, terrestrial habitat, ecosystem productivity) can be difficult because of different functional units (catchments versus ecoregions), scales, and relationships associated with each environmental theme. In some cases it may be better to apply specific models and metrics to each environmental theme and then spatially integrate the results to assess multiple theme conditions for any specific area (Jones et al., 2001a).

Finally, results presented in this study are based on relative values for metrics and indices (e.g. summary of PC scores). As such they can be used to compare catchments and grid cells, but not to predict actual conditions (e.g. a prediction of the impairment due to N and P export). The results are hypotheses of the range in environmental conditions and change relative to environmental themes and need to be tested and validated through finer-scale studies. Results from empirical studies and landscape modeling efforts can be used to establish cut-off and threshold values for many of the metrics used in our assessment. Several of the projects undertaken by the NATO CCMS Landscape Plot Study reported in this volume will help establish cut-off or threshold values for metrics. However, empirical studies quantifying relationships between landscape metrics and species habitat quality, water quality and quantity, and ecosystem productivity over large geographic extents are extremely limited. Few spatially extensive monitoring networks exist over large enough areas to develop many of these relationships. Moreover, Europe is biophysically diverse, which might lead to regional differences in relationships (type of variables and scales). Sampling designs of monitoring programs, and model development, need to consider these differences if meaningful cut-off and threshold values are to be used across extensive areas (Jongman et al., 2006).

For all of the reasons discussed above, we decided on an analysis of relative metrics results rather than one incorporating specific thresholds or cut-off points.

4.3. LANDSCAPE ANALYSIS UNITS – DOES ONE SIZE FIT ALL?

Analysis of finer-scale, spatial units, such as the 64 km^2 grid cells used in our assessment, produced more detailed patterns of landscape conditions and change than did analyses of catchments. Analysis of conservation and protected areas, including World Heritage Sites, would likely produce results similar to grid cells, although it would also be possible to establish different-sized buffer zones that would broaden the scale of the analysis. This would permit an analysis of the vulnerability of conservation areas based on a range of spatial scales.

The ATtILA tool used in this study can generate landscape metrics on any analysis units for which a GIS shapefile exists (Ebert and Wade, 2004). However, the user must consider the sensitivity of certain landscape metrics to the number of pixels or grid cells in the analysis, especially in conducting change analyses. Very small or fine-scale analysis units have fewer pixels than larger units and this may lead to highly variable change estimates resulting from small changes from one date to another (Jones et al., 2001a). For these reasons, change estimates of smaller analysis units are more likely to be influenced by errors in the data, rather than real changes in the landscape. Very large analysis units, such as ecoregions, tend to have many thousand pixels and these types of units tend to be insensitive to changes that may be important for specific environmental themes (Jones et al., 2001a). Therefore, decisions about the size and extent of the analysis units are critical in the design of change analysis studies.

Analysis units also are selected based on their functional relationships to specific environmental themes. For example, catchments are often selected to address water-related issues because of functional relationships to important hydrologic processes. However, catchments used in our analyses were fairly broad and based on 1 km resolution Digital Elevation Models (DEM) and, therefore, captured only course-scale relationships between landscape metrics and hydrologic processes and conditions. Finer-scale DEMs (10–30 m), such as those generated for the Yantra River Basin (Knight et al., 2002), will permit generation of finer-scale catchments and improve the delineation of functional catchment boundaries. These data also will improve delineations of stream and river networks and improve watershed models and estimates of riparian habitat conditions.

A number of ecoregion classifications have been used to stratify areas relative to land cover change (Sohl et al., 2003), biological diversity (Jongman et al., 2006), ecosystem productivity (Bailey, 1983), and terrestrial conditions that affect water quality (Rohm et al., 2002). A primary aim of using these types of analysis units is to reduce variation in estimates of stress (pressures), condition, change, and potential impact. They also are used to reduce variance in potential response to management and policy actions (Jongman et al., 2006). However, recent studies have raised doubts about the ability of broad ecoregion classifications to reduce variance in land cover change and nutrient concentrations in streams and rivers, and in their ability to capture variation in multiple environmental themes. Riitters et al. (2006) found that ecoregions captured relatively small amounts of variability in landscape change. They hypothesized that most land cover change is driven by pressures at the local scale. Wickham et al. (2005) found that ecoregions were not an effective way to partition variation in nutrient concentrations in surface water. They found that land cover composition and pattern were more effective in capturing potential nutrient loads to surface waters.

Another potential analysis approach is the application of sliding windows. This approach involves calculating metric values for individual pixels or grid cells based on the landscape context at one to several scales (one to many sized sliding windows) surrounding the individual pixel. This type of approach has been used to evaluate habitat quality (Riitters et al., 1997), forest fragmentation (Riitters et al., 2000), and disturbance (Zurlini et al., 2006), and may provide for finer-scale analyses of landscape conditions and reveal important scales that influence landscape conditions at local scales.

In summary, it is unlikely that one classification system will capture important aspects of all environmental themes. However, grid cells similar to those used in our assessment may provide a way to capture fundamental landscape features needed to assess a wide range of environmental themes. They also can be reaggregated to larger-size analysis units related to water and terrestrial habitat conditions.

4.4. THE IMPORTANCE OF SPATIAL INTERSECTION AND INTEGRATION OF METRICS

The spatial intersection of biophysical data permitted generation of landscape metrics that may be more closely aligned to processes associated with land, water, and habitat degradation. This differs from analyses that involve metric calculation from single digital maps such as land cover, including those analyses used to assess habitat quality and extent (e.g. see Cushman and McGarigal, 2004). Models relating landscape- and catchment-scale biophysical conditions to water quality use a variety of different biophysical data, but they tend to be more complex and harder to apply over broad geographic areas. Similarly, models that estimate potential soil loss also use combinations of spatial data on slopes, vegetation or land cover, soil erosivity, and precipitation (Dickinson and Collins, 1998), but tend to be applied across broader areas. In all of these cases, there is a tradeoff between complexity (many variables and metrics that utilize relatively fine-scaled spatial and temporal data) and the spatial extent over which the model can be applied (generally fewer variables that are calculated from courser, but more spatially extensive data, Van Rompaey and Govers, 2002). However, in most cases, model prediction is improved by incorporation of finer-scale spatial data that more accurately represent the spatial intersection of important environmental conditions (e.g. erosive soils on steep slopes). Some of the metrics used in our analyses captured important aspects of process models (riparian metrics, cropland on steep slopes, simple N and P export models, marginal land use), but they are computationally less complex and can be applied at multiple scales over extensive areas (Jones et al., 1997, 2001; Wickham et al., 2000). However, because of their relative simplicity, these metrics are most useful for environmental targeting and relative comparisons of geographic areas rather than for specific predictions (e.g. sediment loadings to streams, terrestrial species declines, etc.).

The summary scores from the PCA allowed us to compare overall landscape conditions for catchments and grid cells across Europe. This type of approach has been used to evaluate environmental vulnerability over broad geographic areas (Wickham et al., 1999; Bradley and Smith, 2004; Smith et al., 2006), and fits within the concept of fragility analysis (Zurlini et al., 1999, 2004). Relatively accurate spatial intersection and pattern of metrics and models related to drivers, pressures (stresses), conditions, and impacts (including observed change) are needed to determine an area's fragility or vulnerability (after Zurlini et al., 1999; Bradley and Smith, 2004). These approaches help establish how sensitive areas are to natural and anthropogenic conditions.

4.5. THE DPSIR INDICATOR PARADIGM

Landscape metrics used in this study relate to important aspects of the Driver Pressure State Impact Response (DPSIR) indicators paradigm (OECD, 1998; EU, 1999; Jones et al., 2005). A number of the landscape metrics generated in our study related to pressures, including agriculture on steep slopes, marginal land use, population density, urban density, nitrogen and phosphorus export, and cropland and urban land uses within the riparian zone. Another set related to basic environmental conditions (states), including the amount of natural vegetation at the analysis unit scale and within the riparian zone. Finally, we used a set of metrics that measured changes in states of the environment. These types of metrics may provide important insights into drivers of and potential impacts to important environmental themes (Sohl et al., 2003). Moreover, improved spatial resolution of biophysical and climate data should improve our ability to distinguish between the relative roles of anthropogenic versus natural drivers of pressures and impacts (Plieninger, 2006). Finally, many of the landscape metrics used in this study can help formulate and evaluate catchment- and basin-level land management alternatives (response) to improve environmental quality and increase environmental security, provided that finer-scale spatial data are available (Weber and Hall, 2001; Baker et al., 2004; Kronvang et al., 2005; Kepner et al., 2007, this volume).

4.6. NEXT STEPS AND RECOMMENDATIONS

There are a number of activities that would improve the quality of landscape analyses across Europe. Of primary importance is to acquire or assemble finer-scale core databases for larger areas across Europe (Jones et al., 2005). The following is a list of core databases that will help improve the quality and extent of landscape analyses. Some of these activities may already be under way or completed.

1. Expand 100 m Corine land cover data to other areas of Europe and use pre-classification change methodologies to improve change estimates, including the relabeling of the 1990s data based on the 2000 Corine classification approach
2. Expand agriculturally limited land classification to Turkey and drive the scale down from 8 km to 1 km or less;
3. Acquire 30 m or better DEMs for all of Europe. The DEM will improve slope estimates, agriculturally limited land classification, sub-catchment delineation, and stream and river networks
4. Develop 1 km scale soil properties coverage
5. Acquire climate data on 10 km or finer-scale grid cells. These data will improve a wide range of models, as well as ecosystem classification
6. Acquire in situ and field-scale monitoring data on environmental themes or values to improve model and landscape metric interpretation
7. Develop spatial databases with point source pollution estimates (see Schreiber et al., 2003)
8. Develop spatial databases on human demographic change.

In order to evaluate the degree to which the broader-scale landscape analyses presented in this study capture finer-scale conditions (e.g. effectiveness as a course filter), additional catchment-level studies are needed. These studies would be similar to the analysis on the Yantra River Basin (Nikolova et al., 2007, this volume), but across a wide range of European landscape gradients. The ecological classification system proposed by Jongman et al. (2006) might be an efficient way to identify landscape and biophysical gradients upon which to base additional studies. Additionally, results from this study should be distributed to a number of organizations throughout Europe to determine how well the broader-scale analyses capture regional- and catchment- scale patterns. The most effective way to do this would be to set up a web-based application with the ability to compare individual and multiple sample units and indicators (e.g. radar plots).

A comprehensive survey of spatially distributed models is needed to determine the degree to which threshold or cut-off values can be established for many of the metrics used in this study. Use of model-based thresholds and cut-off values will improve interpretation of landscape metrics relative to specific environmental themes. In some cases, new empirical studies involving gradient analyses and field data collection will be needed to establish important relationships. These studies should include evaluation of remote sensing approaches that more directly measure and represent ecological processes and conditions (Anselmi et al., 2004).

Other statistical approaches should be explored to improve integrated assessment results, and our understanding of relationships between landscape conditions, change, and environmental drivers. Regression Tree Analysis (RTA) helps determine how relationships between environmental variables change across extensive areas (O'Connor et al., 1996; Lawrence and Wright, 2001; Jones et al., 2006). Distinguishing area-specific relationships among environmental variables is an important step forward in conducting broadscale environmental assessments.

Spatial filtering approaches (Riitters et al., 1997; Zurlini et al., 2006) should be explored as an alternative to predefined analysis units, and as a way to capture important patterns and scales of disturbance that potentially influence environment condition and landscape change.

Finally, it will be important to incorporate other environmental themes and values into landscape analyses. We provided a demonstration of the approach by featuring landscape level analyses of terrestrial habitat, water quality, and ecosystem productivity. Future efforts should include themes such as food production (agriculture), energy and water balance, and social and economic factors affecting sustainability and quality of life. European landscapes have unique cultural values that have and will continue to influence landscape conditions and change (Haines-Young, 2005). Moreover, maintaining "cultural landscapes" is viewed as an important environmental management objective and theme in landscape ecology (Pedroli et al., 2006). Therefore, future analyses should include landscape metrics that track status and trends in these important landscape features. Additionally, it will be important to revisit threshold and metric cut-off values of existing metrics so that they reflect sustainable levels of cultural landscapes and other important environmental themes.

Acknowledgments

We thank Christine Estreguil, Centre for Earth Observation, Space Applications Institute, Joint Research Centre, Ispra, Italy, and Roger Sayre, US Geological Survey, Geography Discipline, Reston, Virginia USA, for acquisition of important landscape data used in the analyses. We thank the European Environmental Agency for acquisition of several important landscape databases, including Corine land cover, and the Oak Ridge National Laboratory for European-wide population data. We also thank our colleagues and friends involved in the NATO CCMS Landscape Pilot project for many of the ideas used in this study, and to those who hosted and sponsored each of the five NATO CCMS workshops that resulted in a great exchange of ideas and concepts. Finally, we thank Deniz Yuksel-Beten and NATO CCMS for the opportunity of a lifetime to work with so many fine colleagues from diverse backgrounds and perspectives.

References

Anselmi, S., Chies, M., Giannini, M., Manes, F., and Maselli, F., 2004. Estimation of Mediterranean forest transpiration and photosynthesis through the use of an ecosystem simulation model driven by remotely sensed data. Global Ecology and Biogeography 13, 371–380.

Australian State of the Environment Committee, 2001. Australia state of the environment 2001. Independent Report to the Commonwealth Minister for the Environment and Heritage, CSIRO Publishing on Behalf of the Department of the Environment and Heritage, Canberra, Australia, 135 p.

Bailey, R.G., 1983. Delineation of ecosystem regions. Environmental Management 7, 365–373.

Baker, J.P., Hulse, D.W., Gregory, S.V., White, D., Van Sickle, J., Berger, P.A., Dole, D., and Schumaker, N.H., 2004. Alternative futures for the Willamette River Basin, Oregon. Ecological Applications 14, 313–324.

Behrendt, H., 1996. Inventories of point and diffuse sources and estimated nutrient loads – a comparison for different river basins in central Europe. Water Science Technology 33, 99–107.

Borin, M., Vianello, M., Morari, F., and Zanin, G. 2005. Effectiveness of buffer strips in removing pollutants in runoff from a cultivated field in north-east Italy. Agriculture, Ecosystems and Environment 105, 101–114.

Brauch, H.G., 2006. Desertification: a new security challenge for the Mediterranean? In: W.G. Kepner, J.L. Rubio, D.A. Mouat, and F. Pedrazzini (Editors), Desertification in the Mediterranean Region: A Security Issue. Springer, The Netherlands, pp. 11–85.

Bradley, M.P. and Smith, E.R., 2004. Using science to assess environmental vulnerabilities. Environmental Monitoring and Assessment 94, 1–7.

Burkhard, B., Kumpula, T., and Müller, F., 2003. Landscape assessment for a sustainable reindeer husbandry in northern Scandinavia. EcoSys Bd. 10, 166–124.

Burkhard, B., Müller, F., and Kumpula, T., 2004. Integrative assessment of different reindeer management strategies in North Finland. EcoSys Suppl. Bd. 42, 60–66.

Council of European Communities (CEC), 1992. Directive 1992/43/EC on the conservation of natural and semi-natural habitats and of wild fauna and flora. Council of the European Communities, Brussels.

Council of European Communities (CEC), 2000. Directive 2000/60/EC of the European Parliament and of the Council: establishing a framework for community action in the field of water policy. Journal of European Communication, L327.

Council of European Communities (CEC), 2001. A sustainable Europe for a better world: a European Union strategy for sustainable development (Commission's proposal to the Gothenberg European Council). Commission of the European Communities, Brussels, Communication 264.

Council of European Communities (CEC), 2002. Towards a thematic strategy for soil protection. Commission of European Communities, Brussels, Communication 179.

Cushman, S.A. and McGarigal, K., 2004. Patterns in the species-environmental relationships depend on both scale and choice of response variables. Oikos 105, 117–124.

Dickinson, A. and Collins, R., 1998. Predicting erosion and sediment yield at the catchment scale. Soil Erosion at Multiple Scales. CAB International, pp. 317–342.

DISMED (Desertification Information System for the Mediterranean), 2006 http://www.dismed.eionet.europa.eu/facilities/dismed_products.

Dodson, J.E., Bright, E.A., Coleman, P.R., Durfee, R.C., and Worley, B.A., 2000. Landscan: a global population database for estimating populations at risk. Photogrammetric Engineering and Remote Sensing 66, 849–857.

Domenikiotis, C., Spiliotopoulos, M., Tsiros, E., and Dalezios, N.R., 2004. Early cotton production assessment in Greece based on a combination of the drought Vegetation Condition Index (VCI) and the Bhalme and Mooley Drought Index (BMDI). International Journal of Remote Sensing 25, 5373–5388.

Donohue, I., McGarrigle, M., and Mills, P., 2006. Linking catchment characteristics and water chemistry with the ecological status of Irish rives. Water Research 40, 91–98.

Ebert, D.W. and Wade, T.G., 2004. Analytical Tools Interface for Landscape Assessments (ATtILA): Users Manual. U.S. Environmental Protection Agency Report # EPA/600/R-04/083, Washington, DC.

EPA (U.S. Environmental Protection Agency), 2003. Draft report on the environment 2003. U.S. EPA. Report Number 260-R-02-006, Washington, DC, 167 p.

ESRI, 1996. Introduction to ArcView GIS. Environmental Systems Research Institute, Redlands, CA.

EU (European Union), 1999. Towards environmental pressure indicators for the EU, 1st edition. Communication from the Commission to the Council and the European Parliament on Directions for the EU on Environmental Indicators and Green National Accounts (COM 94/670).

EEA (European Environment Agency), 2003. Europe's environment: the third assessment. Office for Official Publications of the European Communities, Luxembourg, 341 p.

EEA (European Environmental Agency), 2006. The thematic accuracy of Corine land cover 2000 – assessment using LUCAS. Technical report No 7/2006.

Folland, C.K., Karl, T.R., Christy, J.R., Clarke, R.A., Gruza, G.V., Jouzel, J., Mann, M.E., Oerlemans, J., Salinger, M.J., and Wang, S.W., 2001. Observed climate variability and change. In: J.T. Houghton, Y. Ding, D.J. Griggs, M. Noguer, P.J. van der Linden, X. Dai, K. Maskell, and C.A. Johnson (Editors), Climate Change 2001: the Scientific Basis. Contribution of Working Group I to the Third Assessment Report of the Intergovernmental Panel on Climate Change. Cambridge University Press, Cambridge, 881 pp.

Galleo, K.P., Elvidge, C.D., Yang, L., and Reed, B.C., 2004. Trends in night-time city lights and vegetation indices associated with urbanization within the conterminous USA. International Journal of Remote Sensing 25, 2003–2007.

Groten, S.M.E., 1993. NDVI-crop monitoring and early yield assessment in Burkina Faso. International Journal of Remote Sensing 14, 1495–1515.

Haines-Young, R., 2005. Landscape pattern: context and process. In: J.A. Wiens and M.R. Moss (Editors), Issues and Perspectives in Landscape Ecology. Cambridge University Press, Cambridge, pp. 103–111.

Heinz Center (The H. John Heinz III Center for Science, Economics and the Environment), 2002. The state of the nation's ecosystems. Cambridge University Press, New York, 288 p.

Herrero, S.T., 2006. Desertification and environmental security: the case of conflicts between farmers and herders in the arid environments of the Sahel. In: W.G. Kepner, J.L. Rubio, D.A. Mouat, and F. Pedrazzini (Editors), Desertification in the Mediterranean Region: A Security Issue. Springer, The Netherlands, pp. 109–132.

Hooge, P.N. and Eichenlaub, B., 1997. Animal movement extension to arcview. Ver. 1.1. Alaska Biological Science Center, U.S. Geological Survey, Anchorage, AK (http://www.absc.usgs.gov/glba/gistools/animal_mvmt.htm).

Jennings, D.B. and Jarnagin, S.T., 2002. Impervious surfaces and stream flow discharge: a historical perspective in a Mid-Atlantic subwatershed. Landscape Ecology 17, 471–489.

Jones, K.B., Walker, J., Riitters, K.H., Wickham, J.D., and Nicoll, C., 1996. Indicators of landscape integrity. In: J. Walker and D.J. Reuter (Editors), Indicators of Catchment Health: A Technical Perspective. CSIRO Publishing, Victoria, pp. 155–168,

Jones, K.B., Riitters, K.H., Wickham, J.D., Tankersley, R.D. Jr., O'Neill, R.V., Chaloud, D.J., Smith, E.R., and Neale, A.C., 1997. An Ecological Assessment of the United States Mid-Atlantic Region: A Landscape Atlas. EPA/600/R-97/130.

Jones, K.B., Neale, A.C., Wade, T.G., Wickham, J.D., Cross, C.L., Edmonds, C.M, Loveland, T.R., Nash, M.S., Riitters, K.H., and Smith, E.R., 2001a. The consequences of landscape change on ecological resources: an assessment of the United States Mid-Atlantic Region, 1973–1993. Ecosystem Health 7, 229–242.

Jones, K.B., Neale, A.C., Nash, M.S., Van Remotel, R.D., Wickham, J.D., Riitters, K.H., and O'Neill, R.V., 2001b. Predicting nutrient and sediment loadings to streams from landscape metrics: a multiple watershed study from the United States Mid-Atlantic Region. Landscape Ecology 16, 301–312.

Jones, K.B., Walker, J., Zaccarelli, N., Kepner, W.G., and Zurlini, G., 2005. Towards a core data set for landscape assessments. EcoSys Bd. 11, 48–65.

Jones, K.B., Neale, A.C., Wade, T.G., Cross, C.L., Wickham, J.D., Nash, M.S., Edmonds, C.M., Riitters, K.H., O'Neill, R.V., Smith, E.R., and Van Remortel, R., 2006. Multiscale relationships between landscape characteristics and nitrogen concentrations in streams. In: J. Wu, K.B. Jones, H. Li, and O.L. Loucks (Editors), Scaling and Uncertainty Analysis in Ecology: Methods and Applications. Springer, The Netherlands, pp. 205–224.

Jongman, R.H.G., Bunce, R.G.H., Metzger, M.J., Mucher, C.A., Howard, D.C., and Mateus, V.L., 2006. Objectives and applications of a statistical environmental stratification of Europe. Landscape Ecology 21, 409–419.

Kattan, G.H., Alvarez-Lopez, H., and Giraldo, M., 1994. Forest fragmentation and bird extinctions: San Antonio eighty years later. Conservations Biology 9, 138–146.

Kepner, W.G., Zurlini, G., and Müller, F., 2006. Introduction: desertification and security. In: W.G. Kepner, J.L. Rubio, D.A. Mouat, and F. Pedrazzini (Editors), Desertification in the Mediterranean Region, Springer, The Netherlands, pp. 3–9.

Kepner, W.G., Hernandez, M., Heggem, D.T., Semmens, D.J., and Goodrich, D.C., 2007. Watershed assessment of landscape change and impacts to aquatic resources: a multi-scale case study of Oregon (USA) with future scenario analysis. Chapter V.2, this volume.

Knight, C.G. and Hristov, T.N. (Editors), 2002. River environmental knowledge and assessment (REKA) applied to the Yantra River Basin. National Coordination Center for Global Change, the Bulgarian Academy of Sciences, Sofia, Bulgaria, ISBN 954-90485-3-5.

Kondratyev, S., 2007. Influence of catchment land cover on phosphorus balance for large freshwater systems. Chapter V.4, this volume.

Koopowitz, H., Thornhill, A.D., and Anderson, M., 1994. A general stochastic model for the prediction of biodiversity losses based on habitat conservation. Conservation Biology 8, 425–438.

Kronvang., B., Bechmann, M., Lundelkvam, H., Behrendt, H., Rubaek, H., Schoumans, O.F., Syversen, N., Andersen, H.E., and Hoffmann, C.C., 2005. Phosphorus losses from agricultural areas in river basins: effects and uncertainties of targeted mitigation measures. Journal of Environmental Quality 34, 2129–2144.

LADAMER, 2003. Land degradation assessment in Mediterranean Europe. Contribution to the Global Monitoring for Environmental Security (GMES) Full Details Report, 69 p.

Lawrence, R.L. and Wright, A., 2001. Rule-based classification systems using Classification and Regression Tree (CART®) Analysis. Photogrammetric Engineering and Remote Sensing 67, 1137–1142.

Liotta, P.H., 2006. The Poseidon prairie. In: W.G. Kepner, J.L. Rubio, D.A. Mouat, and F. Pedrazzini (Editors), Desertification in the Mediterranean Region: A Security Issue. Springer, The Netherlands, pp. 87–108.

Mattikalli, N.M. and Richards, K.S., 1996. Estimation of surface water quality changes in response to land use change: application of the export coefficient model using remote sensing and geographic information system. Journal of Environmental Management 48, 263–282.

McCollin, D., 1993. Avian distribution patterns in a fragmented wooded landscape (North Humberside, U.K.): the role of between-patch and within-patch structure. Global Ecology and Biogeography Letters 3, 48–62.

Minor, T.B., Landcaster, J., Wade, T.G., Wickham, J.D., Whitford, W.G., and Jones, K.B., 1999. Evaluating change in rangeland condition using multitemporal AVHRR data and geographic information system analysis. Environmental Monitoring and Assessment 59, 211–223.

Mucher, C.A., Bunce, R.G.H., Hennekens, S.M., and Schaminee, J.H.J., 2004. Mapping European habitats to support the design and implementation of a Pan-European ecological network: the PENNHAB project. Alterra-rapport 952, Wageningen, The Netherlands.

Nash, M.S., Wade, T.G., Heggem, D.T., and Wickham, J.D., 2006. Does anthropogenic activities or nature dominate the shaping of the landscape in the Oregon pilot study area for 1990–1999? In: W.G. Kepner, J.L. Rubio, D.A. Mouat, and F. Pedrazzini (Editors), Desertification in the Mediterranean Region: A Security Issue, Series C: Environmental Security, Vol. 3, Springer, The Netherlands, pp. 305–323.

Nikolova, M., Nedkov, S., Semmens, D., and Iankov, S., 2007. Environmental quality and landscape risk assessment in the Yantra River Basin. Chapter V.3, this volume.

Noss, R.F., LaRoe, E.T., and Scott, M.J., 1995. Endangered ecosystems of the United States: a preliminary assessment of loss and degradation. U.S. National Biological Survey Report 28, U.S. National Biological Survey, Washington, DC.

O'Connor, R.J., Jones, M.T., White, D., Hunsaker, C., Loveland, T., Jones, B., and Preston, E., 1996. Spatial partitioning of environmental correlates of avian biodiversity in the conterminous United States. Biodiversity Letters 3, 97110.

OECD (Organization for Economic Co-Operation and Development), 1998. Towards sustainable development – environmental indicators. OECD, Paris.

Ojima, D.S., Galvin, K.A., and Turner, B.L.II., 1994. The global impact of land-use change. BioScience 44, 300–304.

Pedroli, B., Pinto-Correia, T., and Cornish, P., 2006. Landscape – what's in it? Trends in European landscape science and priority themes for concerted research. Landscape Ecology 21, 421–430.

Peterjohn, W.T. and Correll, D.L., 1984. Nutrient dynamics in an agricultural watershed: observations on the role of a riparian forest. Ecology 65, 1466–1475.

Peters, D.P.C., Gosz, J.R., Pockman, W.T., Small, E.E., Parmenter, R.R., Collins, S., and Muldavin, E., 2006. Integrating patch and boundary dynamics to understand and predict biotic transitions at multiple scales. Landscape Ecology 21, 19–33.

Plieninger, T., 2006. Habitat loss, fragmentation, and alteration – quantifying the impact of land-use changes on a Spanish dehesa landscape by use of aerial photography and GIS. Landscape Ecology 21, 91–105.

Reynolds, C.A., Yitayew, M., Slack, D.C., Hutchinson, C.F., Huete, A., and Petersen, M.S., 2000. Estimating crop yields and production by integrating the FAO Crop Specific Water Balance model with real-time satellite data and ground-based ancillary data. International Journal of Remote Sensing 21(18), 3487–3508.

Riitters, K.H., O'Neill, R.V., and Jones, K.B., 1997. Assessing habitat suitability at multiple scales: a landscape-level approach. Biological Conservation 81, 191–202.

Riitters, K.H. Wickham, J.D., Volgelmann, J.E., and Jones, K.B., 2000. Global patterns of forest fragmentation. Conservation Ecology 4(2), 3 http//www.consecol.org.vol.14/iss2/art3.

Riitters, K.H., Wickham, J.D., and Wade, T.G., 2006. Evaluating ecoregions for sampling and mapping land-cover patterns. Photogrammetric Engineering and Remote Sensing 72, 781–788.

Rogan, J. and Chen, D.M., 2004. Remote sensing technology for mapping and monitoring land-cover and land-use change. Progress in Planning 61, 301–325.

Rohm, C.M., Omernik, J.M., Woods, A.J., and Stoddard, J.L., 2002. Regional characteristics of nutrient concentrations in streams and their application to nutrient criteria development. Journal of American Water Resources Association 38, 213–239.

SAS/EDT, 1999. User's Guide, SAS Institute, Cary, NC.

Saunders, D.A., Hobbs, R.J., and Margules, C.R., 1991. Biological consequences of ecosystem fragmentation: a review. Conservation Biology 5, 18–32.

Schonwiese, C.D. and Rapp, J., 1997. Climate trends atlas of Europe based on observations 1891–1990. Kluwer, Dordrecht, 228 pp.

Schreiber, H., Behrendt, H., Constantinescu, L.T., Cvitanic, I., Drumea, D., Jabucar, D., Juran, S., Pataki, B., Snishko, S., and Zessner, M., 2003. Nutrient emissions from diffuse and point sources into the River Danube and its main tributaries for the period of 1998–2000 – results and problems. Proceedings of the Diffuse Pollution Conference, Dublin, Ireland, pp. 9–18 – 9–24.

Scott, J.M., David, F., Csuti, B., Noss, R., Butterfield, B., Groves, C., Anderson, S., Caicco, S., D'Erchia, F., Edwards, T.C. Jr., Ulliman, J., and Wright, G., 2003. Gap analysis: a geographic approach to protection of biological diversity. Wildlife Monographs 123.

Slonecker, E.T., Jennings, D.B., and Garofalo, D., 2001. Remote sensing of impervious surfaces: a review. Remote Sensing Reviews 20, 1231–1242.

Smith, E.R., McKinnis, P., Tran, L.T., and O'Neill, R.V., 2006. The effects of uncertainty on estimating the relative environmental quality of watersheds across a region. Landscape Ecology 21, 225–231.

Sohl, T.L., Gallant, A.L., and Loveland, T.R., 2003. The characteristics and interpretability of land surface change and implications for project design. Photogrammetic Engineering and Remote Sensing 70, 439–450.

Swiaczny, F., 2005. International demographic trends and environmental services. EcoSys Bd. 11, 122–133.

Ten Brink, B., Hosper, S., and Collijn, F., 1991. A quantitative method for description and assessment of ecosystems: the AMOEBA-approach. Marine Pollution Bulletin 23, 265–270.

Tran, L.T., Knight, C.G., O'Neill, R.V., and Smith, E.R., 2004. Integrated environmental assessment of the Mid-Atlantic Region with analytical network process. Environmental Monitoring and Assessment 94, 263–277.

Tuchel, D., 2004. Romanians warn of Danube Delta ecological disaster. Environmental News Service, 28 September.

Van Rompaey, A.J.J. and Govers, G., 2002. Data quality and model complexity for regional scale soil erosion predictions. International Journal of Geographical Information Science 16, 663–680.

Vogelmann, J.E., 1995. Assessment of forest fragmentation in southern New England using remote sensing and geographic information systems technology. Conservation Biology 9, 439–449.

Walker, J., Veitch, S., Dowling, T.D., Braaten, R., Guppy, L., and Herron, N., 2002. Assessment of catchment condition: the intensive land use zone in Australia. CSIRO, Canberra. www.affa.gov.au/catcon/

Weber, J.L. and Hall, M., 2001. Towards spatial and territorial indicators using land cover data. European Environmental Agency, Technical Report 59, Copenhagen.

White, D., Minotti, P.G., Barczak, M.J.,Sifneos, J.C., Freemark, K.E., Santelmann, M.V., Steinitz, C.F., Kiester, A.R., and Preston, E.M., 1997. Assessing risks to biodiversity from future landscape change. Conservation Biology 11, 349–360.

Wickham, J.D., Jones, K.B., Riitters, K.H., O'Neill, R.V., Tankersley, R.D., Neale, A.C., Smith, E.R., and Chaloud, D.J., 1999. Characterizing cumulative environmental impacts on mid-Atlantic watersheds. Journal of Environmental Management 24, 553–560.

Wickham, J.D., Riitters, K.H., O'Neill, R.V., Reckhow, K.H., Wade, T.G., and Jones, K.B., 2000. Land cover as a framework for assessing risk of water pollution. Journal of American Water Research 36, 1417–1422.

Wickham, J.D., O'Neill, R.V, Riitters, K.H., Smith, E.R., Wade, T.G., and Jones, K.B., 2002. Geographic targeting of increases in nutrient export due to future urbanization. Ecological Applications 12, 93–106.

Wickham, J.D., Riitters, K.H., Wade, T.G., and Jones, K.B., 2005. Evaluating the relative roles of ecological regions and land cover composition for guiding establishment of nutrient criteria. Landscape Ecology 20, 791–798.

Young, S.S. and Harris, R., 2005. Changing patterns of global-scale vegetation photosynthesis, 1982–1999. International Journal of Remote Sensing 26, 4537–4563.

Zurlini, G., Grossi, L., and Rossi, O., 1999. A landscape approach to biodiversity and biological health planning: the Map of Italian Nature. Ecosystem Health 5, 296–311.

Zurlini, G., Zaccarelli, N., and Petrosillo, I., 2004. Implementing an integrated framework for ecological risk assessment at the landscape scale. EcoSys Suppl. Bd. 42, 4–15.

Zurlini, G., Riitters, K., Zaccarelli, N., Petrosillo, I., Jones, K.B., and Rossi, L., 2006. Disturbance patterns in a socio-ecological system at multiple scales. Ecological Complexity 3, 119–128.

PART V INTRODUCTION

ASSESSMENTS OF HUMAN-ENVIRONMENTAL SYSTEMS IN LANDSCAPES

RALF SEPPELT
Department for Computational Landscape Ecology Helmholtz-Centre for Environmental Research UFZ, Germany

BENJAMIN BURKHARD
Ecology Centre, University of Kiel, Kiel, Germany

NICOLA ZACCARELLI
Landscape Ecology Laboratory, Department of Biological and Environmental Sciences and Technologies, University of Salento, Lecce, Italy

Land use and land cover change are two of the major factors affecting environmental conditions and many ecosystem services at multiple scales. Understanding of the consequences and implications of changes in land cover and land use is a fundamental part of addressing any environmental security issue and for planning for sustainable development. Processes and causes driving these changes cannot be tracked down to few simple rules and agents and they can not be addressed without the coupling of social-economic and environmental systems. Recent studies are taking into account climate change, which has to be transferred to regional inputs, global market and agricultural production, where for example agricultural subsidized systems are modified, and dynamics in human demographic transitions like aging and migrations. Further potential drivers could be social and ethical norms and values that might change between generations, human societies or due to particular events. Or, at the individual scale, people's response to market opportunities, as mediated by institutional factors, and people's perception of social and economic security.

Thus, the question rises what are the changing impacts of humans to landscapes knowing, that climate, economy but also demography will change? Which consequences are most probable, which consequences are less probably but lead to tremendous consequences, e.g. have a high risk?

A systemic approach is needed to model human-environmental systems and to investigate the complexity of linkages of various environmental and socioeconomic factors and their causal relationships at different hierarchical levels and scales. One weak point of many studies is a frequently static view on dynamic adaptive systems. But rather the historical profile of the system reveals a great deal about current system dynamics and how the system might respond to future shocks. This sets the frame for the studies presented in this part of the book with the special focus on demography and environmental security. The following sections present different state-of-the-art techniques like scenario development, Geographical Information Systems (GIS), remote sensing, interviews, indicator applications, and scale transitions in multifaceted case studies, including Mediterranean coastal areas, central European urban and suburban regions and subarctic tundra landscapes.

An introducing conceptual view on demographic impacts on cultural landscapes is given in the first chapter from Swiaczny. A focus on transportation and infrastructure is laid by Sabbagh and Neef. Several case studies have been collected for detailed study. These are consequences of demographic changes for eastern Germany by Haase et al., indication of state of the environment within the case study Mulfingen (Lenz et al.) and landscape, demographic developments, biodiversity and sustainable land use strategy: a case study on Karaburun peninsula Izmir, Turkey (Nurlu et al.). Zurlini et al. as well as Petrosillo et al. aim at an integrative analysis using the concept of ecosystem services first with comparing a subjective and objective analysis and by analyzing scale mismatches. Human-environmental interrelations in traditional reindeer husbandry in the sparsely populated remote regions of northern Fenno-Scandia are assessed in the chapter of Kumpula et al.

All contributions provide appealing insights into the complex system of human-environmental relations in different landscapes and give new impulses for interdisciplinary researches in the context of environmental security.

DEMOGRAPHIC IMPACTS ON LANDSCAPE CHANGE – A CONCEPTUAL VIEW OF GLOBAL DEMOGRAPHIC TRENDS

FRANK SWIACZNY*
Federal Institute for Population Research, Friedrich-Ebert-Allee 4, 65185 Wiesbaden, Germany

Abstract. Demographic changes will not only impose new challenges on developing regions, developed countries, and countries in transition alike but also will face changes in population without precedence during the next decades. In the latter the population will age and in some cases even decline by number. Related to the ongoing demographic change more and more rural and especially remote areas will furthermore lose population by out migration to urban areas, often leading to an uneven sex and age distribution of the people leaving behind. The demographic change will influence the economic activities in rural areas, such as, e.g. traditional agricultural production patterns and will therefore modify the usage of formerly agricultural landscapes. The same applies to urbanized areas where more and more agricultural landscapes will be modified by incorporating them into the growing urban agglomerations, e.g. for purposes of housing, industry or transport. For this reason the knowledge of the future population developments and their related impacts on the change of landscapes is crucial in understanding and predicting processes of landscape change at a whole. Thus the paper presents a conceptual view on the relationship of demographic and landscape changes based on key figures on the major aspects of the projected future population development, contrasting the population processes of developed countries and countries in transition with those of developing regions.

Keywords: Demographic change; demographic transition; society–nature interactions; landscape change

* Corresponding Author: Dipl.-Geogr. Frank Swiaczny, Federal Institute for Population Research, Friedrich-Ebert-Allee 4, 65185 Wiesbaden, Germany; e-mail: frank.swiaczny@destatis.de

I. Petrosillo et al. (eds.), Use of Landscape Sciences for the Assessment of Environmental Security, 319–328.

1. Conceptual view on the relationship between population and landscape change

One of the aims of this paper is to give a brief introduction into the problem of conceptualizing the impact of current demographic changes on processes of landscape change in order to shape future research in this field of social science and demography.

Figure 1 shows a model of interaction between a socioeconomic and an environmental subsystem (see Swiaczny, 2005b). Both subsystems have their own systems of regulation and respective demand–supply relationships and are linked to each other by the concept of ecosystem services. Ecosystem services are crucial in the production of economic and social well being and there is a contextual framework of economy, law, technology, institutional settings, etc. regulating the interaction of the model. If an ecosystem is no longer able to provide ecosystem services according to the demand – as a result of changes on the demand and/or the supply side of the system – this constitutes an environmental threat.

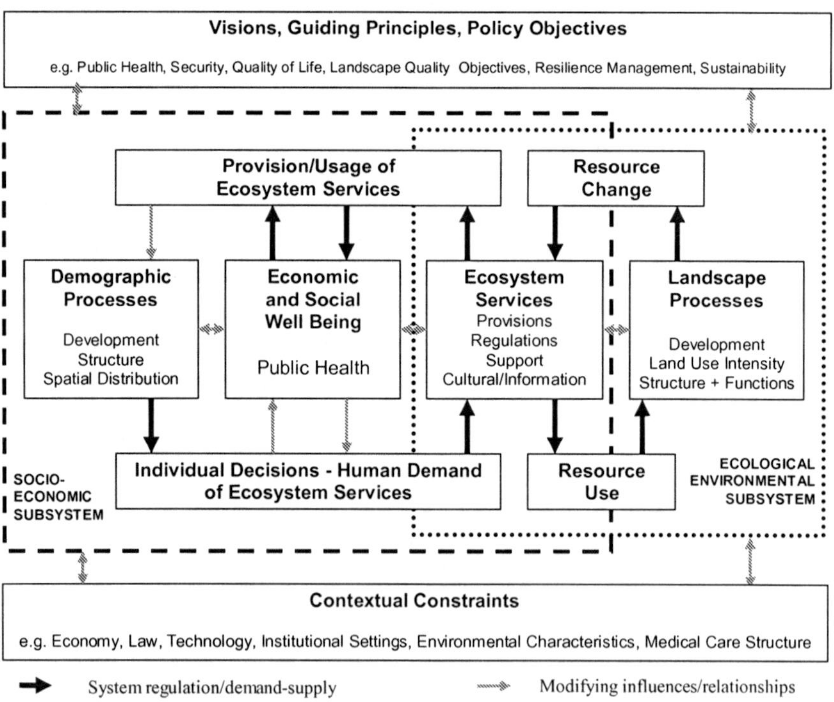

Figure 1. Conceptualization of demographic and environmental processes. (Modified after Sabbagh, Seppelt, Swiaczny, and Zurek. Source: Swiaczny, 2005b.)

The system regulation of the model can be changed by demographic processes. Landscapes are shaped by human use of (and therefore impacts on) the ecosystems of this landscapes. Changes in the shaping factors on the socioeconomic side of the model will result in changes on the ecological–environmental side of the model and may affect the provision of ecosystem services. This means for example that unsustainable systems of production (use of landscapes) and/or unsustainable demand and consumption of ecosystem services may challenge the ability of landscapes to maintain the provision of the respective services and may result in unsustainable landscape changes.

The demand of ecosystem services in this model is conceptualized by human action as a function of individual needs and decisions (cf. Coleman, 1986). The relationship between processes of demographic change – a model on the macro scale – and landscape change – also a model on the macro scale – is in this conceptualization only possible by using human action as a link on the micro level (Figure 2). Only human action – individual or collective – as a sociological concept can cause impacts in the physical sphere of this models view.

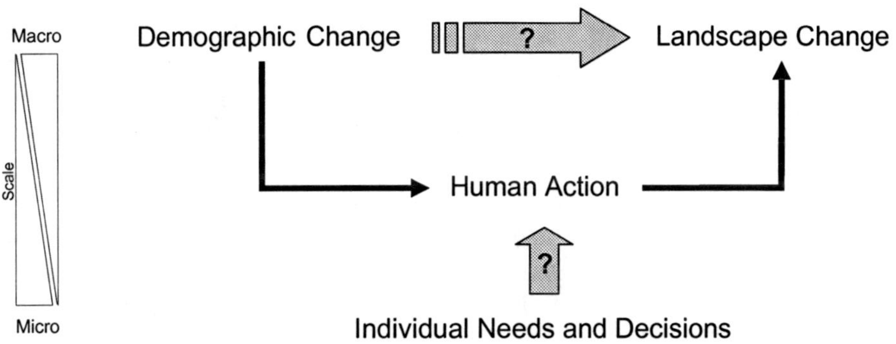

Figure 2. Micro–macro model of human action (Adapted from Coleman, 1986)

For purposes of applied empirical research it is important to enquire which demographic processes will occur and which will have the most impact on the landscapes. This is easily to be made by analyzing the correlation between both (macro) processes, e.g. like "urbanization will transform formerly agricultural land." This example makes the intention of the conceptualization on the micro level clear. Urbanization is a demographic process but it is associated with several other social and economic processes. One has to look which individual interaction affecting the landscape is originally a result of people's needs and decisions correlated with, e.g. a growing or declining population size or a demographically ageing society and what is the result of other processes. Without a more in depth knowledge of the working of those relationships on the spatial and conceptual micro level it is not possible to identify a set of suitable demographic indicators to match the already established sets of environmental indicator for a coherent future monitoring to prevent environmental threats.

2. Future population dynamics

In the previous section of this paper the relationship between processes of demographic change and the environment was shown in a conceptual way. The following section focuses on the processes of demographic change itself, which are to be dealt with in the future. The paper therefore refers to a large amount of – often controversial – scientific writing on the prospective future development of the world population (cf., e.g. Ahmad, 1997; Bongaarts, 2002; Vallin, 2002) which is not to be discussed here in detail due to limitations of the article length (see Schulz and Swiaczny, 2005; Swiaczny, 2005a for an extended list of publications). The figures of this paper are exclusively based on the medium variant projections of the latest World Population Prospects provided by the UN Population Division (UN, 2004b, 2005). As a result of this comprehensive population projections there are three major demographic processes to be identified which affect the relationship between man and ecosystems. Due to the size of the respective population dynamics involved the impacts of these processes on the landscape will be intense and form a potential threat to the environmental security (cf. AAAS, 2000; UN, 2001c; UNDP, 2005; UNFPA, 2006; World Bank, 2003). The relevant processes are (1) the population growth and the respective growth of the population density in developing regions and vice versa in developed and transition countries; (2) the ageing of the population in (nearly) all countries due to a declining fertility; and (3) a changing spatial population distribution, e.g. by differential natural growth rates and processes of spatial mobility (internal and international migration).

According to the UN Population Divisions World Population Prospects the world population growth will remain high in terms of absolute numbers for the coming decades (UN, 2004a); even so the average annual rate of population growth is already declining for some time. The world population will probably exceed the threshold of 9 billion before the year 2050. As a result, the population density will increase considerably, especially in regions with still high population growth rates like in some parts of Asia and Africa (Figure 3). Compared to this global pattern the population numbers and densities of the developed countries in North America and Europe will stay (nearly) stable or even decline (Council of Europe, 2004), leading on a large-scale perspective in some regions already to the problems caused by out migration losses and ageing described in the abstract.

This predicted development is a result of the UN expectation that the number of children per women (Total Fertility Rate, TFR) will globally decline considerably till 2050, except in some of the least developed countries (Bulato and Elwan, 1985). Fortunately they do not belong to the group of the most populous countries, so their effect on the global population development is limited. But locally population stress will remain high in the affected regions. In contrast fertility is currently already below the replacement level of 2.1 children per woman in some developed and most transition countries; leading to aging and declining societies there. The UN calculates their projections with a TFR growing to 1.85 in 2050 for all countries with a lower TFR today. The projections of the UN may be over optimistic, if the TFR remains low in the Industrialized countries (cf. Van de Kaa, 1987).

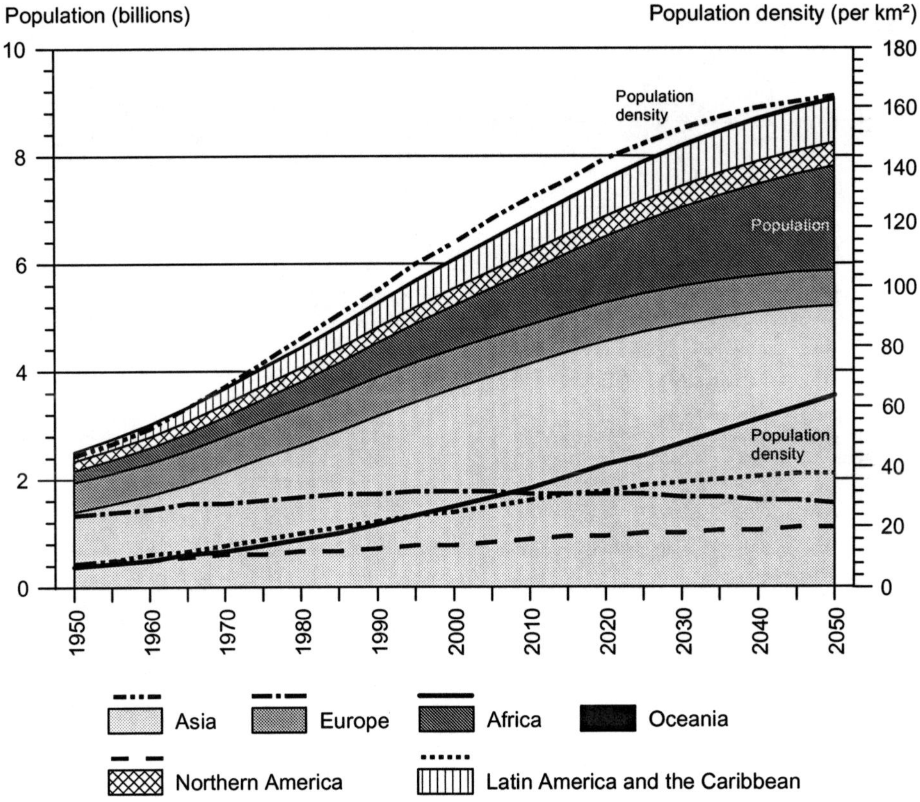

Figure 3. Population (billions, left scale) and population density (inhabitant per km^2, right scale) from 1950 to 2050 by regions according to the 2004 UN World Population Prospects (medium variant).

The result of a declining fertility is always a reduced population growth rate – at least in the long run. An average of less than 2.1 children per woman will in addition lead to a declining population in absolute numbers. Nevertheless, every decline of the TFR will trigger a process called "demographic dividend" (Bloom et al., 2003); first the demographic dependency ratio (Figure 4) between the working and the not working parts of the population will decline and open a "window of opportunity" for economic development. But in the end the dependency ratio will grow again. The result will be an older society than before the period of the "demographic dividend" (Figure 5; UN, 2002b). The ageing of the global population will be attributed with considerable changes in the field of economic activities and therefore impose a changing demand on ecosystems for e.g. industrial and agricultural purposes.

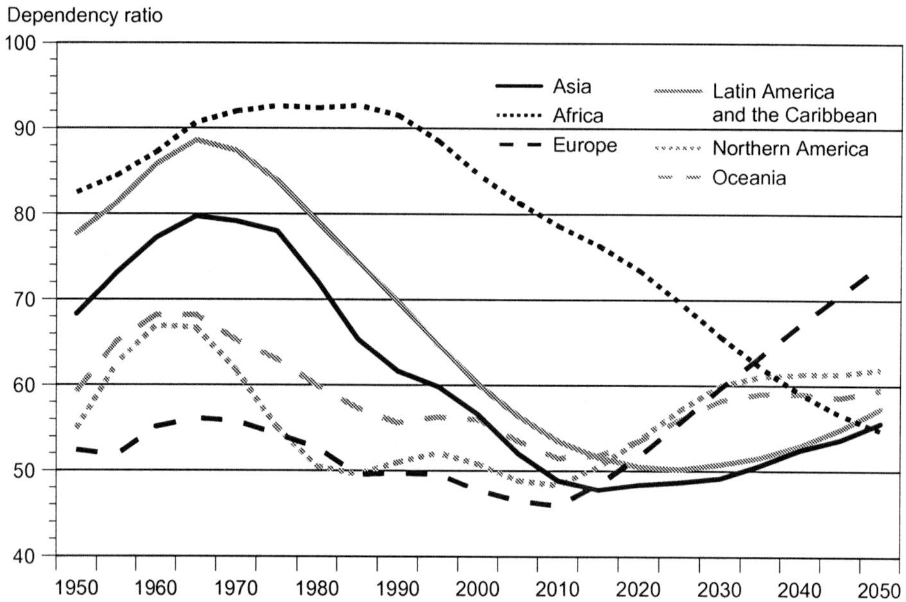

Figure 4. Dependency ratio from 1950 to 2050 by regions according to the 2004 UN World Population Prospects (medium variant).

The UN further expects the urban population to reach a share of more than 50% of the global population by 2008 (Figure 6; UN, 2004b). According to the UN projections all future population growth will take place in the urban areas; resulting in a net gain of approximately 3 billion people in urban or urbanizing areas over the next 45 years. This process is not only a result of the natural population growth but also of intense rural to urban migration patterns in the developing countries (UN, 2001b). It also means leaving behind rural areas which have lost their younger and better-educated parts of the population and so are likely to alter the traditional agricultural activities (Brockerhoff, 1999, 2000). Most population growth will take place in the smaller and medium sized cities in the developing countries. They will grow considerably into the surrounding countryside, transferring formerly agricultural land into "city."

Migration in its international, regional and local forms is also becoming more important in the future (Champion, 1994; UN, 2001a, 2002a; OECD, 2005). The more uneven the demographic development and the economic opportunities are spread spatially, the more likely migration will increase. But, migration patterns are not to be explained by a simple push-pull model matching supply and demand (cf. Massey et al., 1993) as often done in the past. Migration is today highly dependent on perspectives of future living conditions or strategies to improve and secure family income as a whole. Thus migration is easily to be influenced by the perception of (future) income opportunities and living conditions. Migration will probably vary considerably in the future and by doing so migration will also react on the consequences of landscape change, e.g. the lack or decreasing quality of environmental services available in a region. This shows that the relationship between demographic processes and the environment is indeed bidirectional and that people will in fact react on changes in the environment by adopting their locational choices – personally and for economic purposes – accordingly.

This short introduction into the future demographic processes proofed based on UN data that changes of population growth, population density and age distribution will be intense around the world. On smaller scales demographic changes will be even more prominent due to the selective effects of migration and urbanization on the spatial population distribution and the age and sex composition of local populations.

In respect to the landscape change and related environmental threats it is important to note that demographic processes – growing or shrinking and ageing populations – do not have any direct influence on the environment in the proposed conceptualization. More people in itself are no better or worse than less people. Crucial is, e.g. the development of the demand for land or the production and consumption patterns – sometimes driven by the same reasons than population processes (e.g. modernization and individualization of societies) sometimes independent of the forces driving the demographic change (e.g. growing productivity or rationalization of industrial and agricultural production). Unfortunately it is not easy to distinguish driving forces and effects of the demographic change and other current processes affecting landscapes, as it is the case, e.g. with the urbanization in developing countries (population growth and high density, industrialization of production, growth of production, productivity and per capital income, changing live styles and consumption patterns, etc.).

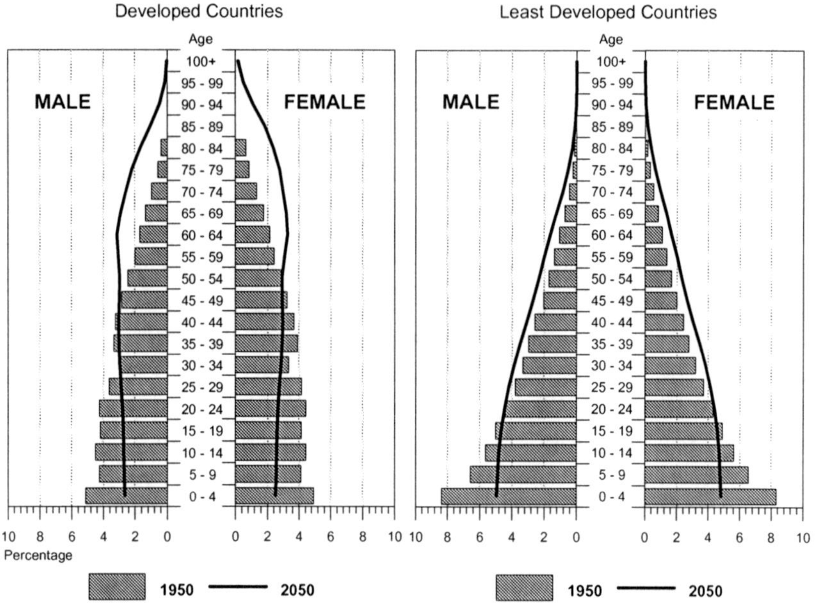

Figure 5. Age structure of the population in developed and least developed countries in 1950 and in 2050 according to the 2004 UN World Population Prospects (medium variant).

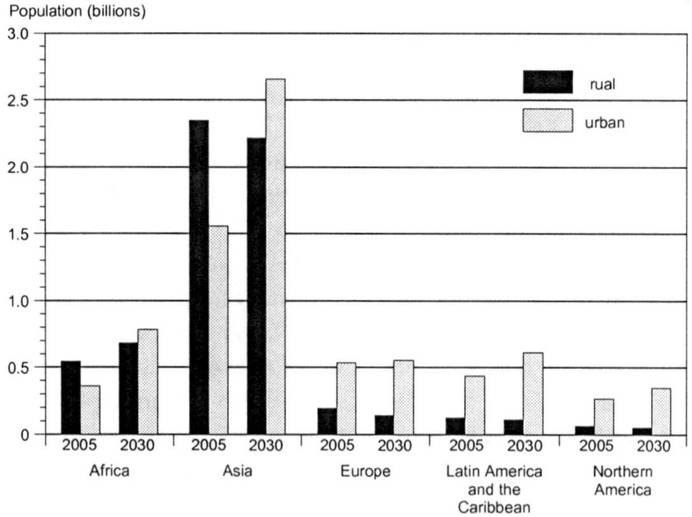

Figure 6. Rural and urban population (billions) in 2005 and in 2030 by regions according to the 2004 UN World Population Prospects (medium variant).

3. Conclusion and outlook

The aim of the paper was to show, that there is a strong relationship between the processes of demographic and environmental changes of landscapes: demography matters!

It was argued that – bearing in mind the observable evidence of these impacts – macro level relationships between demographic changes and landscape changes can only be conceptualized by human action as a sociological concept as a driving force of interaction on a conceptual micro level. Demographic processes as models on the macro level can, nevertheless, be successfully related empirically to respective impacts on landscape changes as well. In this regard more comprehensive research based on regional examples has to be carried out in the future to identify systematical coincidences of demographic and landscape changes in relation to certain environmental characteristics and ecosystem services affected. This will help to identify vulnerabilities and threats imposed by demographic processes on the ecosystem services by means of landscape changes as well as the crucial conceptual micro level notions between demographic and landscape changes.

The section on the future demographic changes to be expected according to the UN projections has shown that all regions of the world will face demographic-environmental threats without precedence – some by anthropogenic stress caused by a growing some by a declining number of people and related population densities, and all will be affected by an ageing of societies and large shares of population living in urban areas. Monitoring and predicting demographically driven threats on environmental security will therefore become more important than ever before.

In order to assess the environmental effects of the intense global demographic changes currently projected for the future, additional empirical and interdisciplinary research is necessary. To enable the implementation of empirical research on the interdependence of the conceptual macro level processes of demographic and landscape change, relevant micro level interactions based on human action have to be conceptualized and empirically established.

References

Ahmad, A., 1997. Demographic transition. The Third World Scenario. Rawat Publishing, Jaipur, India.

American Association for the Advancement of Science, 2000. AAAS Atlas of Population and Environment. University of California Press, Berkeley, CA.

Bloom, D.E., Canning, D., and Sevilla, J., 2003. The Demographic Dividend. RAND, Santa Monica, CA.

Bongaarts, J., 2002. The end of the fertility transition in the developing world, Population Council PRD Working Papers 161. Population Council, Washington, DC.

Brockerhoff, M.P., 1999. Urban growth in developing countries. A review of projections and predictions, Population Council Working Papers 131. Population Council, New York.

Brockerhoff, M.P., 2000. An urbanizing world. Population Bulletin 55(3), 3–44.

Bulatao, R.A. and Elwan, A., 1985. Fertility and mortality transition. Patterns, Projections, and Interdependence. World Bank, Washington, DC.

Champion, A.G., 1994. International migration and demographic change in the developed world. Urban Studies 31, 653–677.

Coleman, J.S., 1986. Social theory, social research and a theory of action. American Journal of Sociology 91, 1309–1335.

Council of Europe, 2004. Recent Demographic Developments in Europe. Council of Europe Publishing, Strasbourg.

Massey, D.S. et al., 1993. Theories of international migration. Population and Development Review 19, 431–466.

OECD, 2005. Trends in International Migration. OECD, Paris.

Schulz, R. and Swiaczny, F., 2005. Bericht 2005 zur Entwicklung der Weltbevölkerung. Zeitschrift für Bevölkerungswissenschaft 30(4), 409–454.

Swiaczny, F., 2005a. Aktuelle Aspekte des Weltbevölkerungsprozesses. Regionalisierte Ergebnisse der UN World Population Prospects 2004, Materialien zur Bevölkerungswissenschaft, Vol. 117, Federal Institute for Population Research. Wiesbaden.

Swiaczny, F., 2005b. International demographic trends. EcoSys Bd. 11, 122–133.

UN, 2001a. Replacement migration. Is it a Solution to Declining and Ageing Populations? UN, New York.

UN, 2001b. The Components of Urban Growth in Developing Countries. UN, New York.

UN, 2001c. World population monitoring 2001. Population, Environment and Development. UN, New York.

UN, 2002a. International Migration Report 2002. UN, New York.

UN, 2002b. World Population Ageing 1950–2050. UN, New York.

UN, 2004a. World Population in 2300. UN, New York.

UN, 2004b. World urbanization prospects. The 2003 Revision. UN, New York.

UN, 2005. World population prospects. The 2004 Revision. UN, New York.

UNDP, 2005. Human Development Report. UN, New York.

UNFPA, 2006. UNFPA Annual Report 2005. UNFPA, New York.

Vallin, J., 2002. The end of the demographic transition. Population and Development Review 28(1), 105–120.

Van de Kaa, D., 1987. Europe's second demographic transition. Population Bulletin 42(1), 1–55.

World Bank, 2003. World development report 2003. Sustainable Development in a Dynamic World. Oxford University Press, Oxford.

LAND USE IMPACTS OF DEMOGRAPHIC CHANGE – LESSONS FROM EASTERN GERMAN URBAN REGIONS

DAGMAR HAASE* AND RALF SEPPELT
Department for Computational Landscape Ecology, Helmholtz-Centre for Environmental Research UFZ, Germany
ANNEGRET HAASE
Department for Urban and Environmental Sociology, Helmholtz-Centre for Environmental Research UFZ, Germany

Abstract. Demographic change has become a major topic regarding the use and stability of European urban regions. It can be seen as the major driving force responsible for "growth" and "non-growth" or "decline" pathways in urban regions for the coming decades. Growing and shrinking urban regions do exist simultaneously next to each other. The trend towards further urban sprawl and dispersion observed in the 1980s in western Europe and the 1990s in East Central Europe accompanying the transition process are about to be replaced by shrinkage and perforation. This is mainly due to the recent decrease in birth rates, ageing and shifting household structures. This chapter analyses the trends and spatial patterns of the impact of demographic changes in urban regions. In the first part different features of demographic change are presented. In the second part, the paper expands on how demographic change affects urban land use, fabric, housing markets, infrastructure, and greenery. Since eastern Germany has been shrinking substantially since 1990, the paper uses this example to show a case in point embedded into the overall European context.

Keywords: Demographic change; shrinkage; perforation; land use impacts; eastern Germany

*Corresponding Author: Dr. Dagmar Haase, Helmholtz-Centre for Environmental Research UFZ, Department for Computational Landscape Ecology, Permoserstr. 15, 04318 Leipzig, Germany, e-mail: dagmar.haase@ufz.de

I. Petrosillo et al. (eds.), Use of Landscape Sciences for the Assessment of Environmental Security, 329–344.

1. Demographic change versus stability – a challenge for landscape sciences

Demographic change has become a major topic for European urban regions and their adjacent rural periphery. Growing and shrinking regions exist simultaneously next to each other within urbanised regions mainly due to low birth rates, an increase in life expectancy, changing household structures and, first and foremost, high dynamics of migration (Kabisch et al., 2006). The trend towards further urban growth and dispersion observed in the 1980s in western Europe and later on in the 1990s in East Central Europe is increasingly accompanied by processes such as urban shrinkage and perforation. Moreover, data on inner-urban migration allow us to assume a certain return of the compact city (Buzar et al., 2007). This paper argues that many of the spatial developments and land use changes in urban regions are effects of demographic change. They are closely related to both the balance of supply and demand in changing consumer relations between space, natural resources, and population.

Whereas social scientists and economists intensively discuss the impact of demographic change upon social structure, cohesion, and the labour market, only a few studies refer to the spatial and land use impacts that are caused by demographic change (Heiland et al., 2006). What are the spatial effects of demographic change in urban regions? Does a population decline automatically mean a reduced number of consumers of urban ecosystem services (water, energy consumption; urban green) or a decrease in urban land consumption? Or will population decline simply lead to an increase in unused and waste land, and low density urban areas (cf. Berg et al., 1982)?

To answer these questions, we need first of all to shed light on the spatial distribution of population, age spectra, households, and migration in urban regions. Second, we have to identify on the relationship between population change, residential mobility and the demands on residential land, urban infrastructure, and recreation areas. Referring to stability and security aspects, in particular the long-term stabilisation of urban regions under demographic decline in terms of compactness of settlements and consumption of natural resources is of interest. Set against this background, it is the purpose of the chapter to analyse the trends and spatial pattern of the impact of demographic changes in urban regions.

Although demographic changes affect many parts of Europe (Kazepov, 2005; Couch et al., 2005), eastern Germany is something like a "fore-runner" in terms of rapid urban shrinkage due both natural population decrease and out-migration (Kabisch, 2005). Therefore, the paper illustrates its pre-assumptions and findings using representative examples from eastern German urban regions and embeds them into the overall European context. To start, in the following section the specifics of eastern German urban development are briefly discussed.

1.1. THE "FORE-RUNNER" ROLE OF EASTERN GERMANY IN POPULATION DECLINE

Currently, large parts of Europe are undergoing a considerable demographic change (Antrop, 2004; Cloet, 2003; Lutz, 2001). In Germany, as far as a general population development is concerned, we find in the coming 20 years a simultaneity of population growth and decline (cf. Figure 2). The population is declining in a long-term perspective (Klingholz, 2004): according to the recent prognosis of the German Federal Agency for Population and Spatial Development (BBR, 2005), the German population will decrease from 82.56 million (2002) to 80–82 million inhabitants up to the year 2015. Up to 2025 a further decline to 77–79 million is expected, followed by an ongoing negative trend up to 2050 (Figure 1). The main reasons therefore are demographic ones: low birth rates (birth rate per woman is about 1.4 in Germany), an increasing number of one-person and childless households, postponement of child-bearing, ageing and interregional migration.

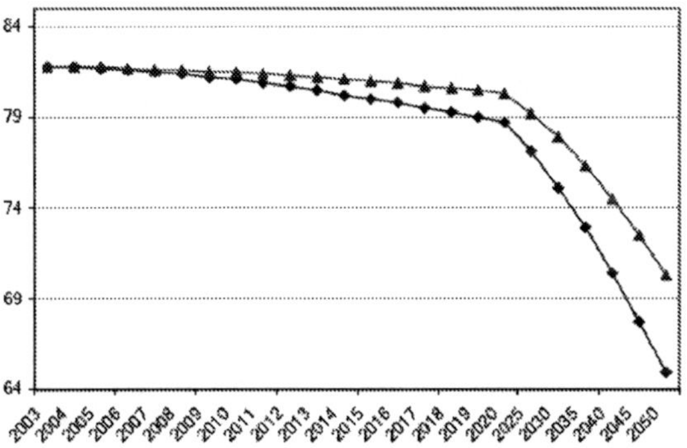

Figure 1. Population prognosis for Germany 2003–2050 in million inhabitants: ▲ = optimistic; ■ = pessimistic estimation. (StBA, 2000.)

In particular eastern Germany has been faced with a dramatic loss of population since the reunification in 1990 due to the above mentioned fall in birth rates (to lowest low rates of meanwhile 0.77) and exorbitant outmigration to the western part of the country: with a birth rate of 1.2, the total population of eastern Germany decreased about 6% from 1990–2005 (>2 million including death surplus). In this vein, the most dramatic population losses from 1989–1998 of nearly 10% can be found in the federal states of Saxony-Anhalt and

Brandenburg; cf. Figures 2 and 3). Those numbers exceed by far values reached in other European countries such as Sweden or Finland and other transition states such as the Baltic States and Romania where the population decline from 1995–2001 reaches –5% to –2% in both welfare countries such as (EUROSTAT, 2006). Looking at urban regions such as Leipzig, Halle, Zwickau, or Chemnitz, more than 15% of their population left the city for either western Germany (labour migration) or suburban settlements (residential migration; see here Figures 5 and 6 in Section 1.4). As Figure 2 suggests, the current population prognosis up to 2020 for eastern German federal countries, except Berlin, anticipates a further population decrease of 10 to >15%.

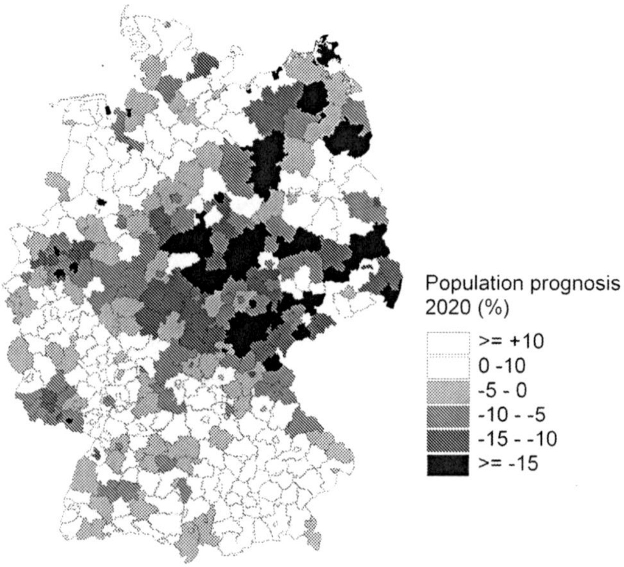

Figure 2. Population prognosis for Germany 2000–2020. (own calculation; modified after BBR, 2005.)

1.2. AGEING

As a result of the decrease in birth rates, the postponement of birth giving and an increase in life expectancy, East German city regions are moreover faced with ageing and an increase in elderly residents (Bösch-Supan et al., 2005).

This trend has a strong spatial component with a dramatic increase in residents aged >75 in urban and rural parts of northeastern Germany, namely Mecklenburg and Brandenburg (Figure 3). In the whole of eastern Germany, the mean age value will grow from 38.2 in 1998 to >50 in 2050. Due to the already existing high share of retired people in most of the urban regions of eastern Germany, the process of ageing is seen as irreversible for the coming decades.

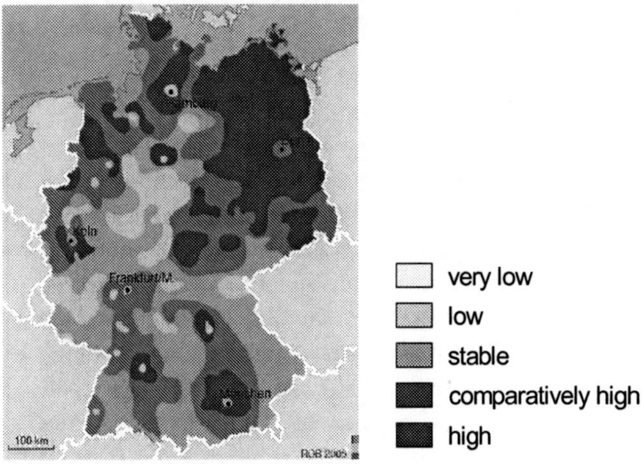

Figure 3. Share of age classes >75 in Germany 2002–2015. (BBR, 2005.)

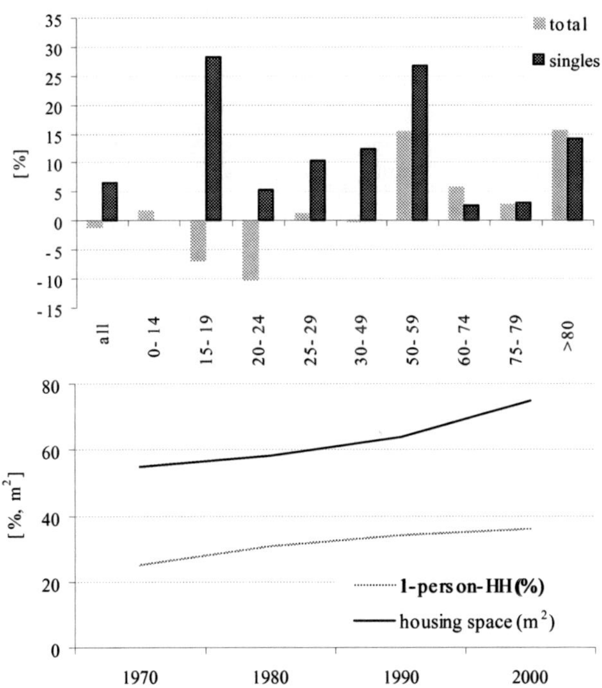

Figure 4. Increase in one-person households in total numbers and distributed over the whole age spectrum in eastern Germany (upper diagram; 1990–2006) and in relation to the housing space requirements. (Data source: EUROSTAT, 2006.)

1.3. DIVERSIFICATION OF HOUSEHOLD STRUCTURES

According to the processes of the *Second Demographic Transition* (SDT; van de Kaa, 2004) the urban population in the whole of Europe is undergoing a further differentiation in terms of a rising diversity of household structures.

Households are becoming smaller and less stable, traditional concepts of the family are losing importance, and non-traditional household types such as one-person households, cohabiting couples, single parents, and younger adults sharing a flat are becoming more widespread (a.o. Buzar et al., 2005).

For eastern Germany, this trend holds particularly true when looking at Figure 4: although the youth quota shows a general decrease, the share of one-person households among them is increasing. The second part of Figure 4 shows that this process is accompanied by an increasing diversity of housing requirements in terms of size, floor plan, and equipment. Smaller households do not necessarily demand for smaller flats. This fact contraries the assumption of lowering land consumption due to population decline.

1.4. SPATIAL PATTERNS OF DEMOGRAPHIC CHANGE

Demographic change, individualisation and related changes in housing preferences combined with a simultaneous rise in the disproportion of existing supply of urban housing, social and technical infrastructure – which is seriously outweighing the demand – challenge the debate on respective land use and landscape effects in urban regions of eastern Germany (Hannemann, 2003). In light of basic knowledge of the correlations between land development processes and population (Ravetz, 2000; Hall, 1992; Himiyama et al., 2005), there have to be assumed considerable spatio-temporal effects of shrinkage such as, generally speaking:

- An increase in *segregation* (spatial separation) between growing and shrinking areas next to each other in a city region
- A small-scale *fragmentation* (spatial separation) of the urban population (fragmented housing geography; Buzar et al., 2007)
- A considerable number of *residential vacancies* in most parts of the urban fabric (Table 1)
- *Large-scale urban brownfields* in both inner-city and suburban areas
- *Outmigration* from the large cities (Figure 5) and related *land abandonment* processes in suburbia, that might have been recently started (Figure 6)

TABLE 1. Number of dwellings and residential vacancies in eastern Germany 1990–2002. (BBR, 2005.)

Federal country	Development of the number of dwellings 1990–2000 (%)	Total residential vacancy 2002 (%)	Residential vacancy in housing cooperatives (%)
Brandenburg	14.2	13.1	14.7
Mecklenburg-Vorpommerania	13.1	11.8	10.7
Saxony	6.9	17.6	19.6
Saxony-Anhalt	7.4	16.9	20.8
Thuringia	6.7	10.2	15.8

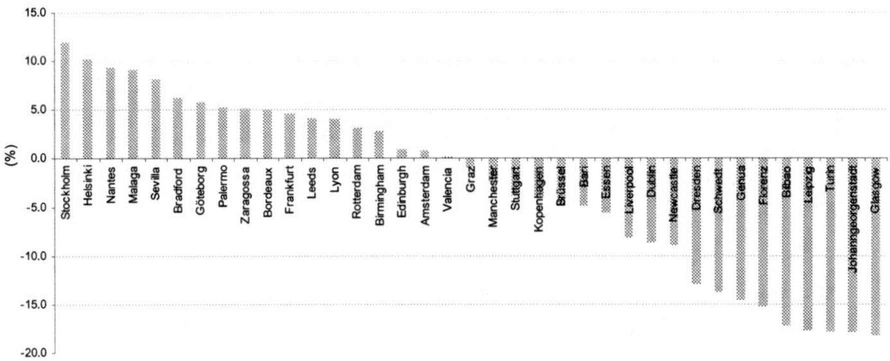

Figure 5. Demographic development of European cities >200,000 inhabitants from 1981–1996. These data prove evidence of the population decline in East German cities and set them into relation to the pan-European change (EUROSTAT, 2006).

- *Perforation* (=patchwork settlement structures without a compact core) of particularly urban landscapes and a respectively related creation of low-density counter settlements (with implications for lower sealing rates, building and population densities)
- *Deconstruction* and large-scale *demolition* of parts of the urban housing stock in cities and possibly suburbia (Haase, D. et al., 2006b)
- A *decreasing* and *modified demand* for technical, transport and social *infrastructure* (demand for schools will decrease compared to the increase in retirement homes and medical infrastructure)
- Remaining trends in *sprawling* land (use) development such as land consumption around cities (2003: 93 ha, 2005: 118 ha, 2020 (prognosis): 104 ha; cf. Nuissl and Rink, 2005)

- New waves of *reurbanisation* that allow one to assume a resurgence of living in the inner city and thus might come along with a revival of the compact city (cf. Figure 6; Buzar et al., 2007; Haase, D. et al., 2006a)

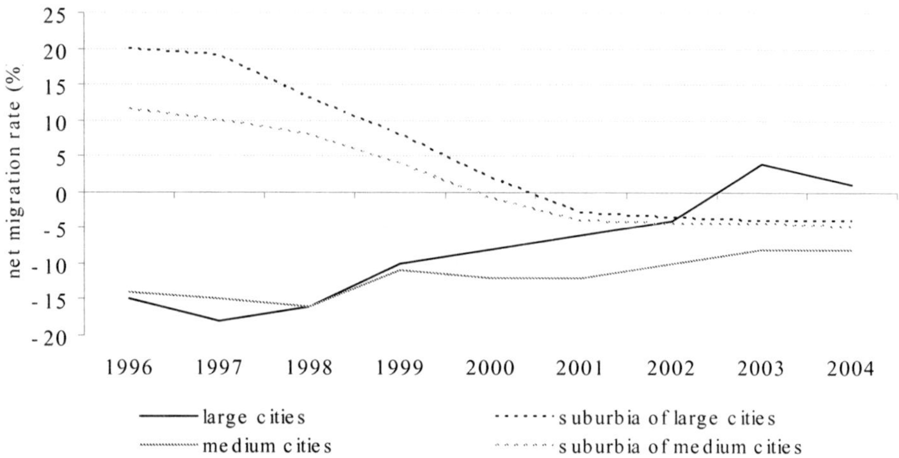

Figure 6. Net migration rate (%) along the rural–urban gradient in eastern Germany 1996–2004. (BBR, 2005.)

The above described land development pathways implicate a major spatial relevance for future land use pattern in eastern German urban regions. It leads to high regional discrepancies between growing and shrinking parts with unstable and differing dynamics along the rural, peri-urban, gradient (Figure 6). Figure 6 clarifies that, compared to the early 1990s, big cities in particular have recently been regaining population from suburbia. However, eastern German urban regions are characterised by decreasing residential densities, perforation into urban clusters and massive residential vacancies on the one hand and, simultaneously, "islands of up-grading" on the other. Is this development comparable to other European city regions? Does this imply a revision of urban policies? For eastern Germany, the described spatial effects of demographic changes in the urban space are under discussion among planners, whereas in major parts of western Europe they are still neglected in the current debate on urban policy (Müller and Siedentop, 2004; Herfert, 2002).

2. New land use patterns?

In this section, selected examples for impacts of demographic change on land use are presented. Although there is a need for more empirical ground, assumptions are made based on causal and logic loops drawn from data explored from demographic and land use statistics for eastern Germany.

2.1. FRAGMENTATION

To improve knowledge regarding environmental effects of demographic change, with particular respect to decline, the spatial allocation of shrinkage (i.e. of restructuring and demolition) at different scales needs to be investigated. Here, old industrialised urban regions in both Western and Central Europe (e.g. Scottish Clydeside, northern England, Lorraine, and Ruhr basin) that have undergone economic and population decline already during the last decades may serve as forerunner examples. Literature reports that shrinkage processes started in most cases with economic decline, the development of brownfields in the inner city and an exodus of population from those areas (Power and Mumford, 1999; Couch et al., 2005). From recent studies on cities such as Liverpool, Manchester, and Newcastle we know, however, that recently there have been started comprehensive, area-based renewal activities that resulted in a revitalisation or – at least – a stabilisation of districts like Manchester-Castlefields and the Ropewalks in Liverpool (Ravetz, 2000; Seo, 2002). Simultaneously, differences between successful and disadvantaged neighbourhoods are increasing (Morrison, 2003).

Eastern Germany is a special case or a “laboratory” to investigate the complex spatial and societal consequences of shrinkage because of the speed of those processes after 1989. Here, deindustrialisation coincided with an enormous outmigration and a demographic “shock” due to the political and societal transition. Evolving housing vacancies of more than 1 million flats are unique in Europe up to present (Haase, A. et al., 2005). The same holds true for the restructuring programme developed by the government for 2002–2009 that foresees subsidies for the demolition of 350,000 flats (Lang and Tenz, 2003). What we find today in eastern Germany is an increasing polarisation between stabilising urban regions such as Leipzig, Dresden, Erfurt, and Jena whose inner cities undergo even processes of household-driven reurbanisation (Haase, A. et al., 2005), and declining cities that keep locked in the shrinkage trap (Halle, Madgeburg, Chemnitz). At the same time, a low-level suburbanisation goes along with the resurgence of inner-city areas. These processes result in a “splintered urban population” (Buzar et al., 2007) at a small scale. Hence, it is fragmentation that will determine the further development of urban space in Europe, and the scholarly as well as planning debate will be on how to focus at best limited financial resources and on giving up declining regions as settlement areas in favour of stabilising others.

2.2. PERFORATION

Undergoing demographic change, eastern German urban regions and respectively their land use patterns might be confronted with an evolving type of urban landscape that "accepts" shrinkage as a development pathway: the *perforated urban landscape* which is less dense and more heterogeneous (in terms of land use and the mixture of open and sealed land), possesses a higher share of typical peripheral or suburban land uses (such as open land, single family houses or commercial units) than densely built up core cities from former times. More people live in higher dense clusters in suburban housing estates than in the inner and the core city. The term of *perforated* cities had been created during the debate on massive residential vacancies and the resulting "scattering" of the remaining urban fabric after demolition activities (Sander, 2006; Lütke-Daldrup, 2001).

Still, there exists no clear idea of how perforation might look like, but initial ideas of this urban type in terms of residential land uses have been already developed (cf. Figure 7): for example, the urban core could be divided into two settlement cores, the development of a poly-centric structure with less dense or even empty areas between tiny cores is imaginable as well as to "laissez faire" the current development what will lead to a fragmented urban body.

Comparing with a western European example, initial ideas of a cohesive but less dense core area of the city had been developed by the Newcastle City Council (NCC, 2000). Here, planners defined "green", "red", and "amber" areas according to their economic growth and quality of life for identifying priority demands of (e.g.) demolition or regeneration.

Figure 7. Forms of demolition in Leipzig-Grünau (left) and urban fabric perforation (upper map: continuous grey colour) in the eastern part of Leipzig (right) (Data source: authors' elaboration.)

Perforation implies that urban areas with strong demographic decline are however sprawling. Thus a heterogeneous mosaic of growing, stabilising and declining urban structures is developing. Formerly compact settlements, old and new built up districts actually change into a mixed urban fabric comprising very different house types, sizes, roof heights, and surrounding open space. Moreover, non-growth and shrinkage are bringing forward residential segregation, further spatial differentiation and small-scale fragmentation ending up possibly in an extreme perforation. To put it differently: high density areas are intertwined with areas of abandonment. Up to present it remains unclear if such structures will lead to an increase or decrease of the demand on natural resources (recreation space, water supply, clean air).

Assuming that urban perforation appears more frequently, do we thus lose the typical compact city and its spatial connectivity? Are we then facing a transport-network-determined land use development between perforated housing and open space clusters? Perforation definitely needs to be further investigated in terms of its structural, density and functional aspects in order to clarify its impact on land use and environmental resources.

2.3. ADAPTATION OF URBAN INFRASTRUCTURE

Besides changes of the urban fabric, infrastructure-related impacts have to be considered as a consequence of demographic change, first and foremost in shrinking city regions. In eastern Germany, we register a considerable underutilisation of the technical infrastructure for water supply and canalisation, which has already begun to underpin efficient operation and increases costs. Moreover, underutilisation causes toxic effects and water resources pollution (Koziol, 2004). For the water consumption since 1990 until today, for example, we state a decrease of 45% in all eastern German federal states. This is purported to be the result of two processes after the societal transition in 1990: (a) an increased sense of responsibility as regards water use due to the "capitalist" water price system and (b) population decline. In the years to come, housing service and maintenance costs are likely to rise in shrinking or perforating communities. To avoid or at least limit higher ancillary costs and the consequent worsening of local conditions in the municipalities affected, low-cost urban renewal strategies of the spatial adaptation of the technical urban infrastructure need to be set on the agenda (examples are given in Table 2). The aspect that a change of household structures and sizes arises (cf. section 1.3) implies changes in water and energy use (Koziol, 2004).

Infrastructure developments in shrinking urban regions are assumed to be determined by a decrease in specific water, heating, and electricity consumption owing to changes in consumer behaviour. Hereby, declining consumption due to the extensive outmigration in eastern German urban regions will lead to a deconcentration of the service area: this requires an adaptation of the infrastructure network (cf. again Table 3).

TABLE 2. Adaptation of infrastructure networks to shrinking or perforating contexts.

Spatial form of adaptation	Land use and transportation impact
Increasing accessibility	Increase in traffic and road network demand
Reduction	Decrease in traffic and road network demand
Centralisation	Decrease in traffic and road network demand
Decentralisation (local networks)	Increase in traffic and road network demand
Temporal-mobile structures	Optimisation of spatial structures
New structures and substitution	No effects for spatial configuration but for total net consumption

Looking at the transportation sector, traffic requirements in the form of road network are supposed to increase by 2015 by approximately 24%, assuming a continuation of the recent trend in transport infrastructure development in eastern Germany. Whereas in compact urban regions travel distances and related noise and air pollution can be limited to a certain extent, spatial segregation and perforation definitely lead to increased travel distances and all its negative effects on urban quality of life (in terms of noise and air pollution). Today, >12% of the total population in eastern Germany need >45 min using the private car for daily commuting compared to 9% in 1995 (INKAR, 2003; BBR, 2005). Subsequently, the decline in population itself does not necessarily contribute to an increase in environmental quality within shrinking urban regions.

2.4. INCREASE IN URBAN BIODIVERSITY?

Turning to the positive side effects of demographic decline, residential and commercial vacancies in inner-city and peripheral areas and related demolition should be understood as a chance for ecological restoration and development of green networks in city regions, too. Mehnert et al. (2005) found a positive correlation between the amount of urban green and the habitat suitability of urban breeding birds (*Picus viridus*). Generally, urban fallow lands are seen as niches for rare specialist species. Thus, demolished sites can be left partially open in order to develop hotspots of urban biodiversity.

However, general positive or negative effects of demographic change for urban ecosystems and biodiversity are not verified yet through empirical studies (Lüthi, 2001). In particular residential vacancy, simple land abandonment (excluding a de-sealing) and perforation are not subsequently followed by positive effects for nature and ecosystems. On the contrary, inefficient solutions could lead to higher environmental impact and more land consumption in shrinking landscapes when, despite an existing urban brownfield, open land at the periphery becomes sealed.

Processes and forms of urban perforation are assumed to have the potential to considerably contribute to structural enrichment and an increase in edge densities (Bolund and Hunhammar, 1999). They let one further think about a *bringing back* of nature to former densely populated and built urban centres. At least *wilderness ideas* (Rink, 2005) in urban landscapes for recreational and educational purposes are in discussion among urban planners and landscape architects. A concept from the eastern German Leipzig suggests newly developed urban greenery in the form of temporary gardens for demolition sites (planned variant) and vacant succession lots at former brownfields (unplanned variant). According to the opinion of the authors, there is potential for more urban green and nature which may retain people in the city and counteract urban sprawl. But going into detail here would exceed the limited space of this chapter.

3. Insights concerning land use impacts arising from demographic change and urban perforation processes

As discussed in this paper, demographic change considerably impacts on recent shrinkage and perforation processes in eastern German urban regions. Except singular empiric results, a comprehensive quantification of these impacts on land use pattern is still to be done. Does, for example, a decreasing population reduce land consumption at the urban periphery in a long-term perspective (Heiland et al., 2006)? Up to present, there is no evidence for a direct relation between the two processes. And further, does shrinkage allow for the improvement or, at least, for a stabilisation of environmental quality in formerly growing, densely built urban areas?

Demographic change means above all a change in the total number of people, age class distribution, a modification in household structures and a rising impact of migration. Whereas population decline in western German urban regions is yet on a moderate level (Ø -1496 inh./a from 1995 to 2004), eastern German urban regions are affected already for 15 years by significant shrinkage processes: low birth rates, high outmigration and ageing rates (Ø - 10365 inh./a from 1995 to 2004; cf. again Figure 1). There, we find a coincidence of seemingly contradictory urban processes: first, although on a low level since the late 1990s, there are traces of prevailing dynamics of suburban growth (single and semi-detached house settlements, new “housing parks”) with adjacent land consumption (commerce and industry) at the urban fringe. Second, at many places, increasing process of depopulation and related shrinkage calculated according to residential vacancy, perforation, demolition and deconstruction in the core city areas can be observed. A thinning out of recent urban fabric and population densities are the consequence. Third, counteracting decline, we find also processes of stabilisation of the housing function as well as increasing population numbers in some inner-city areas.

Our analysis further discussed the fact that demographic change already endangers the infrastructural viability in urban areas that are affected by massive decline. With a declining population density, critical thresholds for the viability of technical, transport, and social infrastructures and in public facilities are reached already now. There are coming up new debates on how to ensure and finance the supply of the remained inhabitants in areas undergoing demolition of housing stock. Due to a thinning out of social and administrative infrastructure and a decrease in individual accessibility, daily travel times might increase although the population decreases. To put it differently: Demographic change on its own does not solve the problem of land and resource consumption, and it does not automatically bring about higher environmental quality for urban regions either.

It is our strong belief that the questions addressed concerning the potential impact of demographic change on urban regions have to be answered jointly by demographic, social, urban and landscape science like the interdisciplinary authorship of this chapter underscores. To what extent and how demands on natural resources and space will really change under the condition of population decline and shrinking urban regions? Do we have to consider a long-term change in demand–supply relations which result either in ecological benefits or in new threats? The question whether future land developments along the rural–urban gradient will be characterised either by polarisation between core city and surrounding areas or by a widespread lowering of densification in the form of perforation has to be left for future comprehensive and comparative analyses of urban regions throughout Europe.

References

Antrop, M., 2004. Landscape change and the urbanization process in Europe. Landscape and Urban Planning 67, 9–26.

BBR, Federal Agency for Spatial development and Construction, 2005. Report of Spatial Development in Germany, Wiesbaden, 2005.

Berg, L., van den, Drewett, R., Klaassen, L., Rossi, A., and Vijverberg, C.H.T., 1982. Urban Europe. A Study of Growth and Decline. Pergamon Press, Oxford.

Bolund, P. and Hunhammar, S., 1999. Ecosystem services in urban areas. Ecological Economics 29, 293–301.

Bösch-Supan, A., Brugiavini, A., Jürgs, H., Mackenbach, J., Siegrist, J., and Weber, G., 2005. Health, Ageing and Retirement in Europe. MEA Mannheim.

Buzar, S., Ogden, P.E., Hall, R., Haase, A., Kabisch, S., and Steinführer, A., 2007. Splintering urban populations: emergent landscapes of reurbanisation in four European cities, Urban Studies add detailed reference (in print).

Buzar, S., Odgen, P., and Hall, R., 2005. Households matter: the quiet demography of urban transformation. Progress of Human Geography 29(4), 413–436.

Cloet, R., 2003. Population Changes 1950–2050 in Europe and North America. Population Statstics.doc 3-03, 1–11.

Couch, C., Nuissl, H., Karecha, J., and Rink, D., 2005. Decline and Sprawl. An evolving type of urban development. European Planning Studies 13(1), 117–136.

Eurostat, 2006. http://europa.eu.int/newcronos/navigation.htm

Gober, P., 1990. The Urban Demographic Landscape. in: Myers, D. (Editors). Housing Demography. Linking Demographic Structure and Housing Markets. Madison, London, pp. 232–248.

Haase, D., Haase, A., Kabisch, S., and Bischoff, P., 2006a. Guidelines for the perfect inner City. Discussing the Appropriateness of Monitoring Approaches for Reurbanisation, European Planning Studies (accepted).

Haase, D., Holzkämper, A., and Seppelt, R., 2006b. Rethinking urban development: Residential vacancy and demolition in Eastern Germany – conceptual model and spatially explicit results. Urban Studies (submitted).

Haase, A., Kabisch, S., and Steinführer, A., 2005. Reurbanisation of Inner-City Areas in European Cities. In: I. Sagan and D.M. Smith (Editors), Society, Economy, Environment – Towards the Sustainable City, Gdańsk, Poznań, pp. 75–91.

Hall, P., 1992. Urban and Regional Planning, Oxford, 1992.

Hannemann, C., 2003. Schrumpfende Städte in Ostdeutschland Ursachen und Folgen einer Stadtentwicklung ohne Wirtschaftswachstum. Aus Politik und Zeitgeschichte 28, 16–23.

Heiland, S., Regener, M., and Stutzriemer, S., 2006. Auswirkungen des demographischen Wandels auf Umwelt- und Naturschutz. Raumordnung und Raumentwicklung 3, 189–198.

Herfert, G., 2002. Disurbanisierung und Reurbanisierung. Polarisierte Raumentwicklung in der ostdeutschen Schrumpfungslandschaft. Raumforschung und Raumordnung 5–6, 334–344.

Himiyama, Y., Mather, A., Bicik, I., and Milanova, E.V. (Editors), 2005. Land Use/Cover Changes in Selected Regions of the World, Vol. IV.

INKAR, 2003. Indicators and maps for spatial development in Germany, Statistical Agencies of Germany and Ministry of Architecture and Regional Development.

Kaa, D. van de, 2004. Is the Second Demographic Transition a useful research concept. Vienna Yearbook of Population Research, pp. 4–10.

Kabisch, S., Haase, A., and Haase, D., 2006. Beyond growth – urban development in shrinking cities as a challenge for modeling approaches. In: A. Voinov, A. Jakeman, and A. Rizzoli (Editors). Proceedings of the iEMSs Third Biennial Meeting: "Summit on Environmental Modelling and Software". International Environmental Modelling and Software Society, Burlington, July 2006. CD-ROM. http://www.iemss.org/iemss2006/sessions/ all.html, ISBN 1-4243-0852-6 978-1-4243-0852-1.

Kabisch, S., 2005. Empirical analyses on housing vacancy and urban shrinkage. In: Y. Hurol, U. Vestbro Dick, and N. Wilkinson, Methodologies in Housing Research, The Urban International Press, Gateshead, GB, pp. 188–205.

Kazepov, Y., 2005. Cities of Europe. Changing Contexts, Local Arrangements, and the Challenge to Urban Cohesion. Blackwell, UK.

Klingholz, R., 2004. Germany 2004 – Breakup to another country. GEO 05, 89–140.

Koziol, M., 2004. Folgen des demographischen Wandels für die kommunale Infrastruktur. Deutsche Zeitschrift für Kommunalwissenschaften 43, 69–83.

Lang, T. and Tenz, E., 2003. Von der schrumpfenden Stadt zur Lean City: Prozesse und Auswirkungen der Stadtschrumpfung in Ostdeutschland und deren Bewältigung, Dortmund.

Lüthi, A., 2001. Eine langfristige ökonomische Perspektive unter demographischen, ökologischen und technologischen Randbedingungen. Norderstedt.

Lütke-Daldrup, E., 2001. Die perforierte Stadt – eine Versuchsanordnung. Stadtbauwelt 150, 40–45.

Lutz, W., 2001. The end of World Population Growth. Nature 412, 543–545.

Mehnert, D., Haase, D., Lausch, A., Auhagen, A., Dormann, C.F., and Seppelt, R., 2005. Bewertung der Habitateignung von Stadtstrukturen unter besonderer Berücksichtigung von Grün- und Brachflächen am Beispiel der Stadt Leipzig. Naturschutz und Landschaftsplanung 2, 54–64.

Morrison, N., 2003. Neighborhoods and Social Cohesion: Experiences from Europe, International Planning Studies 8(2), 115–138.

Müller, B. and Siedentop, S., 2004. Wachstum und Schrumpfung in Deutschland – Trends, Perspektiven und Herausforderungen für die räumliche Planung und Entwicklung. Deutsche Zeitschrift für Kommunalwissenschaften 43, 14–32.

Newcastle City Council (NCC), 2000. Going for growth – a green paper, A city wide vision of Newcastle 2020, Newcastle (http://www.newcastle.gov.uk/gfg2002.nsf/)

Nuissl, H. and Rink, D., 2005. The "production" of urban sprawl. Urban sprawl in Eastern Germany as a phenomenon of post-socialist transformation. Cities 22, 123–143.

Power, A. and Mumford, K., 1999. The Slow Death of Great Cities? Urban Abandonment and Urban Renaissance. YPS.

Ravetz, J., 2000. City region 2020, London.

Rink, D., 2005. Surrogate nature or wilderness? Social perceptions and notions of nature in an urban context, In: I. Kowarik and S. Körner (Editors), Wild Urban Woodlands. New Perspectives for Urban Forestry, Springer, Berlin, pp. 67–80.

Sander, R., 2006. Urban development and planning in the built City: Cities under Pressure for Change – An Introduction. German Journal of Regional Sciences 45, 1.

Seo, J.-K., 2002. Re-urbanisation in regenerated areas of Manchester and Glasgow. New residents and the problems of sustainability. Cities 19(2), 113–121.

Statistisches Bundesamt (StBA; Federal Agency for Statistics in Germany, 2000). Population Development in Germany until 2050, Wiesbaden.

THE CONSEQUENCES OF DEMOGRAPHIC CHANGE IN RHINELAND-PALATINATE: SCENARIOS OF LANDSCAPE CONSUMPTION FOR SETTLEMENT AND TRANSPORTATION AREAS

JAMILL SABBAGH* AND MARCO NEEF
RegioComun – Institute for Strategical Regional Development, Johannes Gutenberg-Universität Mainz, Department of Geography, 55099 Mainz, Germany

Abstract. Land consumption for settlement has been a popular topic for discussion in spatial planning for several decades. But although there is no shortage of declarations and protestations that the reduction of land consumption plays a key role in sustainable development, no really effective countermeasures have been implemented. As a result, little progress can be made in attempts to reduce the rate of land consumption. And yet there are good reasons for giving serious consideration to control options once again. A number of European countries face a drastic change in terms of their demographic and economic structure. The debates over land consumption go along with this phenomenon of predicted dramatic falls in population. Can we prevent the growth of land consumption in view of a shrinking population? Data about changes in population, settlement, and transportation areas are shown for Rhineland-Palatinate region in Germany. Predicted trends for the next 50 years suggest increases in settlement areas despite reductions in the total population. This represents a future crisis in land available for agriculture and environmental values. Planning and policy instruments at the national level are suggested to complement existing measures to promote a sustainable future for German landscapes.

* Corresponding Author: Dr. Jamill Sabbagh, RegioComun – Institute for Strategical Regional Development, Johannes Gutenberg-Universität Mainz, Department of Geography, 55099 Mainz, Germany, e-mail: J.Sabbagh@t-online.de

I. Petrosillo et al. (eds.), Use of Landscape Sciences for the Assessment of Environmental Security, 345–356.

Keywords: Demographic change; population development; population shrinkage; landscape consumption; sustainable land use; Rhineland-Palatinate; settlement and transportation areas

1. Demographic change and sustainable land use – trends in Germany

Germany will face a severe decrease in population during the next 25–50 years, a trend that will intensify beyond 2050. By 2050 the projected total population of Germany is expected to fall from over 82 million to under 70 million accompanied by a fundamental change in demographic structures. For example, the share of people between the ages of 20 and 65, i.e., people who by present standards are of employable age, will then constitute only about 55% of the total population as opposed to 62% at the beginning of the millennium. To examine the nature of the projected population changes, and the consequences for sustainable land use and land consumption we will present and examine several key indicators.

For over a century birth rates (live birth per thousands per year) have been falling in Germany, and for the last 70 years each generation of children has been smaller in numbers than that of its parents. A suitable indicator to reflect this trend is the fertility rate. It is the average number of children a woman will have in her lifetime. Germany's fertility rate has been sufficient to maintain a small increase in population during the 1960s – the time of the "baby boomers." Since then the fertility rate has dropped from over 2.5 (1965) to about 1.3 children per woman in 2004. The fertility rate of every generation must be at 2.1 in order to secure a steady state in the total population.

As populations change and societal values and expectations change, the link between population development and land consumption follow different trajectories. Long-term analyses show that landscape consumption tends to grow faster than population, even if the total population remains static or shrinks. This can be illustrated using a survey of the Deutsche Bank (Just, 2004), which analyzed developments in the German real estate market.

The predictions suggest that the number of private households will increase until 2030. This is mainly due to changing household structures. According to the Deutsche Bank survey the demand for housing area will continue to focus on quality in terms of location, size, and household. The supply of barrier free and apartments suitable for the elderly will be of particular importance to the housing sector. Reconstruction and revitalization will become inherent parts of urban renewal concepts. It becomes evident that the real estate market has to be prepared for new consumer needs and target groups: the growing number of single and two-people households and the growing number of households of elderly people.

In addition regions with a strong economy will show disproportionately high growth of population and a rising demand in housing space. To meet this demand growing regions in western Germany would have to build 275,000 apartments per year of 110 m^2 living area in average – that are 15% more than in 2002. In contrast, landlords and housing societies in other regions will have to adjust to house and apartment vacancies.

Thus social indicators and regional disparities of demographic change can be used to reflect changes in real estate markets and trends in land consumption. Data from Rhineland-Palatinate, a German Bundesland (state), will be used to illustrate population shrinkage and growth in land consumption and consider what this means in order to achieve sustainable land use management.

2. Consequences of demographic change in Rhineland-Palatinate

Population data and the population prognoses are important for spatial planning. They give spatial information about the needs for infrastructure such as kindergartens, schools, and aged-care facilities, etc. Furthermore they show us trends of household structures, in future demands of living area as well as in the labor force potential.

According to various demographic projections of Rhineland-Palatinate will decrease significantly in the period 2000–2050. The demographic projections in Figure 1 show a range of possible total population changes using different assumptions regarding fertility, mortality, migration, and labor-force participation rate. The Bundesland will lose between 314,000 and 1 million inhabitants during the period 2000–2050.

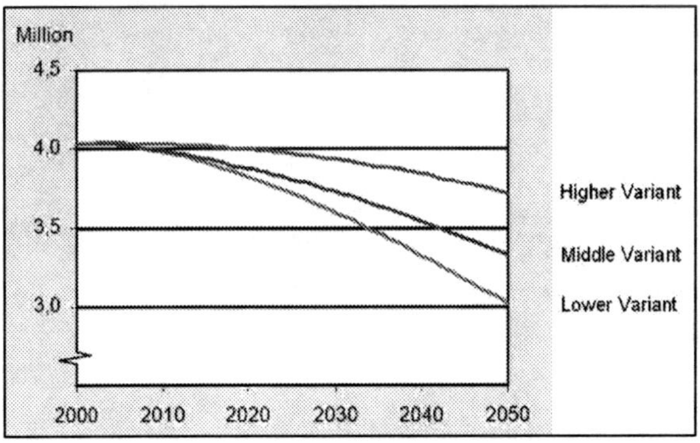

Figure 1. Variants of drop in population in Rhineland-Palatinate 2000–2050. (Source: Statistisches Landesamt Rheinland-Pfalz.)

According to the middle variant, Rhineland-Palatinate will lose 18% of its total population until 2050 and decreases to a population of 3.2 Million. This prognosis is based on the assumptions that fertility rate will be at a constant level of 1.4 until 2050, expectation of life will increase by four years, and that there will be an immigration surplus of 5,000 persons per year. The drop in population varies between districts and cities of Rhineland-Palatinate (Figure 2).

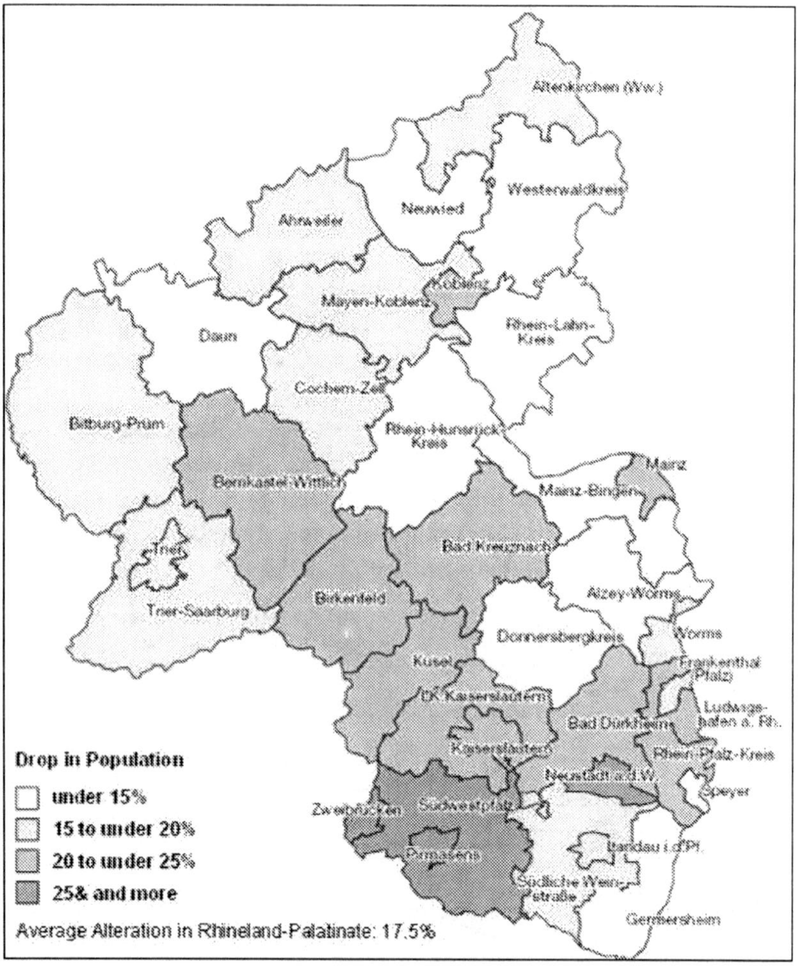

Figure 2. Drop in population in Rhineland-Palatinate per district 2000–2050 (middle variant). (Source: Statistisches Landesamt Rheinland-Pfalz, 2005.)

For policy development and infrastructure projections at the Bundesland and district levels the drop in population will be more important than the age proportion of the people. Two studies by the state office of statistics

(Statistisches Landesamt) in Rhineland-Palatinate come to the following conclusions: firstly, the resident population will age significantly between 2005 and 2040; and secondly, the population in the employable ages decline significantly after 2015.

According to the middle variant, the proportion of the people of 60 years and more will increase from one quarter to one third until 2050. In contrast, the proportion of the population younger than 20 as well as between 20 and 60 years will clearly decline (Figure 3). Looking at the ageing structure Figure 4 and Figure 5 indicate the shift in the population structure.

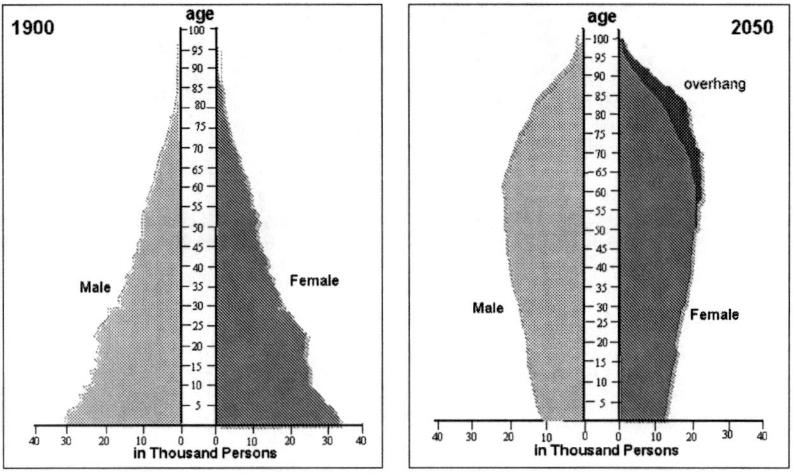

Figure 3. Population pyramide in Rhineland-Palatinate 1900 and 2050. (Own design; Data: Statistisches Landesamt Rheinland-Pfalz.)

3. Landscape consumption for settlement and transportation areas

Since the 1950s, settlement and transportation areas in Germany have approximately doubled (from 7.1% in 1950 to 12.5% in 2005). This trend is also true for Rhineland-Palatinate (Statistisches Landesamt, 2000). One of the major findings of the territorial census of 2001 in Germany was the continuing high rate of land consumption, mainly caused by new infrastructure and housing projects. Moreover, the report stresses, that this trend will not change in the near future.

Landscape consumption has serious impacts on the goods and services provided by the environment. Landscape consumption reduces native vegetation cover and the availability of arable land. The vegetation and soil resources serve us in multiple ways, such as a carbon sink, a purifier of drinking water and a buffer of contaminates. Of further importance are the natural living space for flora and fauna and not least the beauty of the open landscape and its function as ecological corridors.

Soil and landscape consumption seem to be an inevitable consequence of every municipal development, always torn between the society's conflict of interests – social, economic, and ecological demands. Looking at the current rate of landscape consumption (Figure 6), there is little possibility of a sustainable land use future for the Bundesland.

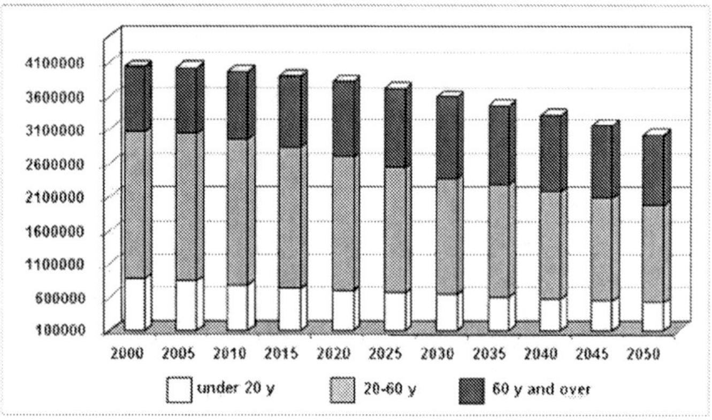

Figure 4. Shift of age structure in Rhineland-Palatinate 2000–2050. (Own design 2005; Data: Statistisches Landesamt Rheinland-Pfalz.)

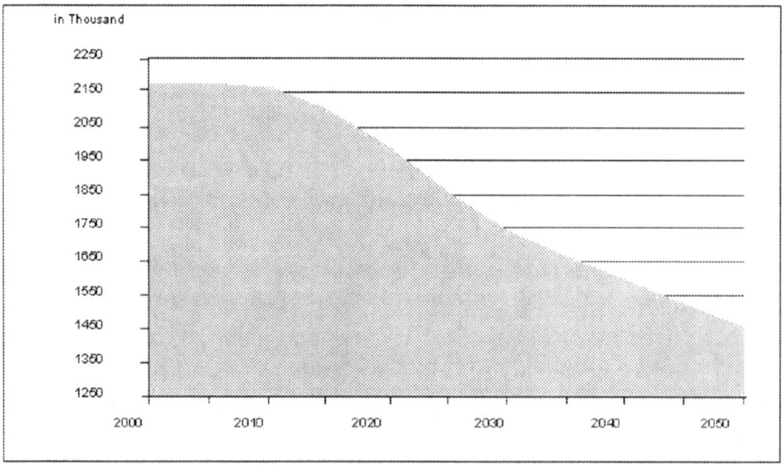

Figure 5. Population of working age in Rhineland-Palatinate. (Own design; Data: Statistisches Landesamt Rheinland-Pfalz.)

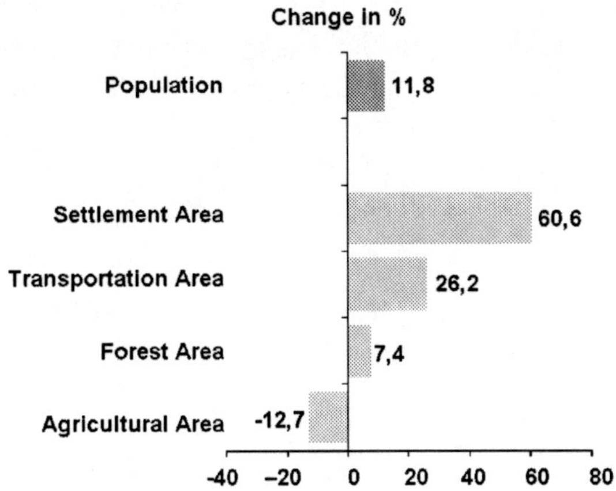

Figure 6. Population development and landscape consumption of selected uses in Rhineland-Palatinate 1978–2004. (Own design; Data: Statistisches Landesamt Rheinland-Pfalz, 2005.)

The main reasons for the enormous extension of settlement and transportation areas are growing wealth and a rising level of mobility. These are linked to the existing ideal of a home in the country, rising requirements of accommodations and false subsidies policy. In 2003, nationwide about 93 hectare of open space were consumed by settlement and transportation area per day – in Rhineland-Palatinate this number was at about 6 hectare per day. The total percentage of settlement and transportation area rose from 9.6% in 1978 to 13.8% in 2004.

4. Increasing landscape consumption: suburbanization and the countryside

The changes in populations and areas used for settlement and transport development for the urban and countryside areas are shown in Figure 7. The main expansion of settlement development took place in the countryside and especially in suburban areas near the cities. This is, because we find an enormous land-price discrepancy between the urban area and the infrastructural well-equipped periphery.

In Rhineland-Palatinate, which has the smallest settlement structure in Germany, most of the people today, like 25 years ago, live in municipalities between 2000 and 5000 inhabitants (15.3%). Here we also find the population density coming closest to the state average of 210 residents per km^2, which is also true for the amount of 665 m^2 of settlement and transportation area per capita (Rhineland-Palatinate: 675 m^2 in 2004). Comparing population development and land consumption in these municipalities, which are likely to be representative, we find that population has grown in average 16.6% between 1981 and 2000, whereas settlement area has grown 38.8%. In the same period the population of the biggest cities stagnated or declined. Thus, making statements regarding settlement sprawl, we have to attach greater importance to the small towns. These findings also document the trend "out of the city, into the countryside."

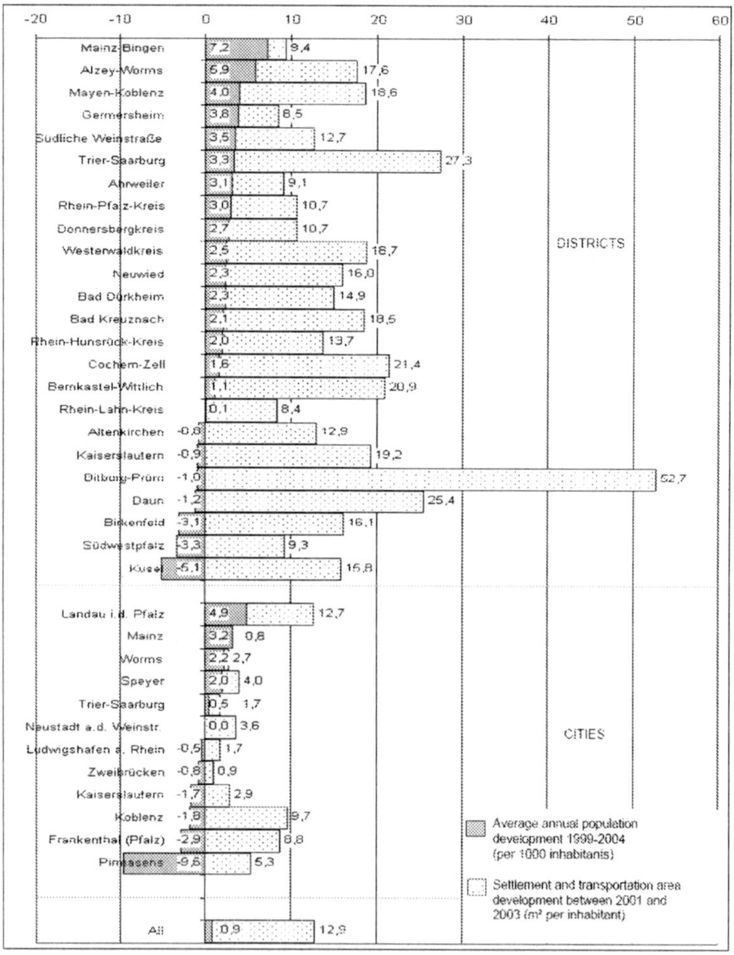

Figure 7. Landscape consumption exceeding population development in cities and periphery – average annual population development 1999–2004 and settlement and transportation area development 2001–2003. (Own design 2006; Data: Statistisches Landesamt Rheinland-Pfalz.)

A reason for the increase of urban area is the long prevailing general principle of functional separation, which has lead to an increasing spatial division between living, working, supplying, and recreation. Furthermore, urbanization results not only in landscape consumption but also in the loss of urban diversity and increasing traffic. Road building has to compensate the division of working and living, associated with increasing commuter traffic and longer commuter distances. Estimates of settlement and transportation area suggest a climb to 13.4% of the total state area until 2010. This equals a loss of about 564,000 hectares since 1997 of potential environmental services including an accelerating loss of biodiversity.

Apart from losses of environmental services landscape consumption by settlement often brings about the disorganization of historical settlement structure – followed by settlement sprawl or even *Siedlungsbrei* (settlement mash). Other adverse consequences are inevitable. With every dedication of new building areas or the improvement of transport infrastructure, we set processes in motion that include so-called feedback effects. For instance, the division of working from living (which then is in the "green" countryside) leads to growing traffic, an impairment of the quality of living in the city, new migration into the suburb and thus to the construction of new circuitous roads.

5. Increasing landscape consumption and changing lifestyle

The reason that landscape consumption clearly outreaches the population development is growing wealth, accompanied by the increase of comfort and quality. Social life in Germany is increasingly characterized by attributes like individualization, anonymity, and even the evanescence of solidarity. This is associated with the loss of traditional family structures, displayed by decreasing number of marriages and increasing number of divorces and single parents. Moreover, job-related mobility leads to frequent changes of location. As a result the number of single households soars and we can find regions, where the total number of households is rising whilst population declines. Living area has almost tripled from 15 m^2 per capita in 1950 to 41 m^2 per capita in 2000 (Iwanow, 2001). The spatial area of workplaces in industry, trade, and service has grown as well.

6. Landscape consumption in the clash of interests between social, economic, and ecological demands

Landscape consumption is also triggered by the fierce economic competition regarding the settlement of industry, business, and housing between municipalities.

Local authorities attract industrial settlements by the means of low land prices and low land purchase taxes (in some cases even self-financing): “To put it bluntly, neither clients nor mayors nor parliamentarians – i.e. those who are finally responsible for the decision as to whether land may be used for development purposes, are not personally interested in using less land. Furthermore, regional planning institutions are wary of quarrelling with communities over the designation of building land. None of these protagonists can nowadays afford to appear as putative inhibitors of growth.” (Schulz and Dosch, 2005: 3)

7. Towards sustainable land use and possible courses of action

In order to overcome the enormous rate of landscape consumption the German Government has set the objective to “reduce daily increase in settlement and transportation area to 30 hectare (for all Germany) until the year 2020” in their national sustainability strategy of April 2002 (Bundesregierung, 2002). Furthermore, the Enquete Commission “Protection of Man and Environment” of the German Parliament calls for a land use reduction to 12 ha per day until 2010. The commission forecasts that, given the current trend, Germany will be completely covered with settlement and transportation areas in about 80 years. Other experts, such as the Wuppertal Institute, call for a zero growth strategy until 2010.

Against this background, the Council on Sustainability has presented its recommendations entitled “More Value for Land: the 30 Hectare-objective for a Sustainable Urban and Rural Development” for the revision of National Sustainability Strategy of the Federal Government in July 2004 (Rat für Nachhaltige Entwicklung, 2004). In this document they call for a new management structure to reach a “30 hectare-objective.” Among other things it is recommended that planning obligations for the objectives of land policy are introduced, that true costs including lost environmental services and plans are considered in a better way and that land recycling is promoted. However, presently it is unclear how a lower trend can be achieved.

Thus, the issue of land consumption is on the list of persistent environmental problems that are characterized by the fact that governmental policy instruments have only limited short-term effects. The reduction of landscape consumption can only be achieved by the quantitative control and qualitative regulation of the further growth of settlement areas. History has shown that a reduction in land consumption rates cannot be achieved by an isolated policy instrument, but only through a mix of coherent policies and instruments from several policy areas, including instruments of police law, economic incentives, and organizational instruments (e.g. urban and regional planning and traffic management). The approaches of instruments are best assigned to five different groups:

- Instruments of *planning law* (e.g. selected extension of planning law; land dedications through focal settlements)
- Instruments of *information* (e.g. increasing awareness; market transparency; effective land registers)
- Instruments of *cooperation* (e.g. regional industrial and commercial land pools; intermunicipal negotiations; vitalization of the regional idea)
- *Economic* instruments (e.g. rearrangement of tax incentives and the finance system of the municipalities; system of tradable land certificates)
- Instruments of *financial* subsidies

According to the findings above those instruments must meet the objectives:

- A reversal of the subsidies policy to redevelopment-oriented planning
- Include the concept of ecosystem service losses in natural resource policy development
- The efficient use of existing settlement structures in due consideration of cost-reducing potentials; that is (re-) utilization prior to new construction, inner development prior to outer development; utilization of building management to scoop area potentials
- The promotion of area-saving building methods and optimal use of urban density; enlargement of existing land use potentials without additional land consumption; subsequent densification
- An active land policy including innovative legal and economic instruments to secure a sustainable land use and intensified cooperation between urban and suburban areas
- Reduction, respectively management of consumption and demand into sound settlement forms

In summary, as much land has been claimed during the past 50 years as in the last 1,000–1,200 years of settlement history. This is caused among other things by the individual availability of motor transport, life style changes resulting in settlements beyond the urban centres – combined with the dissociation of settlement and transportation planning, working places, and housing locations. A stagnating or even shrinking population asks for growing settlement areas. This suggests a pessimistic answer to the question in the beginning, if we can prevent the growth of land consumption in view of a shrinking population. In this context, the conflict over the scarcity of the valuable asset of land is particularly keen. What is needed is political will at all levels and regionally differentiated and integrated concepts of land management.

References

Bundesforschungsanstalt für Landeskunde und Raumordnung, 1998. Nachhaltige Stadtentwicklung/Anforderungen an Städtebau, Wirtschaft und Handel. Kongress in Merseburg am 27. u. 28. Mai 1997 im Auftrag des Bundesministeriums für Forschungsvorhaben des Experimentellen Wohnungs- und Städtebaus. Bonn.

Deutscher Bundestag, 2002. Wohngeld- und Mietenbericht 2002. Unterrichtung durch die Bundesregierung. Drucksache 15/2200. Berlin.

GDW, 2004. Mietwohnungen in Deutschland - ein attraktives und wertbeständiges Marktsegment. Entwicklung, Besonderheiten und aktuelle Trends im internationalen Vergleich. GdW Branchenberichte. Hamburg.

Iwanow, I. et al., 2001. Zukünftige Wohnungsnachfrage und Neubaubedarf in Ost- und Westdeutschland. Eine Forschungskooperation zwischen IÖR und BBR. IÖR-Texte 133. Dresden.

Just, T. (Deutsche Bank Research), 2004. Demographische Entwicklung verschont öffentliche Infrastruktur nicht. Frankfurt.

Rat für Nachhaltige Entwicklung, 2004. Mehr Wert für die Fläche: Das "Ziel-30-ha" für die Nachhaltigkeit in Stadt und Land. Empfehlungen des Rates für Nachhaltige Entwicklung an die Bundesregierung. Berlin.

Schulz, B. and Dosch, F., 2005. New Strategies for Redevelopment-Oriented Urban Planning. In: DISP 160 (1/2005): Urban Sprawl. online: http://www.nsl.ethz.ch/index.php/content/download/1050/6418/file/ (12.02.2006)

Statistisches Bundesamt, 2002. Bautätigkeit und Wohnungen. Mikrozensus - Zusatzerhebung 2002. Bestand und Struktur der Wohneinheiten. Wohnsituation der Haushalte. Fachserie 5/ Heft 1. Wiesbaden.

Statistisches Bundesamt, 2004. Datenreport 2004. Zahlen und Fakten über die Bundesrepublik Deutschland. Schriftenreihe, Band 440. Bonn.

Statistisches Landesamt Rheinland-Pfalz, 2005. Rheinland-Pfalz 2005 – Bevölkerungsentwicklung und Auswirkungen. Mainz.

Statistisches Landesamt Rheinland-Pfalz, 2004. Rheinland-Pfalz 2050, Bd. II. Auswirkungen der Demographischen Entwicklung. Bad Ems.

Statistisches Landesamt Rheinland-Pfalz, 2005. Flächennutzung 2004: Entwicklung wichtiger Nutzungsarten. Press Release 2005-10-12. Mainz.

Statistisches Landesamt Rheinland-Pfalz, 2004. Rheinland-Pfalz 2050 – Zeitreihen, Strukturdaten, Analysen. Mainz.

Statistisches Landesamt Rheinland-Pfalz, 2002. Flächenverbrauch und Versiegelung – Auch ein rheinland-pfälzisches Problem? Statistische Monatshefte 11/55. Mainz.

Umweltbundesamt (Editor), 2003. Berücksichtigung von Umweltgesichtspunkten bei Subventionen -Sektorstudie Wohnungsbau. Texte 31/03. Berlin. online: http://www.umweltdaten.de/publikationen/fpdf-l/2341.pdf (17.3.2006)

LANDSCAPE, DEMOGRAPHIC DEVELOPMENTS, BIODIVERSITY, AND SUSTAINABLE LAND USE STRATEGY: A CASE STUDY ON KARABURUN PENINSULA IZMIR, TURKEY

ENGIN NURLU* AND UMIT ERDEM
Centre for Environmental Studies & Department of Landscape Architecture, Faculty of Agriculture, Ege University, Izmir, Turkey

MUNIR OZTURK
Rtd. Professor, Ege University, Bornova, Izmir, Turkey

AYKUT GUVENSEN
Department of Biology, Faculty of Science, Ege University, Izmir, Turkey

TEVFIK TURK
Verim Mapping and Engineering Services, Izmir, Turkey

Abstract. The variability of natural characteristics, social and historical aspects, and land use in Karaburun Peninsula led us to choose it in order to determine the possible impacts on the biodiversity and landscape fragmentation. More than 50% of the area is covered with forests and maquis. Mainly, remotely sensed data together with ancillary data-*maps, statistics, and local information* were used to refine interpretation and the assignment of the study area into the categories of the CORINE Land Cover Nomenclature. Several ground control points using global positioning systems were determined during field observations. Distribution characteristics of dominant plant communities and accompanying taxa were recorded by using GPS. The objective of this pilot study was to contribute to landscape management for environmental security based on ecological indicators. Attempts have been made to provide guidelines for the inhabitants of the peninsula towards sustainable development on socio-ecological basis.

* Corresponding Author: Dr. Engin Nurlu, Centre for Environmental Studies & Department of Landscape Architecture, Faculty of Agriculture, Ege University, 35100, Bornova, Izmir, Turkey, e-mail: engin.nurlu@ege.edu.tr

I. Petrosillo et al. (eds.), Use of Landscape Sciences for the Assessment of Environmental Security, 357–368.

Keywords: Karaburun Peninsula; land use; CORINE LC classification; environmental security

1. Introduction

The landscape is defined as an area as perceived by people, whose character is the result of the action and interaction of natural and/or human factors (Council of Europe, 2000). They are recognized as a part and parcel of the natural, historical, cultural, and scientific heritage. During the past few decades the landscapes all around the world have undergone severe degradations which ultimately have lead to dramatic changes (McGlade, 2004). Loss of biodiversity, loss of agricultural lands, and land degradation are the results of these changes which are creating threats to environmental security defined by UN as the relative stability of earth's natural ecosystems against human activity. Recently scientists have developed indicators to monitor and classify landscape changes (Banko et al., 2003). These are defined as parameters providing information about, and describing the state of a phenomenon/environment/area, with a significance extending beyond that directly associated with a parameter value (OECD, 2003). The information must be relevant to specific assessment questions, developed to focus monitoring data on environmental management issues (Jackson et al., 2000). The indicators have an important function in the research and decision-making process, contribute a lot on the objectives of sustainable development and can be used at international and national level for environment reporting as well as measurement of environmental performance, thus supporting policy and decision-making on landscape development (OECD, 2003). They can characterize the material dimensions of landscape change together with the mental and cultural aspects of its development. The indicators can demonstrate and provide estimates of actual changes in land use that have occurred over a certain time, with the help of data about the land cover and land use. Interpretation of aerial photos and field surveys are combined in order to record the current situation in the countryside. The interpretation and the surveys are regularly repeated and are a valuable tool for illustrating the changes in the landscape (Tress and Tress, 2003). The indicators for 2007 proposed by EPA on the environment include land indicators and ecological condition indicators (US EPA, 2005) together with species diversity, threatened and protected species and tourism intensity, whereas Environmental Monitoring and Assessment Program gives two generic types of indicators: condition indicators (biotic and abiotic indicators) and stressor indicators (Barber, 1994).

Landscape indicators are a particular category of ecological indicators that are determined for a predefined geographic or a biogeographic area. They are usually based on remotely sensed data or other geographic information, and like ecological indicators, they can be based on a single measure or a combination of measures (Kepner et al., 1995).

Environmental security, which is a very important human issue, is related to watersheds, forests, soil cover, croplands, genetic resources, climate, and other factors such as, protection oriented activities and their management. In this investigation, an ecological indicator based attempt has been made to contribute to landscape management for environmental security and to provide guidelines for the inhabitants of the Karaburun Peninsula towards sustainable development on socio-ecological basis.

2. Study area

The study area, Karaburun Peninsula, covering an area of 436 km^2, lying between 26°21′ –26°38′ longitudes and 38°40′–38°25′ latitudes, is one of the undisturbed sites in the Aegean Region of Turkey (Figure 1). The coastline of the peninsula has high cliffs penetrating into the sea with shores of various dimensions (Figure 2). The elevation in the area varies from sea level up to 1,300 m., with the highest peak Akdag. Steep mountains on most parts of the peninsula have affected the settlements and land use in the area, which embodies a population of 13,546 inhabitants according to Census 2000.

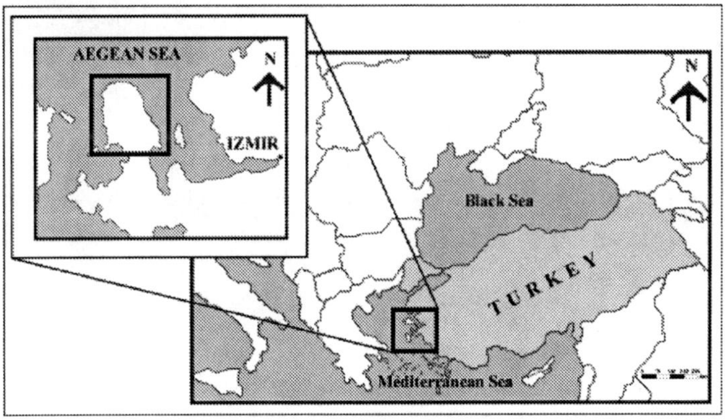

Figure 1. The map showing location of the study area.

The area experiences typical Mediterranean climate with mild rainy winters and dry hot summers. Mean annual temperature ranges between 15°C and 20°C and mean annual precipitation varies between 650 and 750 mm, which is mainly concentrated in the winter months (Anonymous, 2006). Soils are classified within the soil taxonomic unit of non-calcareous brown and red brown Mediterranean, and inland plains are mainly colluvial (Figure 2).

The study area includes forests and maquis (50%), the rest is of grasslands, agricultural areas, rural and urban centers (Nurlu et al., 2003).

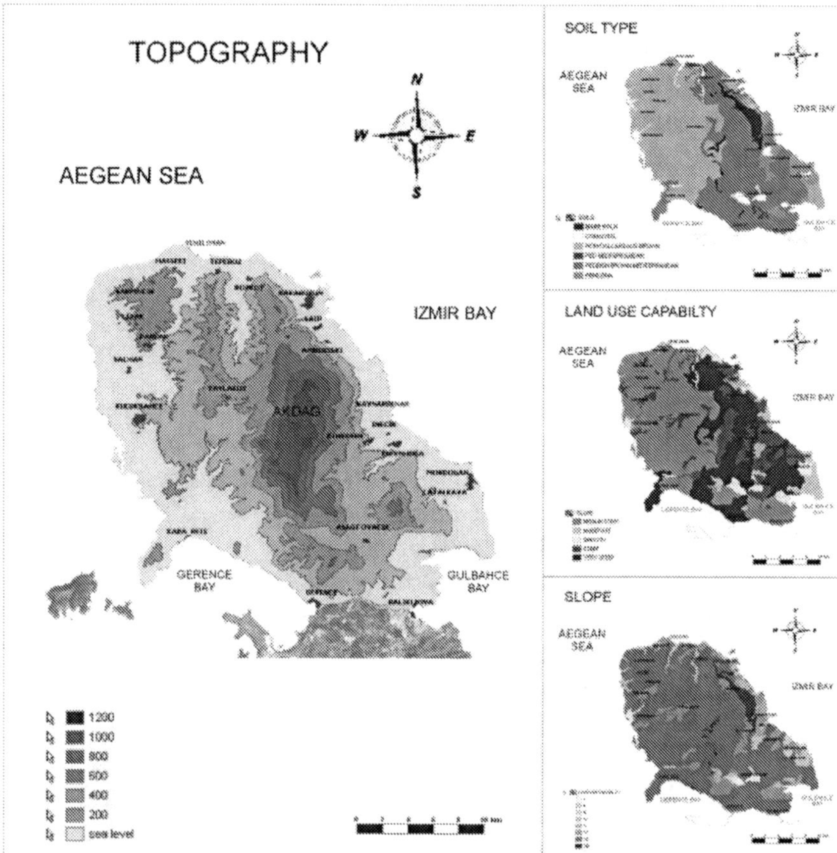

Figure 2. Topography, soil types, and slope characteristics of the study area.

3. Material and methods

Mainly, remotely sensed data together with ancillary data (standard topographic maps, soil maps, environmental land use plans, etc.) were used to create the 1:25,000 scaled land cover map of the study area. Prior to the whole process land covers were defined using Landsat 7 data of September, 2000. Several well-distributed geographical control points obtained from 1:25,000 scale topographic map were used to calculate the geometric transformation. To improve and verify the accuracy of the image, soil maps and environmental land use plan of 1:25,000 scale covering the study area, showing the main landscape features such as relief, soils, forests, maquis, shrubs, wetlands, etc. together with standard topographic maps of 1:25,000 scale showing contours, road networks, and settlements were used. Image Analyst software was used for interpretation of satellite data by using ground control points and ancillary data.

The software Geo Media Professional 5.0 was used for conducting the data acquisition and overlay operations.

The method of this study is based on the sequence of steps required for rural planning as developed by Golley and Bellot (1999), which includes inventory, diagnosis, and evaluation of alternatives. To organize and evaluate ground surface data; remotely sensed information and geographic information systems were used. The methodology included computer-assisted photo interpretation of satellite data by using ancillary data and field observations. Ancillary data were used to refine interpretation and the assignment of the territory into the categories of the CORINE Land Cover Nomenclature (European Commission, 1993). Prior to the mapping of land cover, the satellite imagery was geometrically corrected to the Universal transverse Mercator (UTM) projection. A Root Mean Square Error (RMSE) evaluation was then performed to assess image to map rectification accuracy. The RMSE for the rectified image was <0.5 pixel. After the geometric correction stage, supervised classification algorithm was formed for categorization of different land use/land cover types following the field survey by selecting representative classes. The land cover classes on the created maps were taken according to the CORINE LC classification. Distribution of land cover in different areas was recorded on the basis of artificial surfaces, agricultural areas, forests, semi-natural areas, wetlands, water bodies and other uses. The classification accuracy of each use/land cover type of the image was evaluated using a stratified random sampling design. A confusion matrix was produced, from which the overall accuracy and kappa index of agreement for each land use/land cover category was computed. Distribution characteristics of dominant plant communities and accompanying taxa were recorded by using GPS points. "Flora of Turkey and East Aegean Islands" (Davis, 1965, 1985; Davis et al., 1988) was used for the plant identifications.

4. Results and discussion

4.1. SETTLEMENTS

Owing to its convenient location, Karaburun Peninsula has been an important center of civilization both politically and economically since the ancient era (Isik, 2002). The most important record stating demographic, social, and economical situation of the peninsula is the census results taken in 1831. According to this census, there were 23 villages on the peninsula (Isik, 2002). During early 19th century, economy on the peninsula depended on agriculture. Vineyards and olive plantations were the primary sources of revenue. Topographic conditions and the trade routes being only at the Aegean Sea forced inhabitants to establish their settlements on the coastline (Figure 2). Immigrations, earthquakes, especially the one in 1949, and the rapid development of coastal settlements after 1980s affected the area's settlement distribution. Two types of settlement developments have been observed on the peninsula since 1980. First type is agriculture oriented, people of which are mainly involved in cultivations on the coastal plains. Second type is of summer houses, which are constantly increasing. The population of the peninsula reaches up to 50.000 during the hot season.

4.2. DEMOGRAPHIC DEVELOPMENTS

The demographic structure of the area reveals that it has undergone several changes. While the decrease in the population of peninsula between 1927 and 1975 was caused by immigration to central Izmir, the increase in the population between 1975 and 2000 was caused by second house constructions (Figure 3). Annual population growth rate in the area between 1990 and 2000 was 39.91‰ in total. While urban population growth in the same period was 22.63‰, it was 45.32‰ in rural areas (SIS, 2004). The reason why rural population growth was so high is that a mass second housing took place in coastal villages which are considered rural in documents. According to the census taken in 2000, 4,397 out of the total population of 13,546 is classified rural. Common lifestyle of the rural population includes animal farming, cultivation of ornamental plants, citrus, olive, artichoke, and some other agricultural crops. The local population is dominated by middle-aged people but people from all age groups are observed in summer when they come to spend their summer in their second houses.

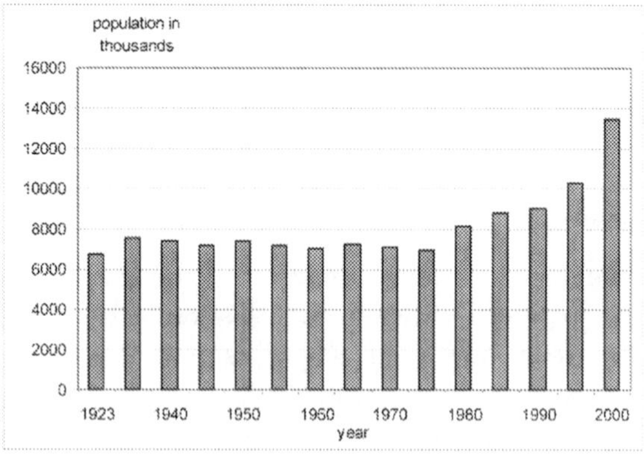

Figure 3. Population change in the study area.

4.3. LAND USE

Land use is one of the most visible effects of humans on the earth and can have effects on both human health and ecological systems. UN environmental security definition notably includes deforestation caused by "clearing" of lands, soil depletion, and desertification caused by intensive monoculture techniques. In our study, functional aspects of land have been considered to be the actual land uses of the area such as; agricultural, forest/heathlands, grasslands, and other uses including urban and nonarable areas. Land uses were obtained from the statistical report of Ministry of Agriculture and Rural Affairs for the years 1995, 1998, 2001, and 2004 (Figure 4). There has not been much land use change in the last decade due to topographic features, and lack of transport facilities such as; roads and public transport.

Agricultural areas occupy less than 10% of the total study area. Over two third of this is covered by olive plantations and the rest is used for growing artichoke and grapes. Olive oil processed at a small scale in local olive oil factories is still a major income source for the area. Growing narcissus as an ornamental crop has recently become an intensive agricultural activity.

The results obtained by using supervised classification showed that 5% of the total study area are covered by artificial surfaces (urban fabric); 15% agricultural areas (arable land, permanent crops, heterogeneous agriculture areas); 75% forests and seminatural areas (coniferous forest, moors and heathland, sclerophyllous vegetation), 1% wetlands (coastal wetlands), and 4% other uses (Figure 5).

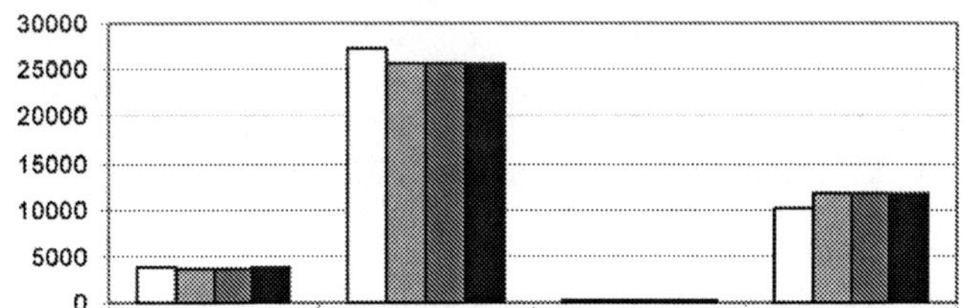

	agricultural areas	forests/heatlands	grasslands	non-arable and other areas
1995	3820	27351	200	10129
1998	3705	25825	200	11770
2001	3705	25825	200	11770
2004	3766	25825	200	11709

Figure 4. Land use changes in the study area.

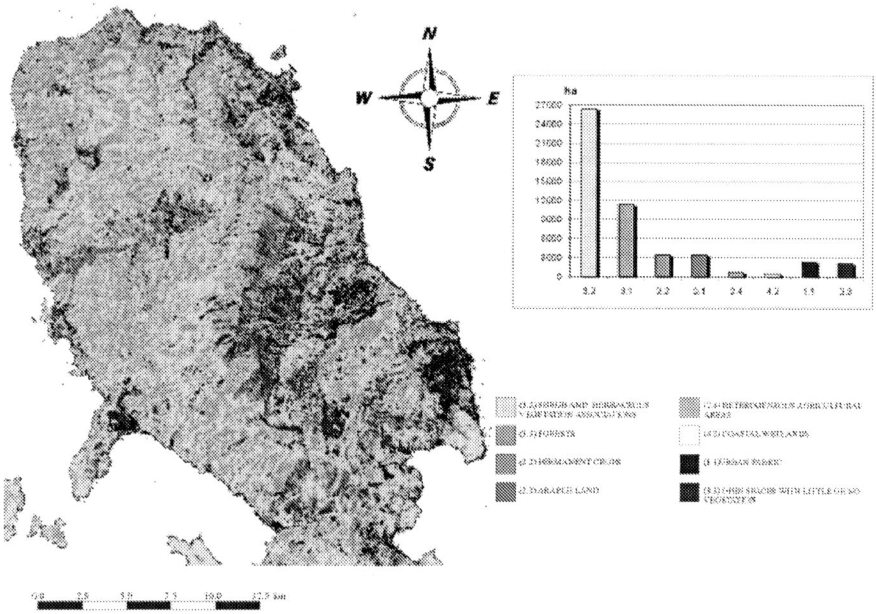

Figure 5. Land cover map of the study area.

4.4. BIODIVERSITY

Over the last 200 years, demographic outburst, overexploitation of natural resources and environmental degradation has resulted in a decline in global biodiversity. The European Commission is seeking an agreement on concrete measures to put a break on this decline and promoting actions to reduce its loss. Experts agree that the backbone of global biodiversity protection is the establishment of a network of protected areas in which threatened species and ecosystems can survive.

Karaburun Peninsula is one of the typical representatives of the Mediterranean phytogeographical region dominated by a rich biodiversity. There are nearly 384 plant taxa distributed in the peninsula, belonging mainly to the 70 families such as Fabaceae (42 taxa), Poaceae (35 taxa), and Asteraceae (30 taxa) (Davis, 1965, 1985; Davis et al., 1988; Bekat and Secmen, 1984). Field surveys have shown that the area is covered with *Pinus brutia* forests which exist at elevations of 200–500 m, shrubs and/or herbaceous vegetation, and many of sclerophyllous species. Floristic surveys were carried out during the field surveys and plant communities were defined digitally using CORINE Land Cover

Nomenclature. The red data book of Turkish plants (Ekim et al., 2000) reveals that seven endemic and four nonendemic taxa from the plant cover are in danger of becoming extinct. These are: *Cyclamen hederifolium* (vulnerable), *Micromeria nervosa* (vulnerable), *Globularia alypum* (vulnerable), *Erysimum pusillum* (vulnerable), *Sideritis sipylea* (lower risk-near threatened), *Stachys cretica* subsp. *smyrnaea* (lower risk-least concern), *S. cretica* subsp. *anatolica* (lower risk-least concern), *Minuartia anatolica* var. *anatolica* (lower risk-least concern), *Trigonella smyrnea* (lower risk-conservation dependent), *Campanula lyrata* subsp. *lyrata* (lower risk-least concern), *Erodium sibthorpianum* subsp. *vetteri* (data deficient).

The dominant species of the coniferous forest is *Pinus brutia*. In some parts coniferous forests are mixed with *Quercus coccifera,* and *Myrtus communis*. The dominant species of moors and heathland is *Sarcopoterium spinosum.* In the sclerophyllous vegetation *Quercus coccifera, Cistus creticus, C. salviifolius, Pistacia lentiscus*, and *P. terebinthus* subsp. *palaestina* dominate the area. Transitional woodland shrubs in the area are: *Juniperus phoenicia, J. oxycedrus* subsp. *macrocarpa, P. brutia* and *Olea europea* var. *sylvestris*. In open spaces with little or no vegetation such as beaches, dunes, sand plains *Juncus maritimus, Euphorbia peplis,* and *Tribulus terrestris* are very common. Species covering salt marshes and saline areas are: *Artrochnemum fruticosum, Halimione portulacoides, Halocnemum strobilaceum, Limonium bellidifolium, Salicornia europaea, Tamarix parviflora*, and *T. smyrnensis.*

The area is quite an important breeding area for more than 200 terrestrial and marine species in particular Audouin's Gull (*Larus audoinii*), Chukar (*Alectoris chukar*), Lesser Kestrel (*Falco naumanni*), Eleonora's Falcon (*Falco eleonorae*) and Yelkouan Shearwater (*Puffinus yelkouan*) (Kilic and Eken, 2004). The area is also an important site for internationally protected sea mammals Eurasian Otter, *Lutra lutra*, and the Mediterranean Monk Seal, *Monachus monachus*. However, there has been a decrease in the population of the latter since 1970s (Veryeri et al., 2002). The reason is that settlements especially second houses are expanding towards the caves.

5. Conclusion

Karaburun Peninsula is one of the least disturbed sites in the Eastern Mediterranean basin. However, several factors related to small scale degradative activities have started coming to the forefront in this area. In order to save the area from a destructive fate, measures should be taken from now on against these activities by making use of landscape indicators and environmental security principals.

The environmental security is safeguarding of water, soil, vegetation, climate, and other components that ultimately underpin all our socioeconomic activities (Brennan, 1999; Carius and Leitzman, 1999). If these resources are degraded or depleted, security of a region declines too. We need to spend much greater effort to tackle environmental problems and best way to cut off is at source. It has not yet been adopted within long-term thinking by the people in Karaburun Peninsula.

Sustainable use of land resources is of vital importance for the study area. Soil is becoming a scarce commodity and object of competition among different sectors using it and this competition is growing. This development will ultimately lead to land degradation and pollution of other resources including ecosystems. Challenges to professionals working in this field are enormous and a holistic and integrated approach is needed. The crucial and most important concerns to be addressed both politically and technically are: how to cope with this situation and to find sustainable solutions. The immediate answer is through drastic changes in the methods of using and managing land resources. Failure to find the appropriate linkage between soil, food, and environmental security derives mainly from mismanagement. When addressing environmental security issues, there is a need to manage land resources more prudently, in quantity as well as quality, and reduce degradative pressures. Management of environmental parameters should be given a priority aimed at better food supply, livelihoods and nature in a sustainable manner. Integrated approaches must take into account not only scientific and technical, but also the socio-economical and environmental aspects. A new generation of efficient land use management systems should be designed while sustaining ecosystems and the environment. New technologies and management techniques will play an important role in meeting the challenge of demographic outburst and increased food demand. A tremendous gap still exists between research and its implementation. New researches should lead to technologies that would conserve the natural resources, in particular land, and is environmentally friendly, technically appropriate, economically viable and socially acceptable. An integrated approach, including ecological, economical and social factors and needs and strategies should be taken.

This will be possible by rectifying management and usage practices of land resources allocated to the agricultural and touristic sectors. An ecological land planning division should be created in the peninsula which can help in restricting the sprawl of second homes on prime quality lands and discourage undue encroachment on arable land. Scientists, sociologists, planners, administrators, and locals should join hands for the protection of virgin areas by international legislations, raise public awareness for environmental protection, and control demographic pressures in order to seek balance and harmony between people and land.

References

Anonymous, 2006. Official Web Site of Karaburun District Government, http://www.karaburun.gov.tr.

Banko, G., Zethner, G., Wrbka, T., and Schmitzberger, I., 2003. Landscape types as the optimal domain for developing landscape indicators. In: W. Dramstad and C. Sogge (Editors), Agricultural Impacts on Landscapes: Developing Indicators for Policy Analysis, Proceedings from NIJOS/OECD Expert Meeting on Agricultural Landscape Indicators, Norwegian Institute of Land Inventory, pp. 317–328.

Barber, M.C. (Editor), 1994. Environmental Monitoring and Assessment Program: Indicator Development Strategy, US Environmental Protection Agency, Office of Research and Development, Environmental Research Laboratory EPA/620/R-94/XXX.

Bekat, L. and Secmen, O., 1984. Akdag-Karaburun (Izmir)'un bitki ekolojisi ve sosyolojisi yonunden arastirilmasi. Bulletin Faculty of Science 7(1), 103–110, Series B, Ege University, Bornova, Izmir.

Brennan, E.M., 1999. Population, urbanization, environment, and security: a summary of the issues. Environmental Change and Security Project 5, 4–14, Woodrow Wilson Center for Scholars, Smithsonian Institution, Washington DC.

Carius, A. and Lietzmann, K.M. (Editors), 1999. Environmental Change and Security: A European Perspective. Springer, Berlin, Germany.

Council of Europe, 2000. European Landscape Convention-Explanatory Report & Convention, Florence.

Davis, P.H., 1965–1985. Flora of Turkey and East Aegean Islands, Vols. I–IX, Edinburgh University Press.

Davis, P.H., Mill, R.R., and Tan, K., 1988. Flora of Turkey and East Aegean Islands, Vol. X, Edinburgh University Press.

Ekim, P., Koyuncu, M., Vural, M., Duman, H., Aytac, Z., and Adıguzel, N., 2000. Red Data Book of Turkish Plants (Pterodophyta and Spermatophyta), Bariscan Ofset, Ankara, p. 246.

European Commission, 1993. CORINE Land Cover: Technical Guide, (Report EUR 12585), Office for Official Publications of the European Communities, Luxembourg, p. 144.

Golley, F.B. and Bellot, J. (Editors), 1999. Rural Planning from an Environmental Systems Perspective, Springer, New York, p. 377.

Isik, S., 2002. Historical Geography of Karaburun Peninsula, Ege Univ. Faculty of Literature Publications: 120, Ege University Press, Izmir, p. 80.

Jackson, L.E., Kurtz, J.C., and Fisher, W.S. (Editors), 2000. Evaluation Guidelines for Ecological Indicators, U.S. Environmental Protection Agency, Office of Research and Development, Environmental Monitoring and Assessment Program, Washington, D.C., EPA/620/R-99/005.

Kepner, W.G., Jones, K.B., Chaloud, D.J., Wickham, J.D., Riiters, K.H., and O'Neill, R.V., 1995. Mid-Atlantic Landscape Indicators Project Plan, Environmental Monitoring and Assessment Program, U.S. Environmental Protection Agency, Office of Research and Development, National Exposure Research Laboratory, Washington, D.C., EPA/620/R-95/003.

Kilic, D. and Eken, G., 2004. Turkiye'nin Onemli Kus Alanlari-2004 Güncellemesi, Doga Dernegi, Ankara, p. 232.

McGlade, J., 2004. EU Conference on Changing Land Use in Europe-Getting the Picture, European Environment Agency, 9–12 November, 2004, The Netherlands.

Nurlu, E., Erdem, U., Guvensen, A., and Yilmaz, O., 2003. Plant cover and land degradation relationship on Aegean Coastal Zone. In: I. Gitas and J.S.M. Ayanz (Editors), Environmental Monitoring in the South-eastern Mediterranean Region Using RS/GIS Techniques, Options-

Mediterraneennes, Series B: Studies and Research Number 44, CIHEAM/MAICh-EU-AIDCO, Greece, pp. 57–68.

OECD, 2003. OECD Environmental Indicators – Development, Measurement and Use, Reference Paper, OECD Environmental Performance and Information Division, Paris Cedex, France.

SIS, 2004. State Institute of Statistics, Census of Population by Administrative Division, Prime Ministry, Republic of Turkey, Ankara, http://www.die.gov.tr.

Tress, B. and Tress, G., 2003. Shaping future landscape: the scenario approach. In: W. Dramstad and C. Sogge (Editors), Agricultural Impacts on Landscapes: Developing Indicators for Policy Analysis, Proceedings from NIJOS/OECD Expert Meeting on Agricultural Landscape Indicators,, Norwegian Institute of Land Inventory, pp. 68–81.

US EPA, 2005. Proposed Indicators for 2007 Report on the Environment (ROE 2007), www.epa.gov/ncea/

Veryeri, O.N., Nurlu, E., and Erdem, U., 2002. A Research on Globalizations and the Mediterranean *Monachus monachus* at Karaburun Peninsula. In: D. Camarda, R. Giordano, and L. Grassini (Editors), Proceedings of Meeting on Local Resources and Global Trades-Environments and Agriculture in the Mediterranean Region, Rabat, CIHEAM-Politecnico di Bari, p. 54.

INDICATION OF THE STATE OF THE ENVIRONMENT WITH GIS AND PEOPLE: A CASE STUDY AND PLANNING TOOL FOR MULFINGEN, A MUNICIPALITY IN SOUTH GERMANY

ROMAN LENZ* AND ANGELIKA JANY
Hochschule für Wirtschaft und Umwelt, Nürtingen, Germany

Abstract. As an outcome of a five years research project and a couple of experiences in other studies we present a case study and planning tool about an indication of the state of the environment of the municipality Mulfingen, South Germany. Municipalities in Germany do have the major planning power and thus steer their development, including socio-ecological resources and loads. Firstly, we discuss indicator sets for sustainable development, focusing on the environment. Secondly, we explain the context and the main case study in the municipality Mulfingen. In a third step we present our indicator set, chosen together with the stakeholders. Fourthly, we describe the use of GIS and data base, and show the advantages of calculations and visualisation. After that, 21 indicators in the categories and so-called action fields: Energy supply, water supply, waste, settlements, environmental protection, agriculture, tourism, and nature conservation are calculated. Four of them indicate good conditions, eight show more or less acceptable situations, and nine indicator values are seriously bad. As a final result, the complete balance of the state of the environment is presented as an amoeba diagram. For the overall approach we conclude, that it is of high importance to define indicators as well as the use of information technology closely with the people involved, in order to optimize the outreach of the approach.

Keywords: Environmental indicators; Geographical Information Systems (GIS); Human-Environment Systems; Landscape assessment

* Corresponding Author: Professor Roman Lenz, Hochschule für Wirtschaft und Umwelt, Schelmenwasen 4-8, Nürtingen, D-72622, Germany.

I. Petrosillo et al. (eds.), Use of Landscape Sciences for the Assessment of Environmental Security, 369–381.

1. Introduction and objectives

Related to the pilot study "Use of Landscape Science for Environmental Assessment," this case study focuses on the use of indicators for environmental assessment. While ecological and/or environmental indication itself is already a widely discussed and complex issue (e.g. Müller and Lenz, 2006), an additional challenge is to involve people and get acceptance by decision-makers – a problem, which was already partly highlighted 25 years ago (e.g. in 1991 by Ten Brink with his amoeba approach), and which finds currently many support by visualization techniques and participatory approaches (e.g. Tress and Tress, 2003).

In this case study, we also highlight a kind of "High-Tech Approach" in using remote sensing data and detailed data bases, all combined and processed with a geographic information system (GIS) and a relational data base. Related to the Landscape Indicator Examples of the pilot study, we made some own suggestions on indicators which were intensively discussed with the stakeholders in the chosen municipality. With a clear participatory component, the working group and round table discussion (see later), including an information system for mapping indicator values and potentially running different scenarios for reduction of environmental impacts, the study contributes to the following objectives:

- To demonstrate the use of ecological indicators for environmental assessment
- To illustrate the utility of this approach in environmental decision-making, planning, and preservation/restoration projects

From these objectives, general approaches, and indicator examples, we rose the following questions:

1. What is a good indicator set – indicating environmental health (Lenz, 1999) and thus helping to assess and achieve environmental security?
2. How can we improve a landscape assessment by information technology?
3. What are the lessons learned for future studies?

2. General method and data

Our method can be briefly described as so-called action research: We offer a problem description via an indicator set to stakeholders and jointly use this or a modified one for the assessment of the state of the environment. The method was developed by Lewis in the fifties (described in Ehret, 1997) for projects in developing countries, and should help to combine expert with stakeholders knowledge for problem identification and solving in an iterative way. It was used in an interdisciplinary project on regional development in Hohenlohe, Germany (Kirchner-Heßler et al., 1999) and its various subprojects (e.g. Beuttler and Lenz, 2003; Lenz and Peters, 2006). Identified relevant data and its interrelations are stored in a GIS and a database system, with additional visualization possibilities.

For environmental balancing we use a Geographical Information System (GIS), because it allows us a spatial as well as consistent relation between basic data and calculation results, and vice versa. Firstly, in the GIS the basic data (ATKIS = digitales Amtliches Topographisch-Karthographisches Informations system, ALK = Automatisierte Liegenschaftskarte, ALB = Automatisiertes Liegenschaftsbuch) are stored and prepared for further overlays. Secondly, we added data provided by additional sources, like interpretations of aerial photographs to delineate land use types, a soil map and information about the fauna. The district administration provided us extensive data on protected areas (nature conservation, water), which can be used by the municipalities. Besides these spatial data, further information is used for calculation procedures of the indicators, which collection and preparation is supported by the specific administrations.

The digital preparation and processing of the data enables an extensive environmental information pool for the municipality. This analysis aims at the detection of weak points as wells as development potentials.

The results are used in future planning processes, and the applicability of the system can be further developed. Some essential aspects of the concept, the case study and the planning tool are described in the following.

3. GIS and indicator set

Since more than a decade, environmental impact assessment, regional or spatial planning, and environmental balancing seem to develop similarities, e.g. joint basic methodological approaches like the use of environmental indicators, the focus on same environmental media like air, water, soil, flora/fauna, etc. (Lenz, 1999). Especially GIS-based software systems show their multiple applications in these fields (Lenz and Beuttler, 2003; Lenz et al., 2004).

Experiences from a set of regional environmental (or eco-) balances show a wide range of advantages as well as disadvantages of the widespread use of GIS-based planning tools. With the background of concepts and examples for spatial eco-balances for the district Pfaffenhofen (Upper Bavaria, Germany; cf. Lenz, 1997) and the municipality Mulfingen (Hohenlohe, Germany) – both related to the concept of environmental indicators of the advisory board of environmental affairs of the Federal Republic of Germany (SRU, 1994) and the

Federal Environmental Agency (UBA, 1995) – we could show GIS-based information systems of a high practical relevancy. On the basis of the GIS software ArcView 3.x, the database management system Access, and html scripts, we developed environmental information systems to balance environmental effects in a map scale of 1:5,000 – 1:50,000, in order to provide the administration with tools for an environmentally sound and sustainable development of their area (Lenz and Beuttler, 2003).

The approach of regional eco-balancing combines the classical landscape planning (predominantly for the protection of environmental compartments and recreation properties as zones in landscapes) with a balancing of a distinct, but – in terms of environmental protection – broad set of environmental indicators for (effect) eco-balances. Hence the latter becomes spatially related due to georeferencing. By taking on board the district administration and establishing an information system, a high practical relevancy and acceptance of the final users can be achieved. Aim is a system to balance environmental impacts in a map scale of 1:10,000–1:50,000, in order to provide the district administration with tools for an environmentally sound and sustainable development of their region.

If there are classical ecological planning units and procedures (e.g. regions, landscapes, watersheds, and related tools like land consolidation) then we should consider the already developed approaches in the field of landscape planning as well as the high importance of having "state of the art" ecological indicators. In the case study Mulfingen we first used 24 indicators suggested by Lenz (1999). The final 21 indicators used in this study – as a result of the round table process – are listed in Figure 1. The further quantification of their status are the basis of the quality of every ecological balancing, besides the tools provided by environmental informatics (see results).

4. Working group and round table

To guarantee an acceptance in the administration and in the population it is important to have a round table accompanying the eco-balance. In the municipality of Mulfingen we founded a working group with representatives of the municipality, municipal councillors, and interested citizens. We presented the action fields and their indicators. At the beginning we underestimated the significance of the discussions about measures. Although we still were in the balancing phase we were talking about measures to improve the environmental situation of Mulfingen. The working group of the eco-balance ran finally into the local agenda process of the municipality where the realisation of the proposed measures took place. For example, they organized a course for ecological driving or an excursion to show applications of solar energy or rain water using in private households.

The main thing of the working group was to combine the local knowledge with the knowledge of the scientists to find solutions of the environmental problems – a typical kind of action research. Hence, the final 21 indicators (see Figure 1) chosen for balancing are a result of this process.

5. Technical design of the planning tool

The technical instrumentation of the eco-balance, developed already in a former project for the district of Pfaffenhofen in Bavaria, is based on a PC concept, using Windows NT and application software as mentioned below (cf. Lenz, 1997). There are three major interlinkages to be established:

GIS (ArcView)–HTML–DBMS (Access)

When starting the ArcView-projects, simultaneously Netscape will be started, and the HTML page, which gives later on the access to the choice of maps, will be written (dynamical generation).

Establishment of new views/layouts/themes: ArcView → DBMS → HTML

When in ArcView one of the documents is elaborated, three steps while storageing will follow:

1. In the DBMS the following inscriptions will be done: author, date of establishment, last change, name of ODB files (name be put together from date and time).
2. Storage of the new document takes place as ODB file (ODB export) with the corresponding name.
3. The HTML page, which contents the register of maps and themes, will be updated with an Avenue (which is the programming language in ArcView) script (as line file of the corresponding HTML syntax).

Call of existing views/layouts/themes: HTML → ArcView → DBMS → ArcView

1. Via Netscape a selection of the view /layout/theme will be carried out. The chosen name will be delivered to a Visual Basic-file (communicate.exe).
2. VB program will call an Avenue-script and delivers the (view/layout/theme)-name.
3. In the DBMS, with this name the ODB file will be searched.
4. The ODB file will be loaded and displayed (ODB import)

Documentation of a theme: HTML → ArcView → HTML

1. In order to get information over an active theme, on the HTML page "documentation" has to be selected. This calls the VB file, which starts an Avenue-script.

2. Via the script the HTML documentation page will be generated dynamically.
3. The HTML page will be displayed.

6. Results

The technical system and planning tool is rather elegant and powerful, and especially in combination with scenarios – in this paper we only use the future area consumption of the legal structure plan as a trend scenario – it started to

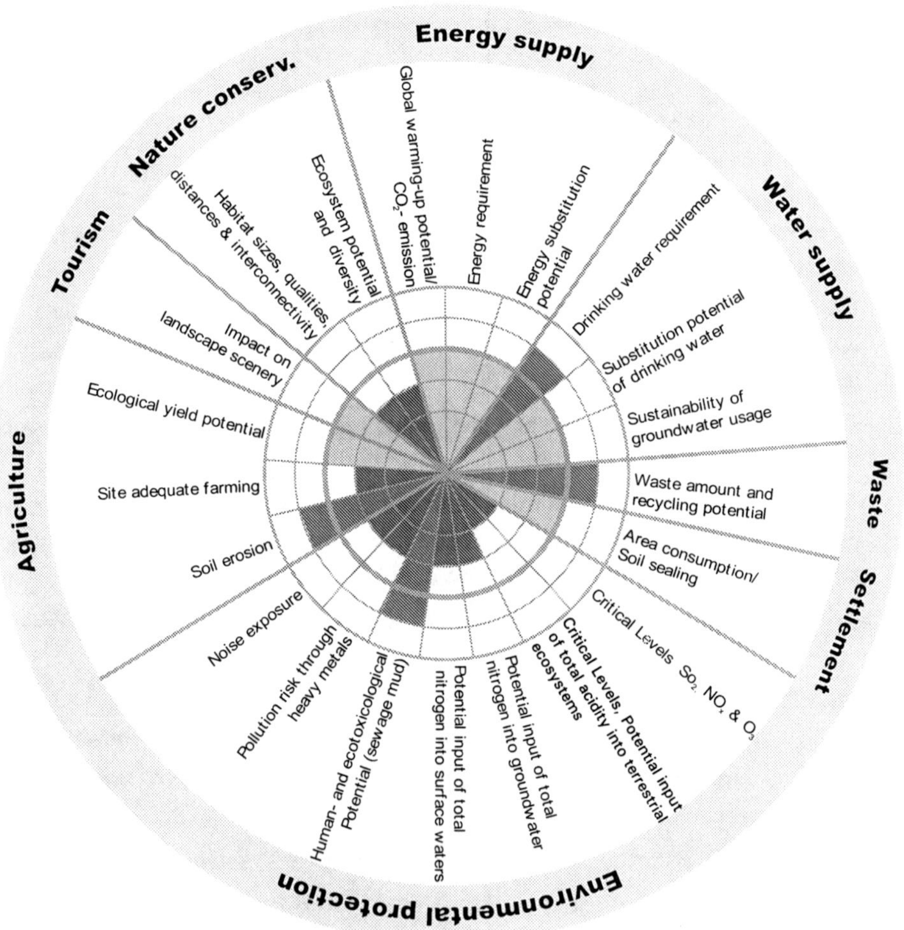

Figure 1. Amoeba diagram with Environmental Indicators and action fields. The darker the gray color, the less good is the environmental status of an indicator in the action field, respectively.

get accepted within the administration. Nevertheless, only a few people in the district administration are able to run the software – although it became a standard one in all districts in several states of Germany – and hence this aspect is still the most relevant restriction of the use of the system. In addition, many other feed backs on the system from our target groups where like: "technically and scientifically very good, but too complicated/oversized for policy and administration…." What all of them appreciated very much, was a graphical illustration of the outcome of the environmental balancing in a very aggregated way (cf. Ten Brink, 1991), as shown in Figure 1 (cf. WG, 2000). Twenty-one indicators in the categories and so-called action fields: Energy supply, water supply, waste, settlements, environmental protection, agriculture, tourism, and nature conservation, were calculated. Four of them indicate good conditions, eight show more or less acceptable situations, and nine indicator values are seriously bad.

The following map (Figure 2) shows the main land use types in the municipality of Mulfingen, where we also applied an eco-balance by support of a GIS. One of the easiest exercises is the calculation of area consumption over past and future times. Municipality planning can therefore be compared and assessed.

The following three examples will illustrate the applications possible for the municipality. They focus on renewable energy, area, and water consumption; all examples, which indicate the status of environmental health and environmental services, and which can be used to enhance environmental security in improving the status of a sustainable self supply.

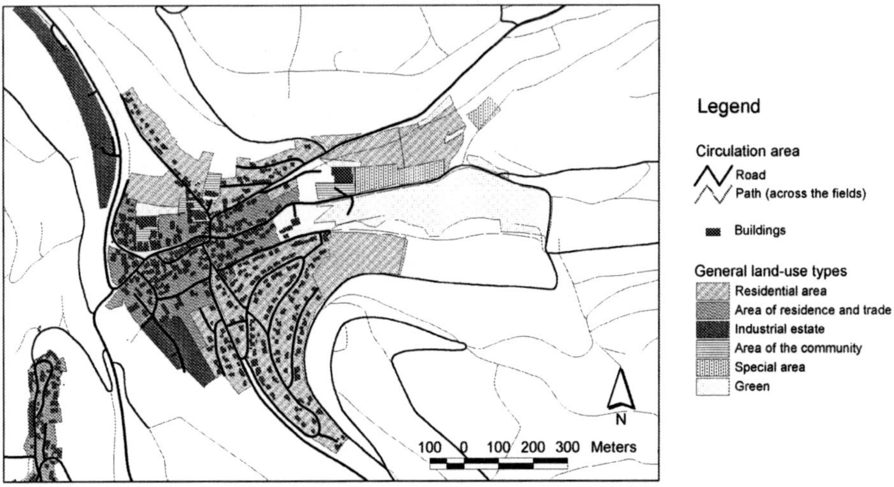

Figure 2. Land use types municipality Mulfingen.

6.1. SOLAR ENERGY POTENTIAL

In using the GIS software ArcView3.2, roof areas can be derived from ALK data. After subtraction of areas with expositions, inclinations, shading, and other restrictions making them unsuitable for solar energy use, the actual roof area can be multiplied with the annual insolation. The result is the amount of regenerative energy from solar radiation, which could be used by the municipality either for heating or for power supply (Figure 3).

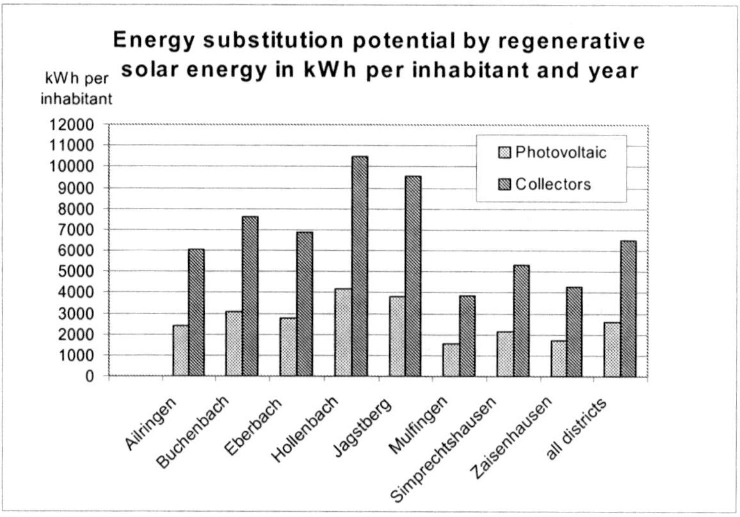

Figure 3. Energy substitution potential by regenerative solar energy in the several parts of the municipality Mulfingen.

Hence, for the whole municipality of Mulfingen we could produce 3.300 kWh per inhabitant by photovoltaic, or 8.200 kWh by collectors.

In comparison: If we take the annual average energy consumption of every inhabitant in Baden-Württemberg in 1998 of about 24.000 kWh (household, traffic, etc.), solar energy could potentially provide 14% or 35% of this consumption, respectively.

A concrete example for the future use of this energy substitution potential by solar radiation is an accepted proposal of the municipality government, in all new building areas to investigate the optimization potential for solar energy!

Furthermore, within the restoration of the primary school, a solar energy collection will be established.

6.2. AREA CONSUMPTION

Table 1 shows, that area consumption in Mulfingen is increasing since 1981–1990, but is still in an average category of "sparingly consuming" compared to many other municipalities in the state Baden-Württemberg.

TABLE 1. Annual area consumption municipality Mulfingen (1981–2010).

Eco-balance municipality Mulfingen Annual area consumption (average)		
1981-1990	legally binding land-use plan (B-Plan)	0,0157%
1991-2000	legally binding land-use plan (B-Plan)	0,0219%
2001-2010	structure plan (FNP)	0,0505%
1981-2010		0,0294%
Assessment of the area consumption		
more than average Nation 1993-1995		> 0,1%
sparingly consumed		< = 0,1%
sustainable Enquete 1997		< 0,01%

In order to ensure a sustainable development of settlement areas, in the long term the annual area consumption must not exceed 0.01% of the total area of the municipality (Enquete-Kommission des Deutschen Bundestages, 1997). This clearly means that the official structure plan of the municipality is not sustainable in the long and short run. For the balance of the past area consumption we took the information of the legally binding land use plan. For the future area consumption the structure plan delivers the necessary information.

6.3. DRINKING WATER CONSUMPTION

For another detection of eventually weak points, the indicator of "drinking water consumption" will be illustrated. With the aid of annual consumption data and number of inhabitants in the several parts of the municipality, we calculate the daily drinking water consumption per person. The different results indicate increased values in some part compared to reference values in literature (Zeisel, 1998, see Figure 4). For the municipality the tasks remains, to detect the causes of this weak points and to establish measures for a reduction. The indicator "drinking water potential" is suitable – due to its basic data – for an annual balancing. For the environmental balancing in Mulfingen we used the programming language Avenue in ArcView3.2 to develop a user-friendly tool for the calculations. With this user interface, it is easy to update the data annually by the administration (see Figure 4) and "control" the trends. This is an important function – to control the development of the state of the environment – of environmental balancing. A regular data collection enables us to follow the status of indicator values. The digital processing makes the ongoing balancing easier as well as other planning in the municipality.

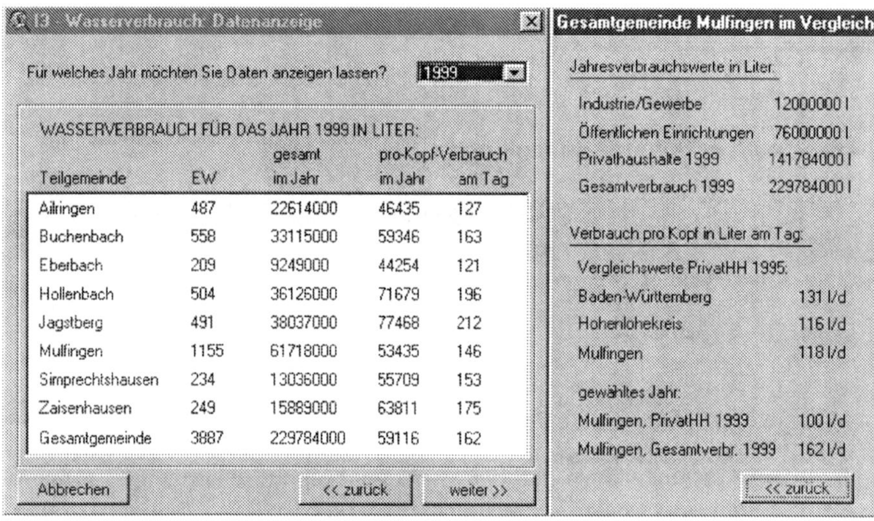

Figure 4. Water consumption of the several parts of the municipality in 1999 compared to reference values from state statistics (screen shot, in German language).

Restrictions in the availability of (digital and others) data is a typical limitation of this approach. Single indicators can only be roughly estimated, because data are missing or can be only collected with high expenses. Especially so-called impact- and reaction indicators (in comparison to status indicators) need extensive data sets for balancing. For example, indication of soil erosion need detailed soil as well as relief and land use data, which are not generally available. On the other hand this indicator only varies when the cultivation will be changed. Noise impact from roads can only be assessed, where countings of traffic had been performed. In Germany every five years the

German Ministry for construction, traffic, and housing performs a national big traffic census of the main roads (so-called Große Verkehrszählung). The data for the waste accumulation are only available on the district level but not on the municipal level. A positive aspect of this fact is that gaps in data can better be detected, and specific fields of problems can be elucidated. In future, more data will be provided by district administrations for better use on lower-level planning like for municipalities, so that data quantity and access is improving (Lenz, 1997, 1999).

One aspect has an influence on the data availability: It is the time period balancing. Some indicators like the drinking water consumption or the waste accumulation should be collected yearly. In contrast soil erosion or area consumption are balanced in longer periods.

7. Discussion

What is a good indicator set – indicating environmental health and thus helping to assess and achieve environmental security?

We discussed indicator sets for sustainable development in the working group and round table, focusing on the environment, and we also explained and presented our indicator set, chosen together with the stakeholders. Twelve of the indicators like soil erosion, global warming-up potential, energy and drinking water requirement, waste amount and recycling potential, critical levels and loads, potential nitrogen input into the ground and surface water, human- and ecotoxicological potential, pollution risk through heavy metals, and noise exposure show a clear relation to environmental health and security aspects. We suggest encouraging the decision makers to focus mitigation strategies on those indicators, which show low values: critical levels and loads, potential nitrogen input into ground and surface water, pollution risk through heavy metals, and noise exposure. To our opinion, environmental health and security can be better assessed when local people are involved already in the choice of indicators and the related definition of problems (what do we want to indicate?).

How can we improve a landscape assessment by information technology?

We described the use of GIS and database, and show the advantages of calculations and visualization. After that, a couple of individual indicator values are highlighted, and also the complete balance of the state of the environment was presented in an amoeba diagram, comprising all environmental media (flora and fauna, soil, water, climate, and air) and categorized as action fields (Energy supply, water supply, waste, settlements, environmental protection, agriculture, tourism, and nature conservation). Information technology used for the planning tool helps a lot in documentation, calculation and visualization. On the other hand, for untrained people it can build a barrier to use it.

Lessons learned for future studies?

Up to now we did not see any other more or less complete balancing of the environmental status, based on an environmental indicator set AND balanced spatially explicit. Especially the spatial reference to all land use activities allows regional planning not only to set priorities in action fields but also relate them priority areas. Hence a powerful GIS software with qualified users, although it still causes limitations in administrative handling, is necessary. It should be assured that the whole eco-balance is operated and maintained regularly.

Such eco-balances represent a chance to communicate environmental issues to citizens in an understandable way. So they should be involved in the realization of measures to improve the balance in the long term. The local agenda offers various possibilities for this process.

The indicators related to the actions field nature conservation represent still a lot of research work which is to be done. Different approaches work which different indicators on various levels. So the discussion must keep going.

Acknowledgment

We would like to thank the Deutsche Bundesstiftung Umwelt (DBU), the German Federal Ministry for Education, Research and Technology (BMBF) and many fruitful NATO workshops for support.

References

Beuttler, A. and Lenz, R. (Editors), 2003. Kommunale Umweltbilanz Mulfingen. Oekom Verlag, München, 100pp. with CD-ROM.

Ehret, W., 1997. Reorientation of extension. A Case Study of Participatory Action Research with a Non-government Organization in Northern Nigeria. Margraf, Weikersheim.

Enquete-Kommission des Deutschen Bundestages, 1997. Konzept Nachhaltigkeit; Fundamente für die Gesellschaft von morgen, Zwischenbericht der Enquete-Kommission "Schutz des Menschen und der Umwelt – Ziele und Rahmenbedingungen einer nachhaltigen zukunftsverträglichen Entwicklung" Deutscher Bundestag, 13. Wahlperiode, Bonn.

Kirchner-Heßler, R., Konold, W., Lenz, R., and Thomas, A., 1999. Ökologische Konzeptionen für Agrarlandschaften – Modellprojekt Kulturlandschaft Hohenlohe – ein Forschungskonzept. Naturschutz und Landschaftsplanung, 31.Jg., H. 9, 275–282.

Lenz, R.J.M., 1997. Design and prototype of an information system for regional eco-balances. In: R. Denzer, D. Swayne, and G. Schimak (Editors), Proceedings of the International

Symposium on Environmental Software Systems, Whistler, B.C., Canada, Chapman & Hall, London, pp. 57–63.

Lenz, R., 1999. Landscape health indicators for regional eco-balances. In: Pykh/Hyatt/Lenz (Editors), Advances in sustainable development: Environmental Indices – Systems Analysis Approach. EOLSS, Oxford, pp. 12–23.

Lenz, R.J.M. and Beuttler, A., 2003. Experiences with GIS-based planning tool for spatial eco-balances. Environmental Modelling & Software 18, 581–585.

Lenz, R. and Peters, D., 2006. From data to decisions. Steps to an application-oriented landscape research. Ecological Indicators 6, 250–263.

Lenz, R., Glemnitz, M., Lausch, A., Müller F., and Sabbagh J., 2004. Functional system analysis of spatial human-environment-interactions. EcoSys Suppl. Bd. 42, 104–107.

Müller, F. and Lenz, R., 2006. Ecological indicators: theoretical fundamentals of consistent applications in environmental management. Ecological Indicators 6, 1–5.

SRU (Rat von Sachverständigen in Umweltfragen), 1994. Umweltgutachten, Metzler-Poeschel, Stuttgart, 380pp.

Ten Brink, B., 1991. The AMOEBA approach as a useful tool for estimating sustainable development, In: O. Kuik and H. Verbruggen (Editors), In Search for Indicators of Sustainable Development, Kluwer Academic, Dodrrecht, pp. 71–87.

Tress, B. and Tress, G., 2003. Scenario visualisation for participatory landscape planning – a study from Denmark. Landscape and Urban Planning 64, 161–178.

UBA (Umweltbundesamt), 1995. Methodik der produktbezogenen Ökobilanzen. Wirkungsbilanz und Bewertung, UBA Texte 23/95, Berlin.

WG (Working Group) Regionale Ökobilanz Pfaffenhofen, 2000. Regionale Ökobilanzen für eine umweltgerechte und nachhaltige Raumnutzungsplanung auf mittlerer Maßstabsebene, Final report to the Deutsche Bundesstiftung Umwelt (in German), unpublished.

Zeisel, J., 1998. Grauwassernutzung; in Innovation Betriebs- und Grauwassernutzung, Schriftenreihe des fbr 3, Frankfurt.

ENVIRONMENTAL SECURITY AS RELATED TO SCALE MISMATCHES OF DISTURBANCE PATTERNS IN A PANARCHY OF SOCIAL-ECOLOGICAL LANDSCAPES

GIOVANNI ZURLINI*, IRENE PETROSILLO,
AND NICOLA ZACCARELLI
Landscape Ecology Laboratory, Department of Biological and Environmental Sciences and Technologies, University of Salento, Lecce, Italy

KURT H. RIITTERS
US Forest Service, Southern Research Station, Research Triangle Park, North Carolina, USA

Abstract. Environmental security, as the opposite of environmental fragility (vulnerability), is multilayered, multi-scale and complex, existing in both the objective realm of biophysics and society, and the subjective realm of individual human perception. For ecological risk assessments (ERAs), the relevant objects of environmental security are social-ecological landscapes (SELs). ERAs, in this case, are less precise than traditional ERAs, but provide results that are more comprehensive and understandable by stakeholders. In this paper, we detect and quantify the scales and spatial patterns of human land use as ecosystem disturbances at different hierarchical levels in a panarchy of SELs by using a conceptual framework that characterizes multi-scale disturbance patterns exhibited on satellite imagery over a four-year time period in Apulia (South Italy). Multi-scale measurements of the composition and spatial configuration of disturbance are the basis for evaluating fragility through multi-scale disturbance profiles, and the identification of scale mismatches revealed by trajectories diverging from the global profile to local spatial patterns. Scale mismatches of disturbances in space and time determine the role of land use as a disturbance source or sink, and may govern the triggering of landscape changes affecting regional biodiversity. This study clarifies the potential roles

* Corresponding Author: Professor Giovanni Zurlini, Landscape Ecology Laboratory, Department of Biological and Environmental Sciences and Technologies, University of Salento, Ecotekne, Prov.le Lecce-Monteroni, 73100 Lecce, Italy. Phone: +390832298886, Fax: +390832298626, e-mail: giovanni.zurlini@unile.it

I. Petrosillo et al. (eds.), Use of Landscape Sciences for the Assessment of Environmental Security, 383–398.

for environmental security of natural areas and permanent cultivations (olive groves and vineyards) in buffering Mediterranean landscape disturbance dynamics and compensating for disturbances across the whole panarchy of Apulia, allowing for potential landscape planning of disturbance.

Keywords: Environmental security; multi-scale disturbance; scale mismatches; social-ecological landscapes

1. Environmental security and ecological risk assessment (ERA)

The major challenge of environmental security concerns the global environmental change, focusing on the interactions between ecosystems and mankind, the effects of global environment change on environment degradation, the effects of increasing social request for resources, and the erosion of ecosystem services and environmental goods. Because land use change by humans is one of the major factors affecting global environmental change (Millennium Ecosystem Assessment, 2003), the question then arises as to how such environmental stresses and the associated risks might vary geographically or evolve over time. Environmental security addresses the risks to, or fragility (vulnerability) of ecosystem goods and services, as well as the subjective perception of those risks (Petrosillo et al., 2006; Zurlini and Müller, in press).

Environmental security, as the opposite of environmental fragility, is multi-layered, multi-scale and complex, existing in both the objective biophysical and social realms, and the subjective realm (Morel and Linkov, 2006). The relevant objects of environmental security are complex, adaptive systems that, in the real geographic world, are social-ecological landscapes (SELs). Those systems are usually designed as made up of two main components: the social, characterized by human intent, and the ecological, arising without intent. However, those two components are often very hard to distinguish because they have been interacting and coevolving historically, and society has always shaped the ecological component of SELs. Therefore, we can address environmental security more appropriately in terms of SEL security.

The subjective perception of security is fundamental at all levels of human organization, from the individual to government entities, and a “threat” is an abstract concept existing in the domains of feelings and cognition. Security is value laden, and related to our normative systems that today recognize concepts like ecosystem functions and services, ecosystem integrity and sustainability as fundamental values for the survival and well-being of mankind.

A fundamental difference between environmental security and ecological risk assessments (ERAs) is that the goal of ERAs is usually restricted to informing risk management decisions in the objective realm, focusing on the relationships between stressors (e.g. a chemical) and ecological effects at different organization levels (EPA, 1998). The animal-toxicity paradigm (Lackey, 1994) is still the most commonly used approach in an ERA, because it is easy to use and to understand, and because a large database exists for many chemicals and biological species. An ERA that estimates likelihoods of specific ecological effects is conceptually equivalent to an assessment of the cancer risk posed by some human health threat (Suter, 1993). This paradigm works best for chemicals (Suter and Loar, 1992) and it assumes that responses of a simple surrogate are adequate to represent responses at the landscape level; it can be precise and reliable, but with a narrow range of inference because of the simple surrogates used.

There is a growing awareness that a much greater ecological realism must be achieved by ERAs for attaining more informed management decisions (cf. Suter, 1995). Thus, for instance, ERAs should make better use of ecological information such as landscape features to generate spatially explicit estimates of exposure to environmental stressors, e.g. invasive species and physical disturbance (Kapustka, 2005). However, even holistic ERAs seldom address the integrated evidence of the entire complex hierarchical pattern and composition of real social-ecological landscapes, in terms of scaling properties of land use and pertinent anthropogenic disturbances (Zurlini et al., 2004).

The actual object of holistic ERAs should be a real-world social-ecological landscape. ERAs should contribute to the objective evaluation of environmental security and provide less precise results but more comprehensive and understandable by stakeholders (Shrader-Frechette, 1998). This would help people to focus on landscape systems instead of surrogates or proxies, and to recognize that SEL systems are open, hierarchically structured, and self-organizing, with historical trajectories, memory and learning capabilities, and with different processes dominating at different scales (Kay, 2000; Gunderson and Holling, 2002).

In summary, the development of new integrated system-specific evaluation and prediction models for environmental security at multiple scales, framed in terms of both subjective and objective observable quantities in the geographical real world domain, is necessary to formulate and evaluate ideas relevant to environmental security in SELs. Towards this goal, we exercise an evaluation framework with real landscape disturbances and demonstrate its interpretive power by examining actual disturbance maps relative to land use for a panarchy of SELs in Apulia, an administrative region in southern Italy. We exemplify concepts and methods with reference to the recent works of Zurlini et al. (2006;

in press) in the framework of potential environmental security evaluation with a view towards understanding how disturbances might impact biodiversity and ecosystem services through land use and habitat modification. Even though we exercise the framework based only on the objective dynamics of land use and land cover, we believe this framework can represent a common basis for assessing security of SELs both objectively and subjectively, at all levels of human organization, by replacing the traditional interpretation of results in ecological terms with an alternate interpretation in terms of environmental security.

2. Panarchy of social-ecological landscapes

SELs are organized in a panarchy of nested levels of organization, which draws on the notion of hierarchies of influences between embedded scales (Gunderson and Holling, 2002). A panarchy is a nested hierarchy of systems where each system follows an adaptive cycle and interacts with other levels through top-down or bottom-up connections. One of the essential features of the panarchy is that it turns hierarchies into dynamic structures. Individual levels have non-linear multi-stable properties while can be stabilized or destabilized through critical connections between levels.

Understanding environmental security in SELs requires understanding how the actions of humans as a keystone species (*sensu* O'Neill and Kahn, 2000) shape the environment across a range of scales in a panarchy of SELs that take into account the scales and patterns of human land use as ecosystem disturbances (e.g. Figure 1). Decision hierarchies of social systems are intertwined with the hierarchies found at the ecosystem or landscape level (Gunderson and Holling 2002). Anthropogenic disturbances such as changes in land use are determined by the social components of SELs which consist of groups of people organized in a hierarchy at different levels (e.g. household, village, county, province, region, and nation). Within this panarchy (Gunderson and Holling, 2002), the participants have differing views as to which system states are desirable at each level. Any given land use system in the panarchy is likely to overlap multiple ownership and jurisdictional boundaries, and fall under at least three levels of administrative decision and control (e.g. Figure 2). Social-ecological systems may have different dynamics when compared to the ecological component alone because the social domain contains the element of human intent. Thus, management actions can deliberately avoid or seek the crossing of actual and perceived thresholds (Walker et al., 2006). It is not clear whether a common framework of system dynamics could be used to examine and explain both social and ecological systems. Europe is a good place to test models because European landscapes are the result of consecutive reorganizations of the land for a long time to adapt uses and spatial structures to meet changing societal demands (Antrop, 2005). Human influence dominates landscape dynamics in space and time (O'Neill and Kahn, 2000), thus defining limiting constraints at "higher scales" and altering the detailed functioning of ecological processes at "lower scales". Land use decisions affect both ecological and social structures and processes, and vice versa.

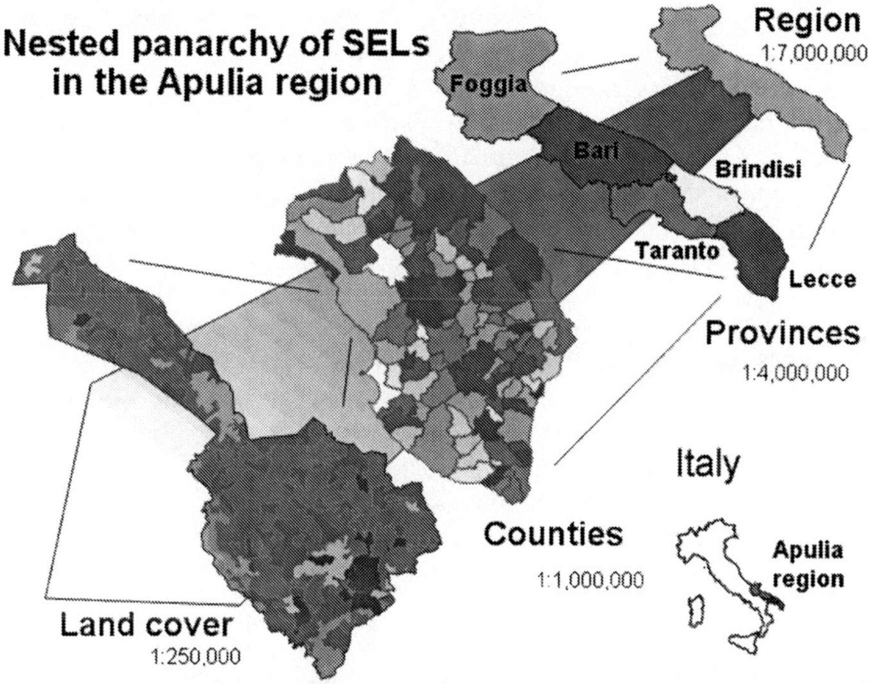

Figure 1. An example of a panarchy of nested SELs for Apulia, an administrative region in southern Italy. Three main levels of governance hierarchy can be identified (one region, five provinces and 258 counties) embodying different social, economic, and cultural constraints. The entire region and each subregion can be described in terms of their unique social-ecological landscapes based on land use composition.

We hypothesize that the characteristic scales of particular phenomena like anthropogenic changes should entrain and constrain ecological processes, and be related to the scales of human interactions with the biophysical environment. If the patterns or scales of human land use change, then the structure and dynamics of SEL as a whole can change accordingly, leading to transitions between alternative phases, when the integral structure of the systems is changed (Kay, 2000; Li, 2002).

In human-driven landscapes, evaluating the disturbance patterns of land use at multiple scales clearly has potential for quantifying and assessing environmental condition, processes of land degradation, subsequent impacts on natural and human resources in SELs, and their consequences on environmental security.

3. Disturbance of what and to what

A fundamental difficulty with social-ecological systems is that their complexity makes it difficult to forecast the future with any sense of reliability. One way of dealing with this problem is to look retrospectively at the observed trends of effects caused by past exposure to stressors and, on this basis, to create future scenarios, taking into account the anticipated changes of the driving forces at work and of their consequent disturbances. The future system trajectories can at least be compared to each other to assess whether management scenarios have more or less effect on trajectories, that is, whether proposed actions will move the system in expected directions at expected rates.

We use land cover change as a measure of disturbance and historical stress. Disturbances have been defined as "any relatively discrete event in space and time that disrupts ecosystem, community, or population structure and changes resources, substrates, or the physical environment" (Pickett and White, 1985). Land cover change is a disturbance because converting forest to agriculture land, or vice versa, alters soil biophysical and chemical properties and associated animal and microbial communities, and agricultural practices such as crop rotation or fire alter the frequency of these disturbances. New land cover types can be juxtaposed and shifted within increasingly fragmented remnant native land cover types, and changes in the structure of the landscape can disturb nutrient transport and transformation (Peterjohn and Correll, 1984), species persistence and biodiversity (Fahrig and Merriam, 1994; With and Crist, 1995), and invasive species (With, 2004).

To detect change, we applied a standardized differencing change detection technique based on the use of the NDVI "greenness" index (Normalized Difference Vegetation Index; Pettorelli et al., 2005; Zurlini et al., 2006). From a set of Landsat TM 5 images for June 1997 and June 2001, after registration, calibration, and atmospheric correction, we derived NDVI values for each pixel and calculated the standardized difference NDVI image. A pixel is considered to be "changed" or "disturbed" whenever it falls within the upper or lower percentile of 5% of the empirical distribution of the standardized difference values (Zurlini et al., 2006). In other words, we define disturbance as *any* detectable alteration of land cover reflecting even tiny and relatively frequent vegetation changes which are mainly assignable to fast human-driven processes. This perspective is different from classical land cover mapping that would ignore, for instance, crop rotation because agricultural fields can be fallow one year and planted the next, and still be labeled as "agricultural fields." In this study, a change in a farming practice is like the use of a prescribed fire which most ecologists would agree is a disturbance even if it did not change the land cover. In the context of environmental security, the justification is that observed changes in NDVI can clearly demonstrate that not only agricultural fields could be more dynamic than other types of land cover systems, but also that agricultural practices like, for instance, fire could spread disturbance agents in the landscape to other neighboring land uses like natural areas or permanent cultivations.

Thus, land uses and covers within SEL mosaics not only might be disturbed by various agents, but also might act as a "source" or a "sink" as to the potential spread of disturbance to neighbor areas, as it may occur because of disturbance agents like, for instance, fire, pests, disease, alien species, urbanization. In Apulia, typical contagious disturbances are related to land use or land cover and reflect changes associated with urban sprawl, conversion of grasslands to cultivation fields, new olive grove tillage, and farming practices such as fire, grazing, and crop rotation. Unlike other disturbances such as storms and hurricanes, or clear cutting, the extent and duration of contagious disturbance events in Apulia are dynamically determined by the interaction of the disturbance with the landscape mosaic.

4. Disturbance patterns at multiple scales

Patterns can be measured in many ways, but many authors have suggested focusing on a few key measures (Li and Reynolds, 1994; Riitters et al., 1995). Li and Reynolds (1995) suggest that the two most fundamental measures of pattern are composition and configuration. Therefore, we characterize landscape patterns of disturbance in terms of the amount (composition) and spatial arrangement of disturbance (configuration or connectivity).

We make use of moving windows to measure composition (*Pd*, the proportion of disturbed pixels within a window) and configuration (*Pdd*; contagion as the proportion of shared edges between disturbed pixels on changed pixels edges within a window) of disturbance patterns at multiple scales (i.e. window sizes), as detected on satellite imagery. The measurements were made for each pixel at multiple scales by using 10 square arbitrary chosen window sizes in pixel units of 3, 5, 9, 15, 25, 45, 75, 115, 165, and 225 thus the window area ranges from 0.81 ha to 5852.25 ha. For each pixel a profile of *Pd* or *Pdd*

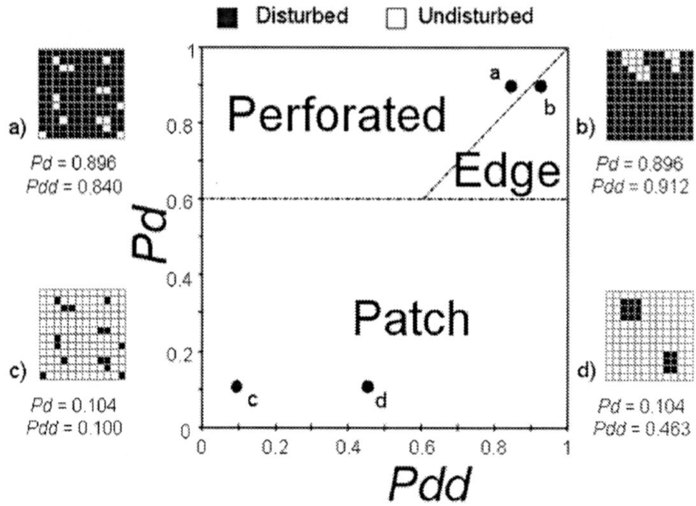

Figure 2. The graphical model used to identify disturbance categories from local measurements of *Pd* and *Pdd* in a fixed-area window. *Pd* is the proportion of disturbed and *Pdd* is disturbance connectivity (modified after Riitters et al., 2000). Four simple examples of binary landscapes (a, b, c, d) are presented by the side of the [*Pd, Pdd*] space for different combinations of composition and configuration: (a) highly disturbed but perforated by undisturbed areas (perforated disturbance), (b) highly disturbed but with clumped undisturbed areas (edge disturbance), (c) low-level and highly fragmented disturbance (spread disturbance), and (d) low-level and clumped disturbance (patchy disturbance). (Modified after Zurlini et al. 2006.)

is defined by the set of values measured at different window sizes. Profiles were aggregated (i.e. averaged) and a mean profile derived applying a broad land use type classification spanning the whole SEL mosaic except for urban regions. We considered four classes roughly coincident to the second level of the European CORINE classification (Heymann et al., 1994) and in particular: arable lands (CORINE code 2.1), permanent cultivations (CORINE code 2.2), heterogeneous agricultural area (CORINE codes 2.3 and 2.4) and natural areas (CORINE codes 3.1, 3.2, 3.3, and 4.1). In contrast to our earlier work (Zurlini et al. 2006), we include land use composition of the SEL by developing at multiple scales the mean accumulation disturbance profiles of each land use from each location.

The [*Pd, Pdd*] phase space (Figure 2) and the use of a convergence point (CP; an asymptotic point for a window exactly equal to the entire study region) to represent SELs (Zurlini et al., 2006), can be very useful to provide the appropriate dynamic representation of different SELs in the panarchy, as traced by their recent disturbance history. For any given location (pixel) in each land use, the trajectory converging to the CP in [*Pd, Pdd*] space describes the accumulation profile of disturbance pattern at increasing scales surrounding that location. If trends in [*Pd, Pdd*] space were similar for two different locations, then both locations have experienced in their surrounding landscapes the same "disturbance profiles" as characterized by the amount and configuration. For example, at a given geographic location, the trend in *Pd* with increasing window size can be interpreted with respect to the disturbances experienced by that location at different spatial lags (Figure 3). A small window with high *Pd* combined with a large window with low *Pd* implies a local heavy disturbance embedded in a larger region of lighter disturbance. Locations characterized by constant *Pd* over window size experience equal amounts of disturbance at all spatial scales.

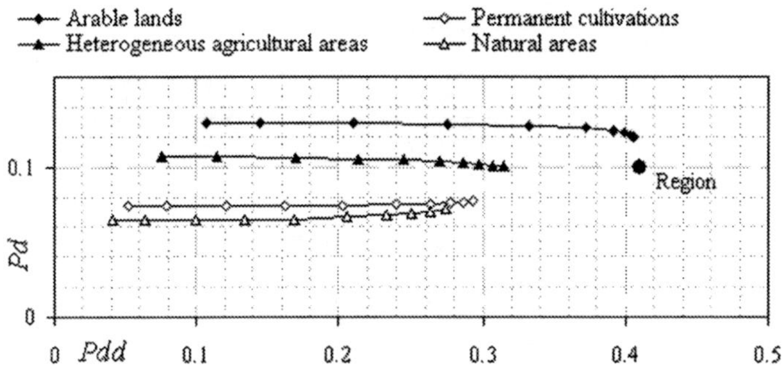

Figure 3. Trends of disturbance profiles at multiple scales (10 window sizes in increasing order from left to right), and relative convergence point (black dot) for four broad land use classes at the regional hierarchical level (see text).

The trajectories of disturbance accumulation profiles at multiple scales on the [*Pd, Pdd*] state space also indicate whether and where land use disturbances might act as a "source" or a "sink" across scales respect to their potential spread to neighbor areas. If a mean profile is always larger than the CP of reference and has a convex trend downwards to the CP (e.g. arable lands, Figure 3), land use acts as a potential disturbance source to the neighbor mosaic because of local heavy disturbance embedded in a larger region of fewer disturbances. Conversely, if a mean profile of a land use is below the CP with a concave trend upwards to the CP (e.g. natural areas, Figure 3), land use locations can be potentially affected by neighbor disturbances (sink) because of local low disturbance embedded in a larger region of heavy disturbances. Disturbance profiles at multiple scales for the four land uses in three different provinces of Apulia region, and province convergence points (CP) are shown in Figure 4.

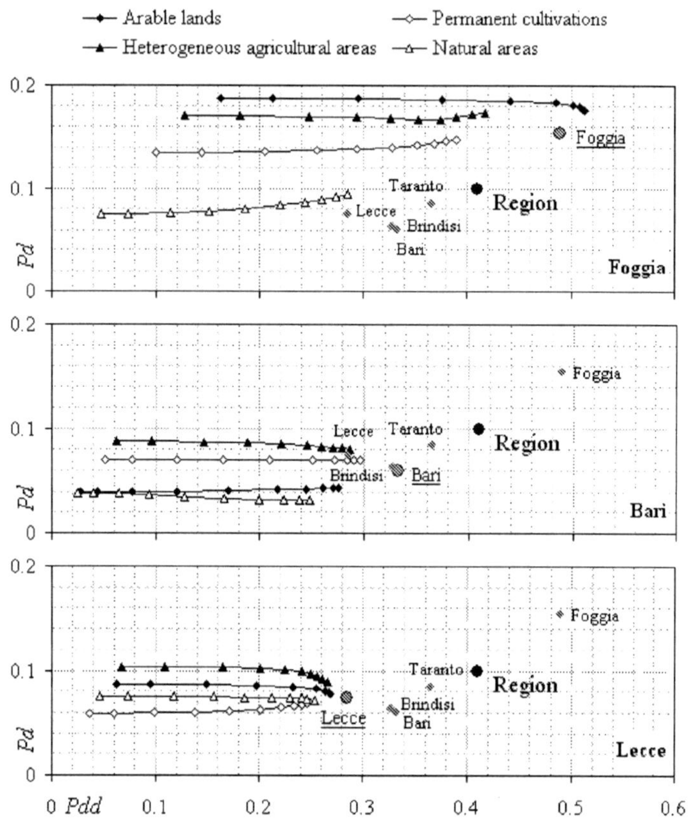

Figure 4. Trends of disturbance profiles at multiple scales (10 window sizes in increasing order from left to right) of the four land uses for three different provinces of Apulia are presented to show their reciprocal source–sink role. Convergence points for the five provinces and for the Apulia region are shown for comparison.

Theoretically, spatial "mismatches" are expected when the spatial scales of management and the spatial scales of ecosystem processes are not aligned, possibly leading to disruptions of the SEL, inefficiencies, and/or loss of important components of the ecological system (Cumming et al., 2006).

In practice, within SEL mosaics, each land use and land cover has its own disturbance due to human management, thus spatial scale mismatches in [*Pd, Pdd*] space can occur for differences in both disturbance accumulation profiles related to the management of different land uses and accumulation rate of disturbance clumping at different spatial lags. Any two geographic locations with the same accumulation trajectory in [*Pd, Pdd*] space experience the same multi-scale disturbance profile with no spatial scale mismatches, as it might occur in some cases for permanent cultivations and natural areas (Figures 3 and 4). Conversely, dissimilar trends imply differences in spatial profiles of disturbance with consequent scale mismatches of disturbance (Figures 3 and 4). Social processes that can lead to mismatches are primarily inherent in land occupancy, which constitutes the hierarchy of social institutions that run the allocation, use, and management of land resources (Figure 1).

The differences of *Pdd* values between window points tell an interesting story about the cross-scale spatial accumulation rate of disturbance clumping of each land use. Such differences are more pronounced and range from natural areas to arable land (Figure 3), meaning that fields have been merged and enlarged to enhance farming efficiency, resulting in almost homogeneously farmed landscapes (e.g. Foggia, Figure 4).

Arable lands and heterogeneous areas (source) generally show at the same scales not only higher disturbance composition (*Pd*), but also cross-scale contagion accumulation increments in disturbance higher than those for permanent cultivations and natural areas (sink).

Distances in the [*Pd, Pdd*] state space between two land use profiles at the same window size (scale; Figure 4) draw directly the attention to spatial scale mismatches of disturbance among land use that can lead to their reciprocal potential role as disturbance source or sink at the same and cross scales, with possible consequent changes in the structure and dynamics of SELs.

5. Discussion: environmental security at multiple scales

Because disturbances are inflicted at multiple scales, various species could be differentially affected by disturbances in the same place, and a potentially useful way to appreciate these differences is to look at how disturbances are patterned in space at multiple scales (Zurlini et al., 2006).

All land use disturbance trajectories in Apulia panarchy are located near the lower left corner in the [*Pd, Pdd*] pattern space (Figures 3 and 4), with a certain invariance of disturbance composition (*Pd*) at increasing disturbance clumping (*Pdd*). Land uses have distinct disturbance profiles at multiple scales with paths fairly parallel to the *Pdd* axis almost up to the CP value of entire region, and with increasing disturbance composition (*Pd*) usually ranging from natural areas to arable land (Figures 3 and 4).

For an environmental security interpretation of the [*Pd, Pdd*] space, we have to look not only at the disturbance accumulation profiles at multiple scales (context) of land use and cover locations, but also at the role those profiles might play as “source” or “sink” across scales within SEL land use mosaics respect to the potential spread of disturbance to neighbor areas.

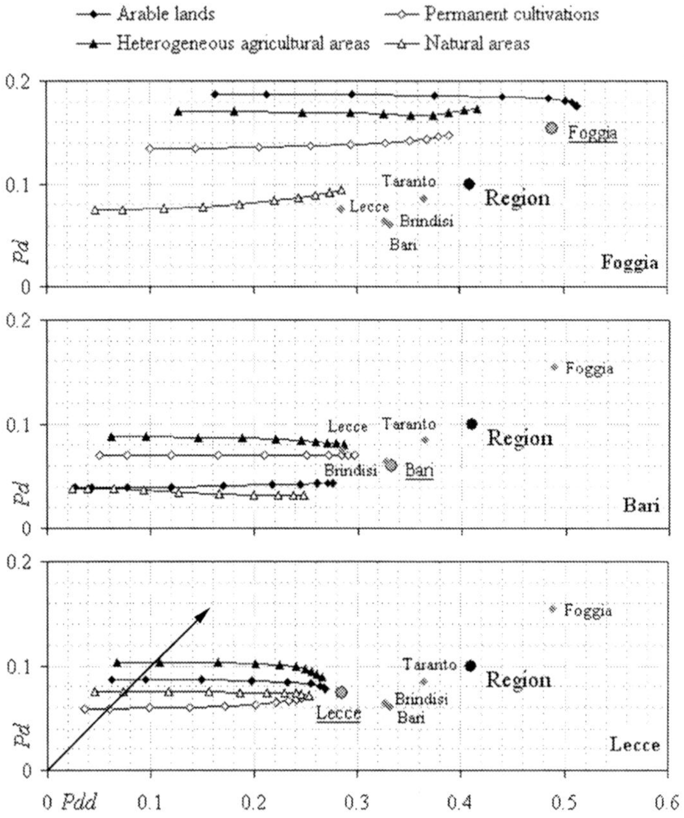

Figure 5. Fragility estimates for the five different provinces spanning Apulia as indicated by convergence points, and comparison of disturbance accumulation profiles at multiple scales (10 window sizes in increasing order from left to right) of the same land use within the same province. The CP for the entire Apulia region is shown for comparison. The arrow indicates the direction of fragility (see text).

As to the first side, the [*Pd, Pdd*] pattern space has already been interpreted in terms of fragility (Zurlini et al., 2006), where fragility is highest for scale domains where disturbance is most likely and clumped for trajectories of location clusters, independently of single location membership to a definite land use. We can identify a gradient of fragility, with fragility increasing with both *Pd* and *Pdd* from the lower left corner to the upper right corner in Figure 2.

The same interpretive framework can be used to compare portions of the SEL such as two provinces in the [*Pd, Pdd*] space (Figure 4), as to their CP, given by its overall *Pd* and *Pdd* values, without taking into account single disturbance patterns of the four land uses. In this way, provinces can be ranked according to relative fragility, and the province of Foggia turns out to be the most fragile (Figure 4). We can also compare the fragility of each single land use at multiple scales among different provinces by looking at its disturbance profiles (Figure 5). In this case, differences in disturbance due to traditional, low-intensity, local land use practices of agriculture and forestry can be revealed, which have greatly promoted habitat diversity in the European human-dominated landscapes during the last centuries.

However, we cannot use the [*Pd, Pdd*] space to interpret and compare disturbance profiles of single land uses within the region or a province, because further factors integral to each specific land use other than disturbance patterns determine its fragility like, for instance, habitat sensitivity (Zurlini et al., 1999), ecosystem service and natural capital values (Costanza et al., 1997).

Natural areas and permanent cultivations are usually thought to have higher natural capital value, and higher potential for regulating landscape dynamics and compensating for disturbances in the SELs of Apulia. Consequently, in an environmental security framework, natural areas and permanent cultivations must be considered intrinsically more fragile (sink) than arable lands which generally act as potential source that could affect neighboring land uses (Figures 3, 4, and 5).

6. Conclusions

This study points out that management of disturbance in the study region will primarily depend more on broader-scale than local-scale patterns of the drivers of disturbance (Figure 3), and clarifies how natural areas and permanent cultivations (olive groves and vineyards) will act in the interplay of disturbance patterns within SELs, regulating landscape mosaic dynamics and compensating for disturbances across scales in South Italy. Both land uses act as buffering mechanisms for land use disturbance, thus providing essential indirect ecosystem services, with consequences likely for regional biodiversity management which requires ecological knowledge of both natural areas and their surroundings.

The [*Pd, Pdd*] space helps to draw attention to spatial scale mismatches among land uses for disturbance accumulation profiles which can determine their reciprocal role as disturbance source or sink at across scales, because of their potential spread to neighbor areas with possible consequent changes in the structure and dynamics of SELs. The reading of [*Pd, Pdd*] space in terms of fragility gradients (or its reverse, environmental security), where fragility is highest anywhere disturbance regime is most likely and clumped, is justified by evidence coming, for instance, from metapopulation simulations which show that increasing spatial aggregation of the disturbance regime always decreases habitat occupancy of species, increases extinction risk, and expands the threshold amount of habitat required for persistence, with more marked effects on species with short dispersal distances (Kallimanis et al., 2005). This is particularly central also to the dispersal of alien species and therefore to the spatial distribution of risk of competition from alien species. Poor dispersers spread more in landscapes in which disturbances are concentrated in space ("contagious" disturbance), whereas good dispersers spread more in landscapes where disturbances are small and dispersed ("fragmented" disturbance) (With, 2004).

However, we acknowledge that agricultural land use intensification might not only mean a decrease in habitat occupancy with consequent higher extinct-tion, but it could also make occasionally more resources available to enhance populations of some species, since the higher productivity of land use compared with generally less productive natural systems may provide more resources such as vegetation biomass, and fruits for birds, mammals and butterflies (Tscharntke et al., 2005).

Current approaches to conserving biodiversity may benefit by incorporating greater understanding of how people and nature interact within complex adaptive systems (Gunderson and Holling, 2002) like SELs, so that scale mismatches of different land uses in land tenure and thresholds of potential concern for environmental security can be identified and managed for a key set of ecological response variables. That could be the basis for intentionally planning and managing the adaptability of the SEL, which is arguably the key to human management of environmental security

References

Antrop, M., 2005. Why landscapes of the past are important for the future. Landscape and Urban Planning 70, 21–34.

Costanza, R., d'Arge, R., de Groot, R., Faber, S., Grasso, M., Hannon, B., Limburg, K., Naeem, S., O'Neill, RV., Paruelo, J., Raskin, R.G., Sutton, P., and van den Belt, M., 1997. The value of the world's ecosystem services and natural capital. Nature 387, 253–260.

Cumming, G.S., Cumming, D.H.M., and Redman, C.L., 2006. Scale mismatches in social-ecological systems: causes, consequences, and solutions. Ecology and Society 11(1), 14. [online] URL: http://www.ecologyandsociety.org/vol11/iss1/art14/

Environmental Protection Agency, US (EPA) (Editors), 1998. Guidelines for ecological risk assessment. EPA/630/R095/002F. Risk Assessment Forum, Washington, DC.

Fahrig, L. and Merriam, G., 1994. Conservation of fragmented populations. Conservation Biology 8, 50–59.

Gunderson, L.H. and Holling, C.S. (Editors), 2002. Panarchy: Understanding Transformations in Human and Natural Systems. Island Press, Washington, DC.

Heymann, Y., Steenmans, C., Croissille, G., and Bossard, M., 1994. Corine Land Cover. Technical Guide. Office for Official Publications of the European Communities, Luxembourg.

Kallimanis, A.S., Kunin, W.E., Halley, J.M., and Sgardelis, S.P., 2005. Metapopulation extinction risk under spatially autocorrelated disturbance. Conservation Biology 19(2), 534–546.

Kapustka, L.A. 2005. Assessing ecological risks at the landscape scale: opportunities and technical limitations. Ecology and Society 10(2), 11. [online] URL: http://www.ecologyandsociety.org/vol10/iss2/art11/

Kay, J.J., 2000. Ecosystems as self-organizing holarchic open systems: narratives and the second law of thermodynamics. In: S.E. Jørgensen and F. Müller (Editors), Handbook of Ecosystem Theories and Management, CRC Press–Lewis Publishers, pp. 135–160.

Lackey, R.T., 1994. Ecological risk assessment. Fisheries 19(4), 14–18.

Li, B.L., 2002. A theoretical framework of ecological phase transitions for characterizing tree-grass dynamics. Acta Biotheoretica 50, 141–154.

Li, H. and Reynolds, J.F., 1994. A simulation experiment to quantify spatial heterogeneity in categorical maps. Ecology 75, 2446–2455.

Li, H. and Reynolds, J.F. 1995. On definition and quantification of heterogeneity. Oikos 73, 280–284.

Millennium Ecosystem Assessment (Editor), 2003. Ecosystems and Human Well-being. Island Press, Washington, DC.

Morel, B. and Linkov, I., 2006. Environmental Security and Environmental Management – the Role of Risk Assessment. Springer, Dodrrecht, supported by Nato Security through Science.

O'Neill, R.V. and Kahn, J.R., 2000. Homo economicus as a keystone species. Bioscience 50, 333–337.

Peterjohn, W.T. and Correll, D.L., 1984. Nutrient dynamics in an agricultural watershed. Observations on the roles of a riparian forest. Ecology 65, 1466–1475.

Petrosillo, I., Zurlini, G., Corlianò, M.E., Zaccarelli, N., and Dadamo, M., 2006. Tourist perception of recreational environment and management in a marine protected area. Landscape and Urban Planning 79, 29–37.

Pettorelli, N., Vik, J.O., Mysterud, A., Gaillard, J.-M., Tucker, C.J. and Stenseth, N.C., 2005. Using the satellite-derived NDVI to assess ecological responses to environmental change. Trends in Ecology and Evolution 20, 503–510.

Pickett, S.T.A. and White, P.S. (Editors), 1985. The ecology of natural disturbance and patch dynamics, Academic Press, Orlando, FL.

Riitters, K.H., O'Neill, R.V., Hunsaker, C.T., Wickham, J.D., Yankee, D.H., Timmins, S.P., Jones, K.B., and Jackson, B.L., 1995. A factor analysis of landscape pattern and structure metrics. Landscape Ecology 10, 23–29.

Riitters, K.H., Wickham, J.D., O'Neill, R.V., Jones, K.B., and Smith, E.R., 2000. Global-scale patterns of forest fragmentation. Conservation Ecology 4(3). [online] URL: http://www.consecol.org/Joumal/vol4/iss2/art3/

Shrader-Frechette, K.S., 1998. What risk management teaches us about ecosystem management. Landscape and Urban Planning 40, 141–150.

Suter, G.W., II (Editor), 1993. Ecological Risk Assessment. Lewis Publishers, Boca Raton, FL.

Suter, G.W., II, 1995. Adapting ecological risk assessment for ecosystem valuation. Ecological Economics 14, 137–141.

Suter, G.W. and Loar, J.M., 1992. Weighing the ecological risks of hazardous waste sites: the Oak Ridge case. Environmental Sciences and Technologies 26(3), 432–438.

Tscharntke, T., Klein, A.M., Kruess, A., Steffan-Dewenter, I., and Thies, C., 2005. Landscape perspectives on agricultural intensification and biodiversity – ecosystem service management. Ecology Letters 8, 857–874.

Walker, B.H., Gunderson, L.H., Kinzig, A.P., Folke, C., Carpenter, S.R., and Schultz, L., 2006. A handful of heuristics and some propositions for understanding resilience in social-ecological systems. Ecology and Society 11(1), 13. [online] URL: http://www.ecologyandsociety.org/vol11/iss1/art13/

With, K.A., 2004. Assessing the risk of invasive spread in fragmented landscapes. Risk Analysis 24(4), 803–815.

With, K.A. and Crist, T.O., 1995. Critical thresholds in species' responses to landscape structure. Ecology 76, 2446–2459.

Zurlini, G., Amadio, V., and Rossi, O., 1999. A landscape approach to biodiversity and biological health planning: the map of the Italian Nature. Ecosystem Health 5, 296–311.

Zurlini, G., Zaccarelli, N., and Petrosillo, I., 2004. Implementing an integrated framework for ecological risk assessment at the landscape level. EcoSys Suppl.-Bd. 42, 4–15.

Zurlini, G., Riitters, K.H., Zaccarelli, N., Petrosillo, I., Jones, K.B., and Rossi, L., 2006. Disturbance patterns in a social-ecological system at multiple scales. Ecological Complexity 3(2), 119–128.

Zurlini, G. and Müller, F., in press. Environmental security and system ecology. In: S.E. Jørgensen (Editor), Encyclopedia of Ecology.

FOSTERING ECOSYSTEM SERVICES' SECURITY BY BOTH OBJECTIVE AND SUBJECTIVE ANALYSES: THE CASE OF A NATURAL PROTECTED AREA IN SOUTHERN ITALY

IRENE PETROSILLO*, NICOLA ZACCARELLI, TEODORO SEMERARO, AND GIOVANNI ZURLINI
Laboratory of Landscape Ecology – Department of Biological and Environmental Sciences and Technologies – University of Salento, Lecce, Italy

Abstract. The maintenance of ecosystem services is the basic guarantee of environmental security that, in an objective sense, aims to evaluate the level of threats to actual acquired goods and services and, in a subjective sense, the level of consciousness and fear that such services will be attacked and possibly lost. To this purpose the aim of this research is: (1) to assess the temporal dynamics of land use and land cover mosaics, and indirectly of ecosystem services, using the economic valuations as surrogates; (2) to verify if the environmental conservation policies can foster ecosystem services; and, since it is still necessary to foster users' perception of ecosystem services in order to reduce their fragility, (3) to compare the results coming from objective and subjective analyses. This is overriding in the case of tourism, where the attarctiveness of tourist destinations will depend on the maintenance of recreational ecosystem services based on both natural and cultural heritages. This research highlights the need for a dynamic and continuous inter-comparison between objective and subjective analyses in order to reduce progressively their possible discordance and, consequently, increase environmental security for a more effective adaptive management of ecosystem goods and services.

Keywords: Ecosystem goods and services; environmental security; socioeconomic valuation

* Corresponding Author: Dr. Petrosillo Irene, Laboratory of Landscape Ecology – Department of Biological and Environmental Sciences and Technologies Ecotekne – University of Salento, Prov.le Lecce-Monteroni, 73100 Lecce (Italy), Phone: +39 0832 298896, Fax: +39 0832 298626, e-mail: irene.petrosillo@unile.it

I. Petrosillo et al. (eds.), Use of Landscape Sciences for the Assessment of Environmental Security, 399–411.

1. Environmental Security: a historical perspective

The relation between the environment and the security of humans and nature has been the object of much research and the subject of many publications in recent decades, but it is only lately becoming an important focus of international environmental policy. The reason is due to the overview provided by environmental security observing that: (1) the environment is the most transnational issue, and its security is an important dimension of peace, national security, and human rights, and (2) over the next 100 years, one third of current global land cover will be transformed, with the world facing increasingly hard choices among consumption, ecosystem goods and services, conservation and management.

The early beginnings of environmental security research are rooted in the discussions of population and the scarcity of resources that began 200 years ago, when Thomas Malthus identified the population growth as the main cause of scarcity and famine (Malthus, 1798). Malthus' essay gave rise to a still ongoing debate between the "Malthusians" or "Neo-Malthusians" (Ehrlich and Ehrlich, 1990), and the "Optimists" (Simon, 1981). The first is concerned about population growth and scarcity that it will bring for a change in mankind's patterns of behavior, whereas the second discusses that human life has the potential to adapt and always find new solutions to scarcities. Afterwards, this debate was placed within a framework of environmental degradation and then explicitly linked to violent conflicts, as a consequence of the general broadening of the concept of security.

McNamara (1968) linked security to the economic development, demonstrating that human security is related directly to the security of the developing world; whereas in 1987 the Brundtland Commission in its report stated that the whole notion of security, as traditionally understood, must be expanded to include the growing impacts of environmental stress at local, regional, national, and global levels (WCED, 1987).

The debate on environmental security began in the late 1980s and has always been quite intense, especially since 1994 (Dabelko, 2004) when it has received an increasing political interest. The debate on these issues is still ongoing, although there is no complete agreement on the way the environment can have a causal role. In recent years, environmental security issues have received increasing worldwide interest from governments, scientific institutions, intergovernmental organizations, and nongovernmental groups (UNEP, 2004).

Over the last decade, scholars have used case studies and other approaches to identify the ways in which environmental change can affect human security.

A useful representation is the "security diagram" (Alcamo et al., 2001). Data points in this diagram represent, for a particular country, the level of environmental pressures, mainly due to human activities, and corresponding environmental sensitivity, as intrinsic feature, in terms of ecosystem services' provision (Figure 1).

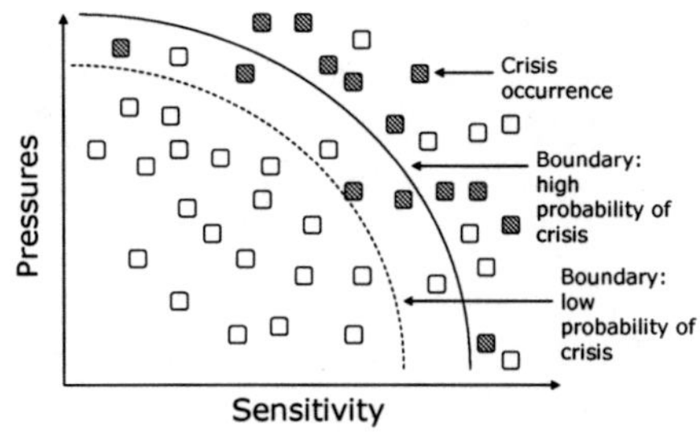

Figure 1. "Security diagram" where the data points represent the level of environmental pressures, due to human activities, and corresponding environmental sensitivity of a particular country, crisis events (data points shared gray) are crowded into the upper right part and a clear boundary between crisis and noncrisis zones is not expected. (Modified after Alcamo et al., 2001.)

Data points that are shaded gray indicate that an environmental crisis, due to the decline of ecosystem services, occurred at this time and in this country. The theoretical diagram shows that the higher the pressure, the more frequent is the occurrence of crisis events. Likewise, the higher the sensitivity, the more frequent will be the crises, because a lower pressure is required to cause a crisis. In the diagram crisis events would be crowded into the upper right part, as shown in Figure 1, however a clear boundary between the crisis and noncrisis zones is not expected.

The crisis state is conceptually similar to the fragility state provided by the conceptual model in Zurlini et al. (1999), where fragility at time t depends on the environmental sensitivity (α), as intrinsic feature of the system, and stresses (U), given by disturbance agents. So, where fragility is higher, there is also a higher likelihood that a crisis event occurs, and therefore a higher risk to environmental security.

2. Environmental security

Human well-being is complex and value-laden, context- and situation-dependent, and reflects social and personal factors such as age, gender, and culture. Despite this diversity and subjectivity, there is wide agreement that human well-being has five key, reinforcing components: the basic material needs for a good life, health, good social relations, security, and freedom and choice (Millennium Ecosystem Assessment, 2003). In particular, security includes the ability to gain access to natural resources, and to safely retain personal safety and physical property (Bloom and Canning, 2001), while freedom and choice includes the ability to acquire, experience, and select what someone likes, including ecosystem goods and services. There can be an asymmetric relationship between freedom and choice and the other components of well-being. For example, it is possible to have enough material goods to survive comfortably and yet to feel far from free. It is also possible to feel secure, to enjoy good social relations and health, and yet not be free (Sen, 1999). On the other side, making choices about the management of ecosystem services is an inherent feature of all human societies.

Since there have been strong linkages and reciprocal influences between the environment and mankind for a long time, we can address environmental security more appropriately in terms of security of socio-ecological systems (SESs; Walker et al., 2002). In the framework of security, we have to consider also the often-neglected regulatory aspect of social learning that should help avoid the mistakes of the past. For instance, on the security diagram (Figure 1) the likelihood of crisis boundary might shift, as society learns how to face and prepare for crises. Therefore, social learning can play an important role helping society to cope better with emergencies (Figure 2).

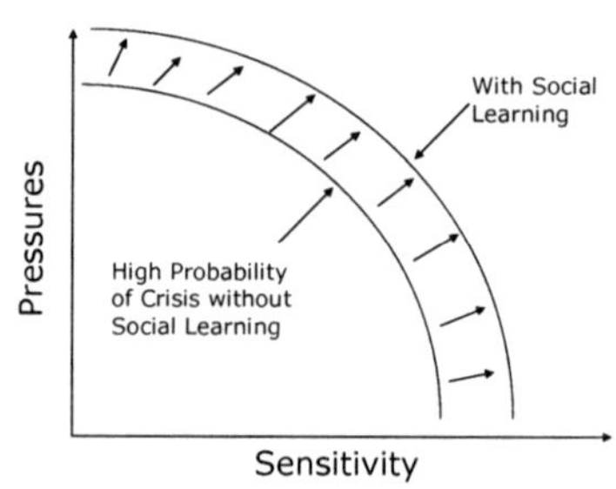

Figure 2. "Security diagram" with and without the effect of social learning showing that crisis boundary might shift as society learns how to face crises. (Modified after Alcamo et al., 2001.)

3. The vital role of ecosystem services

In literature several definitions of ecosystem services exist. In 1997 Daily defined ecosystem services as "the conditions and processes through which natural ecosystems, and the species that make them up, sustain and fulfill human life…", while concurrently Costanza et al. (1997) stated that they "are the benefits humans derive from ecosystem functions...". Afterwards, they have been considered as "the attributes of ecological functions that are valued by humans…" (Scott et al., 1998), and, more recently, as "ecological processes that benefit people directly (e.g. food) or indirectly (e.g. pollination)…" (Luck et al., 2003).

Ecosystem services are so fundamental to life that they are easy to take for granted, and so large in scale that it is hard to imagine that human activities could destroy them. Nevertheless, ecosystems are severely threatened because: (1) people generally are not well informed about the benefits that come from ecosystems and assume ecosystem services to be endlessly regenerating (Daily, 1997); (2) many ecosystems are public, meaning that private markets do not consider them (Heal, 2000); (3) the economic systems used in most countries emphasize values and preferences of individuals (consumer sovereignty) more than the values of communities (Costanza and Folke, 1997); (4) many ecosystem services are not considered at risk, so marginal losses are not given high importance; (5) there is a widespread assumption that ecosystem services can be replaced cost-effectively by technological alternatives; (6) there are few mechanisms or incentive for investment in ecosystem services (Heal, 2000).

Many human activities disrupt, impair, or reengineer ecosystems every day including: the application of pesticides, fertilizers, and animal wastes; pollution of land, water, and air resources; introduction of nonnative species; overharvesting fisheries; destruction of wetlands; erosion of soils; deforestation; and urban sprawl. Although a substantial understanding of many ecosystem services and the scientific principles underlying them already exists, there is still much to learn. The tradeoffs among different services within an ecosystem, the role of biodiversity in maintaining services, and the effects of long- and short-term disturbances are just some of the questions that need to be further explored. The answers to such questions will provide information that is critical to the development of management strategies that will protect ecosystems and help maintain the provisions of the services upon which we depend (Daily, 1997; Ecological Society of America, 1997).

Ecosystem goods and services play an important role in the context of SES security because they are the final product of a wide range of conditions and processes through which natural ecosystems, and the species that are part of them, help sustain and fulfill human life. Those services maintain the production of ecosystem goods, such as seafood, wild game, forage, timber, biomass fuels, natural fibers, and many pharmaceuticals, industrial products, and their precursors. The harvest and trade of these goods represent important and familiar parts of the human economy. In addition to the production of goods, ecosystem services support life through, for instance, purification of air and water, generation and preservation of soils and renewal of their fertility, partial stabilization of climate, provision of aesthetic beauty, and intellectual stimulation that lift the human spirit (Daily, 1997).

As a matter of fact, biodiversity plays an essential role in sustaining ecosystem goods and services, and avoiding social and economic crises that threat SES security. This places more emphasis on the protection of natural areas that serve the preservation of biodiversity and, indirectly, evade crisis and SES insecurity.

Ecosystem service maintenance is the basic guarantee of SES security that, in an objective sense, aims to evaluate the level of threats to actual acquired services (risk assessment) and, in a subjective sense, the level of awareness and fear that such services will be attacked and possibly lost (risk perception).

A comprehensive risk assessment of ecosystem services, in the objective sense, is very difficult to realize in practice given the number and kinds of ecosystem services provided, and because of their complexity that makes it difficult their evaluation and to forecast their future in any meaningful way. However, risk assessment has been and it is still one of the foremost themes underlying environmental and landscape science research.

On the other hand, the risk perception to ecosystem services is also one of the primary aspects to analyze, because people interact based on the meanings they assign to the world around them, and individuals respond to each other in terms of their perceptions. Even if people are "mistaken" in their interpretation of a situation, such interpretations have, nevertheless, *real* consequences. Following this idea, since the late 1960s a substantial body of literature relating to people perception of risk has developed, even if most of the studies refer to perception of natural hazards (Rohrmann, 1999). Much of the early work on risk perceptions concentrated on trying to develop linkages between perception and response in the belief that this would assist the development of risk communication tools; thus making it possible to "educate" the public by changing their perception of risk, bringing them closer to expert perceptions (Gough, 2000), therefore reducing insecurity.

People's perceptions of the risks depend upon a number of factors that may not seem relevant to experts, making technical assessments of the same risks, on how individuals and communities respond according to their perceptions and understanding of the risk; though the links may be complex (Rogers, 1997).

4. The economic valuation of ecosystem services

The main reason to worry about the loss of ecosystem services is that they diminish as ecosystems degrade. The question then is how valuable are these services and the answer to this question is compulsory to inform ecosystem management choices.

Economic valuation usually aims to measure ecosystem services in monetary terms, in order to provide a common metric by which the benefits of a variety of services provided by ecosystems can be quantified. The essence of most of the work on economic valuation of environmental and natural resources has been to find ways to measure benefits that do not enter the market, and so do not have any directly observable monetary comparison.

The economic approach focuses on the cost-benefit analysis of changes in ecosystem services resulting from management decisions, and it is directly relevant for policy. Such approach has been used to derive the total value of ecosystem services at a given time (e.g. Costanza et al., 1997) and to simulate the value of ecosystem services in an integrated Earth system model (Boumans et al., 2002). Efforts to estimate the total value of the services being provided by ecosystems, if conducted properly, can provide useful information on their contribution to well-being. At any given time, an ecosystem provides a specific flow of services, depending on the type of ecosystem, its condition in terms of resources, how it is managed, and its socioeconomic context. Assessing change in the economic value of ecosystem services, resulting from, for instance, two different kinds of management over time can be performed by estimating and comparing, in a spatially explicit way, the changes in value of ecosystem services under the current and the past management regimes.

To this purpose the aim of this research is: (1) to assess the temporal dynamics of land use and land cover mosaics, and indirectly of ecosystem services, using the economic valuations as surrogates; (2) to verify if the environmental conservation policies can foster ecosystem services; and, since it is still necessary to foster users' perception of ecosystem services in order to reduce their fragility, (3) to compare the results coming from objective and subjective analyses.

5. A case study: economic value of Torre Guaceto protected area

The application of the most recent economic approaches to estimate ecosystem services' values are based on the total economic value per hectare of ecosystem services for each biome, as proposed by Costanza et al. (1997), with the procedures developed by Kreuter et al. (2001) and Zhao et al. (2004).

In this study, we developed this approach for the protected area of Torre Guaceto, in southern Italy (Figure 3).

We compiled a time series of land use and land cover maps for the month of July for the years 1954, 1987, 1997, and 2004 by photo interpretation of 1 meter by 1 meter orthorectified aerial photos. We identified 16 terrestrial land cover classes to describe both spatial and temporal landscape dynamics as well as to assess changes in ecosystem service values based on Costanza et al.'s (1997) ecosystem services valuation biomes model. The most representative biome was used as a proxy for each land cover class (Table 1).

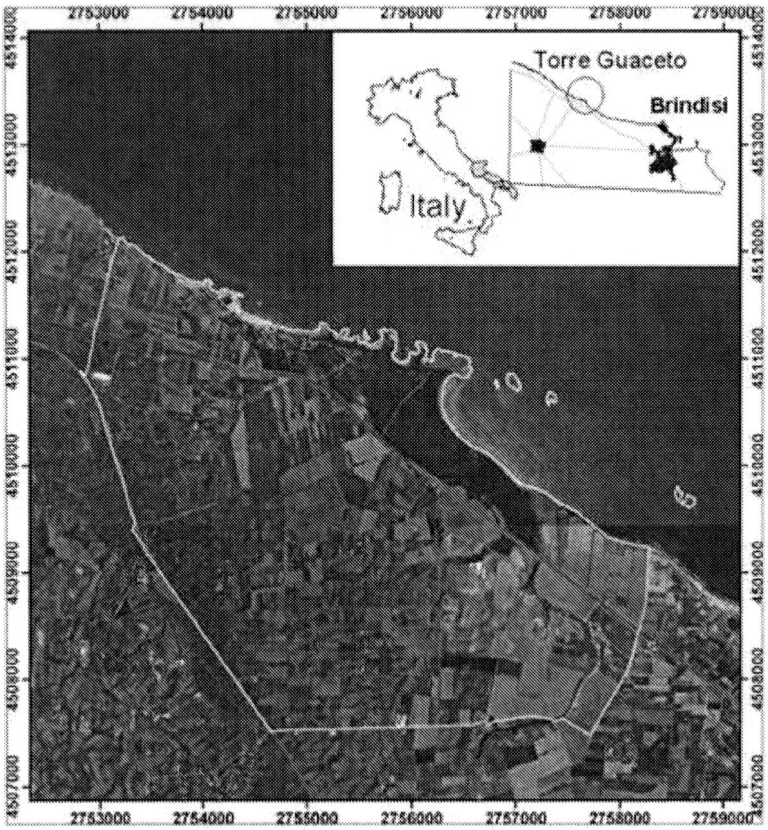

Figure 3. Study area: protected area of Torre Guaceto, southern Italy.

We estimated the Ecosystem Service Value (ESV) at time T using the following relationship:

$$ESV_T=\sum A_k x VC_k \quad (1)$$

where A_k is the area (ha); and VC_k is Costanza's Value coefficient ($\$ha^{-1}$ per year) for land use class "k". Change in ESV is estimated over time by calculating, for each land cover class, the difference between the estimated values for couples of years with reference to 1954, 1987, 1997, and 2004.

The results of change in the economic value of land use classes, for the overall period from 1954 to 2004, are shown in a spatially explicit way in Figure 4. More often, change seems to have affected boundary areas, except for the case of a large central area. The extension of change totals 352 ha (23% of the total area). Changes have been distinguished in gains and losses and, overall, they have caused a reduction in economic value of 138,416 $/year (6% of starting economic value).

TABLE 1. Land cover classes identified in the study area; the most representative biome used as a proxy for each land cover class, and the corresponding ecosystem services coefficient ($\$\times ha^{-1}$ per year), according to the table in Costanza et al., 1997, are also given.

Land cover class	The most representative biome	Ecosystem service coefficient ($\$\times ha^{-1}$ per year)
Urban Railroad Street	Urban	0
Rocky coast	Rock	
Sowable ground	Cropland	92
Uncultivated ground Grassland	Grassland	232
Mediterranean scrub Almond land Olive groves Reforestation area Tamarisk Juniper	Forest	969
Coastal beach	Coastal Beach	4,052
Periodically under inundation land Wetlands	Wetlands	14,785

Looking at the main land use classes involved, changes have been mainly associated with the reduction of "wetlands", converted into "forest" and "cropland", and with the increase of "forest" because of the conversion of "cropland". Even if, it would be noteworthy to present and discuss all the changes observed in each temporal range considered, those results are out of the scope of this paper.

The change in the total economic value of the protected area highlights a general decrease from 1954 to 1997, and an increase from 1997 to 2004. Despite during 1987 the protected area has been declared international important zone according to the Ramsar Convention, this has not apparently produced any relevant positive effect on the total economic value (Figure 5). Conversely, the increase in the total economic value from 1997 to 2004 could be probably attributed to a "conservation" effect determined by the institution of the protected area of Torre Guaceto in 2000 (Figure 5).

Figure 4. The results of spatially explicit change of classes' economic value from 1954 to 2004.

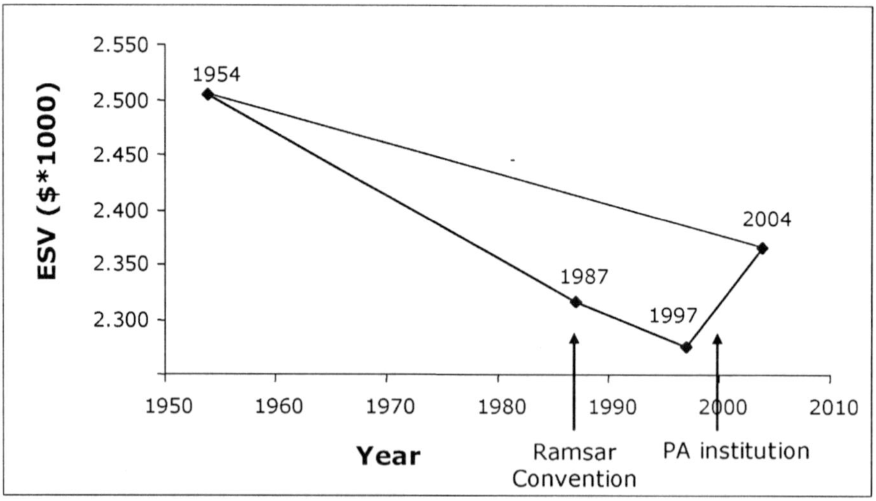

Figure 5. Changes in total economic value for each temporal range considered, highlighting the conservation effect linked to the institution of the protected area of Torre Guaceto in 2000.

6. Discussion and conclusions

The economic estimates of land uses and land covers can represent an effective operational surrogate for evaluating ecosystem goods and services in a comprehensive and objective way. Such surrogates are very useful when there are aspects of SES complexity that are not directly observable and measurable (Carpenter et al., 2005), like ecosystem services. A fundamental difficulty with ecosystem services is that their complexity makes it difficult to forecast their future in any meaningful way. Retrospective analyses help to better understand the trajectory of the SES given by, for instance, a protected area.

Besides, a spatially explicit assessment of losses in economic value determined by the relevant losses in land use and land cover, helps highlight the most fragile zones. Fragility is inversely related to environmental security and depends on factors integral to each specific land use and land cover, habitat sensitivity (Zurlini et al., 1999), ecosystem services and natural capital values (Costanza et al., 1997), but, for instance, also on local land use practices of agriculture and forestry.

The approach developed here allowed identifying zones where ecosystem services are at risk to be lost (fragile), mainly for the anthropogenic pressures exerted on the area. This has been, often, the result of environmental conservation policies that put attention to the last more than to the first disappearing hectare of ecosystem.

It is also important to foster the awareness relative to the perception of ecosystem services in order to include social learning into the environmental assessment. This is paramount in the case of tourism, where the attractiveness of tourist destinations will depend on the maintenance of recreational ecosystem services based on both natural and cultural heritages. For this purpose a survey, based on questionnaires, was carried out to investigate tourists' perception of sensitivity (i.e. natural value of landscape, quality of seawater, etc.) and pressures (i.e. production of waste, beach crowding, etc.), threatening ecosystem services. The results put in evidence that people were aware of the sensitivity of the protected area, but uninformed about their own pressures to the area (Petrosillo et al., 2007).

Thus a clear discordance between objective and subjective analyses has been apparently detected, that can be attributed to a not fully developed social awareness and learning in that area. Low levels of social learning (regulatory mechanism) could be dangerous for ecosystem services security, because can cause an increase of human pressures on the protected area.

However, in the protected area of Torre Guaceto the constraints imposed by the management authority to human activities apparently have already started to reduce human pressures, as exemplified in Figure 5 in terms of "conservation" effect.

An integrative risk assessment approach is essential to better guarantee environmental security at all the levels of SES organization. This study underlines the operational usefulness of surrogates in the context of environmental security assessment in SESs. The economic value approach is a holistic approach that, even though less precise and less reliable than others, provides results more comprehensive and understandable by stakeholders and decision-makers. The next step can be, hopefully, the development of integrated assessment frameworks addressing both economic multi-scale assessments of ecosystem services and social regulatory mechanisms such as social learning, to foster environmental security.

References

Alcamo, J., Endejan, M.B., Kaspar, F., and Rösch, T., 2001. The GLASS model: a strategy for quantifying global environmental security. Environmental Science & Policy 4, 1–12.

Bloom, D. and Canning, D., 2001. Cumulative causality, economic growth, and the demographic transition. In: N. Birdsall, A.C. Kelley, and S.W. Sinding (Editors), Population Matters: Demographic Change, Economic Growth, and Poverty in the Developing World, Oxford University Press, Oxford, pp. 164–197.

Boumans, R., Costanza, R., Farley, J., Wilson, M.A., Portela, R., Rotmans, J., Villa, F., and Grasso, M., 2002. Modeling the dynamics of the integrated earth system and the value of global ecosystem services using the GUMBO model. Ecological Economics 41, 529–560.

Carpenter, S.R., Westley, F., and Turner, M., 2005. Surrogates for resilience of social-ecological systems. Ecosystems 8, 941–944.

Costanza, R. and Folke, C., 1997. Valuing ecosystem services with efficiency, fairness and sustainability as goals. In: G.C. Daily (Editor), Nature's Services: Societal Dependence on Natural Ecosystems. Island Press, Washington, DC, pp. 49–70.

Costanza, R., d'Arge, R., de Groot, R., Farber, S., Grasso, M., Hannon, B., Limburg, K., Naeem, S., O'Neill, R.V., Paruelo, J., Raskin, R.G., Sutton, P., and van den Belt, M., 1997. The value of the world's ecosystem services and natural capital. Nature 387, 253–260.

Dabelko, G.D., 2004. The next steps for environment, population, and security. Environmental Change and Security Project Report 10. Woodrow Wilson International Center for Scholars, Washington, DC, pp. 3–7.

Daily, G.C. (Editor), 1997. Nature's Services, Societal Dependence on Natural Ecosystems. Island Press, Washington DC.

Ecological Society of America, 1997. Ecosystem Services: Benefits Supplied to Human Societies by Natural Ecosystems. Issues in Ecology N. 2, http://www.esa.org/sbi/issue2.htm.

Ehrlich, P. and Ehrlich, A. (Editors), 1990. The population explosion. Touchstone. Simon & Schuster, New York.

Gough, J., 2000. Perception of risk from natural hazards in two remote New Zealand communities. The Australasian Journal of Disaster and Trauma Studies, Vol. 2, http://massey.ac.nz/trauma/issues/2000-2/gough.htm.

Heal, G., 2000. Valuing ecosystem services. Ecosystems 3, 24–30.

Kreuter, U.P., Harris, H.G., Matlack, M.D., and Lacey, R.E., 2001. Change in ecosystem service values in the San Antonio area, Texas. Ecological Economics 39, 333–346.

Luck, G.W., Daily, G.C., and Ehrlich, P.R., 2003. Population diversity and ecosystem services. Trends in Ecology and Evolution 18, 331–336.

Malthus, T. (Editor), 1798. An essay on the principle of population, as it affects the future improvement of society with remarks on the speculations of Mr. Godwin, M. Condorcet, and other writers. Printed for J. Johnson, in St. Paul's ChurchYard, London.

McNamara, R. (Editor), 1968. The Essence of Security. Harper & Row, New York.

Millennium Ecosystem Assessment (Editors), 2003. Ecosystems and Human Wellbeing. Island Press, Washington, DC.

Petrosillo, I., Zurlini, G., Corlianò, M.E., Zaccarelli, N., and Dadamo, M., 2007. Tourist perception of recreational environment and management in a marine protected area. Landscape and Urban Planning 79, 29–37.

Rogers, G.O., 1997. Dynamic risk perception in two communities: risk events and changes in perceived risk. Journal of Environmental Planning and Management 40(1), 59–79.

Rohrmann, B. (Editor), 1999. Risk Perception Research – Review and documentation. Research Center Juelich, RC Studies 68.

Scott, M.J., Bilyard, G.R., Link, S.O., Ulibarri, C.A., and Westerdahl, H.E., 1998. Valuation of ecological resources and functions. Environmental Management 22(1), 49–68.

Sen, A.K. (Editor), 1999. Development as Freedom. Oxford University Press, Oxford.

Simon, J. (Editor), 1981. The Ultimate Resource. Princeton University Press, Princeton, NJ.

UNEP (Editors), 2004. Understanding environment, conflict and cooperation. UNEP, Division of Early Warning and Assessment (DEWA), UNEP, Nairobi. Available from http://wwics.si.edu/topics/pubs/unep.pdf

Walker, B., Carpenter, S., Anderies, J., Abel, N., Cumming, G.S., Janssen, M., Lebel, L., Norberg, J., Peterson, G.D., and Pritchard, R., 2002. Resilience management in social-ecological systems: a working hypothesis for a participatory approach. Conservation Ecology 6(1), 14. [online] URL: http://www.consecol.org/vol6/iss1/art14/

WCED (Editors), 1987. Our Common Future. Oxford University Press, New York.

Zhao, B., Kreuter, U., Li, B., Ma, Z., Chen, J., and Nakagoshi, N., 2004. An ecosystem service value assessment of land use change on Chongming Island, China. Land Use Policy 21, 139–148.

Zurlini, G., Amadio, V., and Rossi, O., 1999. A landscape approach to biodiversity and biological health planning: the Map of Italian Nature. Ecosystem Health 5, 296–311.

ENVIRONMENTAL ASSESSING OF REINDEER HERDING IN CHANGING LANDSCAPES ON DIFFERENT SCALES

TIMO KUMPULA*
Department of Geography, University of Joensuu, Finland
BENJAMIN BURKHARD AND FELIX MÜLLER
Ecology Centre, University of Kiel, Kiel, Germany

Abstract. This paper is based on an interdisciplinary analysis of reindeer management systems in the subarctic and boreal regions of Fennoscandia. The aim was to develop sustainable strategies for the future utilization of the limited natural resources in these northern regions. Here, reindeer herding is competing with powerful land use forms like forestry, energy conversion, mining, and tourism. Problems arise because the ecological conditions of the natural pastures are basic prerequisites for reindeer husbandry, and because the welfare of the reindeer herders is strongly connected to the socioeconomical circumstances in the area. Intensive land use can cause a deteoriation of natural pastures and thus, lead to losses in reindeer husbandry. The described investtigations took place in the reindeer husbandry region of northern Finland with the two herding districts of Näkkälä and Lappi, where reindeer herding traditionally has been the main form of land use. The integrative assessment consisted of comprehensive field studies, applications of remote sensing techniques, data analysis, expert interviews, and modeling. Results are designed to support sustainable decision-making in landscape management.

Keywords: Reindeer husbandry; modeling; indicators; remote sensing; sustainability

* Corresponding Author: Timo Kumpula, Department of Geography, University of Joensuu, Yliopistokatu 7, 80101 Joensuu, Finland, Phone: +358-13-2514563, Fax: +358-13-2513454, e-mail: Timo.Kumpula@Joensuu.fi

I. Petrosillo et al. (eds.), Use of Landscape Sciences for the Assessment of Environmental Security, 413–427.

1. Introduction

Environmental security is related to the risks of losing environmental values, expressed as ecosystem goods and services. These risks are correlated with the dynamics of land use structures that provide basic variables of sustainable landscape management. Often there is a dynamic competition between different land use types, whereby the outcome is influenced by social, economic, and environmental factors as well as features of regional traditions. In the presented case study, respective arguments have been investigated with reference to one traditional, potentially well-adapted land use type; reindeer herding in the northernmost part of Europe.

Wild reindeer (*Rangifer tarandus*) have been part of the northern Fennoscandian ecosystems since the deglaciation about 9,000 years ago. The domestication of reindeer developed in the 17th century when the wild reindeer population declined because of intensive hunting. The first domesticated reindeer were used for transport means and as decoy animals in hunting (SSR, 1999; Pollmann, 1988; Bronny et al., 1985). In the following period, reindeer herding was carried out as part of a nomadic subsistence livelihood for a long time. The last four decades have changed reindeer husbandry drastically. Today, reindeer herding has transformed from a nomadic livelihood to a meat producing enterprise. Herders have settled in villages and adapted to a subsidized market economy, modern equipments like snow mobiles, all terrain vehicles, mobile phones or even helicopters and a more sedentary life style characterize the modern times for reindeer herders. The adaptation to these circumstances causes social, environmental, and economic problems and can be seen as one reason for the increased dependence of the herders on subsidies and the need for higher meat production. On the other hand, these requirements also function as driving forces for the development of higher quality meat products, tourism-related reindeer ranching and other nontraditional forms of income.

The presented research is based on the integrated EU project "The Challenges of Modernity for REiNdeer MANagement (RENMAN, see Forbes et al., 2006)." The focal object of RENMAN was an interdisciplinary and integrative analysis of the reindeer management systems in Fennoscandia. The aim was to provide tools and data to derive strategies for a sustainable use of the natural resources in future. A special focus was put on participation processes of local people and an integration of their indigenous knowledge into the research processes. During the model and tool development, reindeer herders were interviewed and participated in the discussions to focus the research to relevant topics. Data were collected and analyzed with techniques like remote sensing, process modeling, and geographical information systems (see, e.g. Burkhard et al., 2003, 2004). Furthermore, models of socioeconomic and ecological interactions were developed and suitable indicators considering an evaluation system for the ecological, social, and economic sustainability of landscape management regimes on different scales had to be found.

The change of landscape after World War II has been drastic in northern Finland and especially in its coniferous forest areas. The state owns most of the forests in northern Finland, and a big portion of the Finnish post-war growth was based on northern forests and rivers that were harnessed for the hydro power industry. Large clear cuts and heavy ploughing methods were used from 1960s to 1980s. During this period, reindeer herding and the corresponding livelihood had to adapt to the new conditions. In the late 1980s, when discussions about the degradation of pastures were raised, the main blame was put on the numbers of reindeer and on herding practices. Nowadays it is admitted that all land users provide impacts on landscape quality, thus the discussion has become more open.

To promote a sustainable use and management of pastures, there is a need for spatial knowledge about their conditions and quantities. Remote Sensing (RS), Geographical Information Systems (GIS) and systems analytical methods linked with field pasture evaluations provide a methodology for producing reliable inventories of extensive areas such as the reindeer herding districts. Numerous studies in northern America and Fennoscandia have used RS and GIS for pasture inventories of arctic ungulates successfully (George et al., 1977; Ferguson, 1991; Colpaert et al., 1995; Arseneault et al., 1997; Hansen et al., 2001; Theaun et al., 2005; Johansen and Karlsen, 2005). It can be said that the aim of the investigations in the 1990s was to study large areas, e.g. the whole reindeer husbandry area of Finland or Norway. But, in the new millennium, research is focused to a more detailed scale like individual herding districts. Multidisciplinary approaches are increasingly supported and the *Traditional Ecological Knowledge* (TEK) of herders is concerned as valuable information (Rees et al., 2003; Forbes et al., 2006; Burkhard and Müller, 2006; Kitti et al., 2006; Kumpula, 2006; Kumpula et al., 2006).

In this text we will describe different methods that were used to analyze the key problems of contemporary reindeer management in northern Fennoscandia with special focus on the two Finnish reindeer herding districts (*paliskunta* in Finnish) Näkkälä and Lappi (Figure 1). Whilst in the forest-dominated region the problem of land use competition is a dominant factor (Hukkinen et al., 2002), in the rather remotely located tundra ecosystems the conditions of the lichen pastures have been of main interest (Olofsson et al., 2004). A set of 32 indicators for an integrative assessment of reindeer herding systems was developed, and different reindeer management strategies were compared including socioeconomic, ecological, and land use components.

2. Research areas

The analyzed districts have strong Sámi (the indigenous people of this northern area whose culture is strongly connected to reindeer) reindeer herding traditions (Paliskuntain Yhdistys, 2002). Russia (Finland belonged to Russia during that period) and Norway closed their borders in the late 19th century. In 1852 Sámi herders were forbidden to cross the national border with their reindeer and their traditional migration routes with a summer-winter and coastal-inland pasture rotation system were destroyed. As a consequence, the usage of pastures has been altered especially on the Finnish side of the border. In the mid 1950s, a fence was built along the borders to prevent reindeer from crossing it (Pennanen, 2003; Näkkäläjärvi, 2003).

Both herding districts differ as Lappi has been more affected by forestry and the building of artificial lakes for energy generation. Additionally, tourism has a significant impact in Lappi. The competing land use dynamics in the Lappi area were studied as pressures on reindeer herding while in the Näkkälä area, structural dynamics mainly result from pressures caused by reindeer herding itself, namely affecting the quality of winter pastures. In both research areas, representative sites were selected for intensive field studies.

The size of the Näkkälä herding district is 3,539 km^2. Here, 8,700 reindeer were counted in 2002 (2.5 reindeer per km^2, Paliskuntain Yhdistys, 2002). The northern parts of the study region, where the intensive research site Jauristunturit is located, belong to the hemiarctic zone with treeless tundra ecosystems (Oksanen and Virtanen, 1995). The southern parts of the Näkkälä district belong to the northern boreal forest zone. In the northern part of the Näkkälä district, reindeer herding has remained the most important land use form while in the southern regions, forestry has been realized in a non-extensive degree. In Näkkälä, the tourist volume is relatively low. The actual problems of reindeer herding can be summarized to the sufficiency of pastures in order to sustain economically profitable herd sizes, and an aging population structure of the herding communities.

The Lappi herding district has a total size of 4,453 km^2. In 2002 there was a reindeer stock of 8,521 animals, which results in about 1.9 animals per km^2 (Paliskuntain Yhdistys, 2002). Lappi belongs to the northern boreal forest zone, with integrated tundra ecosystems in the north eastern part of the district (Saariselkä mountain area).

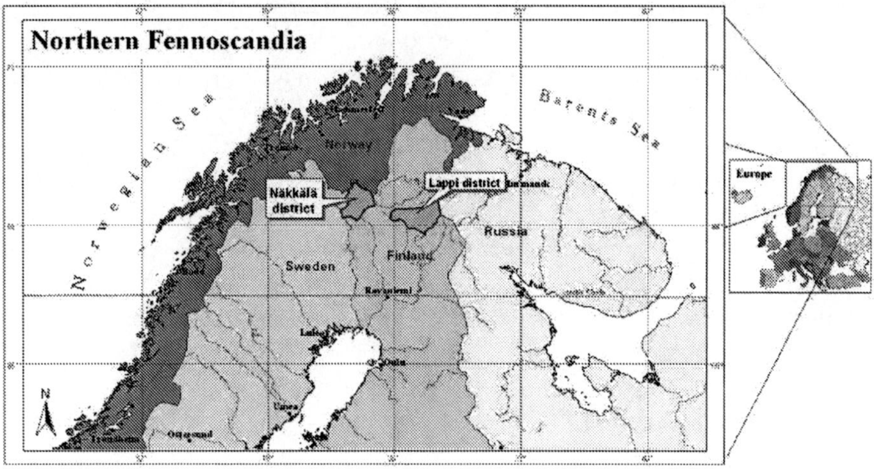

Figure 1. Research areas.

The intensive research sites were located in a forest near the village of Vuotso in the central part of the Lappi district. Additionally, Aapa fens cover large parts of the Lappi district. The district has suffered from heavy forestry, including broad clear cuts, since World War II. Furthermore, the district has lost 12% of the pastures when large water reservoirs were built in 1960s and 1970s. In the eastern parts of the herding district, large nature protection areas cover about 1,800 km^2. Tourism plays a major role for the economy of the area, provoking strong land use conflicts.

3. Materials and methods

Collecting and building the comprehensive database for the integrative assessment was a challenging task. A great amount of field data and information was provided by the project partners (see Forbes et al., 2006). Different GIS data layers (like *Digital Elevation Model* DEM, water bodies, land use types, infrastructural elements, settlements, etc.) were purchased from the Finnish National Land Survey, and additional GIS layers were derived by digitizing different features from satellite images and field investigation points that were marked with GPS.

3.1. RESEARCH SCALES

The environmental assessment of reindeer herding requires the application of data on different scales of the reindeer husbandry system which has a range from national to individual herder's scale. Reindeer herding districts in Finland usually cover 1,000–4,000 km^2. For example, to cover the Näkkälä area with most detailed resolution (40–60 cm high-resolution aerial photographs) it would require about 140 aerial photographs (which usually cover 5 * 5 km). Landsat TM images cover 185 * 185 km (34,000 km^2). ASTER TERRA images cover 60 * 60 km (3,600 km^2) with 15 m spatial resolution (Figure 2). As the sizes of the Finnish herding districts range from 587–5,708 km^2, many of them are divided into subdivisions (*tokkakunta* in Finnish). Individual reindeer herders are most familiar with their own *tokkakunta's* landscape which mostly covers less than 2,000 km^2. Even inside the district, the management styles can vary from one *tokkakunta* to the other.

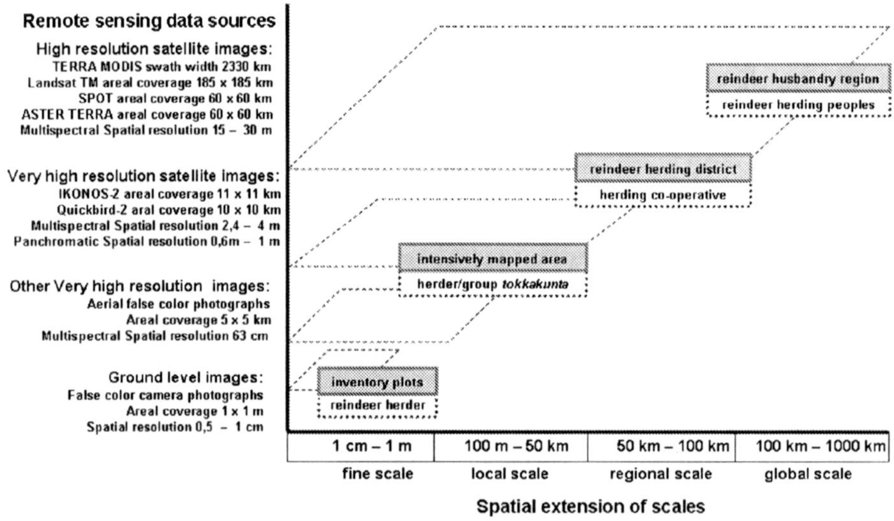

Figure 2. Scales of remote sensing data.

In this research, Landsat TM images were used for the herding district scale. Additionally focus was put on more detailed scales on Näkkälä's late spring and early summer pasture, using very high resolution IKONOS-2 images.

3.2. REMOTE SENSING

For the quantification of the different land use and pasture types, Landsat TM data were used to cover the two targeted herding districts on a general scale.

Three Landsat TM images were used to cover Näkkälä and two Landsat TM images to cover Lappi. For the intensive research area in Näkkälä (10 * 10 km), very high resolution IKONOS-2 satellite imagery data were used (detailed scale). Images were analyzed with ERMapper 6.3 and ArcGIS 8.3 softwares.

Certain vegetation and land use types caused significant classification errors. For example, the spectral reflectance of the forest clear cuts and barren mountain areas were very similar. To avoid these misclassifications images were divided into data sets with or without clear cut areas. Data sets were then classified separately with or without clear cuts and barren mountain classes. After that data sets were merged into one data set. Another group of classes with severe misclassifications were tree growing peat bogs and forest classes. Misclassifications of forest classes and tree growing peat bogs were corrected by using a special peat land mask (GRID 25 * 25 meter) produced by the Finnish National Survey (Kumpula et al., 2005).

3.3. MODEL APPLICATIONS

A conceptual model was developed to show the basic elements, inputs and outputs and the interrelations in contemporary reindeer husbandry. This model was used to promote the understanding of the complex reindeer husbandry system and as a starting point for the subsequent analyses (Burkhard, 2004). For the simulation of water and matter cycling on differently managed reindeer pastures, the simulation model WASMOD (Reiche, 1996) was applied. Thus, effects of different intensities of reindeer grazing and trampling on important ecosystem features and processes were analyzed. Changing land use patterns and dynamics in the socioeconomic systems components were modelled by the development of different future scenarios that were evaluated by expert interviews using the derived indicators (Burkhard, 2004).

3.4. INDICATOR DERIVATION AND SCENARIO APPLICATION

A set consisting of 32 indicators related to the land use pattern, social and economic welfare and ecological integrity (ecosystem state) was developed in interdisciplinary cooperation in the first year of the RENMAN project.

Table 1 gives an overview of these indicators and the availability of the respective data. In the column "Reindeer herder" it is noticeable that information about herders was available for almost all of the 32 indicators.

The set was applied for the evaluation of three different future scenarios related to varying intensities of reindeer herding in the whole Finnish reindeer

TABLE 1. Indicator set and scales of available data.

	Indicators	Herder	Scales of available data		
			Intensive research sites	Herding district	Finnish herding area
Land use	Reindeer herding	O	O	O	O
	Forestry	O	O	O	O
	Agriculture	O	O	O	O
	Energy conversion	O	O	O	O
	Tourism	O	O	O	O
	Mining	O	O	O	O
	Nature protection	O	O	O	O
	Other activities	O	O	O	O
Economic welfare	Reindeer herding	O		O	O
	Employment	O		O	O
	Spending power	O		O	O
	Accommodation	O		O	
	Efficiency & autonomy	O			O
	Reindeer	O		O	O
	Logistic & infrastructure	O		O	O
	Progress	O			O
Social welfare	Personal well-being	O			O
	Health & nutrition	O			O
	Leisure	O			O
	Social security	O			O
	Ethnic/regional identity	O		O	O
	Women's situation	O			O
	Education & culture	O		O	O
	Demography	O		O	O
Ecological integrity	Exergy capture		O		
	Exergy dissipation		O		
	Storage capacity		O		
	Biotic water flow		O		
	Metabolic efficiency		O		
	Nutrient loss		O		
	Biotic diversity	O	O		
	Abiotic diversity	O	O		

husbandry region and in the two case study areas. The aim was to simulate different situations in the Finnish reindeer husbandry region until the year 2030.The scenarios referred to:

- Intensification of reindeer herding: "Agriculturisation" of reindeer husbandry with intensive supplementary feeding, fencing, and machine usage. Special focus is put on meat production and economic output while the number of reindeer herders is decreasing.
- Reduction of reindeer herding: There is no more support for reindeer herding at all, e.g. no more subsidies and subventions, no more special areas for reindeer herding. Reindeer herding is left as tourist attraction and historical remain.
- Business as usual development: All processes, institutions, and connections in future reindeer husbandry are treated in the same way as nowadays.

By simulating these different development paths, the resulting effects on the reindeer management system were analyzed. For the assessment of land use, economic and social welfare variables, various experts with different backgrounds were consulted to explain their opinion on the scenarios and to make quantitative statements using the indicators suggested. For the quantification of the ecological parameters, data from the field campaigns on differently managed reindeer pastures (including intensively, medium, and little grazed areas) and simulation model outcomes were applied.

4. Results

The integrative systems analysis revealed that the key concerns in contemporary reindeer husbandry are very complex and based on multiple factors and interrelations. The derived set of indicators was used to compare different reindeer management strategies and to evaluate them regarding their social, economic, and ecological sustainability. The investigations show, that most of today's problems have socioeconomic or political backgrounds whereas the ecological components did not seem to be critical, at least in the two research areas, at the moment (Burkhard, 2004). This can probably be related to the current number of reindeer which is rather low compared with the large stocks in Finland at the end of the 1980s (Paliskuntain Yhdistys, 2002; Kumpula, 2001).

The results of the remote sensing image classifications (Landsat TM and IKONOS-2) gave an opportunity to investigate pastures at finer and coarser scales (Kumpula et al., 2005; Kumpula, 2006). Landsat TM classifications accuracy was significantly improved by dividing images into regions with or without clear cuts and barren mountains. Also using a peat land mask to correct tree peat bogs and forest classes' mixtures was successful (Kumpula et al., 2005). To conduct detailed studies of ecological integrity both field measurements and very high resolution images are required. This limits the size of the studied area, because of cost and logistic factors. The combination of these techniques made possible to distinguish several grazing intensities from IKONOS-2 image on lichen grounds (Kumpula, 2006) of Nunnanen *tokkakunta*. But already the herding district is such a large area that the collection of data to study ecological integrity with a similar resolution is very difficult. In these cases the methodology has to be modified and modeling has to rely on sources like national or regional environmental, economical, or social databases (Table 1).

Figure 3 gives an overview of a "business as usual" simulation for the Finnish reindeer husbandry system. The basic assumption within this scenario was that no changes take place in the management system within the next 25 years. As a consequence, the land use distribution is altered in the form of decreased reindeer herding and agriculture whereas all other forms of land use, mainly tourism, are increasing. Based on the supposed future decrease of reindeer herding and its impacts on the pasture areas, effects on the ecological systems components were decreasing. Rather high alterations were obvious concerning the abiotic diversity and exergy capture. Regarding the economic welfare of the reindeer herders, "business as usual" has negative effects especially on the employment situation, the spending power, the efficiency, and economy. Moreover, the social welfare components personal well-being and demography are affected negatively. These points have to be taken into account in future management strategies. Some respective insights concerning this point, the results of the other scenarios (reduction and increase of reindeer herding) and the proposed outcome of a sustainable development – scenario can be found in Burkhard and Müller (2006) and Burkhard et al. (2003, 2004).

5. Discussion

The key issues that are challenging the modern reindeer husbandry in northern Finland are the competition for limited natural resources by different land users and the resulting conditions of the natural pastures. The competition for natural resources is mainly connected to socioeconomical circumstances because forestry, mining, hydropower production, tourisms, and reindeer husbandry are mainly concerned with profitability. One sample outcome is the nontraditional supplementary feeding strategy. Although reindeer herding is based on natural pastures, recently in most Finnish herding districts supplementary feeding, at least in late winter, is practised. The changes of natural pastures, e.g. due to

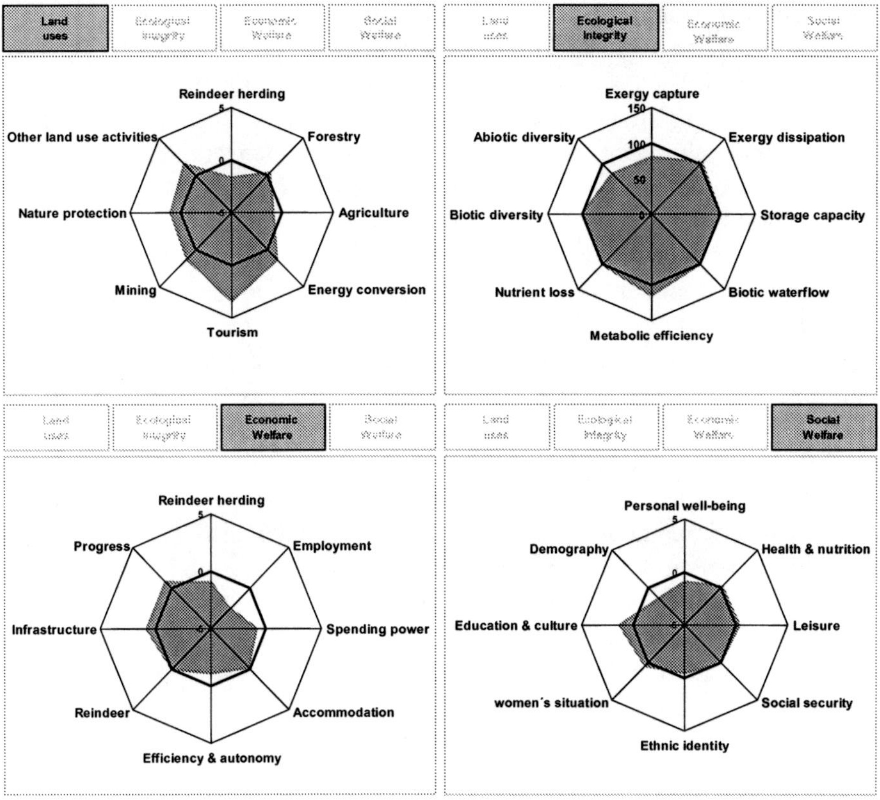

Figure 3. Business as usual scenario simulation for the Finnish reindeer husbandry system. The reference values 0 (for land use, social and economic welfare) resp. 100% (for ecological integrity) represent the situation in the year 2003.

forestry activities, hydropower industry, road building, tourism, and high reindeer numbers have affected the condition and quality of pastures negatively. For example, the loss of old forest arboreal lichen pastures (due to clear cutting forestry) has increased a need of supplementary fodder in spring.

The 32 indicators related to the land use patterns, social and economic welfare, and ecological integrity (ecosystem state) were developed in interdisciplinary cooperation. The quality of comparable integrative ecological assessments is always dependent of the scales of investigations, available data sets, measured parameters, etc. In the future, more efforts have to be paid to the transitions between different spatial and temporal scales and on standardized and reproducible methods of data acquisition. Nevertheless, the chosen methods, models and – last but not least – the results proved to be suitable for the accomplishment of integrative assessments of complex systems and the provision of applicable management recommendations. But, compared to developments in reality, the derived scenarios follow rather deterministic pathways. More dynamic and stochastic modeling procedures that take into account unexpected changes and developments could provide additional interesting results.

Field measurements, statistical data, GIS and remote sensing provide data for modeling. Attention has to be paid to the resolution of the remote sensing and GIS data. Can certain results really be derived from available or used data? For example, the modeling of ecological integrity requires specific, difficult to obtain data at high resolution that is most reliable at the detailed *tokkakunta* scale. Herding district scale or national scale modeling can be conducted with data from environmental and social statistical databases supplemented with information from qualitative and quantitative interviews, but as a consequence modified models have to be used; thus a cross scale strategy of security indicators, connected with a cross-scale modeling procedure should be a focal target of future methodological activities. It can be also argued that is it generally possible to study Finnish reindeer herding husbandry area as a whole because environmental conditions, local practises, traditions and local significance vary drastically from the southern parts of the area to the northern parts. However in general discussion reindeer herding is usually treated as one and only issue.

Practically, if reindeer husbandry investigations are conducted, they should start from smallest reindeer herding unit namely *tokkakunta* level to the coarser district and whole national reindeer management area level with appropriate remote sensing data. Bottom-up approaches may also integrate local actors easier into the research process. Very high resolution satellite data allow most detailed analyses of pastures and land use but, in most cases, the research is aiming to cover large areas which prevents their utilization because of high costs. This fact should not exclude the use of very high resolution data. They could be used to study and analyse certain spatially smaller scale hot spot areas and also to be used as reference data in the classification of coarser resolution images as, e.g. ASTER and Landsat. The basic mapping of Finnish reindeer husbandry area with coarser scale satellite data has been done. In near future GIS databases collected by several institutions (Forest Government, National Land Survey, environmental centers etc.) will become more important in further analysis of changing landscape of reindeer herding. Then very high resolution satellite and digital aerial photographs can be used in updating changes in the landscape, e.g road construction, forestry activities, and tourism-related constructions.

When studying land use with an important historical background as reindeer herding where the long-term survival has required an overall, complete understanding of landscape structures and the interrelated processes, the local knowledge should not be underestimated. The most challenging task is to find a common language among scientists and local actors. This requires the will and ability to explain the respective expertise in an understandable manner from both sides.

6. Conclusions

The main focus of this paper was put on environmental assessment of reindeer herding in northern Finland. Reindeer herding has gone through drastic changes from nomadic life form to meat producing husbandry during last 50 years. Meanwhile landscape changes in boreal coniferous forests have been enormous. Vast wilderness areas have been fragmented by forestry and forestry-related road constructions. The condition and quality of pastures have declined because of forestry activities, hydropower industry, road building, tourism, and also because of high reindeer numbers. Therefore, the long-term provision of focal ecosystem services is endangered, and thus the level of landscape security has been decreasing steadily.

From a scale-based, methodological viewpoint, it can be stated that the environmental assessment of reindeer herding requires multidisplinary approaches. In this research, 32 indicators related to land use patterns, social and economic welfare, and ecological integrity (ecosystem state) were developed. GIS databases and remote sensing images have been coupled with detailed field measurements and statistical data. This methodological pool provided necessary inputs for modeling, whereby the availability of suitable data (in scale and quality) can be seen as a limiting factor of model and scenario building. Very high-resolution satellite imagery (e.g. IKONOS-2 Quickbird-2) is usable in detailed scale investigations and expanding GIS databases will improve modeling and scenarios accuracy in near future. Due to the specificity of certain levels of resolution, we recommend to use bottom-up approaches for investigations of reindeer husbandry, from the smallest reindeer herding unit *tokkakunta* to the national husbandry area scale.

References

Arseneault, D., Villeneuve, N., Boismenu, C., Leblanc, Y., and Deshaye, J., 1997. Estimating lichen biomass and caribou grazing on the wintering grounds of northern Quebec: an application of fire history and Landsat data. Journal of Applied Ecology 34, 65–78.

Bronny, H., Hemmer, N., and Sokki, T.H., 1985. Samische Rentierwirtschaft – Reliktform oder Wachstumsbranche. – Geographische Rundschau 37, H. 10. Braunschweig, pp. 529–536.

Burkhard, B., Kumpula, T., and Müller, F., 2003. Landscape assessment for a sustainable reindeer husbandry in northern Scandinavia. EcoSys Bd. 10, 116–124.

Burkhard, B., 2004, Ecological assessment of the reindeer husbandry system in northern Finland. EcoSys Suppl. Bd. 43, 117 pp.

Burkhard, B., Kumpula, T., and Müller, F., 2004. Integrative assessment of different reindeer management strategies in northern Finland. EcoSys Suppl. Bd. 42, 60–66.

Burkhard, B. and Müller, F., 2006. System analysis of Finnish reindeer management. Ecological Studies 184. In: B.C. Forbes, M. Bölter, L. Müller-Wille, J. Hukkinen, F. Müller, N. Gunslay, and Y. Konstantinov (Editors), Reindeer Management in northernmost Europe. Ecological Studies 184, Springer, Berlin, pp. 341–364.

Colpaert, A., Kumpula, J., and Nieminen, M., 1995. Remote sensing – a tool for reindeer rangeland management. Polar Record 31, 177, 235–244.

Forbes, B.C., Bölter, M., Müller-Wille, L., Hukkinen, J., Müller, F., Gunslay, N., and Konstantinov, Y. (Editors), 2006. Reindeer management in northernmost Europe. Ecological Studies 184. Springer, Berlin.

Ferguson, R.S., 1991. Detection and classification of Muskox habitat on Banks Island, Northwest Territories. Using landsat thematic mapper data. Arctic 44 Supp. 1, 66–74.

George, T.H., Stringer, W.J., Preston, J.E., Fibich, W.R., and Scorup, R., 1977. Reindeer range inventory in western Alaska from computer-aided digital classification of Landsat data. Proceedings 28th Alaska Science Conference, Anchorage, Alaska, pp. 33–42.

Hansen, M.J., Franklin, S.E., Woudsma, C.G., and Peterson, M., 2001. Caribou habitat mapping and fragmentation analysis using Landsat MSS, TM, and GIS data in the North Columbia Mountains, British Columbia, Canada. Remote Sensing of Environment 77, 30–65.

Hukkinen, J., Jääskö, O., Laakso, A., Müller-Wille, L., and Raitio, K., 2002. Poromiehet puhuvat. Helsinki University of Technology.

Johansen, B. and Karlsen, S.R., 2005. Monitoring vegetation changes on Finnmarksvidda, Northern Norway, using Landsat MSS and Landsat TM/ETM+ satellite images. Phytocoenologia 35(4), 969–984.

Kitti, H., Gunslay, N., and Forbes, B.C., 2006. Defining the quality of reindeer pastures: the perspectives of Sámi reindeer herders. In: B.C. Forbes, M. Bölter, L. Müller-Wille, J. Hukkinen, F. Müller, N. Gunslay, and Y. Konstantinov (Editors), Reindeer management in northernmost Europe: linking practical and scientific knowledge in social-ecological systems. Ecological Studies 184, Springer, Berlin, pp. 141–165.

Kumpula, J., 2001. Productivity of the semidomesticated reindeer (Rangifer t. tarandus L.) stock and carrying capacity of pastures in Finland during 1960–1990s. Acta Universitatis Ouluensis, Scientiae Rerum Naturalium A.

Kumpula, T., 2006. Very high resolution remote sensing data in reindeer pasture inventory in northern Fennoscandia. In: B.C. Forbes, M. Bölter, L. Müller-Wille, J. Hukkinen, F. Müller, N. Gunslay, and Y. Konstantinov (Editors), Reindeer Management in northernmost Europe. Ecological Studies 184. Springer, Berlin, pp. 167–185.

Kumpula, T., Burkhard, B., and Müller, F., 2005. Can you see it all in the image? EcoSys Bd. 11, pp. 13–23.

Kumpula, T., Forbes, B.C., and Stammler, F., 2006. Combining data from satellite images and reindeer herders in Arctic Petroleum Development: the case of Yamal, West Siberia. In: O. Heikkinen, K. Suorsa, M. Anttonen, and M. Mönkkönen (Editors), Nordia Yearbook 2006 Theme Issue on Northern Nature and Human Activities. Nordia Geographical Publications 35(2), pp. 17–30.

Näkkäläjärvi, K., 2003. The siida, or Sámi Village, as the basis of community life. In: J. Pennanen and K. Näkkäläjärvi (Editors), Siidastallan. From Lapp Communities to Modern Sámi Life. Siida Sámi Museum Inari. Pohjoinen. Gummerrus, pp. 114–121.

Oksanen, L. and Virtanen, R., 1995. Topographic, altitudinal and regional patterns in continental and suboceanic heath vegetation of northern Fennoscandia. In: L. Oksanen and R.Virtanen (Editors), Geographical ecology of northernmost fennoscandia. Acta Botanica Fennica 153, 1–61 + 10 app.

Olofsson, J., Stark, S., and Oksanen, L., 2004. Reindeer influence on ecosystem processes in the tundra. Oikos 105, 386–396.

Paliskuntain yhdistys, 2002. Information about the reindeer husbandry. Rovaniemi.

Pennanen, J., 2003. Large-scale reindeer herding brings changes to the siidas. In: J. Pennanen and K. Näkkäläjärvi (Editors), Siiddastallan. From Lapp Communities to Modern Sámi Life. Siida Sámi Museum Inari. Pohjoinen, Gummerrus, pp. 138–139.

Pollmann, K., 1988. Die Rentierwirtschaft im südsamischen Raum Skandinaviens sowie außerhalb der traditionellen Rentierweidedistrikte Norwegens. Bamberger Wirtschaftsgeographische Arbeiten Heft 5. Bamberg.

Rees, W.G., Williams, M., and Vitebsky, P., 2003. Mapping land cover change in a reindeer herding area of the Russian Arctic using Landsat TM and ETM+ imagery and indigenous knowledge. Remote Sensing of Environment 85(4), pp. 441–452.

Reiche, E.W., 1996. WASMOD. Ein Modellsystem zur gebietsbezogenen Simulation von Wasser- und Stoffflüssen. Darstellung des aktuellen Entwicklungsstandes. EcoSys, Bd. 4, 143–163.

SSR Svenska Samernas Riksförbund, 1999. Svensk Rennäring. Umeå. (in Swedish)

Theaun, J., Peddle, D.R., and Duguay, C.R., 2005. Mapping lichen in a caribou habitat of Northern Quebec, Canada, using an enhacement-classification method and spectral mixture analysis. Remote Sensing of Environment 94, 232–243.

PART VI INTRODUCTION

ENVIRONMENTAL APPLICATIONS OF LANDSCAPE ECOLOGICAL METHODS – IMPACT ASSESSMENT

OCTAVIAN LIVIU MUNTEAN
"Babeş-Bolyai" University, Faculty of Environmental Science, Cluj-Napoca, Romania

KINGA KRAUZE
International Centre for Ecology, Polish Academy of Sciences, 3 Tylna Str. 90-364 Łódź, Poland

Even in the absence of man, the natural environment undergoes continual change. Man's increasing "control" of his environment often creates conflicts between human goals and natural processes. In order to achieve greater yields or for other purposes, man deflects the natural flows of energy, bypasses natural processes, severs food chains, simplifies ecosystems, and uses large energy subsidies to maintain delicate artificial equilibrium.

Because natural biogeophysical environments fluctuate with time, we are unable easily to distinguish changes brought about by man. But, in order to understand these changes, it is necessary to know what the conditions would have been like if no development had taken place. It is not easy to measure exactly the present condition, far less to assess the significance of past trends and to project these accurately into the future.

Perceptions about environmental impacts can be rather different in different countries. Where poverty is widespread and large numbers of people do not have adequate food, shelter, health care, education, and environmental security, the lack of development may constitute a greater aggregate degradation to life quality than do the environmental impacts of development.

Within this context, the concept of environmental impact assessment has an important relevance in entirely environmental theory and practice because it can be interpreted as a strong environmental tool (based on a legislative measures and procedures), which must be implemented to reconstruction and/or rehabilitate the environmental components that have been affected by human action and its harmful effects. In this context, the environmental impact is part of sustainable territorial planning and it conditioning the healthy future of environment and community. At the same time, environmental impact assessment, as a procedure, is oriented to the rehabilitation, reconstruction, and conservation of the environmental and landscape components (natural and human components).

In environmental studies, impact assessment is related to "environmental restoration" concept which has an important relevance because it refers to a local environmental policy which is related to conservation, diminution, or reconstruction of the environmental components, which have been affected by environmental impact and its effects. For the experts and policy makers, the concept of environmental restoration, as a policy, is trying to define the principles, applications, directions, targets and identify the criteria of the actions leading to a sustainable development (environmental, landscape, economic, and social) under local and regional circumstances. Viewed in this context, environmental applications of landscape ecological methods become tools towards stability and sustainability on the local and regional scale in terms of environmental change and the economic development. The human impacts and their consequences on landscape, environment, and community besides of some assessment methods and techniques are presented within this chapter as local and regional case studies. Sustainable community development, predictive models of landscape change, environmental security, nuclear safety, environmental impact, and landscape ecological assessment are main topics of this chapter.

THE CONTRIBUTION OF QUALITY ASSESSMENT OF ERODED AGRICULTURAL SOIL ON HILLY-UNDULATING LANDSCAPES TO SUSTAINABLE COMMUNITY DEVELOPMENT

BENEDIKTAS JANKAUSKAS* AND GENOVAITE JANKAUSKIENE
Kaltinenai Research Station of Lithuanian Institute of Agriculture, Lithuania

ALGIMANTAS TIKNIUS
University of Klaipeda, Lithuania

Abstract. Sustainable community development depends on different circumstances and conditions and soil is one of the most important natural resources influencing community development. Some soil quality indicators are exceptionally sensitive on hilly undulating relief. The results from 18 years of field experiments at the Kaltinenai Research Station of the Lithuanian Institute of Agriculture indicate that soil erosion processes under the field crop rotation (containing 17% tillage crops, 33% grasses, and 50% cereal grains), disrupt the top layer of unprotected soil, decrease soil organic matter content, impoverish soil physical–chemical properties and instigate landscape leveling. The mean annual erosion rates under the field crop rotation were 9.9, 23.4, and 32.2 Mg ha^{-1} $year^{-1}$, respectively, from slopes of 2–5°, 5–10°, and 10–14°. Erosion-preventive grass–grain crop rotations (containing 67% grasses and 33% cereal grains), as well as the sod-forming perennial grasses for long-term use, decreased mean annual erosion rates by 75–80% compared with the field crop rotation. Both land use systems ensure increasing soil organic matter content, decreasing soil erodibility and increasing soil carbon sequestration, helping to decrease atmospheric CO_2 concentrations and thus positively influencing global climate changes. These land use systems can be considered as soil conserving, sustainable systems ensuring environmental security on the vulnerable hilly undulating agricultural landscape.

* Corresponding Author: Dr. Jankauskas Benediktas, Kaltinenai Research Station of Lithuanian Institute of Agriculture, Lithuania, e-mail: kaltbs@kaltbs.lzi.lt.

I. Petrosillo et al. (eds.), Use of Landscape Sciences for the Assessment of Environmental Security, 431–451.

Keywords: Soil erosion; soil conservation strategy; land use systems; soil organic matter; energy accumulated; sustainable development

1. Introduction and objectives

Ecological systems are composed of complex biological and physical components that are difficult to understand and measure. However, effective management actions and policy decisions require information on the status, condition and trends of ecosystems. Multiple levels of information are needed to make effective decisions and the ideal indicators for measuring ecosystem integrity will incorporate information from multiple dimensions of the ecosystem. A terrestrial index of ecological integrity would be a useful tool for ecosystem managers and decision-makers. The ideal requirements of the terrestrial index of ecosystem integrity are that it be comprehensive and multi-scale, grounded in natural history, relevant and helpful, able to integrate concerns from aquatic and terrestrial ecology, and that it is flexible and measurable (Andreasen et al., 2001). Sustainable development has become a leading target of scientific research and policymakers. In the context of natural resource management, understanding and evaluating the performance of complex socio-environmental systems has become a challenge, and the design of more sustainable alternatives is an imperative need. In addition, there is a need to understand the general principles of sustainability and to use them for operational definitions and practices (Lopez-Ridaura et al., 2002). According to Flint (2003) sustainable development is the parallel consideration of healthy environments, life and human well-being. This includes issues of population, climate, economic prosperity, energy, natural resource use, waste management, biodiversity, watershed protection, technology, agriculture, safe water supplies, international security, politics, green building, sustainable cities, smart development, community/family relations, and human values.

Soil erosion is one of the world's most serious environmental problems, causing extensive losses of cultivated and potentially productive soil and an enormous annual loss of crop yields (Fullen and Catt, 2004; Morgan, 1995). Highly eroded soils tend to have reduced productivity, degraded soil structure, lower organic matter, and are poor media for root growth (Morgan, 1995). Soil erosion on hilly undulating agricultural landscapes is a complex phenomenon involving the detachment and transport of soil particles, storage and runoff of rainwater, and infiltration. The relative magnitude and importance of these processes depend on many factors, including climate, soil, topography, cropping and land management practices, antecedent conditions, and scale (Römkens et al., 2002). The complexity of soil erosion determines the efficacy of different soil erosion processes, such as water, wind, and tillage erosion. The removal of soil by one erosion process can affect the erodibility of the remaining soil to other erosion processes, and one soil erosion process can act as a delivery mechanism for other erosion processes, by depositing soil where it is more readily removed by other erosion processes. Soil erosion also impacts other processes, such as water contamination with sediments and nutrients, pesticide fate in the soil and the environment, and greenhouse gas production and emission. These interactions can complicate modeling efforts, thus there is a tremendous opportunity to increase the accuracy, coherency, and efficiency of environmental indicator initiatives (Lobb et al., 2003). However, many techniques can be used to conserve soils, each with their own relative advantages and disadvantages. Any successful soil conservation plan is a mix of technical and social objectives. Developing effective and viable soil conservation strategies is one of the most pressing soil management problems we face in the early 21st century. These strategies need to be both cost-effective and socially acceptable (Fullen and Catt, 2004).

The cover crops composing agro-ecosystems play a key role in promoting biodiversity (Vandermeer et al., 1998), therefore, multi-species agro-ecosystems (sod-forming long-term perennial grasses and grass–grain crop rotations) are potential components for both soil conservation and biodiversity strategies. Furthermore, the global data set of soil erodibility values shows much unexplained variance and a contributory factor is often the limited measurement period (Torri et al., 1997). Therefore, long-term studies are essential to both assess changes in soil physical properties and the potential of soil conservation techniques (Chan et al., 2002).

Soil erosion processes disrupt the upper layer of unprotected soil, decrease soil organic matter content, impoverish soil physical–chemical properties and instigate landscape leveling. Therefore, sustainable community development on the hilly undulating landscape of the temperate climatic zone must meet the needs of the present without compromising the ability of future generations to meet their own needs. Sustainability depends on different circumstances and conditions. One land use system can meet the needs of the present community, but also lead to progressive land degradation, and thus will present problems for future generations. Solving these goals is the main objective of this work. Analyzing research data from the slopes of different inclination (steepness) we shall: (1) overview soil losses under different crops and land use systems; (2) analyze changes in soil physical properties; (3) discuss changes in soil organic matter content.

2. Materials and methods

Research data were obtained from the Kaltinenai Research Station of the Lithuanian Institute of Agriculture (KRS of LIA), which is located on the southern central Zemaiciai Uplands (55°34' N, 22°29' E). Dystric Albeluvisols prevail in this region of Lithuania; however, soils become Eutric Albeluvisols due to intensive periodical liming, when lime changes the properties of both Ap and deeper soil horizons. Study sites A, B, and C (Figure 1) are on slopes of 2–5°, 5–10°, and 10–14°, respectively. Field trial plot width was 3.6 m and length was 90 m on sites A and C (slopes 2–5° and 10–14°) and 40 m on B (slope 5–10°). The narrow plot width was to facilitate mechanical cultivation and sowing in one tractor pass, while plot length depends on available slope length.

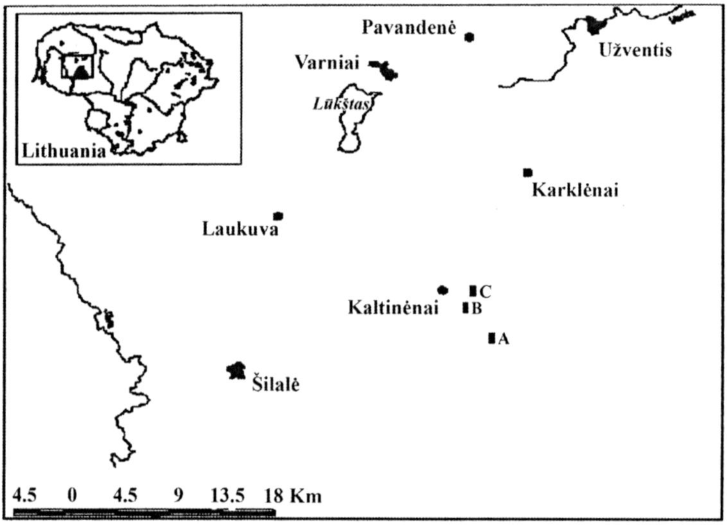

Figure 1. Location of the Zemaiciai Uplands and the long-term field experiments (A, B, and C) in Lithuania.

Field experiments were performed on eroded Eutric Albeluvisol sandy loams (FAO-UNESCO, 1994). Soil was differentially eroded along the slopes, being slightly eroded on 2–5° slopes, moderately eroded on 5–10° slopes, and strongly eroded on 10–14° slopes, with colluvial deposits on basal slopes. Soil erosion was mainly caused by tillage and water erosion under continuous intensive cropping. The agro-chemical properties of Ap horizons (0–20 cm) before field experiments shows topsoil were slightly acid, P-deficient, medium rich in K and contained varying soil organic matter (SOM) contents (Table 1). The highest percentage SOM was on less eroded 2–5° slopes and the least on 10–14° slopes. For historical reasons, soil analytical techniques were mainly former Soviet procedures (Jankauskas and Fullen, 2002). Therefore analytical results differ from those generated by currently internationally accepted protocols (e.g. USDA, NRCS, 1995), but are consistent with former Soviet protocols

(Jankauskas and Jankauskiene, 2003a). Currently, efforts are underway to develop transfer functions between Soviet and international systems (Booth et al., 2003; Jankauskas et al., 2005).

TABLE 1. Mean chemical soil properties of the arable (Ap) horizon (0–20 cm) before field experiments in 1981.

Study sites	Steepness of slope (degrees)	Organic matter ($g\ kg^{-1}$)	Available elements ($mg\ kg^{-1}$)		Exchangeable bases ($cmol(+)kg^{-1}$)	Hydrolytic acidity	pH_{KCl}
			P	K			
A	2–5	28.5	49.8	146.1	119	20.1	5.8
B	5–10	22.0	18.3	127.0	94	24.5	5.3
C	10–14	20.8	29.7	131.2	96	16.7	5.8

Mean annual precipitation in Lithuania is 626 mm, with ~858 mm on the central Zemaiciai Uplands and 750–800 mm on the upland fringe. Annual precipitation during the study period was 635–1075 mm. Plots were deep-ploughed, usually in September, and were bare until spring. Total runoff and erosion from bare soil was measured before the following spring cultivation (usually in mid-April). Plot runoff and erosion were measured on a regular basis, up to weekly during erosive rains, after sowing. Measurements were taken from spring sowing (typically late April or early May) to mid-June for cereals and late August for potatoes.

Long-term field experimental data were collected on the slopes of 2–5°, 5–10°, and 10–14° since 1983. Four crop rotations were compared, specifically:

- The field crop rotation, containing 17% tillage crops (potato), 33% grasses, and 50% cereal grains: 1: winter rye (*Secale cereale* L.), 2: potatoes (*Solanum tuberosum* L.), 3–4: spring barley, 5–6: mixture of clover-timothy (CT) (*Trifolium pratense* L.-*Phelum pratense* L.)
- The grain–grass crop rotation, containing 33% grasses and 67% cereal grains: 1: winter rye, 2–4: spring barley, 5–6: CT
- The grass–grain I crop rotation, containing 67% grasses and 33% cereal grains: 1: winter rye, 2: spring barley, 3–6: CT
- The grass–grain II crop rotation, containing 67% grasses and 33% cereal grains: 1 winter rye, 2: spring barley, 3–6: mixture of orchard grass-red fescue (OF) (*Dactylis glomerata* L.-*Festuca rubra* L.)

Multispecies mixtures of perennial grasses for long-term use as sod-forming grasses (g) were grown on 10–14° slopes instead of the field crop rotation (Table 2). The grass mixture consisted of 20% each of common timothy, red fescue, white clover (*Trifolium repens* L.), Kentucky bluegrass (*Poa pratensis* L.)

and birdsfoot trefoil (*Lotus corniculatus* L). The grass lye replaced the field crop rotation, as tilled crops are not recommended in Lithuania on slopes >10° (Svedas, 2001; Jankauskas, 1996).

Soil management and fertilizer treatments were applied in accordance with measured soil properties and standard regional agricultural practice. Before field experiments, soils were limed with one $CaCO_3$ application (according to hydrolytic acidity). Subsequent liming before each crop rotation enabled pH standardization on all investigated slopes. Chemical fertilizer inputs (ammonium nitrate, granulated superphosphate, and potassium chloride) were used according to plant requirements and soil properties. Mean annual applications of mineral fertilizers were: $N_{60}P_{28.4}K_{66}$ (field crops), $N_{50}P_{26.2}K_{58}$ (grain–grass crops), $N_{70}P_{26.2}K_{66}$ (grass–grain I crops), and $N_{120}P_{26.2}K_{66}$ (grass–grain II crops).

Water erosion rates were assessed by measuring the length and cross-sectional area of rills, to calculate soil loss volume (Chambers et al., 2000; Watson and Evans, 1991; Zaslavskij, 1983). Lost soil volume was calculated using the formula:

$$x = [(\Sigma l_1 p_1 + \Sigma l_2 p_2 + \ldots \Sigma l_n p_n) : n] : y, \tag{1}$$

where x = volume of erosion rills (m^3 ha^{-1}); l_1, l_2, … l_n = rill depth (cm); p_1, p_2, … p_n: rill width (cm, accuracy ±0.1 cm; n: number of rills on the measured plot width; y: measured plot width (m), and Σ: sum of measurements in our case 9) from selected 1 m length segments located at equal distances on the experimental plot. Data were transformed from m^3 ha^{-1} to Mg ha^{-1} using mean soil bulk density from each site.

The significance of differences between treatment means was determined using Fisher's LSD_{05} from the procedures of Dmitriev (1966) and Tonkunas (1957). These procedures included Analysis of Variance (ANOVA) at $P < 0.05$. Mean errors for investigations of over three-year duration (i.e. $n > 3$) were calculated using the formula:

$$s_x = \pm \frac{\sqrt{s_{x_1}^2 + s_{x_2}^2 + \ldots + s_{x_n}^2}}{n} \tag{2}$$

where s_{x_1}, s_{x_2}, s_{x_n} are individual errors of a single (one year) investigation and n is the number of investigations.

Soils were sampled before field experiments commenced in autumn 1981 and every six years (i.e. after each full crop rotation). Representative soil samples (0–20 cm depth) were taken from each field plot. For historical reasons, soil physical–chemical analytical techniques were mainly former Soviet procedures. Soil reaction (pH_{KCl}) was determined in 1M KCl soil sample extracts using a calibrated digital pH meter. Hydrolytic acidity was determined in 1M CH_3COONa

and soil sample extract (ratio sample: extract, 1:25 for mineral soil) by titrating with 1M NaOH (Askinazi, 1975). Exchangeable bases were determined by the Kappen-Hilkovic method, which involves hot titration of 0.1M HCl and soil sample filtrate (ratio sample:extract 1:5) with 0.1 M NaOH (Askinazi, 1975). Ca^{++}, Mg^{++}, K^+, Na^+, and NH_4^+ concentrations were determined on filtrates. Available P and K were extracted with ammonium acetate-lactate (A-L solution pH 3.7; ratio 1:20). Available P was determined by spectrophotometer and K determined by flame photometry (Egner et al., 1960; Vazenin, 1975). Soil organic matter (%) was determined by the Tiurin method (Orlov and Grisina, 1981), which is a wet combustion technique similar to the Walkley-Black method (USDA, NRCS, 1995).

Soil sampling for soil bulk and particle density was performed on the mid-slope, when the same crop was grown in all field experiment treatments (i.e. on perennial grasses in summer before stubble breaking, when preparing soil for winter rye). Soil samples were taken in rings of known volume ($\approx$100 cm^{-3}) and diameter. Samples were collected from 0–5 and 10–15 cm depth and the mean results from both depths are presented. Soil bulk density was determined by removing a block of soil from site, allowing no compaction or crumbling. The block was then dried and weighed and bulk density expressed in grams of oven-dry soil per cubic centimetre (Blake and Hartge, 1986; Motuzas et al., 1996). The soil volume was determined by pycnometer method (Blake, 1965; Motuzas et al., 1996). Total soil porosity and moisture retention capacity are calculated parameters from soil bulk density and particle density (Fedorovskij, 1975; Motuzas et al., 1996):

$$Pr = (1 - Bd : Pd) \times 100 \quad (3)$$

where Pr is soil porosity (%), Bd is soil bulk density (Mg m^{-3}) and Pd is soil particle density (Mg m^{-3}).

$$MRC = Pr : Bd \quad (4)$$

where MRC is moisture retention capacity (%), Pr is soil porosity (%) and Bd is soil bulk density (Mg m^{-3}).

Soil structural stability was analyzed using the Savinov-Bakseev method to determine dry and wet aggregate strength by wet and dry sieving (Fedorovskij, 1975), which involves standard sieves with mesh diameters 7, 5, 3, 1, 0.5, and 0.25 mm. The handset was used for dry sieving and two electro-motor sieve sets for wet sieving, as such sieving precludes operator influence. Aggregate fractions >1.0, 0.25–1.0, and <0.25 mm were used. Practical experience showed these measures enabled an assessment of soil structure for agronomic purposes. Soil aggregate composition is influenced by crop type and stage, liming and irrigation (Fedorovskij, 1975). Therefore, soil aggregate stability was determined on soil samples taken before disking of grass after each crop rotation course in 1988, 1994, and 2000.

Metabolizable energy of crop production was calculated using the equation:

$$ME=0.01746P+0.03123F+0.01365Fb+0.01478N, \quad (5)$$

where ME is the metabolizable energy (MJ kg^{-1}) (DM, dry matter), P the digestible protein, F the digestible fat, Fb the digestible fiber, and N the digestible nitrogen-free extracted components (g kg^{-1} DM) (Jankauskas et al., 2000; Kulpys et al., 1996).

3. Results

We need appropriate indicators to compare the productivity of different crops and crop rotations. "Metabolizable accumulated energy" (MJ ha^{-1}) was selected as the most appropriate indicator. The plant chemical analysis, required for calculating the indicator, permitted us to calculate two additional comparable indicators: quantity of digestible proteins (kg ha^{-1}) and quantity of feed units (f.u. ha^{-1}).

The mean productivity of sod-forming grass mixture in digestible proteins under the described soil management and fertilizer treatments (see "Materials and Methods") was 11.9% higher than the mean productivity of grass–grain crop rotations or 39.4% higher productivity of grain–grass crop rotation on the 10–14° slope. The grass–grain crop rotation exceeds the grain–grass crop rotation mean by 33.5%. The advantage of grass leys in digestible proteins was clearly less on the 2–5° and 5–10° slopes (Table 2). The mean annual amount of metabolizable energy accumulated by the erosion-preventive grass–grain crop rotations was 67.6 GJ ha^{-1} or 10.6% higher productivity than the grain–grass crop rotation on the 10–14° slope, but it was only 3.8% and 2.8% higher than accumulated by the field and by the grain–grass crop rotations, respectively on the 2–5° and 5–10° slopes (Table 2). The productivity of sod-forming grass leys (according to the quantity of feed units) were slightly less than the productivity of field and grain–grass crop rotations on the 2–5°, slope, but productivity of grass leys was higher on the 10–14° slope.

The erosion-protection capability of different crops varied. Soil losses under winter rye were 4.9–13.2 Mg ha^{-1} $year^{-1}$ and under spring barley were 13.9–41.7 Mg ha^{-1} $year^{-1}$ from the slopes of 2–5°, 5–10°, and 10–14°, respectively. Perennial grasses prevented erosion almost completely; with only a small soil loss from the grass–grain I crop rotation due to poor red clover cover in 1992. Potatoes had least erosion-preventive capability, with soil losses from slopes of 2–5° and 5–10° being 8.7 times higher than under winter rye and 3.1 times higher than under spring barley, respectively (Table 3). Erosion largely depended on rainfall intensity when cultivated soil was not protected by plant cover.

TABLE 2. Plant productivity from different land use systems on the slopes of different steepness (1983–2000).

Treatments (crop rotations)	The average 18 years productivity in:		
	Accumulated metabolizable energy (GJ ha^{-1})	Quantity of digestible proteins (kg ha^{-1})	Quantity of feed units (f.u. ha^{-1})
	2–5° slope		
(a) Field	67.9a*	367a	6,500a
(b) Grain-grass	68.2a	357a	6,412a
(c) Grass-grain I	67.0a	402b	6,285b
(d) Grass–grain II	70.9b	466c	6,323b
LSD_{05}	1.08	11.5	145.3
	5–10° slope		
(a) Field	64.2a	361b	6,137a
(b) Grain–grass	65.1a	350a	6,226a
(c) Grass–grain I	67.1b	415c	6,204a
(d) Grass–grain II	69.3c	476d	6,559b
LSD_{05}	1.54	9.2	127.7
	10–14° slope		
(g) Sod-forming grasses**	66.1d	481d	6,194c,d
(b) Grain–grass	61.1a	322a	5,731a
(c) Grass–grain I	65.0b	394b	6,007b
(d) Grass–grain II	70.2c	465c	6,252d,c
LSD_{05}	0.99	12.8	136.5

*Values with the same letter subscript are not significantly ($P < 0.05$) different. (g)** The sod-forming perennial grasses were grown instead of the field crop rotation on the 10–14° slope.

TABLE 3. Influence of plant cover and slope steepness on mean water erosion rates (1983–2000).

Slope Steepness	Soil loss (Mg ha^{-1}) when growing:			
	Perennial grasses	Winter rye	Spring barley	Potato
2–5°	0	4.88±0.399	13.88±1.595	37.27±8.609
5–10°	0	10.32±1.314	29.43±2.569	95.82±15.897
10–14°	0.06 ± 0.002	13.24±2.342	41.72±6.333	*0

*The sod-forming perennial grasses were grown instead of potatoes on the 10–14° slope.

The mean annual erosion rates under the field crop rotation were 9.9, 23.4, and 32.2 Mg ha^{-1} $year^{-1}$, respectively, from the slopes of 2–5°, 5–10°, and 10–14° (Jankauskas et al., 2004). The grass–grain crop rotations decreased mean annual erosion rates by 75–80% compared with the field crop rotation, while under the grain–grass crop rotation it decreased by 23–24%. However, even grass–grain crop rotations could not completely prevent water erosion, with mean annual rates of 7.2–7.4, 4.7–4.9, and 2.4–2.5 Mg ha^{-1} $year^{-1}$ on the slopes 10–14°, 5–10°, and 2–5°, respectively (Figure 2). A comprehensive archive of research data on soil loss under different crops and crop rotations were presented by Jankauskas et al. (2004) and Jankauskas and Tiknius (2004).

Hill altitude gradually decreased under the influences of soil erosion. The annual decreases were: 1–3.22 mm under the field crop rotation, 0.75–2.49 mm under the grain–grass crop rotation and 0.25–0.73 mm under the grass–grain crop rotation (Jankauskas and Tiknius, 2004). It is evident that soil erosion processes instigate landscape leveling. Hence these results may assist modeling landscape change (Müller et al., 2003).

Soil organic carbon (SOC) is a prime soil quality indicator. Carbon is the elemental component of SOM content, which in turn plays a key role in the global carbon cycle (Janzen, 2004; Tan and Lal, 2005). Higher soil losses promote greater SOM loss. Furthermore, various land use systems influence erosion rates and changes in soil physical properties. Erosion-preventive grass–grain crop rotations and perennial grasses for long-term use significantly increased SOM content on 2–5° and 5–10° slopes, compared to field crop rotations. Sod-forming perennial grasses significantly increased SOM content on 10–14° slopes compared with the grain–grass crop rotation (Table 4).

There is evidence of slow but progressive increases in SOM content under the influence of long-term usage of grass leys, compared with SOM content under the field crop rotation. The increase in SOM content was noticeable under the grass–grain I crop rotation only on the 10–14° slope after the first crop rotation, while significant increases in SOM content were evident on all investigated slopes after the third crop rotation.

Change in soil physical properties is slow process. However, there are notable changes in soil dry bulk density under the influence of different land use systems after the 18-year period (Table 5). The highest dry bulk density of soil and lowest soil particle density developed under the field crop rotation. The more favorable conditions for plant growth were under the grass–grain crop rotations on all investigated sites, and under the long-term perennial grasses on the 10–14° slopes.

Even more notable changes occurred in total soil porosity and moisture retention capacity (Table 6). The lowest percentage of total soil porosity and moisture retention capacity was under the field crop rotation. These parameters were significantly higher under the grass–grain crop rotations and especially under the long-term sod-forming perennial grasses.

TABLE 4. Mean contents of soil organic matter under different land use systems on the slopes of different steepness (1983–2000).

Treatments (crop rotations)	SOM (%) after:		
	1st crop rotation, 1988	2nd crop rotation, 1994	3rd crop rotation, 2000
	2–5° slope		
(a) Field	3.47a*	2.73a,b	2.64a
(b) Grain–grass	3.46a	2.54a	2.99b
(c) Grass–grain I	3.08a	3.65b	3.39c
(d) Grass–grain II	3.23a	3.47b	3.46c
LSD_{05}	0.412	0.301	0.284
	5–10° slope		
(a) Field	2.52a	2.37a	2.17a
(b) Grain–grass	2.47a	2.35a	2.01a
(c) Grass–grain I	2.48a	2.27a	2.75 b
(d) Grass–grain II	2.41a	2.31a	2.67b
LSD_{05}	0.287	0.169	0.164
	10–14° slope		
(g) Sod-forming grasses**	2.49a**	2.59b**	2.51b**
(b) Grain–grass	2.42a	2.24a	1.99a
(c) Grass–grain I	2.71b	2.47b	2.45b
(d) Grass–grain II	2.50a	2.39a	2.43b
LSD_{05}	0.232	0.221	0.328

*Values with the same letter subscript are not significantly ($P < 0.05$) different. (g)** The sod-forming perennial grasses were grown instead of the field crop rotation on the 10–14° slope.

TABLE 5. Soil density under different land use on the eroded slopes after three six-course crop rotations, 2000.

Treatments (crop rotations)	Dry bulk density ($Mg\ m^{-3}$)			Soil particle density ($Mg\ m^{-3}$)		
	On the slopes of:					
	2–5°	5–10°	10–14°	2–5°	5–10°	10–14°
(a) Field	1.66a	1.68a	1.37*d	2.58a	2.62a	2.64*c,b
(b) Grain–grass	1.56b	1.68a	1.54a	2.61a	2.63a	2.61a
(c) Grass–grain I	1.47c	1.57a,b	1.49b	2.63a	2.63a	2.63b
(d) Grass-grain II	1.48c	1.58b,a	1.44c	2.62a	2.64a	2.65c
LSD_{05}	0.074	0.118	0.05	0.032	0.038	0.017

* Long-term sod-forming perennial grasses were grown instead of field crop rotation on the 10–14° slope.

TABLE 6. Total soil porosity and moisture retention capacity under different land uses on the eroded slopes after three six-course crop rotations, 2000.

Treatments (crop rotations)	Total soil porosity (%)			Moisture retention capacity (%)		
	on the slopes of:					
	2–5°	5–10°	10–14°	2–5°	5–10°	10–14°
(a) Field	35.9a	35.8a	48.2*c	21.8a	21.4a,c	35.2*c
(b) Grain-grass	40.0b	36.0a,c	41.0a	25.7b	21.5a	26.6a
(c) Grass-grain I	43.9c	40.5b	43.5a	29.8c	25.9b	29.3b
(d) Grass-grain II	43.6c	40.0c	45.7b	29.5c	25.4b,c	31.7b
LSD_{05}	2.69	4.45	2.16	2.72	4.39	2.82

*Long-term sod-forming perennial grasses were grown instead of field crop rotation on the 10–14° slope.

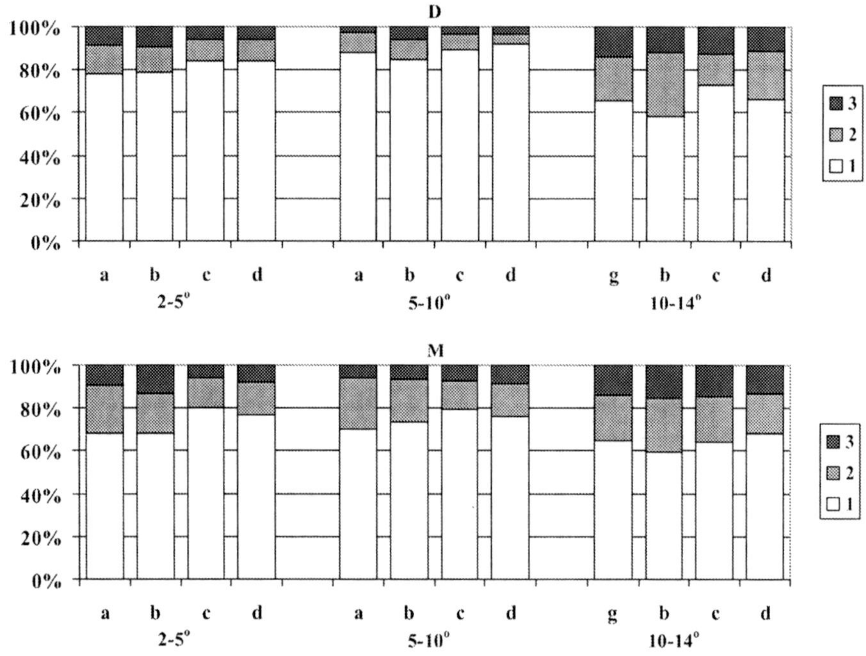

Figure 2. Influence of crop rotations on dry (D) and moist (M) soil aggregate strength after third six-course crop rotations in 2000. Treatments were as follows: D: dry aggregate strength (by dry sieving method), M: moist aggregate strength (by moist sieving method); Crop rotations: a: field, b: grain–grass, c: grass–grain I, d: grass–grain II, g: long-term sod forming perennial grasses. Soil aggregate fractions: (1) >1 mm; (2) 1–0.25 mm; (3) <0.25 mm.

Soil pads exist as complex and dynamic systems of micro- and macro-aggregates and affect many soil physical properties and microbiological processes (Fedorovskij, 1975). Aggregate stability measurements assist soil erodibility studies (Lal, 1979, 1981). According to our investigations the lowest percentage of agronomically most useful soil aggregates (>1 mm) was on the 10–14° slopes: 62–80% by dry sieving and 49–64% by moist sieving (Figure 2). This fraction was 73% and 91% (dry sieving) and 66% and 75% (wet sieving) on the slopes of 2–5° and 5–10°, respectively. The differences in aggregate stability depended on soil texture, which was slightly coarser on the 10–14° slopes.

The influence of different crop rotations was inconsistent in the case of dry sieving after the first course of crop rotations. However, it was evident that grass–grain crop rotations and sod-forming perennial grasses increased the soil aggregate fraction >1 mm in the case of moist sieving on the 10–14° slope (treatments c, d, g). These differences became more notable after the second crop rotation. The positive influence of grass–grain crop rotations on aggregate stability (>1 mm) was evident on all investigated slopes. The most notable increase in soil aggregate stability (dry sieving) under the grass–grain crop rotations and sod-forming perennial grasses was evident after the third crop rotation (Figure 2).

Chenu et al. (2000) and Tisdall (1996) postulated that the largest increase in aggregate stability would be under perennial grasses, including leguminous grass, e.g. red clover (c), rather than cereal grass leys (d). However, the results of three six-course crop rotations (18 years) showed there were no significant differences in soil aggregate stability between grass ley treatments c and d.

Phytocenoses, including sod-forming perennial grasses and grass–grain crop rotations, changed the physical properties of eroded soils: soil bulk density decreased and the percentages of total porosity and moisture retention capacity increased. The grass–grain crop rotations improved soil structure (measured in terms of water stable aggregates) by 6.81% units, compared with the field crop rotation and by 7.28% units compared with the grain–grass crop rotation on slopes of 2–5° and 5–10°, respectively. The grass–grain crop rotations increased the water stable aggregates by 11.03% units and sod-forming perennial grasses increased by 9.86% units, compared with the grain–grass crop rotation on the 10–14° slope. Therefore, grass–grain crop rotations and sod-forming perennial grasses decreased soil erodibility, as soil erodibility is inversely related to aggregate stability.

4. Discussion

Sustainable community development depends on different circumstances and conditions. Sustainable development is a multidimensional way of thinking about the interdependencies among natural, social, and economic systems in our world. It represents a process in which economics, finance, trade, energy, agriculture, industry, and all other policies are implemented in a way to bring about development that is economically, socially, and environmentally sustainable. Thus, the goal of sustainable development is to meet the needs of the present without compromising the ability of future generations to meet their needs (Flint, 2003).

Soil is one of most important natural resources influencing community development. Some soil quality indicators are extremely sensitive on hilly undulating relief. Soil quality is an indication of how well soil performs all of its functions. It cannot be determined by measuring only crop yield, water quality, or any other single outcome. Soil quality is an assessment of how soil performs all of its functions now and how those functions are being preserved for future use.

Fullen and Catt (2004) summarized that developing effective and viable soil conservation strategies is one of the most pressing soil management problems we face in the early 21st century. Spangenberg (2002) wrote that environmental space is a tool for exploring sustainable development benchmarks on a sound scientific basis, and it is helpful to derive indicators of sustainable development for different applications. However, the environmental space concept expresses no preference regarding the structure of the economic system, as long as the environmental and social benchmarks are respected, nor does it suggest specific sustainability indicators (Spangenberg, 2002).

Suitable soil conservation strategies and sustainable community development are essential requirements, satisfying the needs of present and future generations. Evaluating presented research data in terms of the quantity of crop production (Table 2), there is an evident priority for the community to develop areas under perennial grasses on steeper slopes within hilly undulating landscapes. But quantity of crop production is not as indicator for the needs of future generations or even for sustainable community development. Suitable indicators for sustainable land use in our case study were: soil losses due to soil erosion processes (Table 3 and Figure 2), changes in soil carbon content (Table 4), as factors influencing many soil physical–chemical properties: soil nutrient content, soil structural stability (Figure 2), soil bulk density (Table 5), soil porosity and soil water holding capacity (Table 6). These indicators are measurable properties of soil or plants that provide clues about how well the soil is functioning. Useful indicators are easy to measure, are accessible to many users and applicable to field conditions, and are sensitive to variations in management practice. Indicators can be assessed by qualitative or quantitative techniques. After measurements they can be evaluated by studying patterns and comparing results to measurements taken at different times or locations (Suter, 2001).

According to Montenerella et al. (2003), 26 million ha in the European Union suffer from water erosion and 1 million ha from wind erosion. The

Mediterranean region is historically the most severely affected by erosion in Europe. Quite high water erosion rates were estimated and on the hilly–undulating landscape of Lithuania (see Results; Jankauskas et al., 2004; Jankauskas and Fullen, 2006; Jankauskas and Tiknius, 2004). Moreover, in Lithuania tillage erosion must also be considered, along with water and wind erosion (Jankauskas, 2003; Jankauskas and Jankauskiene, 2003b).

High water erosion rates on the Kaltinenai field experiments were estimated starting in the early stages of field investigations. Soil losses were measured after every intensive rainfall and in every spring after snow melt. The essential differences in water erosion rates under different cultivated crops and under different crop rotations as land use systems were estimated after the first and further six-course crop rotations. However, there are uncertainties over the causal mechanisms: changes in physical soil properties (soil bulk density, porosity, water holding capacity, and soil organic matter content) were notable after the first crop rotation. They become more evident after the second and especially after the third six-course crop rotation. Our observations can be supported by results of other investigators. Erosion changes physical soil properties mainly because of the removal of surface soil rich in organic materials and exposure of lower soil layers. Bulk density and hydraulic conductivity of saturated soil increased slightly with erosion level (Arriaga, 2003). Studies generally agree (Arriaga, 2003; Janzen, 2004; Lal, 1981; Tisdall, 1996) that soil organic matter physically and chemically binds primary mineral particles together, thus increasing structural resistance to raindrop impact and splash. The consequent retention of large interstices between pads allows rapid infiltration, hence reducing overland flow (Fullen, 1991). Numerous field and laboratory studies have shown that soils with low organic matter contents are more erodible than more organic soils. Generally, soils with <2% organic matter are highly erodible (Fullen and Catt, 2004).

Water stable aggregate stability data indicate that the grass–grain crop rotations and sod-forming perennial grasses decreased soil erodibility. These results correlate with changes in SOM content. Lado et al. (2004) stated that the low aggregate stability and the high dispersivity of the low organic matter soil allowed the development of a dense and thick crust for all soil aggregate sizes. Conversely, the high aggregate stability and the low dispersivity of the high organic matter soil limited seal formation. Wright and Hons (2004) estimated that management practices, such as no-tillage and high-intensity cropping sequences, have the potential to enhance carbon and nitrogen sequestration in agricultural soils. In agricultural systems, maintenance of SOM has long been recognized as strategy to reduce soil degradation (Mikha and Rice, 2004). Moreover, effects of agricultural land use and land use change on SOC pools play an important role in mitigation of the global greenhouse effect (Leifeld et al., 2005). Increasing SOM content under the grass–grain crop rotations and under sod-forming perennial grass leys accord with this strategy. The conclusion of Dai and Huang (2006) that surface SOM concentration is in general negatively correlated with annual mean temperature and positively correlated with annual mean precipitation and altitude, correspond to our investigations, because higher precipitation predict higher water erosion rates.

SOM content changes in long-term field experiments at the Kaltinenai Research Station illustrate multiple influences of land use systems on SOM dynamics (Table 4). Firstly, the variety of crops as constituents of the rotation can differentially affect carbon sequestration processes (Jankauskas, 1996; Lal et al., 1998). Secondly, different land use systems require different intensities of soil tillage. Consequently, more intense soil tillage stimulates more SOM mineralization, which releases more carbon (C) from the soil store to the atmosphere (Lal, 1999). Thirdly, there were different soil losses due to water erosion under different land use systems: highest losses were under the field crop rotation and the lowest were under grass–grain rotations (Jankauskas et al., 2004). The higher soil losses lead to higher losses of SOM. Furthermore, different land uses influence C-sequestration by changing soil physical properties, such as dry bulk density, total soil porosity and water holding capacity. There were small changes in SOM (%) after both the first and even the second crop rotation (Table 4). However, differences in SOM (%) become more evident after the third crop rotation in 2000. Significantly higher SOM values were found under the grass–grain crop rotations on the 2–5° and 5–10° slopes compared with the field crop rotation, and under the sod-forming perennial grasses on the 10–14° slope compared with the grain–grass crop rotation.

The erosion-preventive capability of crop rotations depended on the erosion-protective properties of constituent crops, and the need for these measures increases with slope gradient. The research data enabled the design of appropriate erosion-resisting crop rotations (Table 7) and these rotations were recommended for erodible soils on 2–10o slopes. Long-term perennial grasses should be grown on slopes >10°. Thus, sod-forming perennial grasses and erosion-resisting crop rotations could assist both erosion control and the ecological stability of the vulnerable hilly undulating relief.

Finally, land use systems, including sod-forming perennial grasses and grass–grain crop rotations, can be considered as sustainable land use systems for Eutric Albeluvisols on the hilly rolling landscape of Lithuania. These land use systems correspond to the needs of both the present community and future generations. The practical recommendations for land users and policy makers contain 17 crop rotation protocols applicable for the different landscape condi-ions. The main attributes of the proposed land conservation and sustainable land use system were the careful selection of optimum erosion-preventive ecosystems (sod-forming perennial grasses or erosion-resisting crop rotations) with high erosion-resisting capabilities. These systems vary in response to slope conditions. Such ecosystems assist erosion control and thus the ecological stability and environmental security of the undulating topography.

TABLE 7. Recommended erosion-resisting crop rotations for fields of varying gradient.

Minimal grass (%)	Group number and composition of crop rotations
	When maximum available slope gradient is 7–10°
80	I. 1: winter grains or spring barley, 2–5: perennial grasses
72	II. 1: winter grains, 2: spring barley, 3–7: perennial grasses
67	III. 1: winter grains, 2: spring barley, 3–6: perennial grasses
63	IV. 1–2: winter grains, 3: spring barley, 4–8: perennial grasses
63	V. 1: winter grains, 2: spring grains, 3: spring barley, 4–8: perennial grasses
60	VI. 1: winter grains, 2: spring barley, 3–5: perennial grasses
	When maximum available slope gradient is 5–7°
57	VII. 1–2: winter grains, 3: spring barley, 4–7: perennial grasses
57	VIII. 1: winter grains, 2: spring grains, 3: spring barley, 4–7: perennial grasses
50	IX. 1–2: winter grains, 3: spring barley, 4–6: perennial grasses
50	X. 1: winter grains, 2: cereal grains with legumes, 3: spring barley, 4–6: perennial grasses
43	XI. 1: winter grains, 2: cereal grains with legumes, 3: winter grains, 4: spring barley, 5–7: perennial grasses
43	XII. 1: winter grains, 2: cereal grains with legumes, 3: spring grains, 4: spring barley, 5–7: perennial grasses
40	XIII. 1: winter grains, 2: spring barley or their mixture with legumes, 3: spring barley, 4–5: perennial grasses
	When maximum available slope gradient is 2–5°
38	XIV. 1: winter grains, 2: spring grains, 3: cereal grains with legumes, 4: winter grains, 5: spring barley, 6–8: perennial grasses
38	XV. 1: winter grains, 2: spring grains, 3: cereal grains with legumes, 4: spring grains, 5: spring barley, 6–8: perennial grasses
33	XVI. 1: winter grains, 2: spring grains, 3: cereal grains with legumes, 4: spring barley, 5–6: perennial grasses
33	XVII. 1–2: winter grains, 3: cereal grains with legumes, 4: spring barley, 5–6: perennial grasses.

5. Conclusions

Effective soil conservation strategies and sustainable community development are vital requirements to satisfy the needs of present and future generations. Land use systems, including sod-forming perennial grasses and grass–grain crop rotations, are sustainable land use systems for Eutric Albeluvisols on the hilly rolling landscape. These land use systems correspond to the needs of both the present and future generations.

The results of long-term field experiments from Kaltinenai Research Station of the Lithuanian Institute of Agriculture allow the following conclusions:

1. The mean annual amount of metabolizable energy accumulated by the erosion-preventive grass–grain crop rotations was 67.6 GJ ha^{-1} or 10.6% higher productivity of grain–grass crop rotation on the 10–14° slopes, but its was only 3.8% and 2.8% higher than accumulated by the field and the grain–grass crop rotations, respectively, on the 2–5° and 5–10° slopes.
2. The mean annual erosion rates under the field crop rotation were 9.9, 23.4, and 32.2 Mg ha^{-1} $year^{-1}$, respectively from the 2–5°, 5–10°, and 10–14° slopes. The grass–grain crop rotations decreased mean annual erosion rates by 75–80% compared with the field crop rotation, while under the grain–grass crop rotation it decreased by 23–24%.
3. Erosion-preventive grass–grain crop rotations and perennial grasses for long-term use significantly increased SOM content on 2–5° and 5–10° slopes, compared to field crop rotations. Sod-forming perennial grasses significantly increased SOM content on 10–14° slopes compared with the grain–grass crop rotation.
4. Percentage of total soil porosity and moisture retention capacity were significantly higher under the grass–grain crop rotations and under the long-term sod-forming perennial grasses compared with these parameters under the field crop rotation.
5. The grass–grain crop rotations increased water stable soil aggregates by 6.81% units compared with the field crop rotation and by 7.28% units compared with grain–grass crop rotation on the 2–5° and 5–10° slopes. The grass–grain crop rotations increased aggregate stability by 11.03% units and sod-forming perennial grasses increased it by 9.86% units compared with grain–grass crop rotation on the 10–14° slope.

Therefore, phytocenoses, including sod-forming perennial grasses and grass–grain crop rotations are recommended to land users and policymakers. These ecosystems vary in response to slope conditions and could assist erosion control, the ecological stability of the vulnerable hilly undulating relief, increasing carbon sequestration, decreasing atmospheric CO_2 concentrations, and positively influencing global climatic changes.

References

Andreasen, J.K., O'Neill, R.V., Noss, R., and Slosser, N.C., 2001. Considerations for the development of a terrestrial index of ecological integrity. Ecological Indicators 1, 21–35.

Arriaga, F., 2003. Soil physical properties and crop productivity of an eroded soil amended with cattle manure. Soil Science 168, 888–899.

Askinazi, D.L., 1975. The methods for determination of liming needs of soddy-podzolic soils. In: A.V. Sokolv and D.L. Askinazi (Editors), The Agro-chemical Methods of Soil Investigation. Nauka, Moscow, pp. 228–243 [in Russian].

Blake, G.R., 1965. Particle density. In: C.A. Black (Editor), Methods of Soil Analysis, Part 1, Agronomy 9. American Society of Agronomy. Madison, WI, pp. 371–373.

Blake, G.R., and Hartge, K.H., 1986. Bulk density. In: A. Klute (Editor), Methods of Soil Analysis, Part 1. Physical and Mineralogical Methods, 2nd edn., Agronomy 9. ASA and SSSA, Madison, WI, pp. 363–375.

Booth, A.C., Fullen, A.M., Jankauskas, B., and Jankauskiene, G., 2003. International calibration of the textural properties of Lithuanian Eutric Albeluvisols. Agricultural Sciences 4, 3–10.

Chambers, B.J., Garwood, T.W.D., and Unwin, R.J., 2000. Controlling soil water erosion and phosphorus losses from arable land in England and Wales. Journal of Environmental Quality 29, 145–150.

Chan, K.Y., Heenan, D.P., and Oates, A., 2002. Soil carbon fractions and relationship to soil quality under different tillage and stubble management. Soil & Tillage Research 63, 133–139.

Chenu, C., Le Bissonnais, Y., and Arrouays, D., 2000. Organic matter influence on clay wettability and soil aggregate stability. Soil Science Society of America Journal 64, 1479–1486.

Dai, W. and Huang, Y., 2006. Relation of soil organic mater concentration to climate and altitude in zonal soils of China. CATENA 65, 87–94.

Dmitriev, E.A., 1966. Use of variance-statistic methods for investigations of soil physical properties. In: S.I. Dolgov (Editor), The Agro-physical Methods of Soil Investigations. Nauka, Moscow, pp. 240–255 [in Russian].

Egner, H., Rehm, H., and Domingo, W.R., 1960. Untersuchungen über die shemische bodenanalyse als grundlage für die beurteilung des nährstoffzustandes der böden. II. Chemische extractionsmethoden zur phosphor und kaliumbestimmung. Uppsala. Sweden [in German].

FAO-UNESCO, 1994. Soil map of the World. Revised Legend with Corrections. ISRIC, Wageningen.

Fedorovskij, D.V., 1975. Methods for evaluation of some physical and water properties of soil in field and pot experiments. In: A.V. Sokolv and D.L. Askinazi (Editors), Methods for Agro-physical Investigation of Soil, Nauka, Moscow, pp. 248–284 [in Russian].

Flint, R.W. 2003. Chapter 4 – Sustainable development: what does sustainability mean to individuals in the conduct of their lives and businesses. In: G.M. Mudacumura and M.S. Shamsul Haque (Editors), Handbook of Development Policy Studies, Marcel Dekker, New York, pp. 67–87.

Fullen, M.A., 1991. Soil organic matter and erosion processes on arable loamy sand soils in the West Midlands of England. Soil Technology 4, 19–31.

Fullen, M.A. and Catt, J.A., 2004. Soil management problems and solutions. Arnold, London.

Jankauskas, B., 1996. Soil erosion. Margi Rastai, Vilnius [in Lithuanian with English abstract].

Jankauskas, B., 2003. Soil erosion monitoring metadata for modelling of landscape change. In: F. Müller, W. Kepner and K. Caesar (Editors), Landscape Sciences for Environmental Assessment, EcoSys 10, 88–97.

Jankauskas, B. and Fullen, M.A., 2002. A pedological investigation of soil erosion severity on undulating land in Lithuania. Canadian Journal of Soil Science 82, 311–321.

Jankauskas, B. and Fullen, M.A., 2006. Lithuania. In: J. Boardman and J. Poesen (Editors), Soil erosion in Europe, Wiley West Sussex, England, pp. 57–66.

Jankauskas, B. and Tiknius, A., 2004. Sustainable and environment-friendly land use system for undulating relief of the Temperate Climate Zone. EcoSys 42, 50–59.

Jankauskas, B., Jankauskiene, G., and Svedas, A., 2000. Comparison of the methods for the calculation of food energy value. Agriculture, Research works of biomedical sciences of Lithuanian institute of agriculture and Lithuanian university of agriculture 72, 239–251.

Jankauskas, B. and Jankauskiene, G., 2003a. Erosion-preventive crop rotations for landscape ecological stability in upland regions of Lithuania. Agriculture, Ecosystems and Environment 95, 129–142.

Jankauskas, B. and Jankauskiene, G., 2003b. Tillage erosion as a predictor of runoff erosion on cultivated slopes. In: M. Bel, J. Benites, B. Boydel et al. (Editors), Soil Management for Sustainability. 16th International Soil Tillage Research Organization Conference, The University of Queensland, Birsbane, pp. 589–595.

Jankauskas, B., Jankauskiene, G., and Fullen, M.A., 2004. Erosion-preventive crop rotations and water erosion rates on undulating slopes in Lithuania. Canadian Journal of Soil Science 84, 177–186.

Jankauskas, B., Jankauskiene, G., Slepetiene, A., Fullen, M.A., and Booth, C.A., 2005. International comparison of analytical protocols for determining soil organic matter content on Lithuanian Albeluvisols. Earth and Environmental Sciences. Acta Universitatis Latviensis 692, 66–77.

Janzen, H.H., 2004. Carbon cycling in earth systems – a soil science perspective. Agriculture, Ecosystems & Environment 3, 399–417.

Kulpys, J., Sederavicius, J., and Monkeviciene, I., 1996. Comparison of the methods of food energy evaluation. Issues of Agronomy and Animal Husbandry Sciences, Kaunas, pp. 462–465.

Lado, M., Paz, A., and Ben-Hur, M., 2004. Organic matter and aggregate size interactions in infiltration, seal formation, and soil loss. Soil Science Society of America Journal 68, 935–942.

Lal, R., 1979. Physical characteristics of soil of the Tropics: determination and management. In: R. Lal and D.J. Greenland (Editors), Soil Physical Properties of Crop Production in the Tropics. Wiley, pp. 3–44.

Lal, R., 1981. Physical properties. In: D.J. Greenland (Editor), Characterization of Soils in Relation to their Classification and Management for Crop Production: Examples from Some Areas of the Humid Tropics, Clarendon Press, Oxford, pp. 144–148.

Lal, R., 1999. Soil management and restoration for C sequestration to mitigate the accelerated greenhouse effect. Progress in Environmental Science 1, 307–326.

Lal, R., Kimble, J.M., Follett, R.F., and Cole, C.V., 1998. The potential of U. S. cropland to sequester carbon and mitigate the greenhouse effect. Ann Arbor Press, Chelsea, MI.

Leifeld, J., Bassin, S., and Fuhrer, J., 2005. Carbon stocks in Swiss agricultural soils predicted by land-use, soil characteristics, and altitude. Agriculture, Ecosystems and Environment 1–2, 255–266.

Lobb, D.A., Lindstrom, M.J., and Schumacher, T.E., 2003. Soil erosion processes and their interactions: implications for environmental indicators. In: R. Francavilga (Editor),

Agricultural impacts on soil erosion and soil biodiversity: developing indicators for policy analysis. OECD Expert Meeting on Soil Erosion and Soil Biodiversity Indicators. Rome, Italy, pp. 325–336.

Lopez-Ridaura, S., Masera, O., and Astier, M., 2002. Evaluating the sustainability of complex socio-environmental systems. The MESMIS framework. Ecological Indicators 2, 135–148.

Mikha, M.M. and Rice, C.W., 2004. Tillage and manure effects on soil and aggregate-associated carbon and nitrogen. Soil Science Society of America Journal 68, 809–816.

Montenerella, L., van Romaey, A., and Jones, R., 2003. Soil erosion risk in Europe. In: R. Francaviglia (Editor), Agricultural inspects on soil erosion and soil biodiversity: developing indicators for policy analysis. OECD Expert Meeting on Soil Erosion and Soil Biodiversity Indicators. Rome, Italy, pp. 87–100.

Morgan, R.P.C., 1995. Soil erosion & Conservation, 2nd edn., Longman, London.

Motuzas, A.J., Buivydaite, V., Danilevicius, V., and Sleinys, R. 1996. Pedology. Mokslo ir enciklop. l-la, Vilnius (in Lithuanian).

Müller, F., Kepner, W., and Caesar, K., 2003. The CCMS pilot study on the use of landscape science for environmental assessment. In: F. Müller, W. Kepner, and K. Caesar (Editors), Landscape Sciences for Environmental Assessment, EcoSys 10, 2–36.

Orlov, D.S. and Grisina, L.A., 1981. Guide in chemistry of humus. MGU, Moscow [in Russian].

Römkens, M.J.M., Helming, K., and Prasad, S.N., 2002. Soil erosion under different rainfall intensities, surface roughness, and soil water regimes. CATENA 46, 103–123.

Spangenberg, J.H., 2002. Environmental space and the prism of sustainability: frameworks for indicators measuring sustainable development. Ecological Indicators 2, 295–309.

Suter, G.W. II, 2001. Applicability of indicator monitoring to ecological risk assessment. Ecological Indicators 2, 101–112.

Svedas, A., 2001. Use of eroded soil for agriculture. In: V.V. Buivydaite, M. Eidukeviciene, I. Eitminaviciute et al. (Editors), Soils of Lithuania. Science and Arts of Lithuania 32, 741–744 (in Lithuanian, with English abstract).

Tan, Z. and Lal, R., 2005. Carbon sequestration potential estimates with changes in land use and tillage practice in Ohio, USA. Agriculture, Ecosystems and Environment 111, 140–152.

Tisdall, J.M., 1996. Formation of soil aggregates and accumulation of soil organic matter. In: M.R. Carter and B.A. Stewart (Editors), Structure and Organic Matter Storages in Agricultural Soils, CRC Press, Boca Raton, pp. 57–96.

Tonkunas, J., 1957. Metodology field experiments. Mokslas, Vilnius [in Lithuanian].

Torri, D., Poesen, J., and Borselli, L., 1997. Predictability and uncertainty of the soil erodibility factor using a global dataset. CATENA 31, 1–22.

USDA, NRCS, 1995. Soil survey laboratory information manual. Soil survey investigations report No. 45, Version 1.0, Lincoln, NE.

Vandermeer, J., Noordwijk, M., Anderson, J., Ong, C., and Perfecto, I., 1998. Global change and multi-species agroecosystems: concepts and issues. Agriculture, Ecosystems and Environment 67, 1–22.

Vazenin, I.G., 1975. The methods for determination of potassium in soil. In: A.V. Sokolv and D.L. Askinazi (Editors), The Agro-chemical Methods of Soil Investigation. Nauka, Moscow, pp. 191–218 [in Russian].

Watson, A. and Evans, R., 1991. A comparison of estimates of soil erosion made in the field and from photographs. Soil & Tillage Research 19, 17–27.

Wright, A.L. and Hons, F.M., 2004. Soil aggregation and carbon and nitrogen storage under soybean cropping sequences. Soil Science Society of America Journal 68, 507–513.

Zaslavskij, M.N., 1983. Erosion Science. Vyshaja skola, Moscow [in Russian].

NUCLEAR SAFETY AND ITS IMPACT ON THE LEVEL OF ENVIRONMENTAL SECURITY IN UKRAIN

VLADIMIR KUZNYETSOV*
Ukrainian Scientific Research Institute of Ecological Problems, Kharkiv, Ukraine

Abstract. Considered in the paper are issues of interdependence between the level of the safety of nuclear facilities operation and level of environmental security in Ukraine. Also discussed is attitude of the population to further development of nuclear energy sector in the country, contradictions between the desire to be energy independent and potential threat to environmental safety.

Keywords: Nuclear safety; environmental security; RAW management; public attitudes

1. Introduction

Environmental security should be considered as a priority criteria for assessment of the level of development of any country. Demographic crisis, high disease rate and mortality of population are results of the system crisis faced by the Ukrainian society. Ignoring the necessity of comprehensive solution of ecological and social problems and lack of due attention to inter-relation between socialeconomic and environmental factors contributes to further degradation of the country (Green Party of Ukraine, 2003).

2. Current threats to environmental security

The most objective assessment of the level of environmental security, including technological, economic, ecological, and social aspects is the assessment of the summary (total) risk or probability of occurrence and further development of unfavorable natural and anthropogenic processes causing subsequent ecological

* Corresponding Author: Dr. Kuznyetsov Vladimir, Ukrainian Scientific Research Institute of Ecological Problems, 6.Bakulin Str, 61166, Kharkiv, Ukraine, e-mail: vladkuz@ukr.net

I. Petrosillo et al. (eds.), Use of Landscape Sciences for the Assessment of Environmental Security, 453–459.

damage. In the process of such an assessment a certain uncertainty arises due to insufficient or inaccurate data on interrelation between a pollutant and man as well as with environment in general.

The National Security Council of Ukraine considers that at the present time Ukraine faces the following threats in the field of environmental security (National Security Council of Ukraine, 1997):

- High anthropogenic pollution load on the natural environment of the country resulting in great damage to it, caused also as a result of the Chernobyl accident
- Inefficient usage of natural resources, wide-scale application of outdated, environmentally dangerous production technologies
- Uncontrolled import to Ukraine of environmentally unsafe technologies, materials, and substances
- Operation of nuclear power stations having outdated designs of nuclear reactors which, probably, poses greatest potential threat to environmental security in Ukraine
- Negative consequences of the previous defense and military activities

The above list of threat shows that for Ukraine potential danger of nuclear power usage for environment and human health still poses the greatest challenge.

3. Current usage of nuclear energy

Nuclear power engineering historically plays important role in Ukraine. Nuclear power plants account for about 53.0% (in 2006) of the total wholesale electricity generation of Ukraine. There are 4 NPS in Ukraine with 15 nuclear reactors, majority of which are of old VVER-1000 type similar to those which were used in Chernobyl NPS. Their design service life is 30 years and most of them are in operation for more than 15 years. Despite extensive safety improvement measures undertaken at these NPS after Chernobyl accident the safety of their operation is still does not meet modern standards (The World Bank, 1995). However, operation of NPS is not only threat to environmental security in Ukraine. Accumulated and currently generated radioactive wastes (RAW) and spent nuclear, the main producer of which are NPS, as well as uranium mining and reprocessing industry (accumulated about 130 millions m^3 of RAW) (Karpan, 2006) also pose environmental threat. Other sources of possible radioactive contamination of environment are medical, scientific, industrial, and other enterprises and organizations which use radioactive substances.

3.1. NPS'S OPERATION SAFETY

Since 2002 National Nuclear Energy Generating Company has been implementing a large-scale "Comprehensive Program" on the safety improvement of operating nuclear power plants. This Program that was approved and continuously monitored by the Government of Ukraine has confirmed once more a paramount attention paid by our State to the NPP safety issues. The Program includes 400 measures on unit equipment and system upgrading aimed at the safety improvement. Implementing the "Comprehensive Program" strengthens the confidence that running units of Ukrainian nuclear power stations will operate safely during their designed service life (30 years). These measures are considered as sufficient enough to improve operational safety of the nuclear facilities and will greatly contribute to increase of the general level of environmental security in Ukraine.

3.2. ATTITUDES TO USE OF NUCLEAR ENERGY

In the recent years, general attitude to nuclear power engineering started to change both worldwide and in Ukraine, in particular. Probably, two most important reasons for such a change of attitude are climate change and increase in price of oil and depletion of its reserves. These phenomena stimulated political debate on nuclear energy usage issues and facilitated a kind of renaissance of interest to nuclear power engineering. Some consider use of nuclear energy as the only practical instrument to control the process of climate change. Concerning Ukraine the issue of climate change plays much lower role in the minds of people. For them accessibility to basic utility services is more urgent need and because Ukraine has very limited oil and natural gas reserves and financial resources are not sufficient to secure reliable availability of needed oil and gas many consider the continuation and moderate expansion of nuclear energy use as, probably, the only alternative available to this country. Also, another reason can be the fact that since the time of the Chernobyl accident (1986) there were no serious accidents at the Ukrainian nuclear stations the scale of which would have been above the acceptable limits used by INES (International Nuclear Event Scale). Reducing the level of dependency of Ukraine on import of oil and gas (mainly from Russia or via Russia) by further development nuclear power engineering potential (for which Ukraine has all required prerequisites , knowledge experience, raw materials, and technological know-how) is considered to be an important tool to achieve greater political and economic independence. Such change of attitudes was probably also a result of the fact that during the period of 1990–2000 production of electric power in Ukraine decreased by 42% To aggravate situation more than 98% of power generating units at TPS of Ukraine have their formal service life (100,000 h) finished and more than half of that are in operation for more than 200,000 h. So, the question is what direction the country should take in future development of its energy sector. Should it rely on traditional TPS requiring coal or oil and natural gas (resources of which are very limited in Ukraine and, therefore, there is a need to import it from Russia or elsewhere). Dependence on energy supply from Russia became a hot political issue in the recent years, especially after "orange revolution" of 2004 when supply of natural gas and oil from Russia became a political tool of influence. It had not only local but international impacts, including those upon some EU countries. These events contributed to initiation of new discussions on the need to further develop nuclear sector of Ukraine by including full nuclear fuel production cycle and establishment of RAW processing and disposal potential in Ukraine. Important factor in change of public opinion was the problem of climate change caused by CO_2 emissions generated at traditional thermal power stations. The possible way to reduce CO_2 emissions would be more heavy reliance on nuclear power generation.

4. Environmental impacts of energy sector

In general, the level of ecological safety of operation of power generation facilities (NPS and TPS) in Ukraine is quite low. Many facilities are prone to failures or technical accidents due to their age, poor maintenance and lack of funding for upgrading. They also generate considerable pollution load. So, in 2002 they accounted for 45% air emissions of pollutants, 25% of total pollutants discharge into water bodies, generated more than 26% of solid industrial wastes and about 65% of total greenhouse gases emissions. Especially serious problem is utilization of RAW (radioactive wastes) generated by NPS.

Historically Ukraine delivered RAW wastes to Russia for processing and burial there. Ukraine also imports nuclear fuel from Russia in exchange for nuclear warheads brought from Ukraine to Russia in the framework of the international agreement on refusal by Ukraine to store and use nuclear weapons. These transactions with Russia make Ukraine also quite dependent and as a result the Ukrainian government strives to develop full-cycle nuclear electric power production, including production of nuclear fuel, its use, and recycling and burial in Ukraine. This effort will require a lot of time and money; therefore, it is planned that by the year 2020 only some elements of such closed-cycle will be achieved in Ukraine.

5. Public perception of nuclear safety and environmental security

In order to facilitate further development of nuclear energy this country should secure required level of safety of nuclear facilities operation and to facilitate wider involvement of population in the decision-making process and gradual rid-off the so-called Chernobyl syndrome under which everything that in this or another way concerns or relates to nuclear energy, materials, and facilities causes in local people fear, mistrust, and psychological stress. Accidents in Chernobyl NPS (1986) and at Three Mile Island in USA (1979) showed what economic, social, environmental as well as health and psychological damage can be inflicted by industrial accidents at such facilities. Large part of the Ukrainian population has negative attitude to nuclear power engineering and unfortunately this attitude is boosted frequently by mass media and various political forces in order to gain some political benefits. In this respect it is interesting to analyze recent survey of how people treat some issues relating to use of nuclear energy in Ukraine. The summarized results are given in Table 1 below. Judging by this data we can see that there are quite a lot of contradictions between the attitudes of the Ukrainian population and those of the authorities to the issues of use of nuclear energy. On one side, there is general unease about usage of nuclear energy, on the other side, many people believe that further development of nuclear energy, provided proper operation safety is secured alongside with due public involvement into consultation and decision-making process, is the way the country should follow in the future in order to reduce its dependence from other countries – suppliers of oil and natural gas.

The results of the survey show that, in general, the issue of environmental security and nuclear safety as its component is not the first priority of concern. Among six questions put to people during the recent public survey the issue of nuclear safety was considered as only the fourth priority concern for majority of the people interviewed (total 2,008 people). This is considerable downward shift since the 1990s. Their principal concerns today are poverty (36% of people get salary below minimum living standard level (poverty line)), unemployment, high prices, and then environmental security. On the other side, poverty makes man more sensitive and susceptible to impact of poor quality environment, because poor people do not have means to protect themselves against such negative impacts.

TABLE 1. Attitudes of the Ukrainian population to issues of using nuclear power (The Razumkov Center, 2005).

No.	Question	Respondents, age 18–40	Respondents, age 40+	Respondents from areas with existing NPS
1.	Are you worried by consequences of the Chernobyl accident	12.8% yes	15.0% yes	17.0 %
2.	Are you worried by plans to build new NPS	6.5 % yes	12.5 % yes	
3.	Which source of energy will dominate in Ukraine in 10 years time:			
	Nuclear energy	26.3 %		36.0 %
	Renewable sources (solar, wind)	14.7 % *(Current share of use – 0.005 %), *though* 31.4 % of respondents could not formulate their answer on this question which means that population is not aware or does not care about renewable sources of energy		
4.	How dangerous are NPS:			
	Very dangerous	24.6 %		
	Rather dangerous	40.3 %		
	Relatively safe	24.1 %		
	Absolutely safe	3.5 %		
5.	Will nuclear PS contribute to higher energy independence of Ukraine	44 %		39.4 %
6.	Are you worried by plans to *prolong* operation of existing NPS	Yes – 55.3 %		
7.	Do you support building of new NPS	No – 54.9 % Yes – 26.8 % Do not know – 18.3 %		

6. Conclusions

Nuclear safety is very important component of general environmental security in Ukraine because of the presence of operational NPS and sad history and consequences of the Chernobyl accident. Considering attitudes of the Ukrainian population to impact of operation of NPS and the level of their safety on the level of environmental security in Ukraine a conclusion can be made that these attitudes differ with age, geographical location, personal association with nuclear accident in Chernobyl, and the level of awareness about the issue of environmental security as such.

Very often these attitudes are contradicting. The future of nuclear power use depends on more wide involvement of population (through national referendums) in decision-making, better information on safety issues relating to use of NPS, improvement of the level of trust to authorities (currently only 7.4% trusts their information or, interestingly enough, to information provided by NGOs!) as well as availability of information on safety, costs, future plans, and environmental consequences of NPS use and development. Measures on eradicating "Chernobyl syndrome" are also necessary (UNDP, 2005).

References

Green Party of Ukraine, 2003. Political Program, Green Party of Ukraine, Kiev, pp. 2–3.

Karpan, N., 2006. Nuclear power engineering cannot wash its hands from the Chernobyl. Zerkalo Nedeli Newspaper, April 4, 2006, Kiev.

National Security Council of Ukraine, 1997. Concept of the national security policy of Ukraine. National Security Council of Ukraine, Kiev, pp. 4–5.

The Razumkov Center, 2005. Nuclear Power Engineering in Ukraine: Problems of Development, Safety and Pub lic Support in the European Integration Context. The Razumkov Center, Kiev.

The World Bank, 1995. Monitoring Environmental Progress. The World Bank, Washington, D.C.

UNDP, 2005. The Human Consequences of the Chernobyl Nuclear Accident: A Strategy for Recovery. UNDP, New York.

ENVIRONMENTAL IMPACT ASSESSMENT AS A TOOL FOR ENVIRONMENTAL RESTORATION: THE CASE STUDY OF COPŞA-MICĂ AREA, ROMANIA

OCTAVIAN-LIVIU MUNTEAN*
"Babeş-Bolyai" University, Faculty of Environmental Science, Cluj-Napoca, Romania

LUCIAN DRĂGUŢ
West University, Department of Geography, Timisoara, Romania
Salzburg University, Z_GIS-Centre for Geoinformatics, Salzburg, Austria

NICOLAE BACIU
"Babeş-Bolyai" University, Faculty of Geography, Cluj-Napoca, Romania

TITUS MAN
"Babeş-Bolyai" University, Faculty of Geography, Cluj-Napoca, Romania

LIVIU BUZILĂ
"Babeş-Bolyai" University, Faculty of Geography, Cluj-Napoca, Romania

IOAN FERENCIK
CISD, Geo-Team, The Hague, Netherlands

Abstract. This paper presents the roots and the consequences of the environmental degradation in Copşa Mică Area (Romania), as well as mitigation measures. The environmental impact assessment (EIA) and territorial analysis were applied on this environmental area. The assessment of the environmental decline was based on four categories of indicators: environmental, economic, social, and landscape structure and quality. We have established a set of

* Corresponding Author: Dr. Octavian-Liviu Muntean, "Babeş-Bolyai" University, Faculty of Environmental Science, Stefan cel Mare Square 4, 400192, Cluj-Napoca, Romania, Phone: +40.264.405316, Fax: +40.264.599444 , e-mail: liviu@enviro.ubbcluj.ro

I. Petrosillo et al. (eds.), Use of Landscape Sciences for the Assessment of Environmental Security, 461–474.

relational environmental indicators, viewed as a tool for community development and enhancement of the environmental security. The methodology and results of the assessment (i.e. assessment matrix and a plan of environmental restoration) will be implemented into the process of local environmental planning.

Keywords: EIA; Copşa Mică; Rapid Impact Assessment Matrix; SWOT analysis; restoration; Romania

1. Introduction

As Barnett (2000) pointed out, the argument that environmental degradation will lead to conflict is a well-established concern of international studies, although skeptics consider it only a minor factor in civil strife (Swart, 1996). On the other hand, the environmental decline impacts upon human security, by triggering negative synergies with extreme events. Thus, the occurrence of extreme events, their superposition with the creeping environmental deteriorations is usually local or regional phenomenon, while the expected consequences are global ones (Bogardi, 2004). Even if not spectacular, processes leading towards environmental degradation are not less important from the perspective of the environmental security. In this context, the research on developing adequate methods to evaluate the environmental impact is more and more important. However, these methods are only efficient when used in copying with the environmental decline. This implies they should be meaningful to the decision-makers, or decision oriented (Pischke and Cashmore, 2006).

Environmental Impact Assessment (EIA) was developed and introduced in the 1960s as a tool to improve and involve the public in decision-making (Kværner et al., 2006). The concept of EIA has a great relevance to the whole environmental theory and practice because it can be interpreted as an environmental tool, which must be oriented to identify the environmental components that have been affected by human impact and its harmful effects (Muntean, 2003).

In this context, EIA has the potential to substantially contribute to sustainable territorial planning and consequently, towards a healthy future development of a given area and its community. At the same time, EIA is oriented to sustain the rehabilitation, reconstruction, and conservation of the environmental components (natural and human components) within a territory. However, EIA needs radical improvements to offer a tool for sustainable planning (Benson, 2003).

The nature and intensity of the human impact upon the environmental components are directly proportional to the natural resources of the territory, the technological level, land use structure, mentalities, and local tradition, etc. The result of this impact is mirrored in the territory by the characteristics and quality of environmental status. Consequently, the assessment of the environmental impact is very complex, taking into accounts the great variety of human actions and their effects upon the environment (i.e. obvious and discrete consequences).

Industrial activities and their extent in time within the study area, together with density and present human pressure have generated various environmental impacts (e.g. direct, indirect, visual, cumulative). The assessment of environmental impact on this area is very important exercise as prerequisite for planning the future evolution of natural and socioeconomic environment. Without getting into details, we can emphasize four types of human impact upon the local area (Muntean, 2003):

- *Direct impact.* Most of the human interventions belong to this type. Their effects were extensively analyzed and assessed by the governmental agencies and by Romanian and foreign experts. These effects are: accumulation of heavy metals in soil, water, and living organisms (Lăcătuşu et al., 1996, Surdu et al., 1998, Bardac et al., 1999), pollution of animals and vegetation (Serdaru et al., 2001), pollution of industrial environment, degradation of the houses, increasing of the unemployment rate and poverty, changes in land use planning, etc.
- *Indirect impact* have produced slow alteration and change of the internal structure of environmental elements: the structure of the soil, increasing in the frequency of professional diseases (Bardac et al., 1999), modification of the albedo, death of the animals (especially horses and fishes), economic stagnation, emergence of pollutants into food (Bratu and Georgescu, 2005), etc.
- *Visual impact.* Aesthetical alteration of the territorial system and landscape of the area is a consequence of the visual impact generated by direct and indirect industrial pollution. Natural, social, and built environments are degraded aesthetically and point out the characteristics of the gloomy landscape of Copşa Mică Area.
- *Cumulative impact* is the result of the territorial and temporal combination of the previous three types and due to its effects, is the most encountered impact in Copşa Mică Area. The importance of its consequences is: severe poverty, decrease in the life expectancy and environmental decline.

The overall goal of this study is to develop a set of relational environmental indicators relevants for the environmental planning process and meaningful for the decision-makers. These indicators were customized to the characteristics of the Copsa Mica Area and then used to elaborate an environmental restoration plan, which would enhance the environmental security within this territory.

2. Material and methods

2.1. STUDY AREA

Copşa Mică Area is located within Târnava Mare Corridor (Transylvanian Tableland) in the central part of Romania (Figure 1). Its environmental decline is due to the industrial pollution and its effects on territorial systems, human health (Pandi et al., 2002), ecosystems, quality of life, and other environmental components (air, soil, water, vegetation, fauna, other human components, etc.). Within the Romanian context, the linkage between environmental decline and poverty was highlighted in some studies (Şimandan, 2006). It was generally accepted as a consequence of inadequate development of heavy metal industry in Romanian communist period and a matter of neglecting the fight against pollution. Copşa Mică is a small town (roughly 5,000 inhabitants) having for more than six decades an important branch of chemical and nonferrous metals industry.

The diffusion of the polluting agents as black smoke and heavy metals (cadmium, zinc and lead) to both air and soil has been generating the most serious environmental problems (Mac and Muntean, 2000). The severe pollution of soils and vegetations with cadmium and lead contributes to decreasing life expectancy of people by 9–10 years (Lăcătuşu et al., 1996). That area is an example for careless environmental management and it may be considered as a representative model in all environmental planning studies (Muntean, 2003).

After 1989 Copşa Mică Area became a subject of environmental concern at both national and international level and was given the legislative status of "less-favored area" in November 2000. Within this legislative context, decision-makers at national, regional, and local level are increasingly involved in the environmental and social rehabilitation of this area.

The present territorial reality reflects the fact that the conflicting environmental relations between the geographical potential and the local socio-economic system have altered in time (Lăcătuşu et al., 1996, Surdu et al., 1998, Bardac et al., 1999, Mac and Muntean, 2000, Serdaru et al., 2001, Pandi et al., 2002, Bratu and Georgescu, 2005). Local industrial strategy had impoverished local environment, the economic viability and people, or as Şimandan said (2006) "people had incomes, but poverty was still everywhere and was induced by this accelerated development." The identification and assessment of the environmental states of this axe of intense human pressure is vital for supporting development policies at local and regional levels.

Figure 1. Geographical location of Copşa Mică Area (Romania).

2.2. THE RAPID IMPACT ASSESSMENT MATRIX (RIAM) FOR EIA

For the time being, the environmental studies are designed according with Romanian legislative framework, which takes into account some indicators used in the elaboration of General Urban Plan and EIA (Environmental Impact Assessment) studies. All environmental and socioeconomic indicators provisioned are based on official data, comprised within the specific chapters and annexes of the above-mentioned studies (i.e. unemployment rate, maximal concentration limit of pollutants, density of buildings in sites, etc.).

Despite of their accuracy, those indicators do not offer a good link between environmental and social information and decision-makers, because many politicians are not skilled enough in reading and assessing them. Moreover, to answer the need for integrated assessment (Aspinall and Pearson, 2000), the indicators should be integrated as well. This is the reason for that we are thinking about a meaningful global indicator to evaluate the environmental decline in this area.

The environmental impact upon the Copşa Mică Area was assessed using the adapted Rapid Impact Assessment Matrix (RIAM), which is a new tool for performing the environmental impact assessment (EIA) (Pastakia and Jensen, 1998; Wang et al., 2006). RIAM uses a structured matrix to allow different judgments (both subjective and based on quantitative data) and provides a transparent and permanent record of these judgments (Pastakia and Jensen, 1998; Al Malek and Mohamed, 2005).

In addition, RIAM is a tool to organize, analyze, and present the results of a holistic environmental impact assessment (EIA). The simple and structured nature of RIAM (Wang et al., 2006) allows for further in-depth analysis of selected components in a rapid and accurate manner (Pastakia and Jensen, 1998).

The main assessment criteria fall into two groups:

- Criteria that are of importance to the local condition, that individually can change the score obtained (A): importance of condition (A1), magnitude of change/effect (A2), and
- Criteria that are of value to the situation, but should not be individually capable of changing the score obtained (B): permanence (B1), reversibility (B2), and cumulative effects (B3)

The sum of the group (B) scores is then multiplied by the result of the group (A) scores to provide a final assessment score (ES) for the condition (Eqs. (1–3), Table 1). The process for the RIAM in its present form can be expressed as:

$$(A1) \times (A2) = AT \quad (1)$$

$$(B1) + (B2) + (B3) = BT \quad (2)$$

$$(AT) \times (BT) = ES \quad (3)$$

The environmental components taken into account were physical, derived and human components (including economic and social aspects). The results of matrix were tied up in an environmental score (ES), which was transformed in range bands (RB) of impacts (from +E = major positive change/impacts to –E: major negative change/impacts). Data concerning quality of life (QOL) in local area were added to RIAM (Table 1).

TABLE 1. Rapid Impact Assessment Matrix (RIAM) of environmental decline and QOL. I – Industry; A – Agriculture; T – Transport; W – Waste management.

Environmental and quality of life components	A1	A2	BI	B2	B3	ES	RB	Impact activities
Physical components								
Geological features	1	–1	3	2	3	–8	–A	I
Landforms	2	–2	2	2	3	–28	–C	I, T, A
Water	3	–3	2	2	3	–63	–D	I, A, T, W
Atmospheric features	3	–2	2	2	2	–36	–D	I, T, W
Derived components								
Vegetation	2	–3	2	2	3	–42	–D	I, A, T
Fauna	2	–1	2	2	2	–12	–B	I, A, T
Soils	2	–2	3	2	3	–32	–C	I, A, T, W
Human components and features								
Buildings	2	–1	3	2	3	–16	–B	I, T, W
Habitat space (residential)	1	–1	2	2	2	–6	–A	I, T, W
Green spaces	1	–1	2	2	2	–6	–A	I, T, W
Economic components	2	+1	1	1	1	+6	+A	I, T
Landscape quality	2	–2	3	2	3	–32	–C	I, T, W, A
Total environmental score	23	–15	27	23	30	–275	–E	
Quality of life								
Human health	2	–2	2	2	3	–28	–C	I, T
Life expectancy	1	–3	3	3	3	–27	–C	I
Rate of unemployment	1	–3	2	2	3	–21	–C	I, A
Educational attainment	1	+1	1	1	1	+3	+A	I
Adjusted real income	1	–3	1	2	3	–18	–B	I
QOL score	6	–13	9	10	13	–91	–E	
Global total score (GTS)	29	–28	36	33	43	–366	–E	

We have considered that human impact generates both environmental decline and decay in quality of life (QOL). It is impossible to list all array of attributes related to the concept of "Quality of Life," but some of them have been described in literature. For instance, in the Human Development Report (UNDP, 2002), Human Development Index measures the achievements of a country in terms of life expectancy, educational attainment, and adjusted real income (standard of life).

We tried to customize those indexes to our case study by adding another two indexes: human health and rate of unemployment. We have also reduced standard of life index to level of incomes, since the first term involves a great number of parameters, which are heavy to cope with.

3. Results and discussion

Global total score (GTS) emerged from RIAM analysis is –366; this value places our area of study into the –E class/major negative changes (accordingly with the range band above mentioned). Although there is a numeric difference between total environmental score (TES, with –275 value) and QOL score (–91), both environmental and QOL components are integrated in the same class of range band (–E). This difference is the consequence of the different number of indicators added (12 for TES and 5 for QOL). The environmental score expresses the environmental decline of the area (maximal negative in our case). In addition, this area may be viewed as an example of environmental decline (Muntean et al., 2003) with consequences on parameters of local human life.

It is easy to remark that only two indicators get positive values (economic components and educational achievements). These positive scores indicate two particular aspects. In the first situation, the economic components have positive value related to the reduction of polluting sources as a result of ceasing activity of the black smoke plant in 1993. In the second case the positive score of educational achievement indicator depicts the educational polarization generated by local employment opportunities (e.g. those related with chemical and nonferrous industries).

The index obtained from RIAM displays a holistic view on the environmental and QOL state of a specific area. As a relational index, it is easily comprehensible to decision-makers and assessors whatever would be their training and skills. It also allows for an objective comparison between different spaces and each component is evaluated according to the range bands.

That factorial assessment allows for establishing priority lists in decision-making process. For instance, in our matrix there are three components (water, atmospheric features, and vegetation) aggregated into the same class (–D) but the greater negative value is attached to water. Indeed, the environmental monitoring studies have been recording frequent values of pollutants higher than maximal limits, both in surface and underground water. The ground water has being polluted since 1972 with lead, zinc, cyanides, and arsenic (Sorocovschi, 1996). Under these circumstances, the quality of water is an important problem of this area and it should be a priority for the local environmental policy.

3.1. EIA AND ENVIRONMENTAL RESTORATION

As Van Cleve et al. (2006) pointed out "regional-scale restoration is a tool of growing importance in environmental management." In the territorial development strategies, the term "restoration" has – from territorial point of view – an important relevance because it means a local environmental policy related to conservation, diminution, or reconstruction of the environmental components, which have been affected by environmental impact and its effects (Muntean, 2003). Thus, the environmental restoration means more than the recovery of degraded lands for the preservation of biodiversity, which is the goal of restoration ecology (Dobson et al., 1997). We consider that the environmental management and restoration seek to improve and protect environmental quality for residents – both by controlling the environmental state, and by segregating activities that are environmentally inconsistent with it. In our case, the environmental management aims at improving the environmental state of Copşa Mică Area and at offering the main environmental protection strategies appropriate to this area.

The preservation of geographical systems requires, besides new ways of thinking the interventions, financial support to allow substantial investments in the environmental protection (Mac and Muntean, 2000). The costs of these actions are estimated at 15 million USD. The consequences of environmental policy in this territory are long-term ones and they should solve the environmental problems.

In this case, the *SWOT (Strengths, Weaknesses, Opportunities, Threats) analysis* allow us to identify strength aspects, weaknesses, opportunities, and threats of an environmental restoration policy (Table 2). Also, we established the four levels of the environmental restoration policy:

- *Scientific level* (based on scientific studies, articles and reports, and participation of the experts, scientists, institutions)
- *Level of public participation* (based on the public participation and consultation of local community, local NGO's, mass media)
- *Decisional level* (based on the decision-making process related to environmental restoration and participation of the national, regional and local authorities and factors)
- *Executive level* (level of implementation) based on implementation and monitoring of policy; the actors at this level are ministries, EPA and local departments of environmental protection)

TABLE 2. SWOT analysis for the environmental restoration.

STRENGTHS (S)	WEAKNESSES (W)	OPPORTUNITIES (O)	THREATS (T)
Favorable geographical location	Historical heavy metal pollution of water, soil, vegetation, air, and fauna	Mitigation measures for industrial pollution	Diminishing of economic interest
Environmental resources (methane, useful rocks)	Environmental impacts (direct, indirect, visual, and cumulative)	Increasing of economic investments in environmental restoration	Unadequate measures of environmental restoration
Environmental public concern	Lack of stakeholders interest	Elaborating of local environmental policy	Rising heavy metal pollution and cumulative effects
Legislative and institutional context	Economic transition	Increasing public participation into the decision-making processes	Unadequate use of environmental funds
EU structural financial support	Degradation of human life's quality	Investments in environmental restoration	Poverty and environmental decline
Status of less-favoured area	Lack of environmental strategies	Economic and financial opportunities	Increasing economic pressure

On the base of the SWOT analysis, territorial integrated approach and EIA, we elaborate an *environmental restoration plan* (Table 3). This is a final result of the area planning and restoration process. Proposals that exist in this plan are prepared for the distribution to experts and decision factors. We conclude that the process of environmental restoration is anticipative, integrative, technical, and participative.

The plan will follow the next steps in territorial implementation: identification of environmental problems in territory; environmental impact assessment and estimated impacts; establish the goals and objectives of environmental restoration; seek the alternative strategies; local planning and the legislative context; implementation of the restoration policy; evaluation and improvement of restoration policy.

These steps must be related to SWOT analysis and integrative territorial approach of the area and then assessed by decision-makers and experts that decide the implementation of the environmental restoration plan.

TABLE 3. Environmental restoration plan of Copşa Mică Area.

ACTIONS	OBJECTIVES	ENVIRONMENTAL TARGETS	TIME HORIZONS
Establishing of restoration area	Mapping of environmental areas and units	Natural, biological, and human components	Short
Integrated environmental approach	Area integrated assessment Biophysical resources EIA, Environmental constraints	Natural , biological, and human components	Short Medium
Natural components restoration	Mitigation of heavy metals pollution	Natural components	Medium Long
Biological components restoration	Natural land cover restoration Mitigation of heavy metals pollution	Biological components (vegetation, fauna, soils)	Medium Long
Landscape restoration	New landscape management (enhance visual quality)	Landscape and land use planning	Medium Long
Build environment restoration	Restoration of buildings and houses	Civilian and industrial buildings and houses	Medium
Social restoration	Reduction of unemployment rate and poverty Improvement of social life	Social environment Communitarian relationships	Long
Economic restoration	Legislative constraints Economic investments Agricultural development	Stakeholders	Medium
Change of collective attitudes toward the sustainability	New environmental behavior Public participation	Decision-making factors Stakeholders	Long

For the experts and policymakers, the concept of environmental restoration could be considered a policy of environmental protection in this area. The environmental restoration is trying to define the principles, directions, targets and identify the criteria of the actions leading to a sustainable development (environmental, economic, and social) of this area under local and regional circumstances. Viewed in this context, "restoration" becomes an attitude towards stability and sustainability on the local and regional scale in terms of environmental change and the economic development.

In Copşa Mică Area, the general problem is to keep fluctuations of environmental parameters within manageable bounds, considering the demands of society on the environmental systems. Within the Copşa Mică Area, environmental restoration is a local policy orientated to four territorial components:

1. Environmental components: improvement of soil fertility, quality of air and water; biodiversity, productivity of ecosystems, natural resources, landscape potential, etc; all these elements are the "environmental support" for local human community and their activities
2. Economical/technological components; these are the main key issues in this territory because of the outdated technology in use for the period of the centralized economy (1960–1989). The sustainable future of the area is likely to be stimulated by an economic shift toward "soft" and "green" technologies
3. Social components; there is an ethnic melting pot in the area (Romanian, Romany, German, Hungarian) with different social attitudes and environmental behavior. A new "common sense" attitude and environmental education are pillars of a new environmental behavior within the area. In addition, the public concern and participation in solving the environmental problems are conditions for a sustainable future of area
4. Legislative and decisional components. Decision-making actors at the local and regional level approve political, legislative and administrative components, environmental reconstruction and planning. In this situation, the new legislative status of "less-favored area" (from 2000 to 2010) of Copşa Mică Area is a key point for the environmental restoration. This status was based on the environmental and social criteria and its objectives promoting the rehabilitation measures of environmental, social, and economic features of area. The legislative aspects of this status offer a beneficial background for environmental planning.

4. Conclusion

The Rapid Impact Assessment Matrix (RIAM) proved to be a powerful tool for Environmental Impact Assessment in the study area. Together with SWOT analysis, it was succesfuly used in elaborating the environmental restoration plan for one of the most polluted areas in Romania. Thus, Copşa Mică Area can be an environmental restoration model for this country. This is a "constructive model" connected with the local community experience in respect with the environment, the level of economic development and the warnings of the scientific research on the state of environment.

This study will be developed and some problems still remain to be solved in the future. The first challenge is the validation of the methodology and results in practice by calibrating the spatial model of Copşa Mică Area (Muntean et al., 2005). Then, it would be interesting to fit our results as function of public preferences and needs, because sometimes scientific and common points of view are different.

As a matter of fact, due to the spatial immobility and environmental attitude many people prefer to work in their neighborhoods even if the plants generate pollution and environmental decline. The meaning of quality of life could be different for people if poverty increases over a specific limit. Thus, for local inhabitants, a high standard of life frequently means to have a work place in spite of threatens upon their health.

We consider that environmental security of area will depend on two main objectives: (a) to mitigate the detrimental impacts and (b) to satisfy the environmental requirements of stakeholders for a better quality of life.

References

Al Malek, S.A. and Mohamed, A.M.O., 2005. Environmental impact assessment of off shore oil spill on desalination plant. Desalination 185, 9–30.

Aspinall, R. and Pearson, D., 2000. Integrated geographical assessment of environmental condition in water catchments: linking landscape ecology, environmental modelling and GIS. Journal of Environmental Management 59, 299–319.

Bardac, D.I., Brezai, C., Florescu, E., Fodor, I., Gurzău, E., Marchian, D., Muntean, C., Obădău, C., Resiga, E., Szasz, A., Selejean, R., and Vulcu, L., 1999. Copşa Mică. Elemente de monografie medicală şi socială. Vol. I. Casa de Presă şi Editură Tribuna, Sibiu, 179 pp.

Barnett, J. 2000. Destabilizing the environment: conflict thesis. Review of International Studies 26, 271–288.

Benson, J.F., 2003. What is the alternative? Impact assessment tools and sustainable planning. Impact Assessment and Project Appraisal 21, 261–280.

Bogardi, J.J., 2004. Hazards, risks and vulnerabilities in a changing environment: the expected onslaught on human security? Global Environmental Change 14, 361–365.

Bratu, I. and Georgescu, C., 2005. Studiu comparativ al contaminării cu metale grele din miere şi sursele florale provenite dintr-o zonă poluată chimic. Revista de Chimie 56, 118–127.

Dobson, A.P., Bradshaw, A.D., and Baker, A.J.M., 1997. Hopes for the future: restoration ecology and conservation biology. Science 277, 515–522.

Kværner, J., Swensen, G., and Erikstad, L., 2006. Assessing environmental vulnerability in EIA – the content and context of the vulnerability concept in an alternative approach to standard EIA procedure. Environmental Impact Assessment Review 26, 511–527.

Lăcătuşu, R., Răuță, C., Cârstea, S., and Ghelase, I., 1996. Soil-plant-man relationships in heavy metal polluted areas in Romania. Applied Geochemistry 11, 105–107.

Mac, I. and Muntean, O.L., 2000. Geosystem, Constraint and Environmental Decline. 29th International Geographical Congress, Abstracts, pp. 194–195.

Muntean, O.L., 2003. A study of environmental planning in Copşa Mică Area (Romania). Würzburger Geographische Manuskripte 63, 163–170.

Muntean, O.L., Baciu, N., and Drăguţ, L., 2003. Environmental decline assessment in Copşa Mică Area (Romania). EcoSys 10, 98–106.

Muntean, O.L., Drăguţ, L., and Baciu, N., 2005. Minimum data sets for landscape indicators using GIS (A Case Study: Târnava Mare Corridor, Romania). EcoSys 11, 24–31.

Pandi, G., Sorocovschi, V., and Vigh, M., 2002. Influence of environment on population health in Copşa Mică. Environmental studies 7, 413–422.

Pastakia, C.M.R. and Jensen, A., 1998. The Rapid Impact Assessment Matrix (RIAM) for EIA. Environmental Impact Assessment Review 18, 461–482.

Pischke, F. and Cashmore, M., 2006. Decision-oriented environmental assessment: an empirical study of its theory and methods. Environmental Impact Assessment Review 26, 643–662.

Serdaru, M., Avram, N., Medrea, N., and Vlădescu, L., 2001. Determination of trace elements content in biological materials – measure of industrial pollution. Analytical Letters 34, 1437–1447.

Şimandan, D., 2006. Marginally Modern. 'Vasile Goldiş' University Press, Arad, p. 264.

Sorocovschi, V., 1996. Podişul Târnavelor. Studiu hidrogeografic. Ed. CETIB, Cluj-Napoca, p. 194.

Surdu, S, Gurzău, A., Gurzău, E.S., Curseu, D., Faraian, D., Bereczki, A., Criotea, C., Bodor, E., and Zeic, A., 1998. Risk assessment/communication in a lead contaminated area of Romania. Central European Journal of Public Health 6, 123–126.

Swart, R., 1996. Security risks of global environmental changes. Global Environmental Change 6, 187–192.

UNDP, 2002. Human Development Report, Oxford University Press.

Van Cleve, F.B., Leschine, T., Klinger, T., and Simenstad, C., 2006. An evaluation of the influence of natural science in regional-scale restoration projects. Environmental Management 37, 367–379.

Wang, Y.M., Yang, J.B., and Xu, D.L., 2006. Environmental impact assessment using the evidential reasoning approach. European Journal of Operational Research 174, 1885–1913.

LANDSCAPE APPROACHES TO ASSESS ENVIRONMENTAL SECURITY: SUMMARY, CONCLUSIONS, AND RECOMMENDATIONS

K. BRUCE JONES*
US Geological Survey, Reston Virginia (USA)

KINGA KRAUZE
Polish Academy of Sciences, Institute of Hyrology, Lodsch (Poland)

FELIX MÜLLER
Ecology Centre, University of Kiel (Germany)

GIOVANNI ZURLINI AND IRENE PETROSILLO
Laboratory of Landscape Ecology – Department of Biological and Environmental Sciences and Technologies – University of Salento (Lecce, Italy)

SERGEY VICTOROV
Russian Academy of Sciences, Institute of Remote Sensing Methods for Geology, St. Petersburg (Russia)

BAI-LIAN LI
Ecological Complexity and Modeling Laboratory, Department of Botany and Plant Sciences, University of California, Riverside (USA)

1. Introduction

Landscape sciences emphasize the importance of spatial pattern and scale in determining the relative degree of environmental security. The primary hypothesis of the landscape sciences is that spatial pattern and distribution of biotic and abiotic attributes of the environment, including people, are important determinants of landscape processes and structures which in turn provide ecological goods and services upon which society depends. Because of the

* Corresponding Author: Dr. K. Bruce Jones, US Geological Survey, Geography Discipline, Reston, Virginia USA, 20192, Phone: +1 703.648.4762, Fax: +1 703.648.5792, e-mail: kbjones@u sgs.gov

I. Petrosillo et al. (eds.), Use of Landscape Sciences for the Assessment of Environmental Security, 475–486.

emphasis on spatial pattern and scale, the landscape sciences also provide insight into how the spatial distribution of human activities influences important landscape processes and structures from which services are derived. As such, the landscape sciences provide a way to determine the continuous interplay between changes in the conditions of human needs (social, economic, cultural) and changes in the environment. The methodologies of the landscape sciences are applied at different scales and hierarchical levels of organization in the context of complex adaptive socioeconomic and ecological systems (Berkes and Folke 1998; Levin 2006).

Taken for granted that the security management of social-ecological landscapes must be based on both objective and subjective assessments, as stated in the introduction, the goal of this book was to demonstrate that there is the need of making full use of the developments of landscape sciences both for improving the objective assessment of environmental security and evaluating and developing approaches and methods that help assess the various aspects of environmental security, including:

- Potential drivers of landscape change (e.g. linkages between economic, social, and economic states and changes, and landscape change)
- Pressures and threats to specific provisional, organizational, and cultural landscape services
- Current conditions
- Changes and trends in landscape conditions
- Changes in services resulting from changes in landscape pattern
- Scaling relationships of different landscape processes and structures
- Linkages (integration) between drivers, pressures, state, impact, and response (the DSPIR framework)
- Landscape properties that indicate potential vulnerability (fragility) or loss of environmental security
- Scenario and management approaches and frameworks that optimize environmental security across a range of environmental themes and scales

These points will be discussed in the following sections with reference to the single papers of this book. Furthermore, the findings will be focused on the demands for adaptive, security-related management and on the idea of *securability*, a landscape feature which describes the potential of an environmental system to provide components of environmental security. In the end, some further developmental steps of landscape sciences are recommended and finally there is a reference to the targets of this book, as they have been formulated in the first chapter.

2. Summary of contributions and findings

A number of conceptual approaches and case studies were presented in the book that dealt with various aspects of environmental security. Many of the contributions dealt with specific types of drivers and pressures, and specific environmental themes and landscape services. Additionally, the book presented conceptual approaches and case studies for a broad range of general environmental themes or values that relate to landscape provisional and organizational/regulatory services. These included terrestrial and aquatic biodiversity and habitat quality, conservation of protected areas, water quality and quantity, terrestrial ecosystem productivity, and human cultural landscape diversity.

Comprehensive conceptual models were presented that link a number of potential drivers to environmental security, including political, social, and economic drivers that can influence wholesale changes in the pattern and extent of land use, each of which can affect a range of services. These conceptual models provide a good summary of our understanding of the causes of landscape change and potential impact on environmental security at different scales. They also provide a basic framework for the types of data and indicators needed to address aspects of environmental security.

A number of chapters in this book analyzed and discussed current and projected changes in human demographic patterns and how they might drive changes in landscape services and environmental security across Europe. These included projections of shrinking populations and population shifts to urban environments (see the chapters by Swiaczny, Haase et al., and Sabbagh and Neef). Both of these likely reflect economic realities; because of financial limitations or changes in social values, people are waiting longer to have children, and better employment opportunities exist within urban areas. Population decline in rural areas reduces the tax base and results in critical infrastructure losses, further accelerating the spatial redistribution of people to urban areas. However, results also suggest that the pattern of population shift may vary considerable across Europe. For example, some urban areas in Germany may experience fragmentation processes (see the chapter by Haase et al.). Larger urban populations and increased spatial extent of urban areas (e.g. urban sprawl) result in a number of landscape changes that potentially will have impacts upon important landscape services. For example, urban expansion can result in loss of forests that provide services such as timber and paper products, removal of air pollutants, hydrological regulation, sequestration of carbon and production of oxygen, filtration of nutrients and sediments from streams and rivers, and habitat for animals and plants. Similarly, urban expansion can result in loss of productive agricultural lands which reduces food production, increases in impervious surfaces that increase the frequency and magnitude of flooding, and increases in pollution load and exposure to people living within and around urban areas. Additionally, loss of people from rural areas can result in accelerated loss of soil, increases in nutrient loads, and increased flooding because people often maintain practices and structures that maintain services. The changes in human demographic patterns described in this book will provide valuable insight into future landscape changes and potential changes in environmental security. A number of chapters dealt with specific *types of pressures or threats*. These included urbanization, agricultural expansion onto lands that are marginal for growing crops or grazing, climate change, and nuclear safety. Each of these chapters provided conceptual frameworks or results from case studies linking specific pressures and threats to changes in landscape states. Many of the chapters dealing with pressures also addressed fundamental changes in landscape condition (state).

Many chapters addressed methods to assess *landscape state* and changes in state. This involved application of metrics, indicators, and models related to landscape structure, and in some cases, function. They also included development and application of sampling strategies that linked in situ measures to spatial data, as well as remote sensing techniques to detect and evaluate change. Two chapters (Zurlini et al. and Li et al.) provide indicators for changes in scaling functions that relate to fundamental changes in ecosystem and landscape processes. In the former, an approach is developed to characterize and interpret the spatial patterns of change as disturbance at multiple scales in a panarchy of nested, bounded, and organized social-ecological landscapes (SELs) like region, province, and county, taking into account the scales and spatial patterns of human land use as source/sink disturbance systems. In the second chapter a new theoretical approach is introduced that links the allometry of energy partitioning among different-sized organisms in ecological community to community stability. The new approach makes it is possible to quantitatively estimate not only the scaling exponents in the dependence of population density and biomass of heterotrophs on body size, but also the absolute values of energy fluxes claimed by organisms of a given size in stable communities, which then can be used as an indicator of an ecosystem state.

A primary conclusion was that there is a paucity of in situ data across and within regions of Europe, and that this deficiency prevented a more thorough interpretation of the consequences of landscape change on provisional and regulating services. Additionally, differences in spatial resolution and extent of existing biophysical coverages across many areas of Europe prevent broad-scale application of spatially distributed landscape models and indicators that would link landscape state and change to provisional and regulatory services.

Nearly all of the case studies that addressed landscape state and change identified potential relationships between state and change, and provisional or regulatory *services*. For example, changes in natural vegetation along streams (riparian vegetation) can be linked to increases or decreases in nutrient and sediment filtering services. In some cases these linkages were quantified from existing empirical studies or models. For example, in the chapters by Kepner et al., Kondratyev, and Nikolova et al., watershed models were used to determine the degree to which landscape state and change influenced provisional services, such as clean water, and regulatory services such as nutrient update, sediment filtering capacity, and flooding abatement. In one chapter (Krause and Wagner), a landscape provisioning service (phosphorus update) was directly measured and linked to a variable of landscape state (vegetation biomass per unit area and percent vegetation cover). Four types of approaches are presented in the book that relate to management or *policy responses* that offer potential to improve or protect environmental security. These include alternative futures scenario analysis, multi-indicator scenario analysis, monitoring designs and systems to assess the effecttiveness of policy and management, and adaptive management strategies. Alternative future scenario analysis is highlighted in the chapters by Kepner et al. and Nikolova. This approach uses spatially distributed watershed or landscape models to link potential landscape management alternatives to provisional and regulatory services. The import feature of this approach is that the spatially distributed models have land cover or land surface parameters such that changes in these parameters can be related to changes in services. Moreover, because of these linkages, it is possible to develop spatially explicit management alternatives (e.g. responses) to improve or protect environmental security. The multi-indicator scenario approach utilizes amoeba diagrams (see the chapter by Kumpula et al.) to evaluate alternative management options. This approach also requires that indicators can be related to specific provisional and regulatory services. A strength of this approach is that multiple indicators and services, and their potential responses to different scenarios, can be displayed in a small set of diagrams. This increases the ability of the decision-maker to interpret complex indicator datasets. A combination of these two approaches would be a valuable next step. A critical component of evaluating the effectiveness of management scenarios and policy changes (e.g. that affect drivers) is development and application of multi-scaled *monitoring strategies*. Such an approach has been applied and used in the United Kingdom (see the chapter by Haines-Young et al.). Finally, the chapter by Petrosillo et al. discusses the concept of *adaptive management*. This concept involves evaluation and reevaluation of management and policy prescriptions and makes management and policy adjustments based on a set of performance indicators. As such, all three of the other approaches described above are important components of the overall adaptive management framework.

3. The landscape sciences approach, environmental securability, and adaptive management

Studies presented in the book showed that an inclusion of landscape sciences into a framework of environmental security is highly recommended. A landscape perspective considering the spatial arrangement of land use types at multiple scales (Turner and Gardner, 1991; Ricketts, 2001) is essential to a more appropriate indication of environmental security. Up to now, however, decision-makers and the public have focused on changes in delivery of ecological services and mitigation of resulting conflicts (e.g. pressures and impacts according to DPSIR indicator framework), rather than on the drivers and sources of disturbances that occur at multiple scales, which jeopardize organizational, cultural, and provisional services. Additionally, important biological components and processes of ecosystems have been largely missing from the objective assessments of environmental security. The majority of proposed solutions have been based on chemical and physical analyses, and remediation and engineering methods and technologies have been implemented on a case-by-case basis at a local scale (bottom-up approach). Yet, many of the drivers and pressures exert their influence from multiple scales. In contrast, the novel landscape approach to environmental security concentrates on identification of hierarchy of, and linkages between processes and components across scales and multiple environmental themes and issues (Figure 1). The landscape approach is prospective and prescriptive rather than the more traditional, reactionary, and retrospective approach (impairment focus, Figure 1), and as such, offers the potential to identify critical drivers and pressures and to avoid future declines in environmental security. It also offers the potential to establish and protect critical landscape provisional, regulatory, and cultural services. Through the concept of ecological hierarchy (Allen and Starr, 1982; O'Neill et al. 1986), and with the recent implementation of the panarchy theory (Gunderson and Holling, 2003), the landscape sciences also provide a framework for addressing multiple-scale interactions between pressures (e.g. anthropogenic drivers, climate change), sensitivity, condition, and potential trajectories of social-ecological landscapes. Because the landscape sciences produce spatially explicit relationships among factors given in the DPSIR framework, it provides a way to evaluate how alternative landscape futures and management scenarios (e.g. response) might increase environmental security. It also provides a novel framework upon which to implement adaptive management strategy.

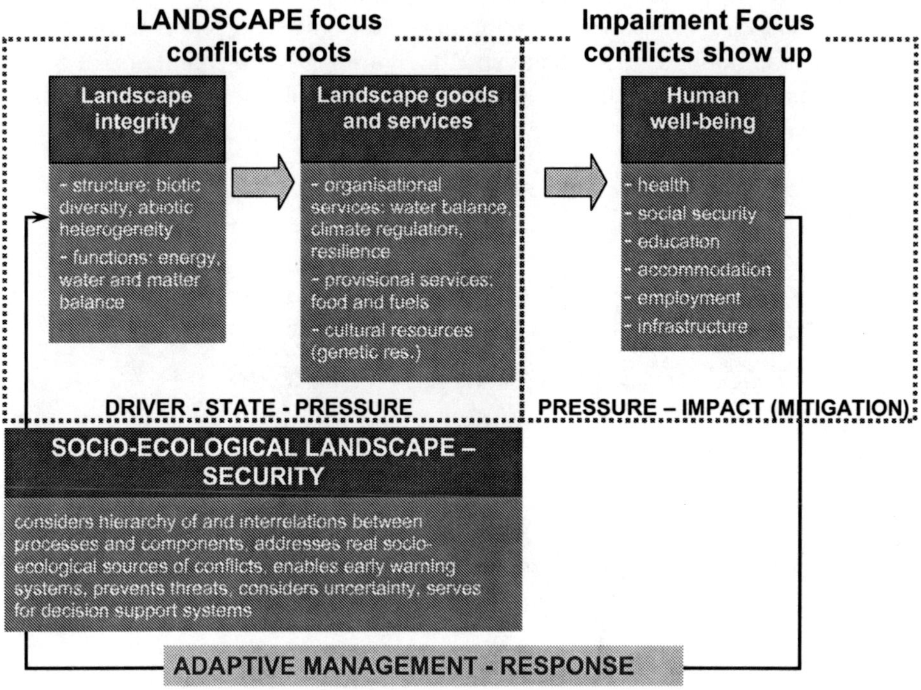

Figure 1. The landscape sciences approach, environmental security, and adaptive management.

Adaptive management is focused on retrieving and improving aspects like social learning and traditional knowledge, through learning by doing. In this context, the landscape sciences provide insights into spatial relationships between factors influencing environmental conditions (e.g. drivers and pressures), the inherent capacity or resiliency of a given areas based on its biophysical attributes (e.g. an areas' sensitivity to different drivers and pressures), and the area's environmental condition (state) and trends (impact and trajectory). When combined information on social and cultural values and expectations, the landscape sciences approach provides a way to evaluate an area's relative environmental *securability*, e.g. the likelihood of being able to change to or maintain a landscape state whereby important landscape services are obtained or preserved over long periods of time.

Such an approach and concept could be an important addition to the concept and the practice of adaptive management (Figure 2). That is, a primary follow-on to adaptive management from the landscape approach which is prospective and prescriptive. Furthermore, possible discordances between objective and subjective analyses can represent an important added determinant of *securability*, in that the likelihood to bridge that gap will contribute to the likelihood of being able to change to or maintain a landscape state in the future whereby important landscape services are obtained or preserved.

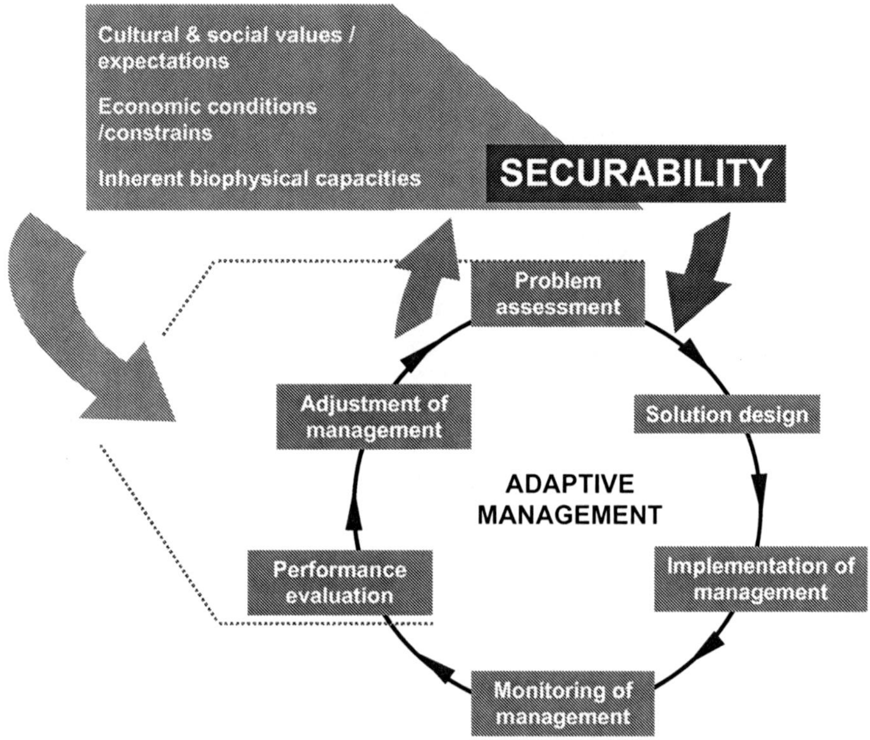

Figure 2. Environmental *securabilty* and its importance in the adaptive management process.

Areas with low levels of *securability* would require greater levels of attention and human and capital investments whereas areas with relatively high *securability* would require less intervention but would still require monitoring.

4. Recommendations

The NATO CCMS landscape pilot study provided a number of important insights and case studies on how the landscape sciences can be used to address environmental security. However, an opportunity exists to expand the application of the findings in this book to a more comprehensive and integrated approach that potentially could address multi-environmental themes and threats at a range of spatial scales. In order to exploit this potential, future studies should be focused on development of reliable, integrated risk assessment methods and quantitative measures from a landscape perspective to evaluate both vulnerability (fragility) and uncertainty associated with environmental security at different scales and hierarchical levels. One immediate goal could be to develop a decision support system, tool, and/or concept that allows users to reduce and avoid conflicts among environmental security decisions, considering also people' perception of threats. This will involve landscape scenario analysis, including an in-deep investigation of linkages between drivers, pressures, states, impacts, and responses (DPSIR framework), with special focus on incorporation of demographic, economic, and social factors including perception of threats. It will also be important to understand the role of humans in maintaining ecosystem services, especially in that humans maintain cultural–ecological landscapes that provide important services (e.g. prevention of soil erosion, landscape mosaics that support biological diversity, etc.).

To apply such a political conception, a number of questions have to be examined, which could also lead to future research activities. The respective fields of problems might be investigated in a potential future project, which should take into account the following steps:

- Select three to five regions in Europe that represent biophysical, cultural, and demographic gradients. These gradients would be used to evaluate metrics and indicators representing the full range of components in the DPSIR framework. Develop these for multiple provisional, organizational, and cultural landscape services.
- Develop methodologies to characterize securability. This would include biophysical (see Jongman et al., 2006), cultural (Potschin and Haines-Young, 2003), sociopolitical (Brauch, 2006), and economic classifications with regards to relative environmental securability.
- Apply securability classifications in each study area at multiple scales (Zurlini et al., 2006) and hierarchical levels in a panarchy of nested, bounded, and organized socio-ecological landscapes (SELs) like region, province, and county;
- Develop an alternative landscape futures decision support tool (see the chapter by Kepner, and Baker et al., 2004) that permits an analysis of alternative management scenarios given different levels of environmental securability.
- Apply the amoeba diagram approach to assess changes in multi-indicator values given different alternative future scenarios.

- Apply the decision support tool in each of the study areas and hold public workshops to get feedback on results.
- Modify the approach and decision support tool based on public workshops and implement the approach as part of a comprehensive adaptive management strategy among NATO CCMS members and partners.

Finally, with regards to the set of questions, tasks, and challenges presented in the introduction of the book, we offer the following conclusions:

- Landscape sciences comprise a very high potential to address and assess environmental security. Due to the typical scales of observation, landscape ecological methodologies are most suitable to link social, physical, and biological sciences, which is an important formal precondition for a successful assessment of environmental security.
- Landscape security is a very young concept which has to be integrated into the canon of environmental developmental targets, such as ecosystem/landscape integrity, health, and sustainability. Environmental security illuminates specific societal demands which can not be enlighten by the corresponding concepts and approaches.
- The articles in this book have shown that there are several significant landscape components related to environmental security. Basing on the knowledge of their organization, it is possible to reduce the number of significant variables to create a manageable set of security indicators.
- On the base of these indicators and a linkage with well-known landscape methodologies, an evaluation of the state of environmental security becomes possible at the landscape level, i.e. if linked with scenarios of different possible futures.

In the articles, it has furthermore been demonstrated that landscape sciences can offer several tools, such as indicators, models, monitoring strategies, and assessment approaches which can be combined to capture security items and to better understand the complex interrelations of of human–environmntal systems. These instruments can be used to perform long-term analyses of landscape potentials as well as the respective risks and threats. Therefore, they are good prerequisites for an integrative, interdisciplinary strategy of adaptive management for environmental security.

In spite of these large methodological potentials many problems still have to be solved. But a rapid application can be carried out already, if the linkages between the components and the respective are optimized. Thus, the development can be accelerated if there is a focus and joint endeavor with the focal target of integration, on a scientific, environmental, and political level.

Acknowledgments

Based on the success of the project, the editors want to make a concluding recommendation. We have been cooperating for a number of years within the pilot study on landscape assessments. The scientific targets of the project have been fulfilled in a degree which really satisfies us, and the productivity of the group in some instances has been overwhelming. There has been a deep exchange of methods and knowledge and several problems of the national working groups were solved due to the assistance of the colleagues. Furthermore, new ideas were realized due to a joint methodological or conceptual development. Therefore, the pilot study format has been a very successful instrument for all of us, and we, therefore, strongly urge the continued incentive to bring scientists from many countries together, who would otherwise not get in contact without the pilot study support mechanisms. Therefore, we are grateful to the colleagues, managers, and supervisors from NATO for giving us this unique chance. We also thank Rick Kutz for initializing this project, and of course we want to thank all the colleagues who participated for their contributions and for the wonderful opportunity to make new friends.

References

Allen, T.F.H. and Starr, T.B., 1982. Hierarchy: perspectives for ecological complexity. University of Chicago Press, Chicago, IL.

Baker, J.P., Hulse, D.W., Gregory, S.V., White, D., Van Sickle, J., Berger, P.A., Dole, D., and Schumaker, N.H., 2004. Alternative futures for the Willamette River Basin, Oregon. Ecological Applications 14, 313–324.

Berkes, F., and Folke, C. (Editors). 1998. Linking Social and Ecological Systems: Management Practices and Social Mechanisms for Building Resilience. Cambridge University Press. Cambridge.

Brauch, H.G., 2006. Desertification: a new security challenge for the Mediterranean? In: W.G. Kepner, J.L. Rubio, D.A. Mouat, and F. Pedrazzini (Editors), Desertification in the Mediterranean Region: A Security Issue. Springer, The Netherlands, pp. 11–85.

Gunderson, L.H. and Holling, C.S. (Editors), 2003. Panarchy: Understanding Transformations in Human and Natural Systems. Island Press, Washington, DC.

Jongman, R.H.G., Bunce, R.G.H., Metzger, M.J., Mucher, C.A., Howard, D.C., and Mateus, V.L., 2006. Objectives and applications of a statistical environmental stratification of Europe. Landscape Ecology 21, 409–419.

Levin, S.A., 2006. Learning to live in a global commons: socioeconomic challenges for a sustainable environment. Ecological Research 21, 328–333.

O'Neill, R.V., DeAngelis, D.L., Waide, J.B., and Allen, T.F.H., 1986. A hierarchical concept of ecosystems. Princeton University Press, Princeton, NJ.

Potschin, M. and Haines-Young, R., 2003. Improving the quality of environmental assessments using the concept of natural capital: a case study from Southern Germany. Landscape and Urban Planning 63, 93–108.

Ricketts, T.H., 2001. The matrix matters: effective isolation in fragmented landscapes. American Naturalist 15, 87–99.

Turner, M.G. and Gardner, R.H. (Editors), 1991. Quantitative Methods in Landscape Ecology. Springer, New York,.

Zurlini, G., Riitters, K.H., Zaccarelli, N., Petrosillo, I., Jones, K.B., and Rossi, L., 2006. Disturbance patterns in a socio-ecological system at multiple scales. Ecological Complexity 3, 119–128.

ACKNOWLEDGMENTS

We want to thank sincerely the reviewers of this book for their great efforts in reviewing the manuscripts and for their kind comments and suggestions. In particular:

Name	Nationality
Baciu Nicolae	Romania
Barson Michele	Australia
Blaschke Thomas	Austria
Bölter Manfred	Germany
Boumans Roelof	USA
Bray Dennis	Germany
Burkhard Benjamin	Germany
Chen Xiongwen	USA
Clagette Peter	USA
Coulston John	USA
Dadamo Marco	Italy
Dragut Lucian	Austria
Ehrlich Daniele	Italy
Forbes Bruce	Finland
Fraschetti Simona	Italy
Godlewska Malgorzata	Poland
Grünbühel M. Clemens	Italy
Kedziora Andrzej	Poland
Kirtland Dave	USA
Klug Hermann	Austria
Krysanova Valentina	Germany
Lang Stefan	Austria
Lenz Roman	Germany
Mancinelli Giorgio	Italy

Name	Nationality
Muchoney M. Douglas	Switzerland
Nash Maliha	USA
Nikolova Maryiana	Bulgaria
O'Malley Robin	USA
Reck Heinrich	Germany
Schöpfer Elisabeth	Austria
Selman Paul	United Kingdom

Semmens Darius	USA
Simandan Dragos	Canada
Slonecker E. Terence	USA
Smith R. Betsy	USA
Smith H. Jonathan	USA
Steinhäusler Friedrich	Austria
Sutton C. Paul	USA
Vadineanu Angheluta	Romania
Viaroli Pierluigi	Italy
Wade Timothy	USA
Wagner F. Paul	USA
Walker Joseph	Australia
White Denis	USA
Wickham James	USA
Windhorst Wilhelm	Germany
Zaccarelli Nicola	Italy

APPENDIX I

LIST OF PROJECT PRINCIPAL INVESTIGATORS

IN THE NATO/CCMS PILOT STUDY ON THE USE OF LANDSCAPE

SCIENCES FOR ENVIRONMENTAL ASSESSMENT

BULGARIA

Stoyan Nedkov
Bulgarian Academy of Sciences
Institute of Geography
Akad. G. Bonchev Str. Bl.3
1113 Sofia, Bulgaria
Phone: +359 2 979 33 71
Fax: +359 2 870 02 04
E-mail: snedkov@abv.bg

Mariyana Nikolova
Bulgarian Academy of Sciences
Institute of Geography
Akad. G. Bonchev Str. Bl.3
1113 Sofia, Bulgaria
Phone: +359 2 870 02 53
Fax: + 359 2 870 02 04
E-mail: mn@bas.bg

GERMANY

Felix Müller (Pilot Study Co-Director)
Ecology Center
University of Kiel
Schauenburgerstrasse 112
D-24118 Kiel
Germany
Phone: +49 431 880 3251, +49 171 643 7418
Fax: +49 431 880 4083
E-mail: fmueller@ecology.uni-kiel.de

Benjamin Burkhard
Ecology Center
University of Kiel
Olshausenstr. 40
D - 24098 Kiel
Germany

Phone: +49 431 880 1230
Fax: +49 431 880 4083
E-mail: bburkhard@ecology.uni-kiel.de

Roman Lenz
Institute of Applied Research
University Nürtingen
Schelmenwasen 4-8
72622 Nürtingen
Germany
Phone: +49 7022 404 192
Fax: +49 7022 404 209
E-mail: roman.lenz@hfwu.de

Jamill Sabbagh
RegioComun – Institute for Strategic Regional Management
University of Mainz
Becherweg 21
55099 Mainz
Germany
Phone: +49 6131 3920218
Fax: +49 6131 3920219
E-mail: J.Sabbagh@geo.uni-mainz.de

Ralf Seppelt
Centre for Environmental Research
Department for Computational Landscape Ecology
Permoser Str. 15
04318 Leipzig
Germany
Phone: +49 341 235 3940
Fax: +49 341 235 3939
E-mail: Ralf.Seppelt@ufz.de

ITALY

Giovanni Zurlini
Department of Biological and Environmental Sciences and Technologies
Landscape Ecology Laboratory
University of Salento
Ecotekne (Campus)
Via per Monteroni
I-73100 Lecce
Italy
Phone: +39 0832 298886

Fax: +39 832 298626
E-mail: giovanni.zurlini@unile.it

LATVIA

Juris Zagars
Venspils International Radio Astronomy Center
101, Inzhenieru Street
LV3600 Ventspils
Latvia
Phone: +371 944 1737
Fax: +371 362 8303
E-mail: yzh@venta.lv

LITHUANIA

Benediktas Jankauskas
Kaltinenai Research Station
Lithuanian Institute of Agriculture
Varniu 17, LT-75451 Kaltinenai,
Silale Distr.
Lithuania
Phone and Fax: +370 449 57141
E-mail: kaltbs@kaltbs.lzi.lt

Algimantas Tiknius
Klaipeda University, Manto 84
LT-92294 Klaipeda
Lithuania
Phone: +370 46 398854
Fax: +370 46 398802
E-mail: algis.tiknius@gmail.com

NORWAY

Wendy Fjellstad
Norwegian Forest and Landscape Institute
P.O. Box 115
N-1431 Ås
Norway
Phone: +47 6494 9704
Fax: +47 6494 8001
E-mail: wendy.fjellstad@skogoglandskap.no

POLAND

Kinga Krauze
International Centre for Ecology
Polish Academy of Sciences
3 Tylna Str.
90-364 Lodz
Poland
Phone: +48 42 681 70 07
Fax: +48 42 681 69 30
E-mail: kingak@biol.uni.lodz.pl

ROMANIA

Mihaela Lazarescu
National Institute for Research and Development in Environmental Protection
Splaiul independentei nr. 294, sector 6, code 77703
Bucharest 78, Romania
Phone: +40 212 210990
Fax: +40 212 203805
E-mail: mihaelalazarescu@yahoo.com

Lucian Drăgut
University of Salzburg
Z_GIS – Centre of Geoinformatics
Kapitelgasse 4-6
A-5020 Salzburg
Austria
Phone: +43 662 8044 0
Fax: +43 662 8044 214
E-mail: lucian.dragut@sbg.ac.at

Octavian-Liviu Muntean
Faculty of Environmental Science
University "BABES-BOLYAI"
P-ta Stefan cel Mare, No. 4
400192, Cluj-Napoca
Romania
Phone: +40 264 405316
Fax: +40 264 599444
E-mail: liviumuntean@yahoo.com, liviu@enviro.ubbcluj.ro

TURKEY

Engin Nurlu
Ege University Faculty of Agriculture

Department of Landscape Architecture
35100 Bornova, Izmir
Turkey
Phone: + 90 232 3884000, + 90 232 3692100
Fax: + 90 232 3881864
E-mail: engin.nurlu@ege.edu.tr

UNITED KINGDOM

Marion Potschin
Centre for Environmental Management
School of Geography
University of Nottingham
Nottingham NG7 2RD, England
Phone: +44 115 84 67398
Fax: +44 115 951 5249
E-mail: Marion.Potschin@Nottingham.ac.uk

Roy Haines-Young
Centre for Environmental Management
School of Geography
University of Nottingham
Nottingham NG7 2RD, England
Phone: +44 115 951 5498
Fax: +44 115 951 5249
E-mail: Roy.Haines-Young@Nottingham.ac.uk

UNITED STATES

William G. Kepner (Pilot Study Co-Director)
US Environmental Protection Agency
Office of Research and Development
944 East Harmon Avenue
Las Vegas, Nevada 89119
USA
Phone: +1 702 798 2193
Fax: +1 702 798 2208
E-mail: kepner.william@epa.gov

K. Bruce Jones
US Geological Survey
516 National Center
12201 Sunrise Valley Dr.
Reston, Virginia 20192
USA

Phone: +1 703 648-4762
Fax: +1 703 648-5792
E-mail: kbjones@usgs.gov

C. Gregory Knight
Professor of Geography and Gosnell Senior Faculty Scholar
College of Earth and Mineral Sciences
302 Walker Building,
Pennsylvania State University
University Park, Pennslyvania 16802
USA
Phone: +1 814 863-8571
Fax +1 814 863-7943
E-mail: cgk@psu.edu

B.-L. Li
Ecological Complexity and Modeling Laboratory
University of California
Riverside, California 92521
USA
Phone: +1 951 827 4776
Fax: +1 951 827 4437
E-mail: bai-lian.li@ucr.edu

EAPC PARTNER COUNTRIES

AUSTRIA
Thomas Blaschke
Z_GIS, Centre for Geoinformatics, University of Salzburg
Hellbrunner Str. 34, A – 5020, Salzburg
Austria
Phone: +43 662 80445225
Fax: +43 662 8044525
E-mail: thomas.blaschke@sbg.ac.at

FINLAND
Timo Kumpula
Department of Geography
University of Joensuu
Yliopistokatu 7
80101 Joensuu
Finland
Phone: +358 13 2514563

Fax: +358 13 2513454
E-mail: timo.kumpula@joensuu.fi

GEORGIA

Gia Sopadze
Tbilisi State University
Geography Department
7 Bakhtrioni Str., Tbilisi
Georgia
Phone: +995 32 36 60 14
Fax: +995 32 94 05 79
E-mail: gia_sopadze@yahoo.com

RUSSIAN FEDERATION

Sergey A. Kondratyev
Institute of Limnology of the Russian
Academy of Sciences
Sevastyanova Str., 9
196105 St. Petersburg
Russian Federation
Phone: +7 812 387 0276
Fax: +7 812 388 7327
E-mail: kondratyev@limno.org.ru

Sergey V. Victorov
Institute of Remote Sensing Methods for Geology
6 Birzhevoy proezd
199034, St. Petersburg
Russian Federation
Phone: +7 812 323 3660
Fax: +7 812 328 3916
E-mail: s1941vic@yahoo.com

UKRAINE

Vladimir Kuznyetsov
Department of International Research Projects
Ukrainian Scientific Research Institute of Ecological Problems (USRIEP)
Ministry of Environment of Ukraine
6 Bakulina Str.
61166 Kharkiv
Ukraine
Phone: +380 50 402 84 28

Fax: +380 57 719 46 92
E-mail: vladkuz@niiep.kharkov.ua; vladkuz5@hotmail.com

Olga Marushevska
Ministry of Environment of Ukraine
35, Uritsky Str., Kiev
Ukraine
Phone: + 380 44 206 31 10
E-mail: marushevska@menr.gov.ua

AUSTRALIA (Non-NATO)

Joe Walker
CSIRO Land and Water
C.S. Christian Laboratory
GPO Box 1666
Canberra, ACT 2601
Australia
Phone: +61 2 62465725
Fax: + 61 2 62465800
E-mail: joe.walker@csiro.au

APPENDIX II

LIST OF PROJECT PRODUCTS FOR NATO/CCMS PILOT STUDY ON THE USE OF LANDSCAPE SCIENCES FOR ENVIRONMENTAL ASSESSMENT

I. International Workshops

(a) Las Vegas, Nevada, USA (Inaugural Meeting); 3–5 April 2002
(b) Salzau, Germany; 19–21 November 2002
(c) Debe, Poland; 1–3 September 2003
(d) Lecce, Italy; 5–9 September 2004
(e) Veliko Turnovo, Bulgaria; 4–8 September 2005
(f) St. Petersburg, Russia; 3–6 July 2006

II. Web site

NATO/Science for Peace and Security web site for Landscape Sciences Pilot Study
(http://www.nato.int/science/pilot-studies/lsea/lsea-index.htm)

III. Published Proceedings

(a) Kepner, W., Kutz, F., Müller, F., and Sajwaj, T., 2002. Final Report. NATO/CCMS Pilot Study Organizational Meeting; Use of Landscape Sciences for Environmental Assessment. 54 pp.
(b) Müller, F., Kepner, W., and Caesar, K. (Editors), 2003. Landscape Sciences for Environmental Assessment. Proceedings from a CCMS Pilot Study Meeting at Salzau Castle, Germany. EcoSys 10, 174 pp.
(c) Müller, F. (Editor), 2004. Landscape Sciences for Environmental Assessment. Proceedings from a CCMS Pilot Study Meeting at Debe, Poland. EcoSys 42, 115 pp.
(d) Petrosillo, I., Zurlini, G., Kepner, W., and Müller, F. (Editors), 2005. Landscape Sciences for Environmental Assessment. Proceedings from a CCMS Pilot Study Meeting at Lecce, Italy. EcoSys 11, 168 pp.

IV. Analytical Tools

(a) Automated Geospatial Watershed Assessment (AGWA) tool
(http://www.epa.gov/nerlesd1/land-sci/agwa/introduction.htm)
AGWA is a GIS-based hydrologic modeling tool that provides an analysis system for environmental managers, natural resource specialists, policy-makers, scientists, and the general public to perform watershed assessments at multiple spatial and temporal scales.

Printed in the United Kingdom
by Lightning Source UK Ltd.
123889UK00006B/64-75/A